Essential Developmental Biology

Essential Developmental Biology

Edited by Leonardo Reeves

MURPHY & MOORE
www.murphy-moorepublishing.com

Published by Murphy & Moore Publishing,
1 Rockefeller Plaza,
New York City, NY 10020, USA
www.murphy-moorepublishing.com

Essential Developmental Biology
Edited by Leonardo Reeves

International Standard Book Number: 978-1-63987-201-5 (Hardback)

Cataloging-in-Publication Data

Essential developmental biology / edited by Leonardo Reeves.
p. cm.
Includes bibliographical references and index.
ISBN 978-1-63987-201-5
1. Developmental biology. 2. Growth. 3. Biology. 4. Ontogeny.
I. Reeves, Leonardo.
QH491 .E77 2022
571.8--dc23

TABLE OF CONTENTS

PREFACE

Developmental biology refers to the process of growth and development of plants and animals. It encompasses the biology and mechanisms of metamorphosis, asexual reproduction, stem cell growth and differentiation, and regeneration. Tissue patterning, tissue growth, and tissue morphogenesis are the chief processes related to the embryonic development of animals. Although plant development shares many similarities with animal development, plant morphogenesis is achieved via differential growth due to immotility of plant cells. Research in developmental biology emphasizes the use of model organisms. This book elucidates the concepts and innovative models around prospective developments with respect to developmental biology. It presents researches and studies performed by experts across the globe. This book is a resource guide for experts as well as students.

This book is the end result of constructive efforts and intensive research done by experts in this field. The aim of this book is to enlighten the readers with recent information in this area of research. The information provided in this profound book would serve as a valuable reference to students and researchers in this field.

At the end, I would like to thank all the authors for devoting their precious time and providing their valuable contribution to this book. I would also like to express my gratitude to my fellow colleagues who encouraged me throughout the process.

Editor

1

Notochord Cells in Intervertebral Disc Development and Degeneration

Matthew R. McCann [1,2] and Cheryle A. Séguin [1,2,*]

Academic Editors: Robert W. Dettman and Andy Wessels

1 Department of Physiology and Pharmacology, Schulich School of Medicine & Dentistry, Western University, London, ON N6A 5C1, Canada; mmccann4@uwo.ca
2 Bone and Joint Institute, Western University, London, ON N6A 5C1, Canada
* Correspondence: cheryle.seguin@schulich.uwo.ca

Abstract: The intervertebral disc is a complex structure responsible for flexibility, multi-axial motion, and load transmission throughout the spine. Importantly, degeneration of the intervertebral disc is thought to be an initiating factor for back pain. Due to a lack of understanding of the pathways that govern disc degeneration, there are currently no disease-modifying treatments to delay or prevent degenerative disc disease. This review presents an overview of our current understanding of the developmental processes that regulate intervertebral disc formation, with particular emphasis on the role of the notochord and notochord-derived cells in disc homeostasis and how their loss can result in degeneration. We then describe the role of small animal models in understanding the development of the disc and their use to interrogate disc degeneration and associated pathologies. Finally, we highlight essential development pathways that are associated with disc degeneration and/or implicated in the reparative response of the tissue that might serve as targets for future therapeutic approaches.

Keywords: animal models; disc degeneration; intervertebral disc; notochord; nucleus pulposus

---

1. Introduction

> "Know from whence you came. If you know whence you came, there are absolutely no limitations to where you can go."—James Baldwin

In trying to understand the cellular and molecular basis of tissue degeneration, it is important to consider the dynamic interplay involving cell-cell communication, gene regulation, and growth factor and cytokine secretion required to form a healthy, functional tissue during normal development. Once we have identified pathways that serve to regulate tissue formation, we can start to investigate alterations in these signals during normal aging, disease, and tissue repair.

Low back pain has become one of the most common causes of disability, affecting overall well-being and work performance, with recent reports indicating a lifetime prevalence as high as 85% in industrialized countries [1]. In fact, according to the recent *Global Burden of Disease* study, low back pain is the most common musculoskeletal disorder and the leading cause of years lived with disability in all developed countries [2]. The prevalence of chronic low back pain has increased remarkably over the last decade for all ages and genders; increases have been reported to be as high as 162% between 1992 and 2006 in middle-aged individuals in the US [3]. Alarmingly, there are no disease-modifying therapies for low back pain and current treatment options do not adequately provide improved outcomes [4]. While several risk factors for back pain have been identified such as, obesity [5], psychological factors [6], age and sex [7], and genetic variants [8], the molecular cause(s) of back pain remain elusive.

Although the etiology of low back pain is unknown, it is often associated with degeneration of the intervertebral disc (IVD). The current lack of disease-modifying therapeutics for disc degeneration is linked to a poor understanding of the biological mechanisms that regulate IVD development, health and degeneration. This review provides an overview of IVD biology, with a specific emphasis on the development of the nucleus pulposus (NP) and how the embryonic notochord plays an essential role in adult IVD homeostasis. Furthermore, we explore the opportunities and challenges associated with the use of animal models to study the IVD.

2. Functions and Structure of the Intervertebral Disc

IVDs are complex connective tissue structures that serve to anchor adjacent vertebral bodies along the spinal column, essential for mechanical stabilization of the spine and load bearing during axial compression [9]. The cartilaginous joints formed by IVDs also provide flexibility and movement to the spinal column. The IVD consists of three distinct, yet interdependent specialized tissues: the central viscous NP, the outer fibrillar annulus fibrosus (AF), and the cartilage end-plates (CEP) that anchor the disc to the adjacent vertebral bones. The diversity in the structure of these tissues is related to the specific organization of their extracellular matrix (ECM), which is produced and maintained by distinct cell populations.

The NP is composed primarily of large aggregating proteoglycans held together loosely by an irregular network of type-II collagen and elastin fibers. The aggrecan monomers are highly sulphated by covalently bound glycosaminoglycans (GAGs) (chondroitin sulfate, keratin sulfate) and cross-linked to hyaluronic acid via cartilage link protein [10]. The GAGs confer a net negative charge to the aggrecan molecules, attracting large numbers of counter ions and therefore water molecules [11,12]. The influx of water to the NP is created by osmotic pressure [13], and the water molecules are retained in the matrix in part by the irregular meshwork of type II collagen [14]. The osmotic pressure provides the stiffness required for the NP to maintain height and turgor against compressive loads. With sustained loading, such as human daily activity, roughly 20%–25% of the water slowly diffuses out of the disc into the paravertebral muscles, resulting in a decrease in disc height [15,16]. Upon removal of the compressive load (*i.e.*, nightly rest), water is reabsorbed into the NP matrix, resulting in a re-expansion of the tissue and recovery of disc height [17–19].

The AF is a fibrocartilagenous structure formed by distinct lamellae composed of bundles of type I collagen fibers oriented at oblique angles, that alternate within each consecutive lamella to form an angle-ply structure (25 lamellae in the AF of a human disc) [20]. This arrangement provides mechanical support to resist the turgor pressure that is applied by the NP. The outer AF contains elastic fibers (elastin, elaunin and oxytalan proteins) that directly associate with the adjacent vertebra and extend into the bone as Sharpey's fibers to anchor the IVD into bone [21]. The inner AF forms a transition zone between the fibrous outer AF and the gelatinous NP, containing higher levels of glycosaminoglycans and type-II collagen in the interlamellar matrix when compared to the outer AF [22].

The disc is anchored superiorly and inferiorly to the adjacent vertebral bodies through the CEPs, thin layers of hyaline cartilage that interface with the inner AF and NP. In humans, the adult NP is the largest avascular tissue in the body and therefore depends on nutrient, metabolic by-products and O_2 gas exchange through the CEPs via passive diffusion from the highly vascularized vertebral bone [23]. The CEP is formed by chondrocytes similar to those found in articular cartilage, which secrete a type II collagen- and proteoglycan-rich extracellular matrix.

3. Intervertebral Disc Development

3.1. Early Embryo and Node Development

Much of our knowledge of mammalian notochord development is derived from experiments using mouse models due to their short gestation period, large litter sizes and ease of genetic manipulation.

As such, the data presented below apply to the murine model and developmental stages, unless otherwise indicated.

During gastrulation the primitive streak is formed along the longitudinal axis of the embryo, establishing longitudinal (anterior/posterior) and bilateral (left/right) symmetry. In the mouse, a cluster of cells located near the anterior end of the primitive streak ingress to form the node (Figure 1). The node is a transient, pit-shaped structure located at the distal tip the early embryo, responsible for establishing the left asymmetry of the body plan. Monocilia localized to the apical surface of node cells beat in a clockwise rotation (when viewed from the ventral side) to drive a leftward flow of extraembryonic fluid containing morphogens, such as Nodal and the Nodal antagonist Cerl-2, secreted by the columnar epithelial cells of ventral node [24–27]. This fluid flow promotes left side specification via two possible mechanisms: either through direct morphogen-induced differentiation on the left side of the embryo, or by inducing a mechanosensitive calcium signalling response to the nodal flow in the perinodal crown cells that promotes cell specification. While the underlying mechanisms remain to be established, the node and resultant nodal flow are necessary for asymmetric differentiation of the left axis of the embryo [28–30].

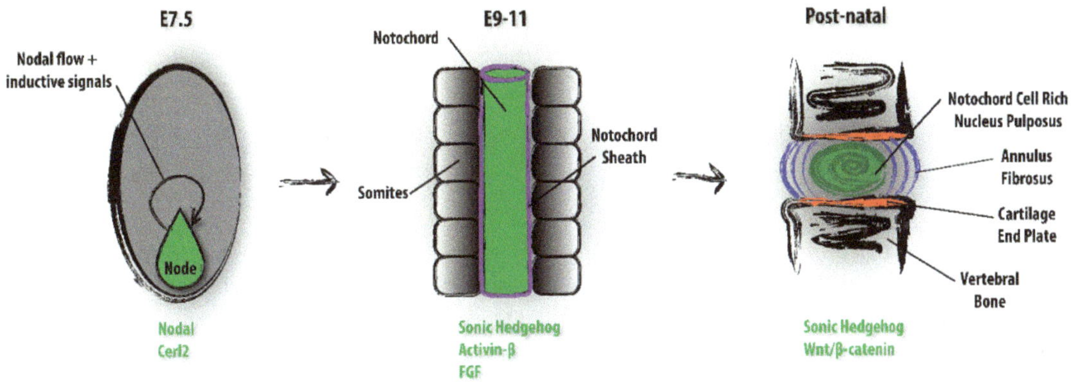

Figure 1. Schematic representation of key events in the transition from notochord to nucleus pulposus development in the murine model. Notochord cells within the mouse embryonic node at embryonic day 7.5 play a fundamental role in left/right patterning of the early embryo by secreting Nodal and Cerl-2. Notochord cells proliferate and migrate to form the embryonic notochord (surrounded by the notochord sheath) between embryonic day 9–11 where they provide instructive cues that pattern the neural tube and sclerotome through secretion of sonic hedgehog. Notochord cells then go on to form the mature nucleus pulposus (NP) in the post-natal intervertebral disc (IVD) (shown here in the sagittal plane), where they continue to secrete inductive factors, including sonic hedgehog and Wnt, which serve to regulate intervertebral disc homeostasis.

3.2. Notochord Development

In the mouse, the first stages of notochord development originate at the notochordal plate between E8.0–E8.5. This plate-like structure lays in the ventral midline of the embryo and is continuous with the dorsal gut endoderm. At E9.0, the notochord plate folds off the gut endoderm to form the anterior notochord, where it rests in a central position in the embryo, flanked by the dorsal ridge of the neural tube (the floor plate) and ventrally by the gut endoderm (the endoderm plate). Laterally, the notochord is flanked by the paraxial mesoderm, which will go on to form the somites and subsequently the AF and vertebrae.

Anatomically, the notochord is a continuous rod-like structure that forms the primitive anterior/posterior axis of the embryo. The notochord is able to resist compressive loads along the axis of the embryo yet is sufficiently flexible to resist buckling [31]. The unique structural and cellular features of the notochord achieve this rigidity. In mammals the notochord is surrounded by an acellular sheath composed of collagens, laminins and proteoglycans [32,33], while in zebrafish the sheath is made up of an epidermal cell layer; both serving to encapsulate the notochord [34]. In zebrafish, alterations in the composition of the notochord sheath, such as loss of *col15a1* or *emilin3*, result in specific phenotypes associated with notochord malformations [35–38]. Furthermore, disruptions in the structural integrity of the sheath result in loss of the embryonic axis and perturbations in spinal column development [33]. In mammals, the cell type responsible for secretion of the notochord sheath has yet to be established, with the extracellular matrix proteins produced by notochord cells, the surrounding mesenchyme cells or a combination thereof. Notochord cells also directly contribute to the mechanical properties of the notochord due to the presence of large intracellular vacuoles that increase cell size and occupy ~80% of the cell volume [39]. The presence of these vacuoles have been described in numerous species including amphibians [40], birds [41], mammals [42] and humans [43]. While conserved across species, their classification has been long debated. Recent work by Ellis *et al.* demonstrated that in zebrafish, notochord cell vacuoles are lysosome related organelles that require late endosomal trafficking and H^+-ATPase-dependent acidification [44]. In contrast to typical lysosomes that have a low pH [45], V-ATPase in notochord vacuoles generates a proton rich gradient that allows water to flow into the vacuole through osmosis [46]. Work from our lab has shown that H^+-ATPases are enriched in the mouse IVD [47], and others have indicated their importance in bovine [48] and canine [39] NP cells suggesting that the processes governing notochord vacuole formation maybe be conserved. The expansion of intracellular vacuoles contributes to longitudinal elongation of the embryonic axis; disruption of vacuolated cell differentiation or integrity in the zebrafish was shown to alter notochord and spine morphogenesis [44]. In mammals, the notochord remains in place until the development of the permanent axial skeleton (e.g., vertebrae, IVDs).

3.3. Notochord Signalling

During early development, the notochord is generally considered a signalling centre, secreting growth factors and morphogens responsible for patterning of surrounding tissues, including the neural tube [49], sclerotome [50], pancreas [51] and aorta [52]. The best characterized is its ability to direct neural tube neurogenesis during dorsal-ventral patterning. Mutations that disrupt notochord formation result in altered floor-plate induction [53]. Conversely, grafting experiments that placed an ectopic notochord adjacent to the neural tube established a secondary floor plate [49,54]. This process is thought to be mediated by notochord-derived SHH and chordin, as transplantation of microcarrier beads likewise resulted in the formation of a secondary floor plate [55].

The notochord also influences mesodermal muscle determination in the sclerotome; experiments have shown that mutations that prevent notochord development disrupt sclerotome formation [56]. In zebrafish, mutation of the *floating head* (*flh*) gene that disrupts notochord formation results in ectopic midline muscle development [57], suggesting that notochord-derived signals inhibit muscle development. This interaction appears to be mediated by pair-box transcription factors, specifically Pax1 induction and Pax-3 and Pax-7 repression [56,58]. Finally, the notochord is also known to contribute to endoderm patterning [59], where notochord-derived activin-βB and FGF2 have been shown to repress SHH in the endoderm thereby promoting pancreas development [60].

While the importance of the notochord in early embryonic development is well established, the postnatal role of notochord cells has been a source of ongoing controversy. Work from our lab and others have shown that in the mouse the notochord disappears in areas where the vertebral bodies form but expands within the perichordal disc, to form the central NP [61,62]. Interestingly, the mechanisms by which the notochord undergoes regression in the presumptive vertebrae are unknown [63,64]. The absence of detectable notochord cell apoptosis in this region or proliferation in the future NP [65] has led to the hypothesis that notochord cells either succumb to pressure from the developing bone mesenchyme and are forced into the future NP, or undergo directed migration to sites of the future NP [64]. Within the newly formed NP, notochord cells produce a glycosaminoglycan-rich extracellular matrix, which separates the original cell mass into a network of cell clusters.

3.4. Role of the Mesenchyme in Intervertebral Disc Development

Development of the axial skeleton is a multi-step process, beginning with notochord formation and requiring fine-tuned coordination and pre-patterning of individual skeletal segments. In the mouse, the paraxial mesoderm undergoes segmentation between E7.5–E11.5 to form the somites, which are metameric structures that exist transiently during vertebrate development [66]. The somites lie adjacent to the neural tube and the notochord. In response to bone morphogenetic protein (BMP) and Wnt signals from the ectoderm and Shh signaling from the notochord and neural tube floorplate, the dorsal aspect of the somites differentiate to form the dermomyotome, which gives rise to the dermis. At E12.5, the ventral portion of the somites differentiates to form the sclerotome that will give rise to the ribs, vertebral bodies and CEP and AF of the IVD [67]. The sclerotome cells migrate and aggregate around the notochord to form a continuous perichordal tube [68]. At E15.5, these mesenchymal cells acquire a metameric pattern of condensed and non-condensed segments along the anterior/posterior axis. This process is regulated by Notch and Wnt signalling to maintain somite polarity [69,70]. The notochord secretes Shh thereby regulating differentiation of the sclerotome in the formation of the axial skeleton. Notably, the vertebral column is not formed in mice lacking *Shh* [71]; however, targeted deletion of *Shh* in the floorplate does not alter IVD development, and deletion of *Shh* in both the floorplate and notochord results in the loss of IVD and vertebral body structures [72]. Lineage tracing using both the *Gdf5*-Cre [73] and *Tbx18*-Cre [74] demonstrated the mesenchymal origin of the AF. The condensed segments will later contribute to the mesenchyme-derived AF, whereas the less-condensed segments form the templates of the future vertebral bones.

3.5. Notochord Cell Contribution to the Nucleus Pulposus

Unlike the cells of the AF and CEP that remain relatively stable throughout life, the cells of the NP undergo drastic changes within the first decade of life in humans. In most vertebrates, including humans and mice, there is a progressive loss of large vacuolated notochord cells immediately after birth and the NP becomes populated by small cartilage-like nucleus pulposus cells [75,76]. Interestingly, the loss of the large vacuolated notochord cells from the NP is associated with the onset of disc degeneration, suggesting that these cells contribute to maintenance of the NP [75,77,78]. Despite this correlation, only recently has there been insight into the underlying biological mechanisms.

There have historically been two conflicting hypotheses regarding the origin of the cartilage-like cells of the NP. It was originally suggested that these cells were of mesenchymal origin, resulting from migration of cells to the NP from either the surrounding CEP [79] or the perichondrium at the periphery of the IVD [80]. During IVD formation, notochord cells were believed to direct mesenchymal cell migration, stimulate matrix synthesis, and then undergo apoptosis or necrosis [81,82]. More recent studies have suggested an alternate model in which notochord cells serve as NP progenitors and undergo terminal differentiation to give rise to the cartilage-like nucleus pulposus cells [77,83,84]. Our group and others conducted lineage-tracing studies in mice that established that during normal development and aging all cells of the NP are notochord-derived (from E15.5 until 21 months of age) . To investigate the cellular composition of the NP during IVD degeneration, a mechanically-induced model of disc degeneration was used in mice where bone marrow[61,62]-derived cells were labeled with GFP. These studies demonstrated an increased number of GFP$^+$ BMSCs in the vertebral bone marrow, vascular canals of the CEP and the site of injury in the outer AF; however, BMSC-derived cells were not detected in the NP [85]. While other genetic approaches and *in vivo* models of disc degeneration or other spine-related pathologies should be used to examine the cellular composition of the IVD in various disease states, the data to date supports the notochord cell maturation model. In fact, recent publications in the field have suggested that mature cells within the NP should instead be referred to as "nucleus pulpocytes" to underscore their distinct developmental origin and function [86].

4. Intervertebral Disc Degeneration

The most common cause of back pain is disc degeneration, the etiology of which is poorly understood. Consequently, there are no specific criteria to distinguish degeneration from the physiological processes of growth, aging or adaptive remodeling [18]. Disc degeneration has perhaps been best defined as a cell-mediated response to progressive structural failure [18]. This process is thought to initiate with changes to the cellular microenvironment within the disc and progress over decades, resulting in structural breakdown and functional deficiency [87,88]. Degeneration of the IVD is associated with increased extracellular matrix breakdown [89], abnormal (fibrotic) matrix synthesis [90], inflammation [91], and in-growth of nociceptive nerves and blood vessels into a typically aneural and avascular tissue [92] (Figure 2).

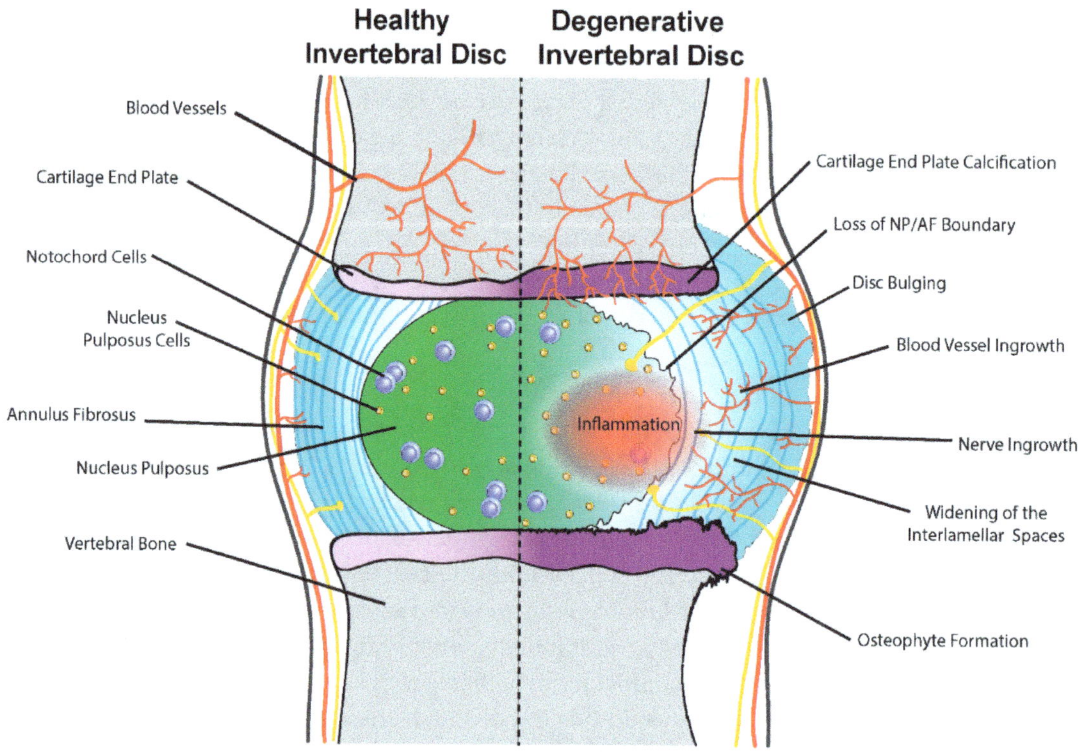

Figure 2. Schematic representation highlighting the hallmarks of human disc degeneration. Compared to healthy intervertebral disc, the degenerative disc has increased inflammation, blood vessel and neuronal ingrowth, loss of the boundary between the NP and annulus fibrosus (AF), and widening of the interlamellar space between collagen bundles in the AF that often results in disc bulging. In advanced degeneration the cartilage endplate calcifies and osteophytes form on the adjacent vertebral bones.

Initiation of the degenerative process is thought to originate specifically in the NP [93,94]. The deterioration of the major structural proteoglycan aggrecan [95,96], along with decreased production of other ECM components in the NP, lead to reduced hydration [97], loss of disc height, and an overall inability to absorb compressive load [98,99]. Compressive forces are therefore distributed to the surrounding AF, which results in altered mechanical properties and structural deterioration including radial and circumferential tears in the AF [100,101]. These tears often precede radial bulges or herniations of the NP substance into the adjacent spinal cord, resulting in pain [88,102]. Currently, there are symptomatic treatments for late stages of degenerative disc disease but no disease-modifying therapeutics [103].

Notochord Cells and Intervertebral Disc Homeostasis

Within the postnatal IVD notochord cells are thought to maintain their role as important signaling effectors regulating IVD cell function. This function was first suggested by co-culture experiments that demonstrated the ability of notochord cells to increase proteoglycan synthesis in mature NP cells [78]. This effect was shown to be independent of direct cell contact, as notochord-condition media could recapitulate the anabolic effects of notochord cells on NP cells [104,105] as well as AF cells [106]. Notochord secreted factors, including the matricellular protein CCN2, directly regulate NP cell function by upregulating anabolic gene expression and downregulating catabolic gene expression [107–109]. Notochord secreted factors have also been shown to inhibit cell death and apoptosis of NP cells and protect NP cells from the degradative effects of cytokine exposure [110].

Since disc degeneration and LBP are associated with nerve and blood vessel in-growth to the IVD [92], studies have investigated the ability of notochord-secreted molecules to prevent axons and/or blood vessels from entering inappropriate territories. During mouse development, secretion of semaphorin 3E from the notochord and lateral plate mesoderm has been shown to regulate dorsal aorta formation [52]. In the IVD, notochord cells have been shown to inhibit angiogenesis by suppressing Vegf expression in endothelial cells [111]. With regards to neuronal patterning, studies of early chick development established that the notochord and dermamyotomes regulate axonal guidance [112]. These studies also identified semaphorin 3A as a notochord-derived axon repellent [113]. Interestingly, the outer AF of degenerative human discs shows reduced expression of semaphorin 3A compared to healthy IVDs, suggesting that IVD-associated nerve repellent pathways maybe be conserved across species [114]. In keeping with the correlation between the loss of notochord cells and neural in-growth to the IVD, expression of Nrp-1, another inhibitor of anxonal growth expressed in the NP, is decreased in degenerative human discs [114] and in aged rat NP [115].

The microenvironment of the NP may also contribute to repelling innervation. The ECM of the notochord differs from that of cartilage tissues in that it lacks keratin sulfate and has high levels of chondroitin sulfate [116]. In zebrafish, chondroitin sulfate has been shown to guide the development of the nervous system by inhibiting the formation of segmental motor neurons [117]. Co-culture studies demonstrated that chondroitin sulfate derived from the notochord or notochord sheath directly induced dorsal root ganglia cone collapse and increased axonal repulsion [118]. The role of notochord-derived chondroitin sulphate in restricting neural in-growth appears to be maintained in the adult IVD; recent studies report that soluble factors derived from porcine notochord cell conditioned media inhibit *in vitro* neurite outgrowth and that this affect was ablated with chondroitin sulfate digestion [119]. Interestingly, a genome-wide association study identified *CHST3* as a susceptibility gene for lumbar disc degeneration [120], and loss-of-function mutation in *CHST3* results in chondrodysplasia with major involvement of the spine [121]. *CHST3* is the gene encoding chondroitin 6-*O*-sulfotransferase, an enzyme required for proper chondroitin sulfate synthesis that is required for proper cartilage formation [122]. While chondroitin 6-*O*-sulfotransferase activity is required for water retention with the IVD [120], it may serve a dual purpose of also inhibiting neural in-growth [123,124]. Conversely, NP tissue is able to cause nerve tissue injury in the absence of mechanical compression [125], indicating that it is most likely chemical in nature.

5. Small Animal Models to Study Intervertebral Disc Development and Degeneration

5.1. Notochord Cells in Animal Models

Due to the difficulties associated with obtaining "healthy" intact human IVD tissue and cells or biological material from the early stages of disease onset, animal models have become invaluable to study the IVD biology [126]. Animal models have been incorporated into studies aimed at characterizing disc development, disease progression and to develop therapeutic interventions. In all mammals, development of the IVD involves common pathways associated with node formation, notochord elongation and aggregation of the mesenchyme around the central notochord (as described above). Although the overall process is similar, one important difference between species is the postnatal persistence of notochord cells in the NP—an important factor to consider when studying NP aging or degeneration. Unlike humans, species such as the rat, pig, cat, and rabbit, retain notochord cells throughout their life [16]. Large animals such as bovine and sheep better resemble humans in that the notochord cells present at birth rapidly decline in number early in life [126].

The use of canine models has revealed insights into the importance of notochord cells in IVD health [127,128]. While there is a large genetic variability in canine breeds, they can be classified as either chondrodystrophic or non-chondrodystrophic. Chondrodystrophia is altered development in the cartilage of the long bones, especially those involved the region of the epiphysial plates, resulting in arrested growth of the long bones [129]. Chondrodystrophic dogs (dachshund, bulldog, and beagle) have a higher prevalence and earlier age of onset of disc degeneration, while non-chondrodystrophic dogs are relatively resistant to disc degeneration in the earlier years of life [130,131]. Strikingly, in chondrodystrophic breeds, notochord cells in the NP are replaced by small nucleus pulposus cells in the first year of life and IVD degeneration proceeds rapidly thereafter [132]. Conversely, in non-chondrodystrophic dogs notochord cells persist in the IVD throughout life [133,134], and while these breeds experience some degree of disc degeneration, it occurs at a much slower rate.

Mouse models are also commonly used to study human disease due to the ease of their genetic manipulation, highly conserved genome, comparable physiology, short gestation period and low housing costs [135,136]. The murine NP is predominantly composed of notochord cells until skeletal maturity, and like humans, there is a progressive decline in notochord cells and a concomitant increase in nucleus pulposus cells with age [137]. An important tool in mouse genetics is Cre/loxP-based conditional mutagenesis. In this system, expression of Cre recombinase can be regulated by tissue-specific and/or temporally-regulated promoters, leading to excision of loxP-flanked ("floxed") genes via intra-chromosomal recombination leading to gain or loss of gene expression. This strategy confers the ability to track specific cell types over time, or assess the function of a specific gene within a given cell type based on the specificity of Cre recombinase expression [138].

5.2. Notochord Associated Cre Mouse Strains

Several genes or gene loci have been targeted to generate Cre expressing mouse strains that demonstrate Cre expression in the notochord, including the T-box transcription factor brachyury, *T*-Cre [139], flat-top (*Cfap126*), a cilia and flagella associated protein, $Fltp^{T2AiCre}$ [140], and type-II collagen *Col2a1*-Cre [141–144]. However, in these strains Cre expression is not notochord-specific, and is detected in the lateral mesoderm, the neural tube floorplate, head mesenchyme and cartilaginous tissues, respectively. Consequently, novel notochord-specific Cre strains were required to trace the developmental lineage of NP cells. The first reported and most commonly used notochord-Cre mouse is the *Shh*-Cre strain [145], which when crossed with a conditional lacZ reporter marked cells throughout the notochord at E10.5, notochord cells during segmentation at E12.5, and the presumptive NP at E16.5 [61]. Importantly, no β-galactosidase-positive cells were detected in the surrounding AF or CEPs. To ensure that detection of Cre expression in the postnatal NP was not associated with the induction of *Shh* expression in mature NP cells, a tamoxifen-inducible $Shh\text{-}CreER^{T2}$ [145] was used to transiently induce Cre expression in the notochord between E8.5–E12.5. This experiment conclusively demonstrated that all the cells the mature NP (in mice ⩾19 months of age) were derived from the *Shh* expressing cells of the node and notochord [61].

The sickle tail (*Skt*) gene has been reported to play a role in the development of the notochord and IVD. The Skt protein is localized to the cytoplasm where it is thought to act as a structural element or a scaffolding protein. *Skt* was an interesting target for the development of a notochord-specific Cre strain since in gene-trapping experiments, intense β-galactosidase expression was detected in the notochord and at birth the mice had kinky tails, compressed IVDs, and disorganized NP tissues [146]. *Skt*-Cre mice were subsequently generated and when crossed to a conditional LacZ reporter, β-galactosidase expression was detected in the notochord at E9.5 and E12.5. At E15.5

β-galactosidase expression was predominantly detected in the NP however, small cell populations were also detected in the primitive AF. In adult mice, β-galactosidase expression was detected in the NP but large populations of AF were also LacZ positive [147], making the *Skt*-Cre mouse not suitable for NP-specific genetic experiments.

Forkhead box protein A2 (FoxA2), also known as hepatocyte nuclear factor 3-β, is a transcription factor necessary for the formation and maintenance of the definitive endoderm [148] and required for node and notochord formation [149]. Both a *FoxA2*-Cre strain [150,151] and an inducible $FoxA2^{mcm}$ Cre strain [150] have been generated, and when crossed to a conditional Rosa26-LacZ reporter mouse, β-galactosidase expression was detected in the endoderm but also in the node, floorplate, and notochord [150]. *FoxA2*-Cre has been used to knockout hypoxia inducing factor-1α, which demonstrated a NP phenotype [152].

Lastly, work form our group targeted the *Noto* locus that encodes a homeobox transcription factor that is transiently but specifically expressed in the node and the posterior notochord between E7.5–E12.5. Noto has been shown to regulate morphogenesis, node ciliogenesis and left-right embryo patterning [153]. Characterization of Cre specificity in this mouse showed specific localization to the node, notochord and NP during development and in adult mice [62], and this mouse strain has subsequently been used to study the role of notochord-derived factors in NP biology [108].

6. IVD Disease and Repair: A Mirror of Development?

Basic research and the use of animal models have demonstrated the role of the notochord as a regulator of disc development, enabling investigations into the ability of notochord-associated signals to modulate disc degeneration. One example is the bone morphogenetic protein (BMP) pathway which is critical for embryonic notochord patterning [154], IVD development, and is also reactivated during disc degeneration in rabbits [155] and rodents [156] localized to the NP and CEP [157]. Consequently, BMPs have been evaluated as potential therapeutic targets to treat disc degeneration, owing to their ability to promote ECM gene expression [158] and their anti-apoptoic effects [159]. However, the use of recombinant BMPs to treat disc degeneration has been called into question due their potent ability to induce osteogenesis leading to spinal fusion [160], and associated complications including swelling in the cervical spine [161]. It remains unclear whether the increased BMP signaling detected within the degenerate IVD is a result of the degenerative cascade or an indication of a reparative tissue response; further studies are required to elucidate the role of BMP in disc degeneration.

The Wnt/β-catenin pathway has been shown to maintain notochord progenitor cell fate during early notochord formation and is required for posterior extension of the notochord [162]. β-catenin is also critical in IVD development. Targeted ablation of β-catenin in type II collagen expressing cells lead to CEP calcification; conversely, ectopic activation of β-catenin in the AF and vertebral growth plate resulted in severe deterioration of both structures [163]. This may be a direct effect of Wnt/β-catenin activation or an indirect effect mediated by altered SHH signaling, given the feedback between these two pathways in the NP [164,165]. Interestingly, in the adult NP expression of both SHH and Wnt/β-catenin are downregulated, as are Wnt targets in the whole IVD [164]. The down-regulation of Wnt/β-catenin signaling in the aged disc has been shown to be reversible and that re-activation of this pathway can induce molecular markers of the IVD that was lost in degeneration [164]. Interestingly, canonical Wnt signalling has also been shown to be upregulated in the NP during early disc degeneration in chondrodystophic dogs [166]. Taken together, these findings suggest a dual role for Wnt/β-catenin in the IVD and its potential as a therapeutic intervention for degenerative disc disease.

Brachyury is a highly conserved transcription factor that is expressed in the developing notochord, required during embryogenesis for proper development of the primitive streak, axial and posterior mesoderm [167]. Brachury is expressed in notochord cells of the NP but not in the nucleus pulposus cells of mature rat discs [115]. Interestingly, increased expression of brachyury, resulting from either gene mutation [168] or gene duplication [169], has been associated with the notochord-derived tumour termed chordoma. Chordoma is a rare invasive cancer responsible for 20% of primary spine tumours, located at clavius of the skull (32%) and sacrococcygeal region (29%), and less frequently in cervical, thoracic and lumbar vertebrae [170,171]. Chordomas are morphologically similar to notochord cells, formed by physaliferous cells that co-express notochord cell genes including cytokeratins 8, 18, and 19 and brachyury [79]. Fate mapping of the notochord demonstrated that during disc formation, not all notochord cells are incorporated in the IVD; some persist in the vertebral bone [61,62]. In humans, notochord cell remnants have been detected in 20% of adult vertebrae but do not develop into neoplasms [172]. The current hypothesis is the chordoma formation results from the activation and proliferation of notochord remnants; while the pathways regulating this process have yet to be elucidated; it is intriguing that malignant transformation is associated with the activation of developmental notochord factors such as brachyury.

7. Conclusions

The IVD is a highly specialized tissue that requires complex developmental patterning to form a complex joint. The notochord and notochord-derived cells play an integral role in the development of the embryo and continue to be important in maintaining IVD homeostasis. Studies to date suggest that as the IVD undergoes degeneration, reactivation of these pathways may not contribute to tissue breakdown but instead reflect the cellular attempt to activate a program of tissue repair. Future studies are needed to understand these processes and exploit them as potential therapeutics for disc degeneration.

Acknowledgments: This work was funded by the Natural Sciences and Engineering Research Council of Canada (371546) and Canadian Institutes of Health Research (MOP-115718 and 115068 to C.A.S.). M.R.M. was supported by a CIHR Doctoral Award and the Joint Motion Program—A CIHR Training Program in Musculoskeletal Health Research and Leadership. C.A.S. is the recipient of a Scholar Award from The Arthritis Society/Canadian Arthritis Network, a CIHR New Investigator Award and an Early Researcher Award from the Ontario Ministry of Research and Innovation.

Author Contributions: M.R.M and C.A.S. were both involved in drafting the article and revising it for intellectual content, and approved the final version to be published.

Conflicts of Interest: The authors declare that they have no competing interests.

References

1. Balague, F.; Mannion, A.F.; Pellise, F.; Cedraschi, C. Non-specific low back pain. *Lancet* **2012**, *379*, 482–491. [CrossRef]
2. Vos, T.; Flaxman, A.D.; Naghavi, M.; Lozano, R.; Michaud, C.; Ezzati, M.; Shibuya, K.; Salomon, J.A.; Abdalla, S.; Aboyans, V.; *et al.* Years lived with disability (YLDs) for 1160 sequelae of 289 diseases and injuries 1990–2010: A systematic analysis for the global burden of disease study 2010. *Lancet* **2012**, *380*, 2163–2196. [CrossRef]
3. Manchikanti, L.; Singh, V.; Falco, F.J.; Benyamin, R.M.; Hirsch, J.A. Epidemiology of low back pain in adults. *Neuromodulation* **2014**, *17* (Suppl. 2), 3–10. [CrossRef] [PubMed]
4. Friedman, B.W.; O'Mahony, S.; Mulvey, L.; Davitt, M.; Choi, H.; Xia, S.; Esses, D.; Bijur, P.E.; Gallagher, E.J. One-week and 3-month outcomes after an emergency department visit for undifferentiated musculoskeletal low back pain. *Ann. Emerg. Med.* **2012**, *59*, 128–133. [CrossRef] [PubMed]
5. Shiri, R.; Karppinen, J.; Leino-Arjas, P.; Solovieva, S.; Viikari-Juntura, E. The association between obesity and low back pain: A meta-analysis. *Am. J. Epidemiol.* **2010**, *171*, 135–154. [CrossRef] [PubMed]
6. Andersson, G.B. Epidemiological features of chronic low-back pain. *Lancet* **1999**, *354*, 581–585. [CrossRef]
7. Hoy, D.; Bain, C.; Williams, G.; March, L.; Brooks, P.; Blyth, F.; Woolf, A.; Vos, T.; Buchbinder, R. A systematic review of the global prevalence of low back pain. *Arthritis Rheum.* **2012**, *64*, 2028–2037. [CrossRef] [PubMed]
8. Mayer, J.E.; Iatridis, J.C.; Chan, D.; Qureshi, S.A.; Gottesman, O.; Hecht, A.C. Genetic polymorphisms associated with intervertebral disc degeneration. *Spine J.* **2013**, *13*, 299–317. [CrossRef] [PubMed]
9. Inoue, N.; Espinoza Orias, A.A. Biomechanics of intervertebral disk degeneration. *Orthop. Clin. N. Am.* **2011**, *42*, 487–499. [CrossRef] [PubMed]

10. Moreland, L.W. Intra-articular hyaluronan (hyaluronic acid) and hylans for the treatment of osteoarthritis: Mechanisms of action. *Arthritis Res. Ther.* **2003**, *5*, 54–67. [CrossRef] [PubMed]
11. Chandran, P.L.; Horkay, F. Aggrecan, an unusual polyelectrolyte: Review of solution behavior and physiological implications. *Acta Biomater.* **2012**, *8*, 3–12. [CrossRef] [PubMed]
12. Nap, R.J.; Szleifer, I. Structure and interactions of aggrecans: Statistical thermodynamic approach. *Biophys. J.* **2008**, *95*, 4570–4583. [CrossRef] [PubMed]
13. Kraemer, J.; Kolditz, D.; Gowin, R. Water and electrolyte content of human intervertebral discs under variable load. *Spine* **1985**, *10*, 69–71. [CrossRef] [PubMed]
14. Mwale, F. Collagen and other proteins of the nucleus pulposus, annulus fibrosus, and cartilage end plates. In *The Intervertebral Disc*; Springer: Berlin, Germany, 2014; pp. 79–92.
15. Matsumura, Y.; Kasai, Y.; Obata, H.; Matsushima, S.; Inaba, T.; Uchida, A. Changes in water content of intervertebral discs and paravertebral muscles before and after bed rest. *J. Orthop. Sci.* **2009**, *14*, 45–50. [CrossRef] [PubMed]
16. Urban, J.P.G.; Roberts, S.; Ralphs, J.R. The nucleus of the intervertebral disc from development to degeneration. *Am. Zool.* **2000**, *40*, 53–61. [CrossRef]
17. Setton, L.A.; Chen, J. Mechanobiology of the intervertebral disc and relevance to disc degeneration. *J. Bone Joint Surg. Am.* **2006**, *88* (Suppl. 2), 52–57. [CrossRef] [PubMed]
18. Adams, M.A.; Roughley, P.J. What is intervertebral disc degeneration, and what causes it? *Spine* **2006**, *31*, 2151–2161. [CrossRef] [PubMed]
19. Watanabe, H.; Yamada, Y.; Kimata, K. Roles of aggrecan, a large chondroitin sulfate proteoglycan, in cartilage structure and function. *J. Biochem.* **1998**, *124*, 687–693. [CrossRef] [PubMed]
20. Cassidy, J.J.; Hiltner, A.; Baer, E. Hierarchical structure of the intervertebral disc. *Connect. Tissue Res.* **1989**, *23*, 75–88. [CrossRef] [PubMed]
21. Smith, L.J.; Fazzalari, N.L. The elastic fibre network of the human lumbar anulus fibrosus: Architecture, mechanical function and potential role in the progression of intervertebral disc degeneration. *Eur. Spine J.* **2009**, *18*, 439–448. [CrossRef] [PubMed]
22. Li, J.; Liu, C.; Guo, Q.; Yang, H.; Li, B. Regional variations in the cellular, biochemical, and biomechanical characteristics of rabbit annulus fibrosus. *PLoS ONE* **2014**, *9*, e91799. [CrossRef] [PubMed]
23. Urban, J.P.; Smith, S.; Fairbank, J.C. Nutrition of the intervertebral disc. *Spine* **2004**, *29*, 2700–2709. [CrossRef] [PubMed]
24. Davidson, B.P.; Tam, P.P. The node of the mouse embryo. *Curr. Biol.* **2000**, *10*, R617–R619. [CrossRef]
25. Hirokawa, N.; Tanaka, Y.; Okada, Y.; Takeda, S. Nodal flow and the generation of left-right asymmetry. *Cell* **2006**, *125*, 33–45. [CrossRef] [PubMed]
26. Brennan, J.; Norris, D.P.; Robertson, E.J. Nodal activity in the node governs left-right asymmetry. *Genes Dev.* **2002**, *16*, 2339–2344. [CrossRef] [PubMed]
27. Marques, S.; Borges, A.C.; Silva, A.C.; Freitas, S.; Cordenonsi, M.; Belo, J.A. The activity of the Nodal antagonist Cerl-2 in the mouse node is required for correct L/R body axis. *Genes Dev.* **2004**, *18*, 2342–2347. [CrossRef] [PubMed]
28. Babu, D.; Roy, S. Left-right asymmetry: Cilia stir up new surprises in the node. *Open Biol.* **2013**, *3*, 130052. [CrossRef] [PubMed]
29. Tanaka, Y.; Okada, Y.; Hirokawa, N. FGF-induced vesicular release of sonic hedgehog and retinoic acid in leftward nodal flow is critical for left-right determination. *Nature* **2005**, *435*, 172–177. [CrossRef] [PubMed]
30. Nonaka, S.; Tanaka, Y.; Okada, Y.; Takeda, S.; Harada, A.; Kanai, Y.; Kido, M.; Hirokawa, N. Randomization of left-right asymmetry due to loss of nodal cilia generating leftward flow of extraembryonic fluid in mice lacking KIF3B motor protein. *Cell* **1998**, *95*, 829–837. [CrossRef]
31. Adams, D.S.; Keller, R.; Koehl, M.A. The mechanics of notochord elongation, straightening and stiffening in the embryo of xenopus laevis. *Development* **1990**, *110*, 115–130. [PubMed]
32. Stemple, D.L. Structure and function of the notochord: An essential organ for chordate development. *Development* **2005**, *132*, 2503–2512. [CrossRef] [PubMed]
33. Choi, K.S.; Harfe, B.D. Hedgehog signaling is required for formation of the notochord sheath and patterning of nuclei pulposi within the intervertebral discs. *Proc. Natl. Acad. Sci. USA* **2011**, *108*, 9484–9489. [CrossRef] [PubMed]

34. Gray, R.S.; Wilm, T.P.; Smith, J.; Bagnat, M.; Dale, R.M.; Topczewski, J.; Johnson, S.L.; Solnica-Krezel, L. Loss of col8a1a function during zebrafish embryogenesis results in congenital vertebral malformations. *Dev. Biol.* **2014**, *386*, 72–85. [CrossRef] [PubMed]
35. Gotz, W.; Osmers, R.; Herken, R. Localisation of extracellular matrix components in the embryonic human notochord and axial mesenchyme. *J. Anat.* **1995**, *186 Pt 1*, 111–121. [PubMed]
36. Corallo, D.; Schiavinato, A.; Trapani, V.; Moro, E.; Argenton, F.; Bonaldo, P. Emilin3 is required for notochord sheath integrity and interacts with Scube2 to regulate notochord-derived hedgehog signals. *Development* **2013**, *140*, 4594–4601. [CrossRef] [PubMed]
37. Pagnon-Minot, A.; Malbouyres, M.; Haftek-Terreau, Z.; Kim, H.R.; Sasaki, T.; Thisse, C.; Thisse, B.; Ingham, P.W.; Ruggiero, F.; le Guellec, D. Collagen XV, a novel factor in zebrafish notochord differentiation and muscle development. *Dev. Biol.* **2008**, *316*, 21–35. [CrossRef] [PubMed]
38. Mangos, S.; Lam, P.Y.; Zhao, A.; Liu, Y.; Mudumana, S.; Vasilyev, A.; Liu, A.; Drummond, I.A. The ADPKD genes pkd1a/b and pkd2 regulate extracellular matrix formation. *Dis. Model. Mech.* **2010**, *3*, 354–365. [CrossRef] [PubMed]
39. Hunter, C.J.; Bianchi, S.; Cheng, P.; Muldrew, K. Osmoregulatory function of large vacuoles found in notochordal cells of the intervertebral disc running title: An osmoregulatory vacuole. *Mol. Cell. Biomech.* **2007**, *4*, 227–237. [PubMed]
40. Waddington, C.; Perry, M.M. The ultrastructure of the developing urodele notochord. *Proc. R. Soc. Lond. B Biol. Sci.* **1962**, *156*, 459–482. [CrossRef]
41. Bancroft, M.; Bellairs, R. The development of the notochord in the chick embryo, studied by scanning and transmission electron microscopy. *J. Embryol. Exp. Morphol.* **1976**, *35*, 383–401. [PubMed]
42. Leeson, T.S.; Leeson, C.R. Observations on the histochemistry and fine structure of the notochord in rabbit embryos. *J. Anat.* **1958**, *92*, 278–285. [PubMed]
43. Babic, M.S. Development of the notochord in normal and malformed human embryos and fetuses. *Int. J. Dev. Biol.* **1991**, *35*, 345–352. [PubMed]
44. Ellis, K.; Bagwell, J.; Bagnat, M. Notochord vacuoles are lysosome-related organelles that function in axis and spine morphogenesis. *J. Cell Biol.* **2013**, *200*, 667–679. [CrossRef] [PubMed]
45. Luzio, J.P.; Pryor, P.R.; Bright, N.A. Lysosomes: Fusion and function. *Nat. Rev. Mol. Cell Biol.* **2007**, *8*, 622–632. [CrossRef] [PubMed]
46. Allen, R.D.; Naitoh, Y. Osmoregulation and contractile vacuoles of protozoa. *Int. Rev. Cytol.* **2002**, *215*, 351–394. [PubMed]
47. McCann, M.R.; Patel, P.; Frimpong, A.; Xiao, Y.; Siqueira, W.L.; Seguin, C.A. Proteomic signature of the murine intervertebral disc. *PLoS ONE* **2015**, *10*, e0117807.
48. Mavrogonatou, E.; Papadimitriou, K.; Urban, J.P.; Papadopoulos, V.; Kletsas, D. Deficiency in the $\alpha 1$ subunit of Na^+/K^+-ATPase enhances the anti-proliferative effect of high osmolality in nucleus pulposus intervertebral disc cells. *J. Cell. Physiol.* **2015**, *230*, 3037–3048. [CrossRef] [PubMed]
49. Yamada, T.; Placzek, M.; Tanaka, H.; Dodd, J.; Jessell, T.M. Control of cell pattern in the developing nervous system: Polarizing activity of the floor plate and notochord. *Cell* **1991**, *64*, 635–647. [CrossRef]
50. Fan, C.M.; Tessier-Lavigne, M. Patterning of mammalian somites by surface ectoderm and notochord: Evidence for sclerotome induction by a hedgehog homolog. *Cell* **1994**, *79*, 1175–1186. [CrossRef]
51. Kim, S.K.; Hebrok, M.; Melton, D.A. Notochord to endoderm signaling is required for pancreas development. *Development* **1997**, *124*, 4243–4252. [PubMed]
52. Meadows, S.M.; Fletcher, P.J.; Moran, C.; Xu, K.; Neufeld, G.; Chauvet, S.; Mann, F.; Krieg, P.A.; Cleaver, O. Integration of repulsive guidance cues generates avascular zones that shape mammalian blood vessels. *Circ. Res.* **2012**, *110*, 34–46. [CrossRef] [PubMed]
53. Clarke, J.D.; Holder, N.; Soffe, S.R.; Storm-Mathisen, J. Neuroanatomical and functional analysis of neural tube formation in notochordless xenopus embryos; laterality of the ventral spinal cord is lost. *Development* **1991**, *112*, 499–516. [PubMed]
54. Basler, K.; Edlund, T.; Jessell, T.M.; Yamada, T. Control of cell pattern in the neural tube: Regulation of cell differentiation by dorsalin-1, a novel TGF beta family member. *Cell* **1993**, *73*, 687–702. [CrossRef]
55. Patten, I.; Placzek, M. Opponent activities of Shh and BMP signaling during floor plate induction *in vivo*. *Curr. Biol.* **2002**, *12*, 47–52. [CrossRef]

56. Goulding, M.; Lumsden, A.; Paquette, A.J. Regulation of Pax-3 expression in the dermomyotome and its role in muscle development. *Development* **1994**, *120*, 957–971. [PubMed]
57. Amacher, S.L.; Kimmel, C.B. Promoting notochord fate and repressing muscle development in zebrafish axial mesoderm. *Development* **1998**, *125*, 1397–1406. [PubMed]
58. Brand-Saberi, B.; Ebensperger, C.; Wilting, J.; Balling, R.; Christ, B. The ventralizing effect of the notochord on somite differentiation in chick embryos. *Anat. Embryol.* **1993**, *188*, 239–245. [CrossRef] [PubMed]
59. Cleaver, O.; Krieg, P.A. Notochord patterning of the endoderm. *Dev. Biol.* **2001**, *234*, 1–12. [CrossRef] [PubMed]
60. Hebrok, M.; Kim, S.K.; Melton, D.A. Notochord repression of endodermal sonic hedgehog permits pancreas development. *Genes Dev.* **1998**, *12*, 1705–1713. [CrossRef] [PubMed]
61. Choi, K.S.; Cohn, M.J.; Harfe, B.D. Identification of nucleus pulposus precursor cells and notochordal remnants in the mouse: Implications for disk degeneration and chordoma formation. *Dev. Dyn.* **2008**, *237*, 3953–3958. [CrossRef] [PubMed]
62. McCann, M.R.; Tamplin, O.J.; Rossant, J.; Seguin, C.A. Tracing notochord-derived cells using a Noto-cre mouse: Implications for intervertebral disc development. *Dis. Model. Mech.* **2012**, *5*, 73–82. [CrossRef] [PubMed]
63. Sivakamasundari, V.; Lufkin, T. Bridging the gap: Understanding embryonic intervertebral disc development. *Cell Dev. Biol.* **2012**, *1*, 103. [PubMed]
64. Lawson, L.; Harfe, B.D. Notochord to nucleus pulposus transition. *Curr. Osteoporos. Rep.* **2015**, *13*, 336–341. [CrossRef] [PubMed]
65. Aszodi, A.; Chan, D.; Hunziker, E.; Bateman, J.F.; Fassler, R. Collagen II is essential for the removal of the notochord and the formation of intervertebral discs. *J. Cell Biol.* **1998**, *143*, 1399–1412. [CrossRef] [PubMed]
66. Forsberg, H.; Crozet, F.; Brown, N.A. Waves of mouse lunatic fringe expression, in four-hour cycles at two-hour intervals, precede somite boundary formation. *Curr. Biol.* **1998**, *8*, 1027–1030. [CrossRef]
67. Brent, A.E.; Tabin, C.J. Developmental regulation of somite derivatives: Muscle, cartilage and tendon. *Curr. Opin. Genet. Dev.* **2002**, *12*, 548–557. [CrossRef]
68. Christ, B.; Wilting, J. From somites to vertebral column. *Ann. Anat.* **1992**, *174*, 23–32. [CrossRef]
69. Zhang, N.; Gridley, T. Defects in somite formation in lunatic fringe-deficient mice. *Nature* **1998**, *394*, 374–377. [PubMed]
70. Barrantes, I.B.; Elia, A.J.; Wunsch, K.; Hrabe de Angelis, M.H.; Mak, T.W.; Rossant, J.; Conlon, R.A.; Gossler, A.; de la Pompa, J.L. Interaction between notch signalling and lunatic fringe during somite boundary formation in the mouse. *Curr. Biol.* **1999**, *9*, 470–480. [CrossRef]
71. Chiang, C.; Litingtung, Y.; Lee, E.; Young, K.E.; Corden, J.L.; Westphal, H.; Beachy, P.A. Cyclopia and defective axial patterning in mice lacking sonic hedgehog gene function. *Nature* **1996**, *383*, 407–413. [CrossRef] [PubMed]
72. Choi, K.S.; Lee, C.; Harfe, B.D. Sonic hedgehog in the notochord is sufficient for patterning of the intervertebral discs. *Mech. Dev.* **2012**, *129*, 255–262. [CrossRef] [PubMed]
73. Mundy, C.; Yasuda, T.; Kinumatsu, T.; Yamaguchi, Y.; Iwamoto, M.; Enomoto-Iwamoto, M.; Koyama, E.; Pacifici, M. Synovial joint formation requires local ext1 expression and heparan sulfate production in developing mouse embryo limbs and spine. *Dev. Biol.* **2011**, *351*, 70–81. [CrossRef] [PubMed]
74. Bruggeman, B.J.; Maier, J.A.; Mohiuddin, Y.S.; Powers, R.; Lo, Y.; Guimaraes-Camboa, N.; Evans, S.M.; Harfe, B.D. Avian intervertebral disc arises from rostral sclerotome and lacks a nucleus pulposus: Implications for evolution of the vertebrate disc. *Dev. Dyn.* **2012**, *241*, 675–683. [CrossRef] [PubMed]
75. Hunter, C.J.; Matyas, J.R.; Duncan, N.A. The notochordal cell in the nucleus pulposus: A review in the context of tissue engineering. *Tissue Eng.* **2003**, *9*, 667–677. [CrossRef] [PubMed]
76. Roberts, S. Disc morphology in health and disease. *Biochem. Soc. Trans.* **2002**, *30*, 864–869. [CrossRef] [PubMed]
77. Boos, N.; Weissbach, S.; Rohrbach, H.; Weiler, C.; Spratt, K.F.; Nerlich, A.G. Classification of age-related changes in lumbar intervertebral discs: 2002 Volvo Award in basic science. *Spine* **2002**, *27*, 2631–2644. [CrossRef] [PubMed]
78. Aguiar, D.J.; Johnson, S.L.; Oegema, T.R. Notochordal cells interact with nucleus pulposus cells: Regulation of proteoglycan synthesis. *Exp. Cell Res.* **1999**, *246*, 129–137. [CrossRef] [PubMed]

79. Vujovic, S.; Henderson, S.; Presneau, N.; Odell, E.; Jacques, T.S.; Tirabosco, R.; Boshoff, C.; Flanagan, A.M. Brachyury, a crucial regulator of notochordal development, is a novel biomarker for chordomas. *J. Pathol.* **2006**, *209*, 157–165. [CrossRef] [PubMed]
80. Henriksson, H.; Thornemo, M.; Karlsson, C.; Hagg, O.; Junevik, K.; Lindahl, A.; Brisby, H. Identification of cell proliferation zones, progenitor cells and a potential stem cell niche in the intervertebral disc region: A study in four species. *Spine* **2009**, *34*, 2278–2287. [CrossRef] [PubMed]
81. Kim, K.W.; Lim, T.H.; Kim, J.G.; Jeong, S.T.; Masuda, K.; An, H.S. The origin of chondrocytes in the nucleus pulposus and histologic findings associated with the transition of a notochordal nucleus pulposus to a fibrocartilaginous nucleus pulposus in intact rabbit intervertebral discs. *Spine* **2003**, *28*, 982–990. [CrossRef] [PubMed]
82. Trout, J.J.; Buckwalter, J.A.; Moore, K.C. Ultrastructure of the human intervertebral disc: II. Cells of the nucleus pulposus. *Anat. Rec.* **1982**, *204*, 307–314. [CrossRef] [PubMed]
83. Pazzaglia, U.E.; Salisbury, J.R.; Byers, P.D. Development and involution of the notochord in the human spine. *J. R. Soc. Med.* **1989**, *82*, 413–415. [PubMed]
84. Liebscher, T.; Haefeli, M.; Wuertz, K.; Nerlich, A.G.; Boos, N. Age-related variation in cell density of human lumbar intervertebral disc. *Spine* **2010**, *36*, 153–159. [CrossRef] [PubMed]
85. Sakai, D.; Nishimura, K.; Tanaka, M.; Nakajima, D.; Grad, S.; Alini, M.; Kawada, H.; Ando, K.; Mochida, J. Migration of bone marrow-derived cells for endogenous repair in a new tail-looping disc degeneration model in the mouse: A pilot study. *Spine J.* **2015**, *15*, 1356–1365. [CrossRef] [PubMed]
86. Colombier, P.; Camus, A.; Lescaudron, L.; Clouet, J.; Guicheux, J. Intervertebral disc regeneration: A great challenge for tissue engineers. *Trends Biotechnol.* **2014**, *32*, 433–435. [CrossRef] [PubMed]
87. Mwale, F.; Roughley, P.; Antoniou, J. Distinction between the extracellular matrix of the nucleus pulposus and hyaline cartilage: A requisite for tissue engineering of intervertebral disc. *Eur. Cell. Mater.* **2004**, *8*, 58–63. [PubMed]
88. Smith, L.J.; Nerurkar, N.L.; Choi, K.S.; Harfe, B.D.; Elliott, D.M. Degeneration and regeneration of the intervertebral disc: Lessons from development. *Dis. Model. Mech.* **2011**, *4*, 31–41. [CrossRef] [PubMed]
89. Freemont, A.J. The cellular pathobiology of the degenerate intervertebral disc and discogenic back pain. *Rheumatology* **2009**, *48*, 5–10. [CrossRef] [PubMed]
90. Peng, Y.; Lv, F.J. Fibrosis in intervertebral disc degeneration: Knowledge and gaps. *Austin J. Orthop. Rheumatol.* **2014**, *1*, 3.
91. Risbud, M.V.; Shapiro, I.M. Role of cytokines in intervertebral disc degeneration: Pain and disc content. *Nat. Rev. Rheumatol.* **2014**, *10*, 44–56. [CrossRef] [PubMed]
92. Freemont, A.J.; Peacock, T.E.; Goupille, P.; Hoyland, J.A.; O'Brien, J.; Jayson, M.I. Nerve ingrowth into diseased intervertebral disc in chronic back pain. *Lancet* **1997**, *350*, 178–181. [CrossRef]
93. Pearce, R.H.; Grimmer, B.J.; Adams, M.E. Degeneration and the chemical composition of the human lumbar intervertebral disc. *J. Orthop. Res.* **1987**, *5*, 198–205. [CrossRef] [PubMed]
94. Sobajima, S.; Kim, J.S.; Gilbertson, L.G.; Kang, J.D. Gene therapy for degenerative disc disease. *Gene Ther.* **2004**, *11*, 390–401. [CrossRef] [PubMed]
95. Roughley, P.J. Biology of intervertebral disc aging and degeneration: Involvement of the extracellular matrix. *Spine* **2004**, *29*, 2691–2699. [CrossRef] [PubMed]
96. Boxberger, J.I.; Sen, S.; Yerramalli, C.S.; Elliott, D.M. Nucleus pulposus glycosaminoglycan content is correlated with axial mechanics in rat lumbar motion segments. *J. Orthop. Res.* **2006**, *24*, 1906–1915. [CrossRef] [PubMed]
97. Buckwalter, J.A. Aging and degeneration of the human intervertebral disc. *Spine* **1995**, *20*, 1307–1314. [PubMed]
98. Barbir, A.; Godburn, K.E.; Michalek, A.J.; Lai, A.; Monsey, R.D.; Iatridis, J.C. Effects of torsion on intervertebral disc gene expression and biomechanics, using a rat tail model. *Spine* **2011**, *36*, 607–614. [CrossRef] [PubMed]
99. O'Connell, G.D.; Guerin, H.L.; Elliott, D.M. Theoretical and uniaxial experimental evaluation of human annulus fibrosus degeneration. *J. Biomech. Eng.* **2009**, *131*, 111007. [CrossRef] [PubMed]
100. Natarajan, R.N.; Williams, J.R.; Andersson, G.B. Modeling changes in intervertebral disc mechanics with degeneration. *J. Bone Joint Surg. Am.* **2006**, *88* (Suppl. 2), 36–40. [CrossRef] [PubMed]
101. Haefeli, M.; Kalberer, F.; Saegesser, D.; Nerlich, A.G.; Boos, N.; Paesold, G. The course of macroscopic degeneration in the human lumbar intervertebral disc. *Spine* **2006**, *31*, 1522–1531. [CrossRef] [PubMed]

102. Nilsson, E.; Nakamae, T.; Olmarker, K. Pain behavior changes following disc puncture relate to nucleus pulposus rather than to the disc injury per se: An experimental study in rats. *Open Orthop. J.* **2011**, *5*, 72–77. [CrossRef] [PubMed]

103. Risbud, M.V.; Schaer, T.P.; Shapiro, I.M. Toward an understanding of the role of notochordal cells in the adult intervertebral disc: From discord to accord. *Dev. Dyn.* **2011**, *239*, 2141–2148. [CrossRef] [PubMed]

104. Korecki, C.L.; MacLean, J.J.; Iatridis, J.C. Dynamic compression effects on intervertebral disc mechanics and biology. *Spine* **2008**, *33*, 1403–1409. [CrossRef] [PubMed]

105. Gantenbein, B.; Calandriello, E.; Wuertz-Kozak, K.; Benneker, L.M.; Keel, M.J.; Chan, S.C. Activation of intervertebral disc cells by co-culture with notochordal cells, conditioned medium and hypoxia. *BMC Musculoskelet. Disord.* **2014**, *15*, 422. [CrossRef] [PubMed]

106. Boyd, L.M.; Chen, J.; Kraus, V.B.; Setton, L.A. Conditioned medium differentially regulates matrix protein gene expression in cells of the intervertebral disc. *Spine* **2004**, *29*, 2217–2222. [CrossRef] [PubMed]

107. Erwin, W.M.; Ashman, K.; O'Donnel, P.; Inman, R.D. Nucleus pulposus notochord cells secrete connective tissue growth factor and up-regulate proteoglycan expression by intervertebral disc chondrocytes. *Arthritis Rheum.* **2006**, *54*, 3859–3867. [CrossRef] [PubMed]

108. Bedore, J.; Sha, W.; McCann, M.R.; Liu, S.; Leask, A.; Seguin, C.A. Impaired intervertebral disc development and premature disc degeneration in mice with notochord-specific deletion of CCN2. *Arthritis Rheum.* **2013**, *65*, 2634–2644. [PubMed]

109. Bedore, J.; Leask, A.; Seguin, C.A. Targeting the extracellular matrix: Matricellular proteins regulate cell-extracellular matrix communication within distinct niches of the intervertebral disc. *Matrix Biol.* **2014**, *37*, 124–130. [CrossRef] [PubMed]

110. Erwin, W.M.; Islam, D.; Inman, R.D.; Fehlings, M.G.; Tsui, F.W. Notochordal cells protect nucleus pulposus cells from degradation and apoptosis: Implications for the mechanisms of intervertebral disc degeneration. *Arthritis Res. Ther.* **2011**, *13*, R215. [CrossRef] [PubMed]

111. Cornejo, M.C.; Cho, S.K.; Giannarelli, C.; Iatridis, J.C.; Purmessur, D. Soluble factors from the notochordal-rich intervertebral disc inhibit endothelial cell invasion and vessel formation in the presence and absence of pro-inflammatory cytokines. *Osteoarthritis Cartilage* **2015**, *23*, 487–496. [CrossRef] [PubMed]

112. Davies, J.A.; Cook, G.M.; Stern, C.D.; Keynes, R.J. Isolation from chick somites of a glycoprotein fraction that causes collapse of dorsal root ganglion growth cones. *Neuron* **1990**, *4*, 11–20. [CrossRef]

113. Anderson, C.N.; Ohta, K.; Quick, M.M.; Fleming, A.; Keynes, R.; Tannahill, D. Molecular analysis of axon repulsion by the notochord. *Development* **2003**, *130*, 1123–1133. [CrossRef] [PubMed]

114. Tolofari, S.K.; Richardson, S.M.; Freemont, A.J.; Hoyland, J.A. Expression of semaphorin 3a and its receptors in the human intervertebral disc: Potential role in regulating neural ingrowth in the degenerate intervertebral disc. *Arthritis Res. Ther.* **2010**, *12*, R1. [CrossRef] [PubMed]

115. Tang, X.; Jing, L.; Chen, J. Changes in the molecular phenotype of nucleus pulposus cells with intervertebral disc aging. *PLoS ONE* **2012**, *7*, e52020. [CrossRef] [PubMed]

116. Mathews, M.B. Comparative biochemistry of chondroitin sulphate-proteins of cartilage and notochord. *Biochem. J.* **1971**, *125*, 37–46. [CrossRef] [PubMed]

117. Bernhardt, R.R.; Schachner, M. Chondroitin sulfates affect the formation of the segmental motor nerves in zebrafish embryos. *Dev. Biol.* **2000**, *221*, 206–219. [CrossRef] [PubMed]

118. Masuda, T.; Fukamauchi, F.; Takeda, Y.; Fujisawa, H.; Watanabe, K.; Okado, N.; Shiga, T. Developmental regulation of notochord-derived repulsion for dorsal root ganglion axons. *Mol. Cell. Neurosci.* **2004**, *25*, 217–227. [CrossRef] [PubMed]

119. Purmessur, D.; Cornejo, M.C.; Cho, S.K.; Roughley, P.J.; Linhardt, R.J.; Hecht, A.C.; Iatridis, J.C. Intact glycosaminoglycans from intervertebral disc-derived notochordal cell-conditioned media inhibit neurite growth while maintaining neuronal cell viability. *Spine J.* **2015**, *15*, 1060–1069. [CrossRef] [PubMed]

120. Song, Y.Q.; Karasugi, T.; Cheung, K.M.; Chiba, K.; Ho, D.W.; Miyake, A.; Kao, P.Y.; Sze, K.L.; Yee, A.; Takahashi, A.; *et al.* Lumbar disc degeneration is linked to a carbohydrate sulfotransferase 3 variant. *J. Clin. Investig.* **2013**, *123*, 4909–4917. [CrossRef] [PubMed]

121. Thiele, H.; Sakano, M.; Kitagawa, H.; Sugahara, K.; Rajab, A.; Hohne, W.; Ritter, H.; Leschik, G.; Nurnberg, P.; Mundlos, S. Loss of chondroitin 6-*O*-sulfotransferase-1 function results in severe human chondrodysplasia with progressive spinal involvement. *Proc. Natl. Acad. Sci. USA* **2004**, *101*, 10155–10160. [CrossRef] [PubMed]

122. Hermanns, P.; Unger, S.; Rossi, A.; Perez-Aytes, A.; Cortina, H.; Bonafe, L.; Boccone, L.; Setzu, V.; Dutoit, M.; Sangiorgi, L.; *et al.* Congenital joint dislocations caused by carbohydrate sulfotransferase 3 deficiency in recessive larsen syndrome and humero-spinal dysostosis. *Am. J. Hum. Genet.* **2008**, *82*, 1368–1374. [CrossRef] [PubMed]
123. Yin, J.; Sakamoto, K.; Zhang, H.; Ito, Z.; Imagama, S.; Kishida, S.; Natori, T.; Sawada, M.; Matsuyama, Y.; Kadomatsu, K. Transforming growth factor-beta1 upregulates keratan sulfate and chondroitin sulfate biosynthesis in microglias after brain injury. *Brain Res.* **2009**, *1263*, 10–22. [CrossRef] [PubMed]
124. Johnson, W.E.; Caterson, B.; Eisenstein, S.M.; Hynds, D.L.; Snow, D.M.; Roberts, S. Human intervertebral disc aggrecan inhibits nerve growth *in vitro*. *Arthritis Rheum.* **2002**, *46*, 2658–2664. [CrossRef] [PubMed]
125. Olmarker, K.; Rydevik, B.; Nordborg, C. Autologous nucleus pulposus induces neurophysiologic and histologic changes in porcine cauda equina nerve roots. *Spine* **1993**, *18*, 1425–1432. [CrossRef] [PubMed]
126. Alini, M.; Eisenstein, S.M.; Ito, K.; Little, C.; Kettler, A.A.; Masuda, K.; Melrose, J.; Ralphs, J.; Stokes, I.; Wilke, H.J. Are animal models useful for studying human disc disorders/degeneration? *Eur. Spine J.* **2008**, *17*, 2–19. [CrossRef] [PubMed]
127. Hunter, C.J.; Matyas, J.R.; Duncan, N.A. The three-dimensional architecture of the notochordal nucleus pulposus: Novel observations on cell structures in the canine intervertebral disc. *J. Anat.* **2003**, *202*, 279–291. [CrossRef] [PubMed]
128. Hunter, C.J.; Matyas, J.R.; Duncan, N.A. The functional significance of cell clusters in the notochordal nucleus pulposus: Survival and signaling in the canine intervertebral disc. *Spine* **2004**, *29*, 1099–1104. [CrossRef] [PubMed]
129. Beachley, M.C.; Graham, F.H., Jr. Hypochondroplastic dwarfism (enchondral chondrodystrophy) in a dog. *J. Am. Vet. Med. Assoc.* **1973**, *163*, 283–284. [PubMed]
130. Hansen, H.J. Some animal bone and joint diseases of comparative interest. *Calcif. Tissue Res.* **1968**, *2*, 23. [CrossRef]
131. Smolders, L.A.; Bergknut, N.; Grinwis, G.C.; Hagman, R.; Lagerstedt, A.S.; Hazewinkel, H.A.; Tryfonidou, M.A.; Meij, B.P. Intervertebral disc degeneration in the dog. Part 2: Chondrodystrophic and non-chondrodystrophic breeds. *Vet. J.* **2013**, *195*, 292–299. [CrossRef] [PubMed]
132. Olby, N.; Harris, T.; Burr, J.; Munana, K.; Sharp, N.; Keene, B. Recovery of pelvic limb function in dogs following acute intervertebral disc herniations. *J. Neurotrauma* **2004**, *21*, 49–59. [CrossRef] [PubMed]
133. Cappello, R.; Bird, J.L.; Pfeiffer, D.; Bayliss, M.T.; Dudhia, J. Notochordal cell produce and assemble extracellular matrix in a distinct manner, which may be responsible for the maintenance of healthy nucleus pulposus. *Spine* **2006**, *31*, 873–882. [CrossRef] [PubMed]
134. Klauser, M.; Forterre, F.; Doherr, M.; Zurbriggen, A.; Spreng, D.; Forterre, S. Evaluation of apoptotic cell death in normal and chondrodystrophic canine intervertebral discs. *Vet. Sci. Dev.* **2012**, *2*, e6. [CrossRef]
135. Rosenthal, N.; Brown, S. The mouse ascending: Perspectives for human-disease models. *Nat. Cell Biol.* **2007**, *9*, 993–999. [CrossRef] [PubMed]
136. Nguyen, D.; Xu, T. The expanding role of mouse genetics for understanding human biology and disease. *Dis. Model. Mech.* **2008**, *1*, 56–66. [CrossRef] [PubMed]
137. Hunter, C.J.; Matyas, J.R.; Duncan, N.A. Cytomorphology of notochordal and chondrocytic cells from the nucleus pulposus: A species comparison. *J. Anat.* **2004**, *205*, 357–362. [CrossRef] [PubMed]
138. Nagy, A. Cre recombinase: The universal reagent for genome tailoring. *Genesis* **2000**, *26*, 99–109. [CrossRef]
139. Perantoni, A.O.; Timofeeva, O.; Naillat, F.; Richman, C.; Pajni-Underwood, S.; Wilson, C.; Vainio, S.; Dove, L.F.; Lewandoski, M. Inactivation of FGF8 in early mesoderm reveals an essential role in kidney development. *Development* **2005**, *132*, 3859–3871. [CrossRef] [PubMed]
140. Lange, A.; Gegg, M.; Burtscher, I.; Bengel, D.; Kremmer, E.; Lickert, H. Fltp(T2AiCre): A new knock-in mouse line for conditional gene targeting in distinct mono- and multiciliated tissues. *Differentiation* **2012**, *83*, S105–S113. [CrossRef] [PubMed]
141. Ovchinnikov, D.A.; Deng, J.M.; Ogunrinu, G.; Behringer, R.R. Col2a1-directed expression of Cre recombinase in differentiating chondrocytes in transgenic mice. *Genesis* **2000**, *26*, 145–146. [CrossRef]
142. Schipani, E.; Ryan, H.E.; Didrickson, S.; Kobayashi, T.; Knight, M.; Johnson, R.S. Hypoxia in cartilage: HIF-1alpha is essential for chondrocyte growth arrest and survival. *Genes Dev.* **2001**, *15*, 2865–2876. [PubMed]

143. Sakai, K.; Hiripi, L.; Glumoff, V.; Brandau, O.; Eerola, R.; Vuorio, E.; Bosze, Z.; Fassler, R.; Aszodi, A. Stage-and tissue-specific expression of a Col2a1-Cre fusion gene in transgenic mice. *Matrix Biol.* **2001**, *19*, 761–767. [CrossRef]
144. Terpstra, L.; Prud'homme, J.; Arabian, A.; Takeda, S.; Karsenty, G.; Dedhar, S.; St-Arnaud, R. Reduced chondrocyte proliferation and chondrodysplasia in mice lacking the integrin-linked kinase in chondrocytes. *J. Cell Biol.* **2003**, *162*, 139–148. [CrossRef] [PubMed]
145. Harfe, B.D.; Scherz, P.J.; Nissim, S.; Tian, H.; McMahon, A.P.; Tabin, C.J. Evidence for an expansion-based temporal shh gradient in specifying vertebrate digit identities. *Cell* **2004**, *118*, 517–528. [CrossRef] [PubMed]
146. Semba, K.; Araki, K.; Li, Z.; Matsumoto, K.; Suzuki, M.; Nakagata, N.; Takagi, K.; Takeya, M.; Yoshinobu, K.; Araki, M.; *et al.* A novel murine gene, sickle tail, linked to the danforth's short tail locus, is required for normal development of the intervertebral disc. *Genetics* **2006**, *172*, 445–456. [CrossRef] [PubMed]
147. Abe, K.; Araki, K.; Tanigawa, M.; Semba, K.; Ando, T.; Sato, M.; Sakai, D.; Hiyama, A.; Mochida, J.; Yamamura, K. A Cre knock-in mouse line on the sickle tail locus induces recombination in the notochord and intervertebral disks. *Genesis* **2012**, *50*, 758–765. [CrossRef] [PubMed]
148. Ang, S.L.; Wierda, A.; Wong, D.; Stevens, K.A.; Cascio, S.; Rossant, J.; Zaret, K.S. The formation and maintenance of the definitive endoderm lineage in the mouse: Involvement of HNF3/forkhead proteins. *Development* **1993**, *119*, 1301–1315. [PubMed]
149. Ang, S.L.; Rossant, J. HNF-3 beta is essential for node and notochord formation in mouse development. *Cell* **1994**, *78*, 561–574. [CrossRef]
150. Park, E.J.; Sun, X.; Nichol, P.; Saijoh, Y.; Martin, J.F.; Moon, A.M. System for tamoxifen-inducible expression of Cre-recombinase from the Foxa2 locus in mice. *Dev. Dyn.* **2008**, *237*, 447–453. [CrossRef] [PubMed]
151. Uetzmann, L.; Burtscher, I.; Lickert, H. A mouse line expressing Foxa2-driven Cre recombinase in node, notochord, floorplate, and endoderm. *Genesis* **2008**, *46*, 515–522. [CrossRef] [PubMed]
152. Merceron, C.; Mangiavini, L.; Robling, A.; Wilson, T.L.; Giaccia, A.J.; Shapiro, I.M.; Schipani, E.; Risbud, M.V. Loss of HIF-1alpha in the notochord results in cell death and complete disappearance of the nucleus pulposus. *PLoS ONE* **2014**, *9*, e110768. [CrossRef] [PubMed]
153. Beckers, A.; Alten, L.; Viebahn, C.; Andre, P.; Gossler, A. The mouse homeobox gene noto regulates node morphogenesis, notochordal ciliogenesis, and left right patterning. *Proc. Natl. Acad. Sci. USA* **2007**, *104*, 15765–15770. [CrossRef] [PubMed]
154. Darras, S.; Nishida, H. The BMP signaling pathway is required together with the FGF pathway for notochord induction in the ascidian embryo. *Development* **2001**, *128*, 2629–2638. [PubMed]
155. Sobajima, S.; Shimer, A.L.; Chadderdon, R.C.; Kompel, J.F.; Kim, J.S.; Gilbertson, L.G.; Kang, J.D. Quantitative analysis of gene expression in a rabbit model of intervertebral disc degeneration by real-time polymerase chain reaction. *Spine J.* **2005**, *5*, 14–23. [CrossRef] [PubMed]
156. Takae, R.; Matsunaga, S.; Origuchi, N.; Yamamoto, T.; Morimoto, N.; Suzuki, S.; Sakou, T. Immunolocalization of bone morphogenetic protein and its receptors in degeneration of intervertebral disc. *Spine* **1999**, *24*, 1397–1401. [PubMed]
157. Kim, H.; Lee, J.U.; Moon, S.H.; Kim, H.C.; Kwon, U.H.; Seol, N.H.; Kim, H.J.; Park, J.O.; Chun, H.J.; Kwon, I.K.; *et al.* Zonal responsiveness of the human intervertebral disc to bone morphogenetic protein-2. *Spine* **2009**, *34*, 1834–1838. [CrossRef] [PubMed]
158. Than, K.D.; Rahman, S.U.; Vanaman, M.J.; Wang, A.C.; Lin, C.Y.; Zhang, H.; la Marca, F.; Park, P. Bone morphogenetic proteins and degenerative disk disease. *Neurosurgery* **2012**, *70*, 996–1002. [CrossRef] [PubMed]
159. Wei, A.; Brisby, H.; Chung, S.A.; Diwan, A.D. Bone morphogenetic protein-7 protects human intervertebral disc cells *in vitro* from apoptosis. *Spine J.* **2008**, *8*, 466–474. [CrossRef] [PubMed]
160. Carragee, E.J.; Hurwitz, E.L.; Weiner, B.K. A critical review of recombinant human bone morphogenetic protein-2 trials in spinal surgery: Emerging safety concerns and lessons learned. *Spine J.* **2011**, *11*, 471–491. [CrossRef] [PubMed]
161. Smucker, J.D.; Rhee, J.M.; Singh, K.; Yoon, S.T.; Heller, J.G. Increased swelling complications associated with off-label usage of rhBMP-2 in the anterior cervical spine. *Spine* **2006**, *31*, 2813–2819. [CrossRef] [PubMed]
162. Ukita, K.; Hirahara, S.; Oshima, N.; Imuta, Y.; Yoshimoto, A.; Jang, C.W.; Oginuma, M.; Saga, Y.; Behringer, R.R.; Kondoh, H.; *et al.* Wnt signaling maintains the notochord fate for progenitor cells and supports the posterior extension of the notochord. *Mech. Dev.* **2009**, *126*, 791–803. [CrossRef] [PubMed]

163. Kondo, N.; Yuasa, T.; Shimono, K.; Tung, W.; Okabe, T.; Yasuhara, R.; Pacifici, M.; Zhang, Y.; Iwamoto, M.; Enomoto-Iwamoto, M. Intervertebral disc development is regulated by Wnt/β-catenin signaling. *Spine* **2011**, *36*, E513–E518. [CrossRef] [PubMed]
164. Winkler, T.; Mahoney, E.J.; Sinner, D.; Wylie, C.C.; Dahia, C.L. Wnt signaling activates Shh signaling in early postnatal intervertebral discs, and re-activates Shh signaling in old discs in the mouse. *PLoS ONE* **2014**, *9*, e98444. [CrossRef] [PubMed]
165. Dahia, C.L.; Mahoney, E.; Wylie, C. Shh signaling from the nucleus pulposus is required for the postnatal growth and differentiation of the mouse intervertebral disc. *PLoS ONE* **2012**, *7*, e35944. [CrossRef] [PubMed]
166. Smolders, L.A.; Meij, B.P.; Riemers, F.M.; Licht, R.; Wubbolts, R.; Heuvel, D.; Grinwis, G.C.; Vernooij, H.C.; Hazewinkel, H.A.; Penning, L.C.; *et al.* Canonical Wnt signaling in the notochordal cell is upregulated in early intervertebral disk degeneration. *J. Orthop. Res.* **2012**, *30*, 950–957. [CrossRef] [PubMed]
167. Showell, C.; Binder, O.; Conlon, F.L. T-box genes in early embryogenesis. *Dev. Dyn.* **2004**, *229*, 201–218. [CrossRef] [PubMed]
168. Pillay, N.; Plagnol, V.; Tarpey, P.S.; Lobo, S.B.; Presneau, N.; Szuhai, K.; Halai, D.; Berisha, F.; Cannon, S.R.; Mead, S.; *et al.* A common single-nucleotide variant in T is strongly associated with chordoma. *Nat. Genet.* **2012**, *44*, 1185–1187. [CrossRef] [PubMed]
169. Yang, X.R.; Ng, D.; Alcorta, D.A.; Liebsch, N.J.; Sheridan, E.; Li, S.; Goldstein, A.M.; Parry, D.M.; Kelley, M.J. T (brachyury) gene duplication confers major susceptibility to familial chordoma. *Nat. Genet.* **2009**, *41*, 1176–1178. [CrossRef] [PubMed]
170. McMaster, M.L.; Goldstein, A.M.; Bromley, C.M.; Ishibe, N.; Parry, D.M. Chordoma: Incidence and survival patterns in the united states, 1973–1995. *Cancer Causes Control* **2001**, *12*, 1–11. [CrossRef] [PubMed]
171. Chambers, K.J.; Lin, D.T.; Meier, J.; Remenschneider, A.; Herr, M.; Gray, S.T. Incidence and survival patterns of cranial chordoma in the united states. *Laryngoscope* **2014**, *124*, 1097–1102. [CrossRef] [PubMed]
172. Yamaguchi, T.; Suzuki, S.; Ishiiwa, H.; Ueda, Y. Intraosseous benign notochordal cell tumours: Overlooked precursors of classic chordomas? *Histopathology* **2004**, *44*, 597–602. [CrossRef] [PubMed]

2

The α-Tubulin gene *TUBA1A* in Brain Development: A Key Ingredient in the Neuronal Isotype Blend

Jayne Aiken [1], Georgia Buscaglia [2], Emily A. Bates [2] and Jeffrey K. Moore [1,*]

[1] Department of Cell and Developmental Biology, University of Colorado School of Medicine, MS8108, 12801 E 17th Ave, Aurora, CO 80045, USA; JAYNE.AIKEN@ucdenver.edu
[2] Department of Pediatrics, University of Colorado Anschutz Medical Campus, Aurora, CO 80045, USA; GEORGIA.BUSCAGLIA@ucdenver.edu (G.B.); EMILY.BATES@ucdenver.edu (E.A.B.)
* Correspondence: jeffrey.moore@ucdenver.edu

Abstract: Microtubules are dynamic cytoskeletal polymers that mediate numerous, essential functions such as axon and dendrite growth and neuron migration throughout brain development. In recent years, sequencing has revealed dominant mutations that disrupt the tubulin protein building blocks of microtubules. These tubulin mutations lead to a spectrum of devastating brain malformations, complex neurological and physical phenotypes, and even fatality. The most common tubulin gene mutated is the α-tubulin gene *TUBA1A*, which is the most prevalent α-tubulin gene expressed in post-mitotic neurons. The normal role of TUBA1A during neuronal maturation, and how mutations alter its function to produce the phenotypes observed in patients, remains unclear. This review synthesizes current knowledge of TUBA1A function and expression during brain development, and the brain malformations caused by mutations in *TUBA1A*.

Keywords: TUBA1A; tubulinopathy; microtubule; neurodevelopment; lissencephaly; polymicrogyria

---

1. Introduction

Brain development is a highly complex process that requires careful coordination of cell proliferation, differentiation, migration, axon and dendrite growth and guidance, and synapse formation. Each of these cellular tasks rely on the proper regulation of microtubules, which are dynamic cytoskeletal polymers that help provide structure and generate force in all eukaryotic cell types. Neurons require elaborate networks of microtubules to support their complex morphologies and to perform a variety of critical functions. In developing neurons, dynamic microtubules are necessary to support essential functions such as morphological changes that underlie polarization and migration [1–3]. In particular, microtubules are key players in the initiation and extension of neurites, axon specification, and neuronal migration. In adult neurons, microtubules provide a stable backbone for established axons and dendrites, acting as essential trafficking routes for microtubule-based motors to carry various cargos from one compartment of the neuron to another. Microtubules also play an important role in synaptic plasticity, where they facilitate the morphological changes to dendritic spines that are thought to underlie synaptic enhancement and depression [4–6]. How microtubules are regulated to support distinct roles in different compartments of the neuron and at different stages of development to adulthood is widely unknown.

An emerging theme in the field is that microtubules may be regulated at the level of their protein building blocks. Microtubules are assembled from protein subunits known as tubulins, which are obligate heterodimers consisting of α- and β-tubulin polypeptides. Nearly all eukaryotes express multiple, different genes encoding α- and β-tubulin, known as isotypes. As many as nine α- and ten β-tubulin isotypes have been identified by genome analysis in humans [7,8]. One isotype, α-tubulin *TUBA1A*, is of particular importance to neural development [9–11], and will be the focus of this review. Throughout the history of tubulin research, the nomenclature describing *TUBA1A* has changed. Various names that have been used to describe the *TUBA1A* and *TUBA1B* isotypes are included in Table 1.

Table 1. α-tubulin protein isotype nomenclature.

	TUBA1A			TUBA1B		
Organism	Human	Mouse	Rat	Human	Mouse	Rat
Aliases	**TUBA1A** [7] **b-α-1** [7,12]	**Tuba1a** [7] **Tuba1** [7] **M-α-1** [10]	**Tuba1a** [7] **Tuba1** [7] **α-T14** [9] **T-α-1** [13]	**TUBA1B** [7] **k-α-1** [7,12]	**Tuba1b** [7] **M-α-2** [10] **Tuba2** [7]	**Tuba1b** [7] **T26** [13]

The strongest evidence for the critical role of tubulin isotypes in brain development comes from a growing number of heterozygous *de novo* missense mutations identified in isotypes of human patients with brain malformations, known as tubulinopathies [11–20]. The tubulin gene most commonly mutated in tubulinopathy patients is *TUBA1A*, the human isoform of α1 tubulin that is highly expressed in post-mitotic neurons [20,21]. Patients with mutations in *TUBA1A* present 5 classes of brain malformations including microlissencephaly, lissencephaly, and polymicrogyria, and a broad spectrum of clinical effects [20]. The spectrum of malformations suggests an important and complex role for TUBA1A in brain development, and mutations may alter TUBA1A protein function in ways that ultimately have drastically different impacts on brain development.

Exploring how TUBA1A contributes to neuronal function will provide important insight into how cells "tune" their microtubule networks using the programmed expression of a specific tubulin isotype. A major outstanding question in the microtubule field is how this tuning is achieved at the molecular level. Distinct tubulin isotypes could lead to outright functional differences due to intrinsic changes to the tubulin protein structure, which could alter microtubule dynamic instability or extrinsic binding of important microtubule-associated proteins (MAPs). Alternatively, carefully timed expression of a specific tubulin isotype could increase the concentration of tubulin present in the cell, supplying a necessary burst of new microtubule assembly during morphological changes such as axon outgrowth. In this review, we will discuss the current understanding of the regulation of microtubule function during neuronal development. We will particularly focus on *TUBA1A*, discussing its regulated expression, mutations associated with brain malformations, and how this α-tubulin might uniquely contribute to normal brain development.

2. Overview of Microtubules and Tubulin Isotypes

2.1. Microtubule Basics

Tubulin, the fundamental protein subunit of the microtubule, is a heterodimer of α- and β-tubulin polypeptides. α- and β-tubulins are conserved across eukaryotic evolution and are related to filament-forming proteins in prokaryotes and bacteriophage [22,23]. These proteins share the common characteristic of polymerizing into dynamic filaments, and this characteristic depends on their ability to bind and hydrolyze GTP. The tubulin heterodimer binds to two molecules of GTP at two separate sites. The so-called "non-exchangeable" or "N-site" is located at the intradimer interface, between α- and β-tubulin [24] (Figure 1A). GTP bound to the N-site is not hydrolyzed and exchanges at a slow rate [25]. In contrast, the "exchangeable", or "E-site", is located at the interdimer interface formed by the β-subunit of one heterodimer and the α-subunit of a neighboring heterodimer [24] (Figure 1A). GTP at the E-site can be hydrolyzed to GDP and subsequently exchanged for a new GTP nucleotide [26]. Importantly, the nucleotide status at the E-site plays a determining role in the conformation of the heterodimer and its interactions with other heterodimers (see description of "maturation" below) [27–29]. These changes alter tubulin activity by either affecting the conformation of the free heterodimer [30] or when it is packed into an microtubule [28,31].

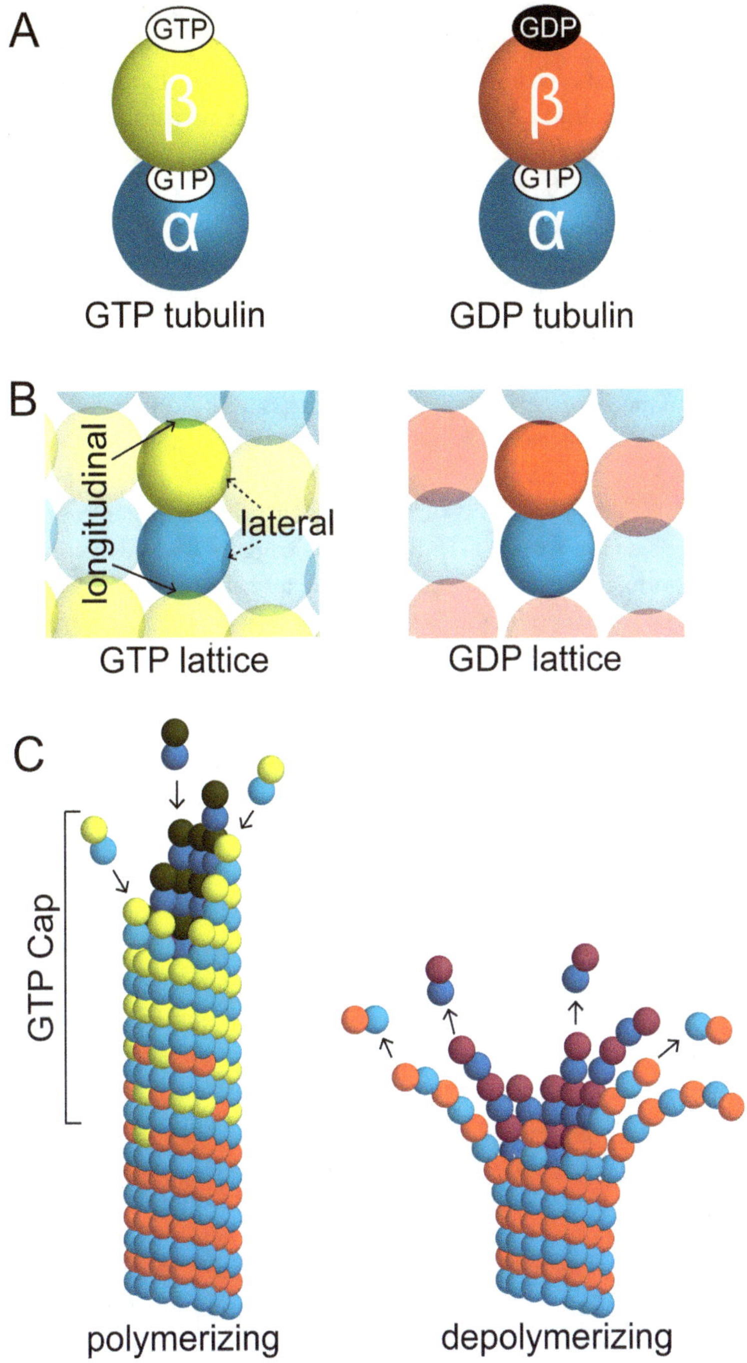

Figure 1. Microtubule structure and dynamics. (**A**) Heterodimer conformation in GTP and GDP states; (**B**) lattice conformation with labeled longitudinal and lateral interfaces; and (**C**) Microtubule conformation during polymerization and depolymerization.

α/β-Heterodimers polymerize into a sheet-like conformation, known as the microtubule lattice (Figure 1B). The lattice consists of longitudinal chains of heterodimers arranged head-to-tail, called protofilaments. Longitudinal interactions involve an extensive hydrophobic interface between α-tubulin of one heterodimer and β-tubulin of the adjacent heterodimer, which completes the E-site [32].

Protofilaments bind along their lateral sides to other protofilaments. Lateral interactions involve α-tubulin of one protofilament binding to the α-tubulins of neighboring protofilaments, and β-tubulin binding to neighboring β-tubulins. In contrast to longitudinal interactions, the interfaces of lateral interactions consist of flexible loop regions of α- and β-tubulin and feature prominent electrostatic contributions [28,32]. The binding energy of lateral interactions is predicted to be much weaker than longitudinal interactions, based on computational modeling [33,34]. In most cells, the lattice consists of 13 protofilaments that close into a cylindrical filament—the microtubule (Figure 1C). The closed cylindrical confirmation of the lattice restricts heterodimer addition or loss (i.e., polymerization or depolymerization) to the ends of the microtubule.

Microtubules are unique among cytoskeletal filaments in that they exhibit a behavior known as "dynamic instability", which is defined as stochastic switching between states polymerization, where heterodimers are added to the growing microtubule lattice, and depolymerization, where heterodimers leave the shrinking microtubule lattice. Dynamic instability is an intrinsic property of microtubules, and is based on changes in the interactions between heterodimers in the lattice. Current models suggest a mechanism involving the "maturation" of tubulin heterodimers in the lattice, which can be depicted in a step-wise manner: (1) an "immature" heterodimer with GTP bound to β-tubulin at the nascent E-site binds to the microtubule end; (2) the next heterodimer arrives at the microtubule end and binds to the exposed β-tubulin of the first heterodimer, completing the E-site and stimulating GTP hydrolysis [26]; and (3) GTP hydrolysis triggers a wave of conformational changes (i.e., "maturation") that compact the heterodimer and allosterically alter interactions with neighboring heterodimers [27,28,30,31]. At this point, the "mature" GDP-bound heterodimer has a weak affinity for the microtubule lattice and favors disassembly; however, the continued addition of new GTP-bound heterodimers at the growing end buries mature heterodimers within the lattice and prevents their escape. This layer of new GTP-bound heterodimers at the growing end is known as the "GTP cap" (Figure 1C). As long as the addition of new GTP-bound heterodimers outpaces the maturation of heterodimers in the lattice, the microtubule will continue to polymerize.

The switch from polymerization to depolymerization is known as catastrophe. Catastrophe is thought to be triggered when the GTP cap is exhausted; that is, when heterodimer addition slows and the cap matures to contain a threshold number of GDP-bound heterodimers. The structural changes in the microtubule that accompany catastrophe are poorly defined. However, cryoelectron microscopy of depolymerizing microtubules shows protofilaments forming "ram's horns" that curl outward from the lattice [30,35] (Figure 1C). The curling of protofilaments away from each other suggests that conformational changes driven by maturation break lateral interactions between protofilaments before breaking longitudinal interactions within protofilaments. Lateral interactions are therefore likely to play an important role in the catastrophe mechanism, and are a key control point for regulating the organization and function of microtubule networks in cells.

Eukaryotic cells harness the dynamic instability of microtubules to construct highly organized networks for transporting cargoes and generating forces across large intracellular distances. The organization of microtubule networks and the generation of movement and force along microtubules is controlled by a diverse array of microtubule-associated proteins (MAPs). For comprehensive discussions on MAPs involved in different neuronal maturation stages, we refer readers to the following reviews [36–39]. Of particular interest for this review are the kinesins and dyneins—motor proteins that drive directional movement along microtubules.

Kinesins are a diverse family of ATPases that use energy from ATP hydrolysis to power directional movement along microtubules. The mechanism of kinesin motility involves the coordination of the head domains, which contain the ATP- and microtubule-binding activities, with adjacent neck linker domains that swing each head forward in an alternating, step-wise manner [40]. Among the 45 kinesin genes identified in mammals, several play important roles during brain development [41–43]. Dyneins represent a structurally and mechanistically different type of microtubule motor. Dynein also uses its ATPase activity to drive directional movement; however, the mechanism is very different from that of kinesin. The dynein mechanism involves a coordinated progression of conformational changes to mechanically link the nucleotide cycle of its AAA + ATPase domains to a microtubule-binding domain that is located ~15 nm away [44]. Furthermore, dyneins predominantly move toward the minus ends of microtubules, while most kinesins move toward the plus ends. Dynein motility is regulated by an assortment of dynein-binding proteins that regulate its speed, level of force production, and cargoes [45–54]. This regulation may explain how a single cytoplasmic dynein gene could be necessary for diverse roles in cells, while kinesins have undergone evolutionary diversification giving rise to 45 genes with specified functions. Because of the differences between dyneins and kinesins, it is likely that these motors use different mechanisms to bind and move along the microtubule surface. However, the contribution of the microtubule surface to motor activity and regulation is poorly understood.

Although much research in the microtubule field focuses on MAPs and motors that regulate microtubule networks by binding and moving along them, it is becoming clear that tubulin heterodimers themselves are intrinsic key regulators. Tubulins exhibit molecular differences that can be genetically encoded through different α- and β-tubulin genes, or conferred by posttranslational modifications. These molecular differences provide cells with a toolkit for changing the properties of microtubule networks.

2.2. Tubulin Isotypes

Nearly all eukaryotes express multiple, distinct genes for α- and β-tubulin, known as tubulin isotypes. Recent analysis of the human genome identified nine α-tubulin isotypes and ten β-tubulin isotypes, along with dozens of pseudogenes [7,8]. In addition to their different chromosomal locations, isotypes can be distinguished by three features: (1) the amino acid sequences they encode; (2) nucleotide sequences of the 3′ untranslated region (UTR); and (3) expression levels in different tissues or developmental stages. We will focus on key differences amongst the α-tubulin isotypes.

The amino acid sequence of α-tubulin is strongly conserved across eukaryotic evolution. For example, the human α-tubulin TUBA1A exhibits 86% sequence identity to α-tubulin in the unicellular eukaryote *Giardia lambia* and 75% identity to the α-tubulin in the budding yeast, *Saccharomyces cerevisiae*. Within the human α-tubulin isotypes (Table 2), there are a small number of amino acid sequence differences. Importantly, these differences are conserved in isotype homologues across species, suggesting that selective pressure may maintain isotype-specific sequence differences (Table 2). The majority of these differences are found in the 15–27 amino acids at the very carboxy-terminus; a region known as the Carboxy-Terminal Tail (CTT). CTTs decorate the outer surface of the microtubule, contain an abundance of amino acids with negatively-charged side chains, and are major sites of post-translational modifications (PTMs), which will be discussed in the next section. Current models propose that the molecular diversity generated by genetically-encoded differences and PTMs in the CTTs act as a "tubulin code" that regulates the activities of MAPs and motors at the microtubule surface [55,56]. Consistent with this model, several studies show that altering the amino acid sequence within the CTT causes changes in microtubule function in vivo [57–61]. Besides the CTTs, human α-tubulin isotypes also exhibit amino acid differences at other positions. For example, the human TUBA1A and TUBA1B isotypes have identical CTTs, but have different amino acids at position 232 and 340. Position 232 is buried deep within α-tubulin, while position 340 lies on the microtubule surface, near the interdimer interface [62]. Whether these amino acid differences lead to functional differences between TUBA1A and TUBA1B has not been investigated. This exemplifies a general deficit in our understanding of α-tubulin isotypes—although we have known of amino acid differences between α-tubulin isotypes for over 30 years, the functional consequences of these differences remain largely unexplored.

Table 2. Isotypes of α-tubulin.

Human Gene	Gene Accession	Protein Accession	Identity to TUBA1A	CTT Amino Acid Sequence	Mouse Gene	Identity to Human Isotype
TUBA1A	NM_006009	NP_006000	-	MAALEKDYEEVGVDSVEGEGEEEGEEY	*Tuba1a*	451/451
TUBA1B	NM_006082	NP_006073	449/451	MAALEKDYEEVGVDSVEGEGEEEGEEY	*Tuba1b*	451/451
TUBA1C	NM_032704	NP_116093	442/451	MAALEKDYEEVGADSADGEDEGEEY	*Tuba1c*	446/449
TUBA3C	NM_006001	NP_005992	440/451	LAALEKDYEEVGVDSVEAEAEEGEEY		
TUBA3D	NM_080386	NP_525125	440/451	LAALEKDYEEVGVDSVEAEAEEGEEY		
TUBA3E	NM_207312	NP_997195	435/451	LAALEKDCEEVGVDSVEAEAEEGEEY		
TUBA4A	NM_006000	NP_005991	432/450	MAALEKDYEEVGIDSYEDEDEGEE	*Tuba4a*	448/448
TUBA8	NM_018943	NP_061816	399/439	LAALEKDYEEVGTDSFEEENEGEEF	*Tuba8*	446/449
TUBAL3	NM_024803	NP_079079	329/444	LAALERDYEEVAQSF	*Tubal3*	369/446

Human genes identified by Khodiyar et al., 2007 are listed, along with accession IDs for DNA and protein. CTT sequences depict the last 15–27 genetically encoded amino acids. Underlined residue is the major site of polyglutamylation [63]. Gray denotes residues that differ from TUBA1A.

In contrast to the highly similar coding sequences of α-tubulin isotypes, the 3′-UTR regions are highly divergent. Vertebrate tubulin isotypes were originally identified from cDNA clones, and it was noted at that time that each isotype contained a distinct 3′UTR sequence [10,64,65]. What makes the differences in the 3′-UTR region especially intriguing is that they are conserved across species, with the 3′-UTR regions of human *TUBA1A* and *TUBA1B* sharing interspecies homology with rat *Tuba1a* and *Tuba1b*, respectively [64]. This conservation of the noncoding region implies selective pressure, and raises the question of how 3′-UTR regions contribute to function. Presumably, the 3′-UTR regions could differentially regulate mRNA stability and/or localization within a cell. A beautiful example of this regulation comes from studies in zebrafish, where the 3′-UTR of the β-tubulin isotype *tubb5* targets the mRNA to axons and distal growth cones during development [66]. This mRNA targeting could provide an appealing mechanism for increasing the supply of tubulin heterodimers at a region of the cytoplasm that is far from the nucleus. Whether the 3′-UTRs of α-tubulin isotypes provide similar regulation awaits discovery.

The third distinguishing feature of tubulin isotypes is their pattern of expression across different tissues and developmental stages. The β-tubulin isotypes have been extensively mapped to different tissues, cell types, and, in some cases, sub-cellular localization [67–71]. We have a comparatively poor understanding of the distributions of α-tubulin isotypes, with the exception of *TUBA1A*. *TUBA1A* is strongly and specifically expressed in the developing nervous system, and provides over 95% of the α-tubulin in the embryonic brain [10,72]. We will extensively discuss *TUBA1A* expression and its regulation in subsequent sections. Spatial and temporal expression data on α-tubulin isotypes remain sparse and may need to be readdressed. Tracking the expression of tubulin isotypes at the protein level is particularly challenging due to the high degree of homology, and will benefit from new approaches to selectively label isotypes without impairing their functions.

Why different cell types express specific tubulin isotypes is a long-standing question. On one hand, the coding and non-coding differences between isotypes could impart functional differences within microtubule networks. The strongest evidence for specific functional roles for α-tubulin isotypes comes from studies of in *Drosophila* and *C. elegans*, which demonstrate isotype-specific requirements for generating proper axonemal structures within cilia and flagella [73–75]. These findings underscore the possibility that isotypes may play specific roles in building complex microtubule architectures and raise the question of whether specific isotypes could be required to build other complex microtubule structures, such as those in neurons. An alternative, but not mutually exclusive explanation for differential isotype expression is that it provides a convenient mechanism to regulate the levels of tubulin protein in a cell. This is a particularly important challenge considering that cells must balance the levels of α- and β-tubulin to form heterodimers, and excess monomer, particularly β-tubulin, can be toxic [76–78]. Studies investigating tubulin isotypes have provided some answers to these questions, but our overall understanding of isotype biology is still in its infancy.

2.3. Tubulin Post-Translational Modifications

In addition to different isotypes, the tubulin subunits can be regulated by diverse PTMs. Various PTMs have been identified on neuronal microtubules, including detyrosination/tyrosination, polyglutamylation, acetylation, and polyamination. In some cases, these PTMs can further amplify genetically-encoded differences between isotypes, since the modified amino acid residues on α- or β-tubulin are only found in a subset of isotypes. In this section, we will briefly summarize the current evidence of major classes of PTMs and their functional impacts on neuronal microtubules. For comprehensive reviews of tubulin PTMs, the reader is referred to several recent reviews [55,79,80].

Detyrosination/tyrosination refers to the enzymatic removal of tyrosine from the CTT of α-tubulin (detyrosination), and subsequent re-ligation (tyrosination). This tyrosine is genetically encoded by six α-tubulin isotypes in humans, including *TUBA1A* (Table 2). Although the enzyme that catalyzes the detyrosination reaction in vivo has not been identified, it is well-established that tyrosination is catalyzed by Tubulin Tyrosine Ligase (TTL), which exclusively acts on free heterodimers to ligate tyrosine onto a detyrosinated α-tubulin [81,82]. Antibodies that selectively bind to either tyrosinated or detyrosinated α-tubulin show that these PTMs can be differentially enriched on specific microtubules within a network, or specific regions of an individual microtubule, and appear to correlate with the age of the microtubule lattice [83–85]. Accordingly, neurons exhibit an enrichment of tyrosinated α-tubulin at regions containing more newly-assembled microtubules (e.g., neurites and the distal ends of axons) while detyrosinated α-tubulin is primarily enriched in regions with older and highly stable microtubules (the axon shaft) [86–88]. Neurons lacking the TTL enzyme undergo aberrant neuronal development, indicating an important role for the cycle of detyrosination/tyrosination [89]. Currently, there is little to no evidence that tyrosination status influences the intrinsic stability of a microtubule. In contrast, tyrosination status does impact the binding of various proteins to the microtubule surface. Cytoskeleton-Associated Protein Glycine-rich (CAP-Gly) domains selectively bind to EEY/F motifs, such as those found in the CTT of α-tubulins, and this binding strongly depends on the aromatic side chain of the terminal Y/F residue [58,90–92]. Detyrosination therefore inhibits the microtubule-binding activity of CAP-Gly-domains. CAP-Gly domains are found in several microtubule-associated proteins, including CLIP170, the tubulin binding co-factors TBCB and TBCE, and the $p150^{glued}$ subunit of dynactin. Recent studies show that the binding of $p150^{glued}$ and CLIP170 to tyrosinated tubulin promotes the initiation of retrograde dynein-dependent transport in vitro and at the distal ends of axons [93,94]. Thus, detyrosination/tyrosination provides a system for local control of microtubule-MAP interactions and transport within a microtubule network.

Polyglutamylation is the most abundant tubulin PTM within the brain and involves the addition of variable-length chains of glutamate residues to genetically-encoded glutamates in the CTT regions of α- or β-tubulin. Polyglutamylation was originally found to extend from α-tubulin peptide sequences that are only present in the TUBA1A and TUBA1B isotypes, with the glutamate chains extending from the γ-carboxyl group of the genetically-encoded glutamate at position 445 [63]. This may represent the primary polyglutamylation site on α-tubulin; however, alternative glutamates in the α-tubulin CTT may also be targeted for modification. The enzymes that catalyze polyglutamylation belong to the Tubulin Tyrosine Ligase Like (TTLL) family [95]. To date, twelve mouse TTLLs have been identified based on sequence homology, although only six of these exhibit polyglutamylase activity in vitro [96]. Among these, TTLLs 5, 6, 11, and 13 selectively modify α-tubulin [95–97]. The removal of glutamate residues is catalyzed by Cytosolic Carboxypeptidase (CCP) enzymes, which can either shorten polyglutamate chains or remove the genetically encoded, penultimate glutamate from detyrosinated α-tubulin, creating a truncated species known as Δ2-tubulin [98]. Similar to detyrosination/tyrosination, polyglutamylation is thought to alter the interactions of MAPs and motors at the microtubule surface. The clearest example is the regulation of microtubule severing by spastin. Here, the length of glutamate chains provides a tunable signal for directing spastin's severing activity [99,100]. Although the mechanistic role of polyglutamylation in neurons is still unclear, it appears to be developmentally regulated and correlate with degeneration. Both in vivo and in vitro experiments show that levels of polyglutamylation vary in different brain regions and increase over development, reaching their highest level in mature neurons [101]. Loss of function mutations affecting the CCP1 enzyme disrupt the normal program of tubulin PTMs in mouse neurons, increasing the amount of polyglutamylated tubulin and decreasing the Δ2-tubulin species [98]. CCP1 disruption causes Purkinje cell degeneration in mice, in a manner that depends on increased polyglutamylation [102–104]. This suggests an important yet poorly understand role for regulated polyglutamylation in neurons.

The presence or absence of specific PTMs can directly affect the stability of the microtubule. For example, the adult brain contains a much higher cold stable population of microtubules than the developing brain, and this increase in the proportion of cold stable microtubules in the brain has been attributed to an accumulation of polyamination on neuronal microtubules [105,106]. Polyamination describes the addition of polyamine to multiple, genetically-encoded glutamine residues in α- and β-tubulin, by the transglutaminase enzyme TG2 [106]. Polyamination is sufficient to stabilize microtubules in vitro, and TG2 protein and activity levels increase postnatally, suggesting a role in neuronal maturation [106]. Acetylation of α-tubulin also correlates with microtubule stability, and staining with antibodies to acetylated tubulin is commonly used as a measure of microtubule stability in cells. However, while long-lived microtubules tend to be acetylated, the direct impact of acetylation on microtubule stability is not well established. Two recent studies indicate that acetylation of α-tubulin may promote microtubule stability through an unexpected mechanism—softening the microtubule lattice and allowing it to withstand bending without breaking and depolymerizing [107,108]. Studies in *C. elegans* demonstrate that microtubule acetylation is necessary to form specialized microtubules with 15-protofilament lattices in touch receptor neurons [109,110]. In addition to direct effects on the microtubule lattice, acetylation has been reported to promote the activities of kinesin-1 motors in vivo [111] and dynein motors in vitro [112]. This is intriguing, since the canonical acetylation site at lysine residue 40 of α-tubulin is located on the luminal side of the microtubule cylinder. How acetylation impacts lattice stability, interactions on the microtubule surface, and the larger implications of microtubule acetylation in vivo remain active areas of research.

3. Roles of Microtubules during Neuronal Development and Adulthood

Microtubules play numerous important roles during brain development, particularly in neurons, where the microtubule cytoskeleton has been intensely studied. Once neuronal progenitors exit the cell cycle to become neurons, diverse microtubule-based maturation stages must occur for the neuron to correctly extend dynamic neurites, migrate to the proper position, form and guide long axons, set-up synapses, and sustain the diverse regions of the neuron. Microtubules are essential for these phases of neuronal maturation and are differentially regulated by numerous MAPs to perform diverse neuronal microtubule-based functions (Figure 2).

As shown in Figure 2, microtubules interact with numerous MAPs during each stage of neuronal maturation to properly perform various cellular tasks. During neurite initiation, dynamic actin forms lamellipodia that become stabilized by invading microtubules (Figure 2A) [113]. Interestingly, the dynamic properties of microtubules at this stage may inform which neurite becomes the future axon. Locally stabilizing microtubules in a particular neurite with stabilizing drugs results in that neurite becoming the axon [114]. Later, as neurites mature, microtubules become stabilized by MAPs such as tau and MAP2 (Figure 2B) [39,115]. The presence of tau helps identify the neurite as the axon, while MAP2 identifies a dendritic fate. At this stage, motors such as kinesin-1 and dynein are important in sorting and pushing microtubules to the end of the neurite [116].

As development proceeds, neuronal microtubules begin to exhibit a distinct polarity and decoration by MAPs depending on which type of process they inhabit. Microtubules within the axon become uniformly oriented with their plus ends directed away from the soma (Figure 2E), while dendritic microtubules retain a mixed polarity (Figure 2H) [117,118]. Dynein has been implicated in establishing and maintaining this plus-end-out (away from the soma) orientation in axons [116].

Plus-end-out polarity is critical for axonal transport, discussed below. The quantity and diversity of MAPs in the neuronal environment, and the interplay between them makes understanding each MAPs' role difficult. For example, XMAP215 directly regulates microtubule plus-end dynamics in vitro, but in the growth cone it regulates the linkage of translocating microtubules to the F-actin network, thereby constraining microtubule growth velocity (Figure 2G) [119]. Future studies are needed to determine how microtubules are regulated throughout each stage of neuronal maturation, with attention given to how microtubule dynamics are regulated by the plethora of MAPs present during each developmental stage.

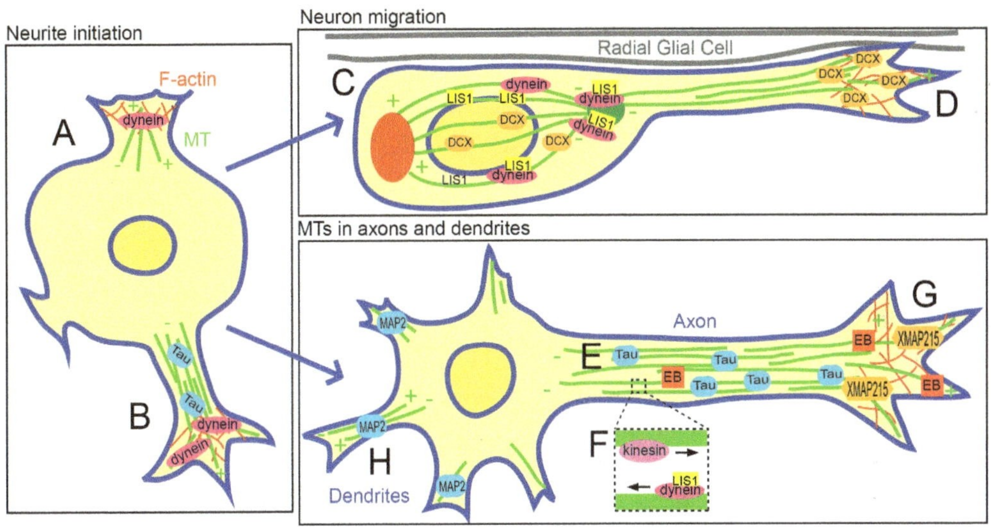

Figure 2. Overview of microtubule tasks during neuron maturation: (**A**) During neurite initiation, microtubules invade nascent lamellipodium; (**B**) microtubules form bundles to stabilize neurites; (**C**) microtubules form a perinuclear cage and provide force for nucleokinesis during neuronal migration; (**D**) in the migrating growth cone, microtubules stabilize and aid leading process growth; (**E**) polarized, bundled microtubules provide structural backbone of axon; (**F**) microtubules act as a transportation track for microtubule motors; (**G**) microtubules support axonal growth cone dynamics; and (**H**) microtubules of mixed polarity provide support to dendrites.

One of the most important functions of microtubules in adult neurons is to facilitate the efficient trafficking of organelles, proteins, mRNAs, and other cargoes across long distances in the axon. To appreciate the importance of efficient transport, consider that the distal regions of an axon can be as far as one meter away from critical protein and mRNA synthesis occurring in the soma. The unique microtubule polarity of the axon that is established early in development organizes transport in the anterograde and retrograde directions, with the help of motor proteins such as kinesin and dynein that move towards the plus or minus ends, respectively (Figure 2F). Kinesins, particularly kinesin-1 and 3 family members, carry cargoes towards the axon terminal by walking towards the plus ends of axonal microtubules [120]. Conversely, dynein carries cargoes back toward the soma by walking towards the minus ends [121]. The combination of uniform microtubule orientation and processive motility by directional molecular motors allows neurons to effectively transport cargoes across large distances.

In addition to establishing the architecture and transport networks within individual neurons, microtubule function is critical for neuronal migration, which is required to form the layers of the cortex and many other structures in the brain. During neuronal migration, microtubules extend into the leading process where they help to steer the protrusive growth cone at the end of the axon (Figure 2D) [122]. Microtubules also generate force to move the nucleus, in a process known as nucleokinesis. Here, microtubules extend back from the centrosome to form a cage-like network around the nucleus [123]. Dynein and its regulator LIS1 generate pulling forces to draw the nucleus toward the centrosome and move the centrosome toward the leading process [124]. Consistent with the important roles of the microtubule network in neuronal migration, mutations in LIS1 and the microtubule regulator doublecortin/DCX are associated with migration disorders that give rise to brain malformations [125].

4. Tubulinopathies Reveal Essential Role of TUBA1A in Brain Formation and Function

Recent studies investigating the genetic cause of brain malformation disorders have revealed that heterozygous, missense mutations to *TUBA1A* and other neuronal tubulin isotypes play an important role in brain development. First shown by Keays et al. in 2007, sequencing of tubulin genes has proven fruitful in uncovering the genetic source of numerous patients exhibiting complex neurological and physical phenotypes with cortical malformations. From these genotype-phenotype analyses, *TUBA1A* is recognized as vital for neurodevelopment based on the devastating effects observed in patients containing heterozygous, *de novo TUBA1A* missense mutations [11–20]. Brain malformation disorders caused by mutations to *TUBA1A* and other neuronally expressed tubulin isotypes are collectively termed "tubulinopathies", and lead to severe cortical abnormalities, mental retardation, and commonly epilepsy and paralysis [11–20]. Patients containing *TUBA1A* mutations exhibit a wide variety of cerebral cortex malformation phenotypes including lissencephaly, pachygyria, microlissencephaly, and polymicrogyria. While known genetic causes of these phenotypes hint at which developmental processes may be disrupted, little is known about how the *TUBA1A* mutations disrupt microtubule functions, or even how these disruptions could cause the larger-scale cellular and tissue problems seen in patients. The wide variety of brain malformations observed in patients leads to the prediction that different missense mutations in *TUBA1A* may disrupt different neuronal maturation phases. Disease-causing mutations to *TUBA1A* therefore provide a valuable opportunity to investigate the numerous neurodevelopmental stages that require TUBA1A, and how microtubules must be regulated for each stage to occur appropriately.

The identification of neurodevelopment disorder-causing *TUBA1A* mutations supports the hypothesis that this particular α-tubulin isotype is essential for neuronal maturation; however how can subtle changes to one α-tubulin protein lead to drastic changes in the formation of the brain? Expression studies of *TUBA1A* mRNA make it clear that *TUBA1A* is by far the most prevalent α-tubulin isotype in the embryonic nervous system, accounting for more than 95% of α-tubulin mRNA [72]. Thus, mutations to *TUBA1A* could greatly affect the available pool of neuronal tubulin. Tubulin mutations can act to alter numerous tubulin/microtubule characteristics and functions, but most of the possible changes fit into three general categories: (A) disrupting tubulin folding and heterodimer formation, thereby depleting the pool of α-tubulin that is competent to assemble microtubules; (B) disrupting polymerization activity and/or microtubule dynamics regulation, which could either deplete the pool of assembly-competent α-tubulin or alter dynamics once mutant heterodimers have formed the microtubule lattice; or (C) assembling appropriately into microtubules but altering microtubule function by disrupting interactions with MAPs and motors (Figure 3). Identifying which category *TUBA1A* mutations fit into will greatly increase our understanding of tubulinopathy disease progression. This task is not easy, however, as structure-function predictions for tubulin residues are rarely straightforward. Tubulin is a complex protein that interacts with numerous binding partners and undergoes long range conformational changes as part of its principle biochemical activity. Therefore, knowing the location of a change in the tubulin sequence does not necessarily give insight into its functional consequences. This underscores the need for functional studies that can test and refine structural predictions. However, despite the growing list of tubulinopathy mutations, few studies provide insight into how individual mutations impact the tubulin protein and its function in vivo. Future studies must be performed to fill in the missing molecular, mechanistic steps between the known *TUBA1A* mutations and the final brain phenotype.

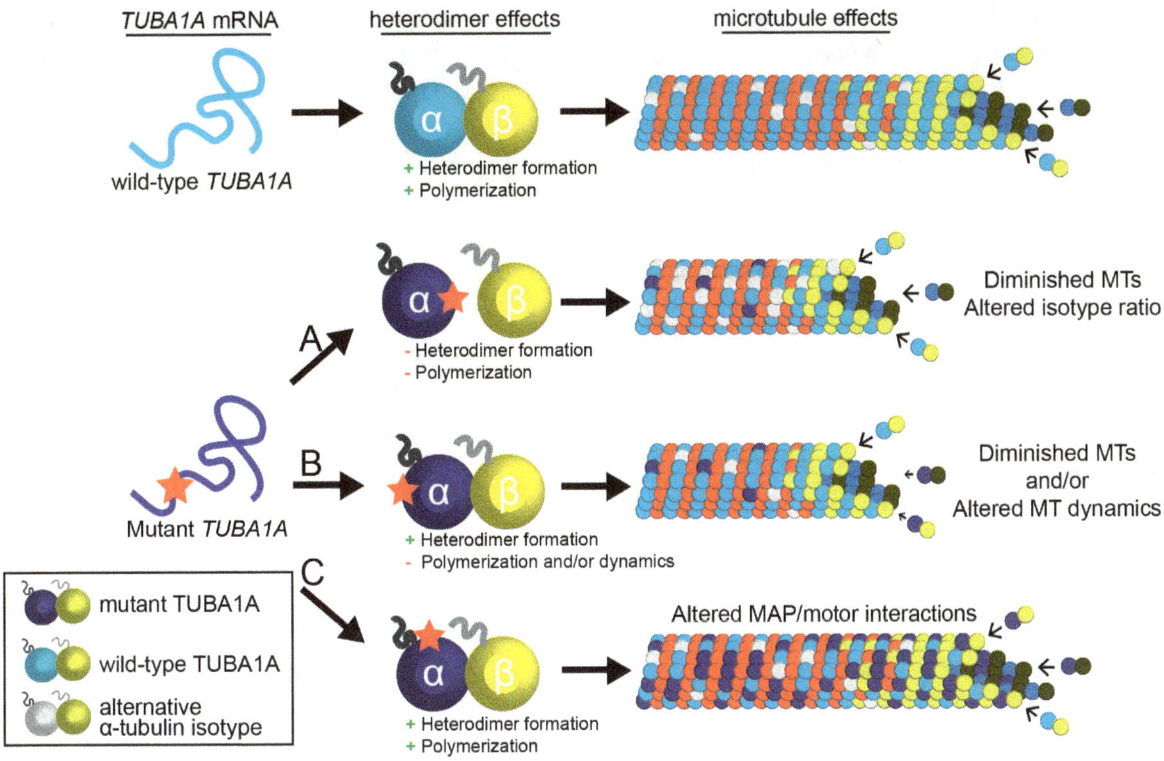

Figure 3. Potential consequences of TUBA1A mutations. TUBA1A mutations may lead to (**A**) protein folding defects or heterodimer instability; or (**B**) altered lattice interactions. Either of these defects may produce haploinsufficiency/loss of function consequences, resulting in fewer polymerization competent tubulin heterodimers available to form microtubules, or changes in microtubule dynamics. This also may lead to changes in the ratio of α-tubulin isotypes present in the microtubule lattice; (**C**) TUBA1A mutations may lead to mutant tubulin heterodimers that appropriately polymerize and cause toxic, gain of function consequences from within the microtubule lattice. Once in the lattice, mutant dimers may intrinsically change microtubule behavior or extrinsically alter MAP binding.

While few studies have specifically addressed how mutations that disrupt TUBA1A cause cortical malformations, understanding the known mechanisms can provide clues into how attributes of TUBA1A contribute to normal brain development. However, there is much more work to be done to elucidate the mechanistic role of TUBA1A. Several different aspects of TUBA1A function could go awry to disrupt cellular processes and ultimately lead to abnormal brain development. For example, interference with neuronal migration can lead to lissencephaly, but not all the aspects of TUBA1A that are important for neuronal migration are known. For example, a mutation to *TUBA1A* could cause haploinsufficiency through many mechanisms, including disruptions to TUBA1A folding, heterodimer formation, or microtubule assembly. This could disrupt neuronal migration due to a pool of incompetent tubulin dimers forming less stable microtubules. Alternatively, specific *TUBA1A* mutants could appropriately assemble into microtubules but then cause dominant changes that "poison" the microtubule network. The dominant disruption could be achieved through changes to microtubule dynamics or disruption of binding of MAPs involved in migration. The following sections attempt to marry the current understanding of cortical malformation progression with how identified *TUBA1A* mutations could potentially act in the established pathway.

4.1. TUBA1A Mutations Linked to Lissencephaly

Lissencephaly describes a set of cortical malformations where at least part of the brain surface appears smooth, lacking the cortical folds that are a hallmark of a healthy human brain [126]. Severe lissencephaly manifests as a complete lack of cortical folds (agyria), while milder forms present as fewer broad folds (pachygyria) or as bands of heterotopic gray matter embedded below the white matter of the cortex (subcortical band heterotopia (SBH)). All of these manifestations of lissencephaly are attributed to neuronal migration errors in the developing cortex. In agyria and pachygyria, neurons fail to reach their appropriate positions leading to a disordered four-layered cortical structure lacking gyri or sulci folds, instead of the normal, six-layered cortex containing folds. In SBH, cortical neurons inappropriately migrate to an area deep to the cortex, forming band-like patterns of grey matter beneath the cortex [127].

In principle, any perturbation to cortical migration could lead to lissencephaly phenotypes, but most identified cases are attributed to mutations to *LIS1* (also known as *PAFAH1B1*), *doublecortin/DCX*, and more recently *TUBA1A* [125]. These genes are encompassed by the cortical migration pathway, and provide the molecular basis for the malformation. LIS1 and DCX play important roles in regulating microtubule-based tasks. LIS1 is an adaptor for the microtubule-motor dynein, acting as a "clutch" that allows dynein to remain attached to microtubules for longer periods of time [128]. This modulation of dynein activity is important during neuronal migration, when dynein is responsible for pulling the microtubule-caged nucleus [129]. While LIS1 modulates dynein behavior, DCX binds directly to microtubules to stabilize and promote polymerization [130]. During migration, DCX facilitates the formation/maintenance of the microtubule cage around the nucleus, as well as stabilizing microtubules in the leading process of the migrating neuron [131]. These proteins are vital to migration, as disruption leads to defective migration, and overexpression of either LIS1 or DCX is sufficient to increase migration rates [131].

As lissencephaly-associated proteins are known to modulate the microtubule network, proper microtubule function must be requisite for correct cortical migration to occur. In fact, lissencephaly is the most prominent malformation attributed to *TUBA1A* mutations, with over 90% of patients exhibiting some form of lissencephaly malformation. This suggests that the consequence of many *TUBA1A* mutants is to disrupt migration in some way. The molecular basis of this disruption, or whether many different changes to tubulin function could lead to migrational defects, remains largely unknown. Table 3 provides the known *TUBA1A* mutations that cause lissencephaly.

Table 3. Tuba1a mutations leading to lissencephaly spectrum phenotypes.

Reference	Mutation	Case Number	Gender/Age	Cortical Phenotype	Corpus Callosum Defect	Other Brain Malformations
[132]	**I5L**	not reported	F/7	perisylvian pachygyria	thin	brainstem mildly hypoplastic
[132]	**I5L**	not reported	F/2	perisylvian pachygyria	splenium hypoplastic	brainstem and cerebellum vermis mildly hypoplastic
[133]	**C25F**	K3373	M/2	lissencephaly, thin cortex	agenesis	poorly differentiated dysmorphic basal ganglia, slightly hypoplastic cerebellar vermis, ventricle dilation
[134]	**E27Q** *	not reported	F/0.5	simplified gyral pattern, diffuse pachygyria	hypoplastic	hypoplastic cerebellar vermis, lateral ventricle dilation
[15]	**E55K**	FR08-D5604	M/3	microcephaly, lissencephaly posterior, pachygyria anterior	partial agenesis	severe vermis hypoplasia, flattened isthmus and pons, hypoplastic hippocampus, trigones and occipital horns dilated
[20,21]	**T56M**	LIS_TUB_003_ foetus18	M/24.3 GW	microcephaly, lissencephaly; absent cortical plate, 2 layered	complete agenesis	severe vermis hypoplasia; hypoplastic basal ganglia, severe hypoplasia of cerebellum and pons, optic nerve hypoplasia
[133]	**R64W** *	NCU_F41	F/3	extremely thin cerebral parenchyma	agenesis	optic nerve hypoplasia, hypoplastic brainstem, agenesis of cerebellum
[19]	**L70S**	not reported	F/2 weeks	lissencephaly, diffuse polymicrogyria-like	absent	cerebellar hypoplasia, enlarged lateral ventricles
[20]	**P72S**	LIS_TUB_012_ foetus22	F/37.8 GW	severe lissencephaly	dysmorphic/hypoplastic	severe vermis hypoplasia
[16]	**L92V**	CM-66	fetus	lissencephaly with cerebellar hypoplasia; microcephaly	absent	small brainstem, cerebellum, and corticospinal tract, severe ventricular dilation
[20,21]	**N101S**	LIS_TUB_079_ foetus25	M/25 GW	microcephaly, lissencephaly; 2–3 layered cortex, poorly differentiated	complete agenesis	severe vermis and hemispheric dysplasia; severe hypoplasia and dysplasia of cerebellum, severe hypoplasia of pons
[20]	**E113K**	LIS_TUB_031	M/11	central pachygyria	normal	normal cerebellum
[135]	R123C *	106115P	M	pachygyria	not reported	not reported
[16]	**V137D**	CM-107	not reported	pachygyria with cerebellar hypoplasia	absent	malformed hippocampus, thin brainstem, severe cerebellar hypoplasia
[135]	**L152Q** *	169451P	F	pachygyria	not reported	not reported
[12]	**I188L**	LIS_TUB_026	F/2	laminar heterotopia	thin, partial agenesis	vermis and brainstem hypoplasia, severe ventricular dilation
[136]	**C200Y**	not reported	F/8	lissencephaly	agenesis	abnormal hippocampus, dysmorphic and hypoplastic basal ganglia and thalamus, hypoplastic cerebellum, enlarged lateral ventricles
[132]	**Y210C**	not reported	M/1.5	lissencephaly anterior, pachygyria posterior	thin	brainstem and cerebellum vermis mildly hypoplastic

Table 3. *Cont.*

Reference	Mutation	Case Number	Gender/Age	Cortical Phenotype	Corpus Callosum Defect	Other Brain Malformations
[137]	R214H	UW168-3	F/4	diffuse irregular gyration and sulcation	partial agenesis	hypoplasia of vermis, asymmetric pons, dysmorphic basal ganglia, cranial nerve hypoplasia, enlarged lateral ventricles
[16]	**D218Y**	LR07-213	not reported	lissencephaly with cerebellar hypoplasia, microcephaly	absent	thin brainstem, hypoplasia of cerebellar vermis
[137]	**I219V**	UW167-3	F/10	diffuse (L > R) irregular gyration and sulcation	partial agenesis	hypoplasia of vermis, asymmetric pons, dysmorphic basal ganglia, cranial nerve hypoplasia, enlarged lateral ventricles
[14]	**I238V**	LIS_TUB_022_ foetus05	M/fetus	two-layered cortex, poorly differentiated	complete agenesis	disorganized hippocampus, internal capsule absent on one side, hypoplastic brainstem and corticospinal tract
[12]	**P263T**	not reported	M/fetus	lissencephaly, microcephaly	agenesis	abnormal hippocampus, cerebellum, vermis and brainstem hypoplasia, severe ventricular dilation
[14]	P263T	LIS_TUB_025_ foetus06	M/fetus	two-layered cortex, poorly differentiated	complete agenesis	disorganized hippocampus, hypoplastic internal capsule, hypoplastic brainstem and corticospinal tract
[11]	**R264C**	not reported	not reported	pachygyria	agenesis	abnormal hippocampus, abnormal vermis, brainstem hypoplasia
[12]	R264C	LIS_TUB_037	M/4	pachygyria	present, abnormal shape	vermis and brainstem hypoplasia, mild ventricular dilation
[12]	R264C	LIS_TUB_036	M/2	pachygyria	present, abnormal shape	vermis hypoplasia, mild ventricular dilation
[13]	R264C	LIS_TUB_041	M/7	perisylvian pachygyria	posterior agenesis	severe dysgenesis of internal capsule
[13]	R264C	LIS_TUB_040	M/1.5	perisylvian pachygyria	mild hypoplasia	moderate dysgenesis of internal capsule
[20]	R264C	LIS_TUB_033	F/1.5	central pachygyria	normal	mild vermis hypoplasia
[20]	R264C	LIS_TUB_034	F/6.5	central pachygyria	hypogenetic	normal cerebellum
[20]	R264C	LIS_TUB_035	M/6	central pachygyria	normal	mild vermis hypoplasia
[138]	R264C	Patient B	M/2	grade 4 agyria, pachygyria (P > A gradient)	present, abnormal shape	hypoplastic basal ganglia
[20,21]	**R264H**	LIS_TUB_002_ foetus20	F/24 GW	microcephaly, lissencephaly; absent cortical plate, 2 layered	complete agenesis	severe vermis hypoplasia; moderate hypoplasia of cerebellum, severe pons hypoplasia
[16]	**A270T**	LR07-244	not reported	pachygyria with cerebellar hypoplasia	absent	malformed hippocampus, thin brainstem, severe cerebellar hypoplasia
[139]	**A270S**	not reported	M/19 mo	mild posterior simplified cerebral gyral pattern	agenesis	severe hypoplastic cerebellar vermis, mildly dysplastic and hypoplastic cerebellar hemispheres, mildly hypoplastic brainstem, dysplastic basal ganglia, thalami, hypoplastic optic nerves, absent olfactory bulbs, lateral and third ventricle dilated
[12]	**L286F**	not reported	M/fetus	lissencephaly	agenesis	abnormal hippocampus, vermis and brainstem hypoplasia, severe ventricular dilation

Table 3. *Cont.*

Reference	Mutation	Case Number	Gender/Age	Cortical Phenotype	Corpus Callosum Defect	Other Brain Malformations
[14]	L286F	LIS_TUB_007_ foetus04	fetus	two-layered cortex, poorly differentiated	complete agenesis	absent hippocampus, olfactory bulb, and internal capsule, hypoplastic brainstem and corticospinal tract
[17]	**V303G**	LIS_TUB_006_ foetus03	fetus	pachygyria, microcephaly	short and thin	thin brainstem, pons and medulla flattened, hypoplastic cerebellum and corticospinal tracts, severe ventricular dilation
[20,21]	**R320H**	LIS_TUB_005_ foetus01	M/25 GW	microcephaly, lissencephaly; absent cortical plate	partial agenesis	severe vermis and hemispheric dysplasia/severe hypoplasia and dysplasia of cerebellum, severe hypoplasia (neuronal over migration) spinal cord anterior horn hypoplasia
[21]	R320H	LIS_TUB_081_ foetus26	M/26 GW	absent cortical plate, 2 layered	complete agenesis	severe hypoplasia and dysplasia of cerebellum, severe hypoplasia of pons
[20,21]	**K326N**	LIS_TUB_004_ foetus08	M/23 GW	microcephaly, lissencephaly; thick 2-layered cortex	partial agenesis; complete agenesis	severe vermis and hemispheric dysplasia/hypoplastic basal ganglia, severe hypoplasia and dysplasia of cerebellum, severe pons hypoplasia
[16]	**N329S**	LR05-388	not reported	lissencephaly with cerebellar hypoplasia, microcephaly	absent	thin brainstem, hypoplasia of cerebellar vermis
[138]	N329S	Patient A	M/0	grade 1 lissencephaly with cerebellar hypoplasia	agenesis	hypoplastic basal ganglia, cerebellum and brain stem
[20,21]	**V371E**	LIS_TUB_080_ foetus24	F/23.3 GW	microcephaly, lissencephaly; 2–3 layered cortex, poorly differentiated	complete agenesis	severe vermis hypoplasia; hypoplastic basal ganglia, severe hypoplasia of cerebellum, severe pons hypoplasia
[140]	**A387V**	not reported	F/5	pachygyria with SBH	thin	simplified hippocampus, highly dysmorphic brainstem, flattened pons, mildly hypoplastic cerebellar vermis
[20]	**A369T**	LIS_TUB_030	M/11	central pachygyria	dysmorphic/hypoplastic	mild vermis hypoplasia
[13]	**L397P**	LIS_TUB_039	M/5.5	perisylvian pachygyria	posterior agenesis	severe vermis dysplasia, severe dysgenesis of internal capsule
[12]	**R402C**	not reported	M/fetus	lissencephaly	abnormally thick	abnormal hippocampus, vermis and brainstem hypoplasia, severe ventricular dilation
[15]	R402C	FR04-D4148	F/11	microcephaly, lissencephaly	thin, rostrum absent, splenium hypoplastic	mild vermis and pons hypoplasia, trigones and occipital horns dilated
[16]	R402C	LP95-073	not reported	lissencephaly	dysmorphic but intact	classic lissencephaly, round hippocampi with rounded rim
[16]	R402C	LR07-008	not reported	lissencephaly	dysmorphic but intact	classic lissencephaly, round hippocampi with rounded rim
[16]	R402C	LR06-210	not reported	lissencephaly	dysmorphic but intact	classic lissencephaly, round hippocampi with rounded rim
[16]	R402C	LR08-035	not reported	lissencephaly	dysmorphic but intact	classic lissencephaly, round hippocampi with rounded rim
[16]	R402C	LR06-064	not reported	lissencephaly	dysmorphic but intact	classic lissencephaly, round hippocampi with rounded rim

Table 3. *Cont.*

Reference	Mutation	Case Number	Gender/Age	Cortical Phenotype	Corpus Callosum Defect	Other Brain Malformations
[20]	R402C	LIS_TUB_019	M/10	moderate lissencephaly	dysmorphic, hypoplastic	mild vermis hypoplasia
[14]	R402C	LIS_TUB_021_foetus07	M/fetus	thick four-layered cortex	abnormally thick and short	disorganized hippocampus, hypoplastic vermis, brainstem, abnormal corticospinal tract
[11,12]	**R402H**	LIS_TUB_023	M/11	lissencephaly	agenesis; thin, partial agenesis	abnormal vermis, brainstem hypoplasia/abnormal hippocampus, vermis and brainstem hypoplasia, severe ventricular dilation
[16]	R402H	LP97-039	not reported	lissencephaly with cerebellar hypoplasia	dysmorphic but intact	moderate cerebellar vermis hypoplasia, classic lissencephaly, round hippocampi with rounded rim
[16]	R402H	LP97-041	not reported	lissencephaly with cerebellar hypoplasia	dysmorphic but intact	moderate cerebellar vermis hypoplasia, classic lissencephaly, round hippocampi with rounded rim,
[20,21]	R402H	LIS_TUB_017_foetus02	M/29 GW	severe lissencephaly	complete agenesis	mild vermis and pons hypoplasia
[20]	R402H	LIS_TUB_014	M/4	severe lissencephaly	hypogenetic	severe hemispheric hypoplasia
[20]	R402H	LIS_TUB_015	M/1	severe lissencephaly	hypogenetic	severe hemispheric hypoplasia
[20]	R402H	LIS_TUB_016	F/1.5	severe lissencephaly	hypogenetic	vermis dysplasia
[141]	R402H	not reported	M/1	lissencephaly	not reported	severe dysplasia of brainstem and cerebellum
[15]	**R402L**	RE07-S1605	M/3	microcephaly, lissencephaly posterior, pachygyria anterior	thin	mild vermis and pons hypoplasia, retrocerebellar cyst, abnormal hippocampus, dilated lateral ventricles, occipital horns, anterior horns
[142]	R402L	not reported	F/1	microcephaly, lissencephaly posterior, pachygyria anterior	mild hypoplasia	severe hypoplasia of cerebellum, mild hypoplasia of brainstem, moderate ventricular dilation
[20,21]	**V409A**	LIS_TUB_011_foetus23	M/32 GW	severe lissencephaly with cerebellar hypoplasia	complete agenesis	severe vermis hypoplasia/severe hypoplasia of cerebellum and pons
[20]	**V409I**	LIS_TUB_032	M/10	central pachygyria	dysmorphic, hypoplastic	normal cerebellum
[12]	**S419L**	LIS_TUB_024	M/18	pachygyria	abnormal shape	abnormal hippocampus, vermis hypoplasia, severe ventricular dilation
[13]	**R422C**	LIS_TUB_042	F/4.5	perisylvian pachygyria	mild hypoplasia	mild vermis hypoplasia, severe dysgenesis of internal capsule
[13]	**R422H**	LIS_TUB_020	F/7	posterior pachygyria	mild hypoplasia	mild vermis hypoplasia, moderate dysgenesis of internal capsule
[15]	R422H	FR05-D4607	M/9	pachygyria with SBH	partial agenesis	severe vermis hypoplasia, dandy-walker malformation, hypoplasia of pons, abnormal hippocampus, dilated lateral ventricles, enlarged 4th ventricle
[15]	R422H	FR07-D5526	F/5	microcephaly, pachygyria with SBH	partial agenesis	moderate vermis hypoplasia, mild pons hypoplasia, abnormal hippocampus, dilated lateral ventricles, enlarged 4th ventricle

Table 3. *Cont.*

Reference	Mutation	Case Number	Gender/Age	Cortical Phenotype	Corpus Callosum Defect	Other Brain Malformations
[16]	R422H	LR05-052	not reported	pachygyria with cerebellar hypoplasia	absent	malformed hippocampus, thin brainstem, severe cerebellar hypoplasia
[16]	R422H	LR08-340	not reported	pachygyria with cerebellar hypoplasia	absent	malformed hippocampus, thin brainstem, severe cerebellar hypoplasia
[20,21]	R422H	LIS_TUB_018_ foetus10	F/28 GW	severe lissencephaly	complete agenesis	mild vermis and pons hypoplasia
[16]	**M425K**	LR08-388	not reported	lissencephaly with cerebellar hypoplasia, microcephaly	absent	thin brainstem, hypoplasia of cerebellar vermis
[20]	**E429Q**	LIS_TUB_001_ foetus09	F/25 GW	microcephaly, lissencephaly	complete agenesis	severe vermis hypoplasia
[21]	E429Q	LIS_TUB_004_ foetus09	F/25 GW	4 layered cortex	complete agenesis	hypoplastic basal ganglia, severe hypoplasia of cerebellum and pons
[13]	**G436R**	LIS_TUB_038	M/7	perisylvian pachygyria	mild hypoplasia	mild vermis hypoplasia, severe dysgenesis of internal capsule

Bold indicates first report of mutation; * Asterisk indicates mutation was identified through whole exome sequencing; Abbreviations: M, male; F, female; GW, gestational weeks SBH, subcortical band heterotopia.

4.2. TUBA1A Mutations Linked to Polymicrogyria

Polymicrogyria describes the cortical malformation characterized by excessive gyration (i.e., multiple, small folds) of the cerebral cortex [143]. The classification of this malformation is complex as cases of polymicrogyria are heterogeneous, with variable pathological results, clinical features, and etiologies. Even key characteristics used to define the malformation are controversial, as some sources point to abnormal cortical lamination as a key feature [144], while others affirm that cortical layering remains normal, and there are simply fewer neuronal populations inhabiting the layers [143]. A number of seemingly unrelated environmental and genetic causes have been implicated as the molecular basis of polymicrogyria, but the details of this developmental disorder remain mysterious.

The major non-genetic causes of polymicrogyria relate to ischemic insults in utero, such as hypoxia, hypoperfusion, congenital infections, and/or inflammation of the microvasculature [144]. Metabolic and mitochondrial diseases have also been implicated, with a mouse model of Zellweger syndrome revealing that polymicrogyria may be caused by defects in glutamate receptor-mediated calcium mobilization during neuronal migration [145]. Additionally, the transcription factor *PAX6,* which is important for neuronal migration and axon guidance, has been implicated in polymicrogyria. Contrary to these few hints implicating neuronal migration in polymicrogyria, the disorder is generally considered a "post-migrational" malformation, with issues occurring after neurons have completed their migratory pathway to form the cortical layers [143,146]. Polymicrogyria seems to be caused by numerous, seemingly unrelated etiologies that all result in an excessive gyration phenotype. This is supported by the identification of many genetic roots of the disorder, including signaling molecules, cytoskeletal elements, and others.

Contrary to lissencephaly, where cortical migration provides a clear culprit for the malformation, the complex etiology of polymicrogyria makes it difficult to predict how *TUBA1A* mutations contribute to the molecular basis of the disease. Polymicrogyria-causing *TUBA1A* mutations are significantly less common than lissencephaly-causing mutations (~13%), suggesting that perhaps only very specific disruptions of tubulin function can lead to polymicrogyria. Understanding how these TUBA1A mutations alter intrinsic properties of microtubules, interactions with MAPs, and consequently cellular functions will provide a window into the cellular progression of the disease, and will help shed light on the "post-migration" vs. "migration" debate. The *TUBA1A* mutations that lead to polymicrogyria are described in Table 4.

Table 4. Tuba1a mutations leading to polymicrogyria phenotypes.

References	Mutation	Case Number	Gender/Age	Cortical Phenotype	Corpus Callosum Defect	Other Brain Malformations
[19]	**L70S**	not reported	F/2 weeks	lissencephaly, diffuse polymicrogyria-like	absent	cerebellar hypoplasia, enlarged lateral ventricles
[21]	P72S	LIS_TUB_012_foetus22	F/37.8 GW	unlayered generalized and asymmetric polymicrogyria	hypoplastic	severe hypoplasia of cerebellum and pons
[20]	**R123C**	LIS_TUB_044	F/3	central polymicrogyria-like cortical dysplasia	normal	vermis dysplasia
[20,21]	**S158L**	LIS_TUB_053_foetus21	F/24.5 GW	unlayered generalized and asymmetric polymicrogyria	complete agenesis	severe vermis and hemispheric dysplasia/hypoplastic basal ganglia, severe hypoplastic and dysmorphic cerebellum, hypoplasia olivary heterotopia
[18]	**Y161H**	LIS_TUB_047	F/11	asymmetrical perisylvian polymicrogyria	moderate hypoplasia	dysmorphic basal ganglia, dysplastic vermis and pons
[20,21]	**R214H**	LIS_TUB_043_foetus11	M/23 GW	central polymicrogyria-like cortical dysplasia; unlayered central and asymmetric polymicrogyria	complete agenesis	normal cerebellum; mild vermian hypoplasia, mild dysplastic olivary nuclei
[18]	**V235L**	LIS_TUB_046	M/7.5	asymmetrical perisylvian polymicrogyria	moderate hypoplasia	dysmorphic basal ganglia
[19]	**A333V**	not reported	M/7	right focal polymicrogyria-like	thin	vermis hypoplasia, mild brainstem hypoplasia, 4th ventricle enlarged
[20]	**V353I**	LIS_TUB_059	M/4	simplified gyral pattern with focal polymicrogyria	partial agenesis	normal
[18]	R390C	LIS_TUB_045	M/1	asymmetrical perisylvian polymicrogyria	severe hypoplasia	dysmorphic basal ganglia, dysplastic vermis, severe hypoplasia of brainstem
[147]	**R390H**	not reported	F/3	focal polymicrogyria	thin and incomplete	hypoplasia of left internal capsule, severe hypoplastic and asymmetric brainstem, vermis dysplasia, enlarged lateral ventricles
[20]	**D396Y**	LIS_TUB_078	F/4	central polymicrogyria-like cortical dysplasia	partial agenesis	severe vermis dysplasia

Bold indicates first report of mutation; Abbreviations: M, male; F, female; GW, gestational weeks.

4.3. *TUBA1A Mutations Linked to Microcephaly*

Microcephaly describes a brain that is significantly smaller than average, typically measured by occipitofrontal circumference (OFC). Different definitions exist as to where the cutoff from small head to microcephaly occurs, with some defining it more stringently as less than 4 SD (i.e., below the 1st centile) below the average [148], while others include a broader range with less than 2 SD (i.e., below the 3rd centile) [149]. However, arguments have been made about the relevance of using such a broad label, as many infants within the −2 SD to −3 SD population will be "normal" [150]. Microcephaly can be categorized into two main divisions: primary microcephaly and secondary microcephaly. Primary microcephaly describes a non-progressive, significantly small head detected prior to 36 gestational weeks (GW), and generally results from reductions in neurogenesis or loss of neural stem cells [150]. The major causes of primary microcephaly include non-genetic, damaging events prior to birth or mutations to genes regulating mitosis in neuronal progenitors [148,150]. In contrast, secondary microcephaly describes when microcephaly progresses postnatally, and is considered a neurodegenerative condition [148]. Causes of secondary microcephaly are numerous and varied, encompassing anything that disrupts orderly development and function of the brain.

As *TUBA1A* is not expressed in neuronal progenitors, microcephaly associated with *TUBA1A* likely fits into the secondary microcephaly classification. Neuronal maturation relies on appropriate regulation of microtubules (Figure 2), so it follows that disrupting microtubule function could alter neuronal development and cause progressive microcephaly. This idea is supported by *TUBA1A* patient cases where head circumference measurements have been taken more than once, such as the patient with p.E27Q whose OFC decreased from within the normal range at birth (−1 SD) to microcephalic (−3.3 SD) by two months of age [134]. Table 5 provides *TUBA1A* mutations linked to microcephaly. In this review, we use the broadest definition of microcephaly and include cases where the OFC is 2 SD below the appropriate mean (i.e., less than the 3rd percentile). Using this classification, ~74% of *TUBA1A* mutations lead to "microcephaly", or smaller than normal head size. It is important to note that microcephaly is never the primary cortical malformation associated with *TUBA1A* mutants, but is a common accompaniment to both lissencephaly and polymicrogyria cortical phenotypes.

Table 5. Tuba1a mutations leading to microcephaly phenotypes. Microcephaly classified as an OFC more than 2 SD below the appropriate mean (i.e., less than the 3rd percentile).

Reference	Mutation	Case Number	Gender/Age	Cortical Phenotype	OFC	Corpus Callosum Defect	Other Brain Malformations
[132]	**I5L**	not reported	F/7	perisylvian pachygyria	−2 SD	thin	brainstem mildly hypoplastic
[134]	**E27Q** *	not reported	F/5	simplified gyral pattern, diffuse pachygyria	−1 SD at birth, −3.3 SD at 2 mo.	hypoplastic	hypoplastic cerebellar vermis, lateral ventricle dilation
[15]	**E55K**	FR08-D5604	M/3	lissencephaly posterior, pachygyria anterior	−7 SD	partial agenesis	severe vermis hypoplasia, flattened isthmus and pons, hypoplastic hippocampus, trigones and occipital horns dilated
[20]	**T56M**	LIS_TUB_003_ foetus18	M/24.3 GW	lissencephaly	microcephaly	complete agenesis	severe vermis hypoplasia
[21]	T56M	LIS_TUB_003_ foetus18	M/24.3 GW	absent cortical plate, 2 layered	<3rd centile	complete agenesis	hypoplastic basal ganglia, severe hypoplasia of cerebellum and pons, optic nerve hypoplasia
[133]	**R64W** *	NCU_F41	F/3	extremely thin cerebral parenchyma	−2.4 SD	agenesis	optic nerve hypoplasia, hypoplastic brainstem, agenesis of cerebellum
[19]	**L70S**	not reported	F/2 weeks	lissencephaly, diffuse polymicrogyria-like	2–9 centile	absent	cerebellar hypoplasia, enlarged lateral ventricles
[21]	P72S	LIS_TUB_012_ foetus22	F/37.8 GW	unlayered generalized and asymmetric polymicrogyria	5th centile	hypoplastic	severe hypoplasia of cerebellum and pons
[16]	**L92V**	CM-66	fetus	lissencephaly with cerebellar hypoplasia	microcephaly	absent	small brainstem, cerebellum, and corticospinal tract, severe ventricular dilation
[20]	**N101S**	LIS_TUB_079_ foetus25	M/25 GW	lissencephaly	microcephaly	complete agenesis	severe vermis and hemispheric dysplasia
[21]	N101S	LIS_TUB_079_ foetus25	M/25 GW	2–3 layered cortex, poorly differentiated	<3rd centile	complete agenesis	severe hypoplasia and dysplasia of cerebellum, severe hypoplasia of pons
[20]	**E113K**	LIS_TUB_031	M/11	central pachygyria	−3 SD	normal	normal cerebellum
[20]	**R123C**	LIS_TUB_044	F/3	central polymicrogyria-like cortical dysplasia	−3 SD	normal	vermis dysplasia
[18]	**Y161H**	LIS_TUB_047	F/11	asymmetrical perisylvian polymicrogyria	3rd centile	moderate hypoplasia	dysmorphic basal ganglia, dysplastic vermis and pons
[12]	**I188L**	LIS_TUB_026	F/2	laminar heterotopia	−4 SD	thin, partial agenesis	vermis and brainstem hypoplasia, severe ventricular dilation
[136]	**C200Y** * reported as C402Y	not reported	F/8	lissencephaly	−3.5 SD	agenesis	abnormal hippocampus, dysmorphic and hypoplastic basal ganglia and thalamus, hypoplastic cerebellum, enlarged lateral ventricles

Table 5. *Cont.*

Reference	Mutation	Case Number	Gender/Age	Cortical Phenotype	OFC	Corpus Callosum Defect	Other Brain Malformations
[132]	**Y210C**	not reported	M/1.5	lissencephaly anterior, pachygyria posterior	−3 SD	thin	brainstem and cerebellum vermis mildly hypoplastic
[137]	R214H	UW168-3	F/4	diffuse irregular gyration and sulcation	<−2.5 SD	partial agenesis	hypoplasia of vermis, asymmetric pons, dysmorphic basal ganglia, cranial nerve hypoplasia, enlarged lateral ventricles
[16]	**D218Y**	LR07-213	Not reported	lissencephaly with cerebellar hypoplasia	microcephaly	absent	thin brainstem, hypoplasia of cerebellar vermis
[12]	**P263T**	not reported	M/fetus	lissencephaly	microcephaly	agenesis	abnormal hippocampus, cerebellum, vermis and brainstem hypoplasia, severe ventricular dilation
[12]	R264C	LIS_TUB_037	M/4	pachygyria	−4.5 SD	present, abnormal shape	vermis and brainstem hypoplasia, mild ventricular dilation
[12]	R264C	LIS_TUB_036	M/2	pachygyria	−4 SD	present, abnormal shape	vermis hypoplasia, mild ventricular dilation
[13]	R264C	LIS_TUB_041	M/7	perisylvian pachygyria	−4 SD	posterior agenesis	severe dysgenesis of internal capsule
[13]	R264C	LIS_TUB_040	M/1.5	perisylvian pachygyria	−4 SD	mild hypoplasia	moderate dysgenesis of internal capsule
[20]	R264C	LIS_TUB_033	F/1.5	central pachygyria	−5 SD	normal	mild vermis hypoplasia
[20]	R264C	LIS_TUB_034	F/6.5	central pachygyria	−4 SD	hypogenetic	normal cerebellum
[20]	R264C	LIS_TUB_035	M/6	central pachygyria	−3 SD	normal	mild vermis hypoplasia
[20]	**R264H**	LIS_TUB_002_ foetus20	F/24 GW	lissencephaly	microcephaly	complete agenesis	severe vermis hypoplasia
[21]	R264H	LIS_TUB_002_ foetus20	F/24 GW	absent cortical plate, 2 layered	<3rd centile	complete agenesis	moderate hypoplasia of cerebellum, severe pons hypoplasia
[139]	**A270S**	not reported	M/19 mo	mild posterior simplified cerebral gyral pattern	microcephaly	agenesis	severe hypoplastic cerebellar vermis, mildly dysplastic and hypoplastic cerebellar hemispheres, mildly hypoplastic brainstem, dysplastic basal ganglia, thalami, hypoplastic optic nerves, absent olfactory bulbs, lateral and third ventricle dilated
[14]	L286F	LIS_TUB_007_ foetus04	fetus	two-layered cortex, poorly differentiated	<3rd centile	complete agenesis	absent hippocampus, olfactory bulb, and internal capsule, hypoplastic brainstem and corticospinal tract
[17]	**V303G**	LIS_TUB_006_ foetus03	fetus	pachygyria	microcephaly	short and thin	thin brainstem, pons and medulla flattened, hypoplastic cerebellum and corticospinal tracts

Table 5. *Cont.*

Reference	Mutation	Case Number	Gender/Age	Cortical Phenotype	OFC	Corpus Callosum Defect	Other Brain Malformations
[20]	**R320H**	LIS_TUB_005_ foetus01	M/25 GW	lissencephaly	microcephaly	partial agenesis	severe vermis and hemispheric dysplasia
[21]	R320H	LIS_TUB_005_ foetus01	M/25 GW	absent cortical plate	<3rd centile	partial agenesis	severe hypoplasia and dysplasia of cerebellum, severe hypoplasia (neuronal over-migration) spinal cord anterior horn hypoplasia
[21]	R320H	LIS_TUB_081_ foetus26	M/26 GW	absent cortical plate, 2 layered	<3rd centile	complete agenesis	severe hypoplasia and dysplasia of cerebellum, severe hypoplasia of pons
[20]	**K326N**	LIS_TUB_004_ foetus08	M/23 GW	lissencephaly	microcephaly	partial agenesis	severe vermis and hemispheric dysplasia
[21]	K326N	LIS_TUB_004_ foetus08	M/23 GW	thick 2-layered cortex	<3rd centile	complete agenesis	hypoplastic basal ganglia, severe hypoplasia and dysplasia of cerebellum, severe pons hypoplasia
[16]	**N329S**	LR05-388	Not reported	lissencephaly with cerebellar hypoplasia	microcephaly	absent	thin brainstem, hypoplasia of cerebellar vermis
[20]	**V353I**	LIS_TUB_059	M/4	simplified gyral pattern with focal polymicrogyria	−2 SD	partial agenesis	normal
[20]	**V371E**	LIS_TUB_080_ foetus24	F/23.3 GW	lissencephaly	<3rd centile	complete agenesis	severe vermis hypoplasia
[21]	V371E	LIS_TUB_080_ foetus24	F/23.3 GW	2–3 layered cortex, poorly differentiated	<3rd centile	complete agenesis	hypoplastic basal ganglia, severe hypoplasia of cerebellum, severe pons hypoplasia
[140]	**A387V**	Not reported	F/5	pachygyria with SBH (subcortical band heterotopia)	<3rd centile	thin	simplified hippocampus, highly dysmorphic brainstem, flattened pons, mildly hypoplastic cerebellar vermis
[18]	R390C	LIS_TUB_045	M/1	asymmetrical perisylvian polymicrogyria	microcephaly	severe hypoplasia	dysmorphic basal ganglia, dysplastic vermis, severe hypoplasia of brainstem
[20]	**A369T**	LIS_TUB_030	M/11	central pachygyria	−2 SD	dysmorphic, hypoplastic	mild vermis hypoplasia
[20]	**D396Y**	LIS_TUB_078	F/4	central polymicrogyria-like cortical dysplasia	−3 SD	partial agenesis	severe vermis dysplasia
[13]	**L397P**	LIS_TUB_039	M/5.5	perisylvian pachygyria	−4 SD	posterior agenesis	severe vermis dysplasia, severe dysgenesis of internal capsule
[15]	R402C	FR04-D4148	F/11	lissencephaly	−4 SD	thin, rostrum absent, splenium hypoplastic	mild vermis and pons hypoplasia, trigones and occipital horns dilated
[20]	R402C	LIS_TUB_019	M/10	moderate lissencephaly	−3 SD	dysmorphic, hypoplastic	mild vermis hypoplasia

Table 5. *Cont.*

Reference	Mutation	Case Number	Gender/Age	Cortical Phenotype	OFC	Corpus Callosum Defect	Other Brain Malformations
[12]	R402H	LIS_TUB_023	M/11	lissencephaly	−3 SD	thin, partial agenesis	abnormal hippocampus, vermis and brainstem hypoplasia, severe ventricular dilation
[20]	R402H	LIS_TUB_014	M/4	severe lissencephaly	−4 SD	hypogenetic	severe hemispheric hypoplasia
[20]	R402H	LIS_TUB_015	M/1	severe lissencephaly	−4 SD	hypogenetic	severe hemispheric hypoplasia
[20]	R402H	LIS_TUB_016	F/1.5	severe lissencephaly	−3 SD	hypogenetic	vermis dysplasia
[141]	R402H	not reported	M/1	lissencephaly	microcephaly		severe dysplasia of brainstem and cerebellum
[15]	**R402L**	RE07-S1605	M/3	lissencephaly posterior, pachygyria anterior	−3.5 SD	thin	mild vermis and pons hypoplasia, retrocerebellar cyst, abnormal hippocampus, dilated lateral ventricles, occipital horns, anterior horns
[142]	R402L	Not reported	F/1	lissencephaly posterior, pachygyria anterior	−4 SD	mild hypoplasia	severe hypoplasia of cerebellum, mild hypoplasia of brainstem
[21]	V409A	LIS_TUB_011_ foetus23	M/32 GW	not available	5th centile	complete agenesis	severe hypoplasia of cerebellum and pons
[20]	**V409I**	LIS_TUB_032	M/10	central pachygyria	−2 SD	dysmorphic, hypoplastic	normal cerebellum
[13]	**R422C**	LIS_TUB_042	F/4.5	perisylvian pachygyria	−3 SD	mild hypoplasia	mild vermis hypoplasia, severe dysgenesis of internal capsule
[13]	**R422H**	LIS_TUB_020	F/7	posterior pachygyria	−4 SD	mild hypoplasia	mild vermis hypoplasia, moderate dysgenesis of internal capsule
[15]	R422H	FR05-D4607	M/9	pachygyria with SBH	−4 SD	partial agenesis	severe vermis hypoplasia, dandy-walker malformation, hypoplasia of pons, abnormal hippocampus
[15]	R422H	FR07-D5526	F/.5	pachygyria with SBH	−3.5 SD	partial agenesis	moderate vermis hypoplasia, mild pons hypoplasia, abnormal hippocampus
[16]	**M425K**	LR08-388	Not reported	lissencephaly with cerebellar hypoplasia	microcephaly	absent	thin brainstem, hypoplasia of cerebellar vermis
[20]	**E429Q**	LIS_TUB_001_ foetus09	F/25 GW	lissencephaly	microcephaly	complete agenesis	severe vermis hypoplasia
[21]	E429Q	LIS_TUB_004_ foetus09	F/25 GW	4 layered cortex	<3rd centile	complete agenesis	hypoplastic basal ganglia, severe hypoplasia of cerebellum and pons
[13]	**G436R**	LIS_TUB_038	M/7	perisylvian pachygyria	−3 SD	mild hypoplasia	mild vermis hypoplasia, severe dysgenesis of internal capsule

Bold indicates first report of mutation; * Asterisk indicates mutation was identified through whole exome sequencing; Abbreviations: OFC, occipitofrontal circumference; M, male; F, female; GW, gestational weeks; SD, standard deviation; SBH, subcortical band heterotopia.

4.4. *TUBA1A Mutations Linked to Cerebellar Dysplasia*

In addition to the cortical malformations described above, *TUBA1A* mutations also commonly cause dysplasia of a variety of other brain regions, most notably the cerebellum. In fact, "lissencephaly with cerebellar hypoplasia" is a common class of *TUBA1A* mutation-induced phenotype [12,16]. In addition, some identified cases hint that tubulin mutations may cause specific cerebellar phenotypes with only subtly disrupted, or normal, cortical folds [137]. Further sequencing of *TUBA1A* in patients without the characteristic cortical lissencephaly phenotype may prove fruitful in uncovering additional *TUBA1A* mutations.

4.5. *Summary of TUBA1A Mutations Associated with Brain Malformations*

In this section, we have merely provided broad categories of brain malformations linked to mutations in *TUBA1A*. Within each category are numerous mutations that lead to the identified cortical malformation, but also other detrimental neurological and physical phenotypes. The increasing number of *TUBA1A* mutations discovered and the growing list of phenotypic consequences call for a greater understanding of the mechanism(s) of tubulinopathy disease progression.

4.6. *Cellular Impact of TUBA1A Mutations*

While numerous studies have identified missense mutations in *TUBA1A* as the genetic basis of brain malformations, few studies have been conducted to determine how these mutations alter tubulin molecularly, or how the mutations alter neuronal maturation/function. Below we describe studies investigating the molecular and cellular impact of *TUBA1A* mutations.

Studies investigating the stability of the mutant tubulin heterodimer and its ability to incorporate into microtubules provide hints at the underlying mechanism of the mutant and point to whether they lead to haploinsufficiency (Figure 3A,B), or dominant disruption of microtubule function from within a polymerized microtubule (Figure 3C), or potentially both. To date, Tian et al. published the most comprehensive and informative study on the consequences of *TUBA1A* mutations on heterodimer formation and stability [151]. TUBA1A mutant proteins were expressed in vitro to perform transcription/translation analysis in rabbit reticulocyte to test tubulin heterodimer yield. They discovered that p.L397P, p.V303G, and p.R402C lead to significantly reduced amount of unstable tubulin heterodimer, and that p.I188L, p.I238V, p.P263T, p.L286F, p.R402H, and p.S419L caused slight reduction to the amount of tubulin heterodimer [151]. These data suggest that mutations such as p.P263T may act more dominantly, while p.L397P, p.V303G, and p.R402C may act more as haploinsufficient mutants. This hypothesis is supported by the discovery that when ectopically expressed, p.P263T can incorporate into microtubules, dampen microtubule dynamics in COS-7 cells, and decrease microtubule growth in neurites of E15.5 primary cortical neurons 1 DIV [16,151]. These data all point to p.P263T acting dominantly, as described by Figure 3C. Alternatively, when p.V303G was ectopically expressed, it caused no change in microtubule dynamics in COS-7 or primary cortical neurons [151], as might be expected from a mutant that leads to few competent tubulin heterodimers entering the microtubule (Figure 3A).

Few *TUBA1A* mutations that lead to lissencephaly have been satisfactorily linked to the migratory pathway. However, inferences can be made about the functions of TUBA1A that may be important for migration based on how MAPs interact with microtubules and on the limited molecular data available on some of the lissencephaly patient *TUBA1A* mutations. For instance, p.R264C mutation occurs frequently in the *TUBA1A* patient population (12.5%), causes the less severe pachygyria phenotype, and is one of the most investigated mutations. Molecular and cellular data interrogating the consequence of p.R264C reveal that the mutation reduces the frequency of heterodimer formation due to compromised folding efficiency [152]. This reduced folding efficiency does not completely cause tubulin heterodimers containing the p.R264C mutant to drop out of the polymerization-competent tubulin pool, as they have been seen to form microtubules in HeLa, COS-7, and P19 cells [16,151]. p.R264C has also been modeled using iPSCs derived from patient cells, revealing that neurospheres generated from the patient iPSCs are capable of normal neurite extension [138]. Taken together, these data reveal that the p.R264C *TUBA1A* mutation probably follows the mutant scenario outlined by Figure 3A, and more importantly that partial haploinsufficiency to *TUBA1A* is sufficient to disrupt cortical migration enough to cause pachygyria. Similarly, cellular studies with p.I188L, which causes laminar heterotopia, show that FLAG-tagged mutant Tuba1a incorporates into microtubules in P19 cells [16], but that the mutation also causes slightly reduced amounts of tubulin heterodimer formation [151].

Three other lissencephaly-causing mutations have been investigated for their effects on cellular function, p.N329S, p.C25F, and p.R64W. Mutation p.N329S causes lissencephaly with cerebellar hypoplasia [16,138], and when neurospheres were generated from patient iPSCs, the mutation caused abnormal neurite extension [138]. While these cellular data suggest that defective neurite extension leads to lissencephaly with cerebellar hypoplasia, how the microtubules are altered by the p.N329S mutation to cause this diminished neurite growth is unknown. Mutation p.C25F leads to lissencephaly/thin cortex, causes decreased microtubule density in mutant transfected COS7 cells, and also caused patient fibroblast microtubules to depolymerize faster in cold [133]. These results are hard to interpret without further molecular evidence, but suggest that p.C25F may cause a shift in the stability in microtubules and that this decrease in stable microtubules is also sufficient to derail cortical migration enough to cause lissencephaly. Mutation p.R64W leads to severe brain malformations, exhibiting extremely thin cerebral parenchyma, agenesis of the cerebellum, and additional devastating defects to other brain regions. Similar to p.C25F, this mutation also causes decreased microtubule density and unstable microtubules [133]. However, because the phenotypes of p.C25F and p.R64W are divergent, with R64W exhibiting much more severe brain malformations, additional data must be collected to distinguish the molecular consequences of these mutations.

Whereas several lissencephaly-causing mutations lead to a loss of microtubules in vivo, there are no examples of mutants acting dominantly to assemble into microtubules and "poison" their function. These gain-of-function effects could be exerted by mutations that alter amino acid residues on the microtubule surface, such as mutations of R402, which cause lissencephaly. No studies have yet investigated how the patient mutations p.R402C and p.R402H impact microtubule function in vivo; therefore, these are high priority for future investigation.

While a few mutations leading to lissencephaly have been examined in terms of molecular and cellular consequence, to our knowledge none of the polymicrogyria-causing *TUBA1A* mutants have been investigated. Understanding how these mutations alter microtubule function will be an important step in understanding the molecular basis of the polymicrogyria malformation.

In addition to the cellular data referenced above, two mouse models that contain *Tuba1a* point mutations have been developed. Mice with *N*-ethyl *N*-nitrosourea (ENU) induced mutations in *Tuba1a* have provide insights into how Tuba1a contributes to neuronal development. For example, an S140G substitution in Tuba1a interferes with efficient GTP binding and heterodimer formation, but heterodimers that include S140G mutant Tuba1a can incorporate into the lattice [11]. One copy of the S140G *Tuba1a* allele is sufficient to cause defective neuronal migration in the hippocampus and the auditory, visual, and somatosensory cortices [11]. This mutation increases neuronal branching and alters the direction of migration in the rostral migratory stream that populates the olfactory bulb [153]. The neuronal migration phenotypes caused by the S140G mutation can be rescued with increased expression of wild-type Tuba1a, indicating that this mutation reduces the function of the protein [11]. Intriguingly, in vitro data suggest that the S140G mutation cannot be rescued by expressing another α-tubulin isotype, Tuba8 [153]. Additionally, presence of the Tuba8 isotype significantly altered microtubule polymerization compared to control or S140G mutant cells. This study provides the first evidence of functionally distinct α-tubulin isotypes in neurons [153]. However, it is important to note that *Tuba8* expression has not been detected in neurons [154].

An N102D substitution in Tuba1a provides different insight into the role of Tuba1a in neural development. Homozygous *Tuba1a* N102D mice have significantly less acetylated tubulin in growth cones, suggesting that microtubules containing Tuba1a N102D are less stable in extending growth cones [155]. On a cellular level, homozygous *Tuba1a* N102D mice have defects in motor axon extension and synapse function [155]. On a tissue level, homozygous *Tuba1a* N102D mice have gross defects in brain development including disorganized cortical layers [155]. The asparagine residue at position 102 is highly conserved across eukaryotes, which allows for the impact of N102D to be isolated and tested in the simple model organism budding yeast. The N to D substitution in yeast α-tubulin interferes with heterodimer stability, impairs its ability to incorporate into the lattice, and increases the frequency of microtubule catastrophes/depolymerization events [155]. Interestingly, yeast cells expressing the mutant allele in one α-tubulin isotype appear to compensate by increasing the cellular levels of an alternative, wild-type α-tubulin isotype. This suggests a novel consequence of *TUBA1A* mutations—cells may respond by shifting the blend of α-tubulin isotypes, which could in turn alter the normal regulation of microtubule networks.

Together, these mouse models demonstrate that Tuba1a is important for axon extension, efficient synapse function, and migration of neurons during development. However, we do not understand how mutations that have been identified in tubulinopathy patients impact microtubule properties or cellular function to disrupt development. Genetic models of patient derived *Tuba1a* mutations would lend much needed insights into the attributes of Tuba1a that are important for its role in neurodevelopment.

5. TUBA1A Expression: A Burst of Tubulin to Fuel Morphogenesis?

To understand the critical role of *TUBA1A* in brain development, it is important to consider which cells express *TUBA1A*, and how expression is regulated.

5.1. TUBA1A Expression Pattern

TUBA1A was originally identified in a cDNA library from human fetal brain tissue [64]. Subsequent northern blot studies in cell lines showed that *TUBA1A* is expressed in neural-derived cells and is particularly abundant in cell lines with characteristic long cytoplasmic processes (such as adherent neuroblastomas), but not in cell lines that exhibit small, round morphologies [9]. Since then, numerous groups have tracked the expression of *TUBA1A* and its vertebrate homologues to neuronal populations after terminal mitosis, as they begin to extend long processes.

A series of classic studies elucidated the expression program of mouse *Tuba1a* mRNA across tissues and developmental stages. The earliest study investigating mouse *Tuba1a* expression via northern blot with probes recognizing the distinct *Tuba1a* 3′-UTR provides the most extensive inquiry into which organs express *Tuba1a* postnatally [10]. This study identified *Tuba1a* mRNA in the adult brain, lung, and at low levels in the testes; however, only alternative α-tubulin mRNAs were detected in the heart, kidney, liver, muscles, spleen, stomach, and thymus. A time course analysis of brain-derived samples revealed that *Tuba1a* is abundant through early postnatal stages, begins to decline by P10, but still is expressed beyond P32 through adulthood [10]. Subsequent studies sought to elucidate the program of *Tuba1a* expression in the brain; to our knowledge there are no other published studies confirming the presence of *Tuba1a* in the adult lung or testes.

Much of our understanding of *Tuba1a* expression in the developing brain comes from studies conducted by Miller and colleagues, who inserted lacZ behind the native *Tuba1a* promoter and used this *Tuba1a*:nlacZ transgenic mouse to monitor *Tuba1a* expression in vivo [156–158]. The expression profiles of mouse *Tuba1a* generated by these approaches, and later a transgenic reporter line using *Tuba1a* promoter-driven EYFP expression [159], are outlined in Tables 6 and 7. These studies show that *Tuba1a* is expressed highly throughout the nervous system over the course of development, starting around E9.5 and continuing through early postnatal stages [10,156,158,159]. This trend holds true through the developing central and peripheral nervous system, with *Tuba1a* induction onset correlating to neurogenesis of numerous neuronal structures including the forebrain, midbrain, hindbrain, spinal cord, retina, and cranial nerves, among others [158]. Primary cultures generated from the transgenic reporter mice revealed that *Tuba1a* is expressed in cells positive for neuronal markers neuron-specific enolase (NSE) and βIII tubulin, but not in cells positive for stem cell or glial markers [156,158–160]. More recent RNA-seq studies from the Barres lab show that *Tuba1a* remains the most abundant α-tubulin isotype mRNA in neurons late in brain development (Postnatal Day 7) [154] (Figure 4). Based on these studies, the current model is that *Tuba1a* expression is induced coincident with or shortly after the terminal cell division that gives rise to neurons, and provides >95% of the α-tubulin mRNA in these cells [72].

Table 6. Studies of *TUBA1A* expression during mouse development.

			Developmental Time Point							
			E9.5	**E10.5**	**E11.5**	**E12**	**E12.5**	**E13.5**	**E14.5**	**E16.5**
Developing CNS	brain	brain, unspecified	[159]		[156]			[156]		
		forebrain	[158]							
		midbrain	[158]							
		hindbrain	[158]							
		telencephalon								[12]
		diencephalon		[156]	[156]			[156]		[12]
		mesencephalon								[12]
		metencephalon		[156]	[156]			[156]		[12]
		myelencephalon		[156]	[156]			[156]		[12]
		developing neocortex					[158]	[156]	[156]	[12]
		developing striatum					[159]			
		developing hippocampus					[159]			
		developing thalamus					[159]			
		developing amygdala					[159]			
		developing hypothalamus					[159]			
		developing cerebellum					[159]			[12]
		developing brain stem								[12]
		spinal cord	[158]	[156]	[156]		[159]	[156]		
		retina				[158]		[156]		
Developing PNS	cranial nerves	cranial ganglia			[156]			[156]		
		trigeminal ganglion	[158]							
		fascioacoustic ganglion	[158]							
		glossopharyngeal ganglia		[158]						
		developing olfactory bulbs			[156]		[159]	[156]		
		nasal epithelium		[158]						
		vomeronasal organ			[158]					
		dorsal root ganglia		[158]	[156]			[156]		[12]
		sensory ganglia			[156]			[156]		
	ANS	sympathetic ganglia		[158]	[156]			[156]		
		parasympathetic ganglia			[156]			[156]		
		heart	[158]							

Table 7. Studies of *TUBA1A* expression in postnatal mouse.

			Postnatal Time Point							
			P0	**P3**	**P6**	**P10**	**P15**	**P22**	**P32**	**Adult**
CNS	brain	brain, unspecified		[10]	[10]	[10]	[10]	[10]	[10]	[10,157]
		olfactory system	[159]							[157,159]
		cortex	[159]			[12]				[157]
		corpus callosum	[159]							[159]
		striatum	[159]							
		rostral migratory stream	[159]			[12]				
		basal ganglia								[157]
		basal forebrain								[157]
		hippocampus	[159]			[12]				[157,159]
		amygdala	[159]							[157,159]
		hypothalamus	[159]							[157,159]
		thalamus	[159]							[157,159]
		subthalamus								[157]
		midbrain	[159]							[157,159]
		pons								[157]
		medulla								[157]
		cerebellum	[159]			[12]				[157]
		brainstem				[12]				[157]
		spinal cord	[159]							[159]
Other		lung		[10]	[10]	[10]	[10]	[10]	[10]	[10]
		testes				[10]			[10]	[10]

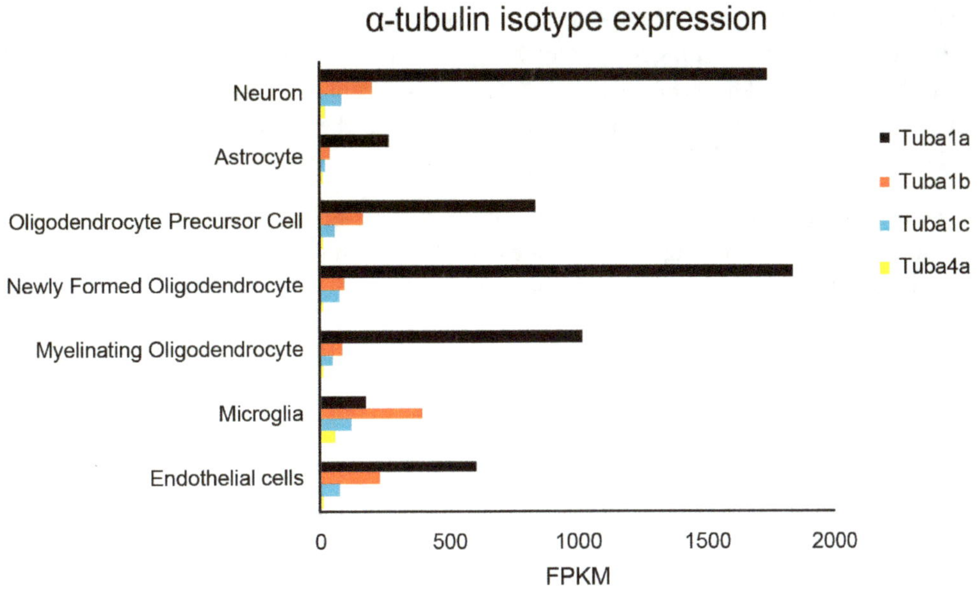

Figure 4. Dataset of RNA-Seq results for α-tubulin isotypes in mouse nervous system cell population. Distribution of α-tubulin isotype mRNA expression in various cell types generated from P7-17 mouse cerebral cortex. OPC population contains 5% microglial contamination. Adapted from [141].

While much of the RNA probing data suggest that *Tuba1a* expression is limited primarily to neurons, the recent RNA-seq dataset contends that *Tuba1a* may be expressed in more cell types than the earlier studies indicate [154]. As stated previously, numerous studies claim that *Tuba1a* mRNA is absent in stem cell, glial, and neuron progenitor populations, limited only to early born neurons [156,158–160]. However, the RNA-Seq data suggest that *Tuba1a* may in fact be expressed in numerous cell types found in the brain, with *Tuba1a* expression equally high in newly-formed oligodendrocytes as it is in neurons at Postnatal Day 7 (Figure 4). As immunohistochemistry double labeling studies suggest that oligodendrocytes in embryonic and adult mouse do not express *Tuba1a* [156,159,160], the contradictory expression of *Tuba1a* in glial populations calls for additional studies to conclusively uncover what cell types contain TUBA1A.

After brain development, *Tuba1a* mRNA levels persist, albeit at lower levels, in the adult mouse brain. Which adult cell populations express *Tuba1a* is somewhat controversial, due to conflicting data. Bamji and Miller found that *Tuba1a* mRNA was panneuronal in the adult mouse brain, with the highest levels found in neuronal populations that have the potential for morphologic growth, such as neurons of the piriform cortex [157] (Table 7). In contrast, Coksaygan and colleagues only detected *Tuba1a* mRNA in some regions of the adult brain (Table 7). How experimental differences might contribute to these discrepancies is not immediately clear.

While much of the *Tuba1a* expression data have been conducted in mouse, studies in rat models yield similar findings. Rat *Tuba1a* mRNA is highly enriched in the embryonic nervous system and is less abundant in adult brain [72]. Additionally, the rat data also support the idea that Tuba1a is induced in neurons undergoing neurite extension, such as in vivo at the cortical plate as well as in vitro in PC12 cells induced to differentiate with NGF treatment [72].

Interestingly, studies of the zebrafish *tuba1a* yield results that overlap with those from mammalian models, but with several important differences. Whereas zebrafish *tuba1a* is dramatically induced in the developing nervous system and declines coincident with the maturation, it is maintained at high levels in progenitor cells in the retinal periphery, lining the brain ventricles, and around the central canal of the spinal cord [161]. When cultured in vitro, these *tuba1a*-expressing cells divide and give rise to new neurons, suggesting that zebrafish may express *tuba1a* in neuronal progenitors as well as newly-born neurons [161]. Consistent with this notion, morpholino-based knock down of zebrafish *tuba1a* suppresses CNS development and leads to fewer differentiating neurons [162]. Intriguingly, Ramachandran et al. generated transgenic fish to conditionally and permanently label *tuba1a*-expressing cells, and found transient *tuba1a* expression in neural progenitors, skeletal muscle, heart, and intestine progenitors [163]. This experiment suggests that *tuba1a* expression is not neuronally limited in zebrafish.

These expression studies strongly suggest that *TUBA1A* is highly induced in neurons as they undergo cellular morphology changes to generate cytoplasmic processes. This suggests a simple model for why neurons, but not other cell types, use the conserved expression program for *TUBA1A*—neurons need significantly *more* α-tubulin at the time when they are maturing to form the long processes necessary to wire the nervous system.

5.2. Mechanisms Regulating TUBA1A Expression

While questions remain regarding which cell populations express *TUBA1A*, it is clear that expression is carefully regulated during development and adulthood. What pathways control this expression program for *TUBA1A*? A variety of factors and conditions that mediate *TUBA1A* mRNA production have been identified, and growth factors involved in neuronal differentiation play a prominent role.

Several studies have shown that growth factors are sufficient to stimulate *Tuba1a* expression. For example, PC12 cells from rats can be induced to differentiate into neurons with the addition of Nerve growth factor (NGF), and consequently *Tuba1a* mRNA levels increase as the differentiated PC12 cells begin to actively extend neurites [72]. Similarly, NGF treatment of neonatal rats causes a 5-10-fold increase of *Tuba1a* mRNA during a developmental period when *Tuba1a* levels normally decrease [164]. In addition to NGF, Fibroblast Growth Factor (FGF) and Platelet-Derived Growth Factor (PDGF) also stimulate *Tuba1a* expression. FGF was first shown to stimulate *Tuba1a* expression in rat sensory neurons, where treatment with FGF-1 increases *Tuba1a* mRNA levels by nearly ten-fold, and causes a concomitant increase in neurite outgrowth [165]. These findings suggest that exogenous stimuli such as growth factors may be sufficient to induce *TUBA1A* expression, at least in certain cell types.

The most detailed evidence of the regulatory pathways that act downstream of growth factors to increase *Tuba1a* transcription comes from a study using mouse primary cultures of E12–E13 cortical progenitors [166]. This study revealed a phosphorylation cascade initialized by FGF and PDGF activation that ends in drastic upregulation of TUBA1A, and ultimately neurogenesis (Figure 5). Inhibiting factors in this pathway, MEK or C/EBP, leads to a significant decrease in *Tuba1a* expression and stops neurogenesis. Conversely, introducing an activated C/EBP phosphomimetic enhances *Tuba1a* expression and promotes neurogenesis. Furthermore, this study identified three C/EBP binding sites in the conserved TUBA1A promoter (Figure 5B), and revealed that direct C/EBP binding to these promoter regions greatly enhances TUBA1A transcription. Interestingly, inhibiting the *Tuba1a* neurogenesis pathway causes the would-be neurons to take on an astrocyte fate. This reveals that the same progenitors can be induced to become either neurons or astrocytes, and that *Tuba1a* activation is an important step in promoting the neuronal fate.

While the studies described above implicate common neurogenesis growth factors in *Tuba1a* expression, the specific requirements for *Tuba1a* activation remain largely unexplored. However, these data provide an additional link between the presence of Tuba1a and neurogenesis, further promoting the idea that Tuba1a may be an important regulator of neuronal fate.

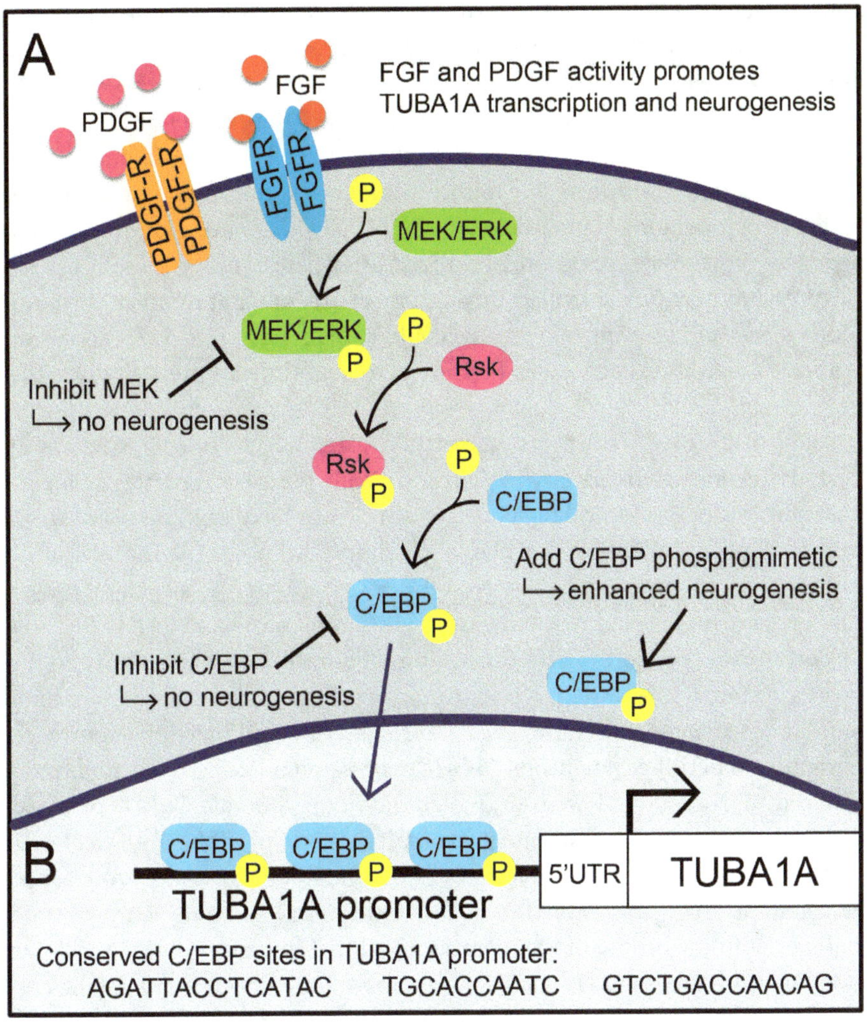

Figure 5. Transcriptional regulation of *TUBA1A* in mouse cortical progenitors. (**A**) FGF/PDGF treatment of cortical progenitors leads to a phosphorylation cascade activating MEK/ERK, Rsk, and C/EBP; and (**B**) in the nucleus, phosphorylated C/EBP directly binds three conserved C/EBP binding sites in the TUBA1A promoter, leading to transcription of TUBA1A and neurogenesis. Based on data from [153].

5.3. *TUBA1A in Regeneration*

The unique temporal expression of *TUBA1A* during brain development indicates a critical, albeit poorly defined role for TUBA1A as a growth-associated neuronal α-tubulin. Interestingly, there is evidence that TUBA1A may also play important roles in remodeling the cytoskeleton during nerve regeneration and repair. Although *Tuba1a* mRNA expression declines after development, it is rapidly induced following nerve injury [72,167–171]. Mathew and Miller proposed that NGF could regulate *Tuba1a* expression following injury like in development [164,172]. The p75 NGF receptor is upregulated following axotomy, providing some support for this model [164,172]. The induction of a developmental α-tubulin, such as *TUBA1A*, during regeneration supports the idea that developing and regenerating neurons may utilize common mechanisms and that TUBA1A is of particular importance to growing neurons, regardless of age.

How might nerve injury trigger the re-emergence of *TUBA1A* expression? Several studies suggest that *Tuba1a* is normally repressed in adult neurons, but if normal processes are disrupted, *Tuba1a* expression is induced. For example, blocking fast axonal transport has been shown to induce *Tuba1a* expression [173,174]. These findings specifically implicate normal axonal transport as part of the mechanism that represses *TUBA1A*. For example, the neuron could be constantly monitoring its axonal integrity using intrinsic cues, such as backup of protein cargoes that would occur within the axon upon injury, or extrinsic cues such as constant uptake of growth factors like CNTF from nearby Schwann cells [167,173]. Alternatively, it has been proposed that repression could be maintained through physical contacts between neurons and neighboring cells [174]. However, current evidence does not support this model, as studies have found that loss of contact with either the basal lamina or the neuron's target structure alone is insufficient to induce *Tuba1a* expression [167,172]. Although the exact mechanisms by which its expression is induced after nerve injury remain unclear, the model that *TUBA1A* expression may be controlled through mechanisms that monitor normal, homeostatic processes in neurons is intuitively appealing. This could explain how *TUBA1A* expression is repressed late in development, when these processes become active, and de-repressed when these processes are disrupted.

The expression of *TUBA1A* during regeneration after nerve injury provides an interesting opportunity to study regenerative capacity in neurons. While peripheral nerves are able to regenerate after injury, most neurons of the central nervous system do not have this capacity [175]. What allows peripheral axons to regenerate, while central axons cannot? Many of the initial studies identifying *Tuba1a* expression in response to injury focused on the peripheral nervous system; however, increases in *Tuba1a* expression have been noted following injury to central nerves as well [170]. This observation provides an opportunity to probe the differences in regenerative capacity between the central and peripheral nervous systems. Interestingly, one study found that *Tuba1a* gene expression after injury was broadly similar between peripheral nerves that regenerate and proximal central nerves that do not [170]. Importantly, in peripheral neurons, *Tuba1a* expression remained elevated until regeneration was complete. In central rubrospinal neurons, which did not regenerate, *Tuba1a* expression remained elevated for as long as two weeks post injury, even after levels of other injury-related cytoskeletal proteins had returned to normal [170]. However, if rubrospinal neurons are transplanted into a peripheral environment, they can regenerate [176]. Other studies have demonstrated that *Tuba1a* expression correlates with the timing of regeneration, and that in general *Tuba1a* will remain elevated until the axon is able to regenerate [172,177]. Thus, these data show that specific central neurons have regenerative capacity, but likely an unfavorable environment prevents complete regeneration. Conversely, study of a different population of central neurons found no evidence of *Tuba1a* upregulation after injury [169]. This may indicate some heterogeneity in regenerative capacity between different populations of central neurons; some neurons may lack the capacity to regenerate, regardless of whether the environment is conducive to regeneration [169]. Interestingly, *Tuba1a* can become upregulated after injury, even in uninjured regions of the nervous system. For example, elevated *Tuba1a* expression was not limited solely to injured peripheral axons, as neighboring uninjured axons can also display increased *Tuba1a* mRNA [172,177]. Additionally, elevated *Tuba1a* expression was found in the red nucleus, a brainstem nucleus that sends axons to the spinal cord, following a spinal cord compression injury [171]. The outlined studies show a selective upregulation of *Tuba1a* in regenerating neurons. The properties of TUBA1A α-tubulin that make it particularly suited to these functions are yet unknown. However, it is interesting to speculate that combinations of different tubulin isotypes could produce microtubules that are preferentially suited for specific functions.

6. Concluding Remarks

In recent years, the α-tubulin gene *TUBA1A* has emerged as a vital component in the development of the nervous system. With the identification of tubulinopathy-causing *TUBA1A* mutations, understanding how TUBA1A must function during neural development becomes an urgent inquiry. Microtubules perform many complex roles throughout development, but it remains unclear how TUBA1A, and the tubulin heterodimers and microtubules it produces, can be regulated to perform such diverse tasks in different neuronal compartments and through developmental and adult stages. The neuronal tubulin gene profile, as well as the poorly understood, yet remarkably complex, regulation of tubulin PTMs and MAPs may provide a starting point for determining how microtubules are tuned throughout neuron development. Brain malformation-causing *TUBA1A* mutations provide an irreplaceable opportunity to explore the role of microtubules at the cellular level of neuronal maturation, and at the tissue level of brain formation. While previous studies have performed the important task of sequencing tubulin genes to uncover underlying genetic mutations, future tubulinopathy studies must focus on the molecular and cellular consequences of these mutations to fully understand how altered TUBA1A function contributes to brain malformation progression. Discovering the molecular mechanisms behind TUBA1A mutations will be an important undertaking in cellular neuroscience, and could bring powerful new insight into the essential roles of microtubules in neural development.

Acknowledgments: Funding: This work was supported by the National Science Foundation Graduate Research Fellowship 1553798 to J.A., the Neuroscience Training Grant T32NS099042 to G.B. and National Institutes of Health 5R01GM112893-03 to J.K.M.

Author Contributions: J.A., G.B., E.A.B. and J.K.M. wrote the paper.

Conflicts of Interest: The authors declare no conflict of interest.

References

1. Dent, E.W.; Kalil, K. Axon branching requires interactions between dynamic microtubules and actin filaments. *J. Neurosci.* **2001**, *21*, 9757–9769. [PubMed]
2. Hu, J.; Bai, X.; Bowen, J.R.; Dolat, L.; Korobova, F.; Yu, W.; Baas, P.W.; Svitkina, T.; Gallo, G.; Spiliotis, E.T. Septin-driven coordination of actin and microtubule remodeling regulates the collateral branching of axons. *Curr. Biol.* **2012**, *22*, 1109–1115. [CrossRef] [PubMed]
3. Lin, S.; Liu, M.; Mozgova, O.I.; Yu, W.Q.; Baas, P.W. Mitotic Motors Coregulate Microtubule Patterns in Axons and Dendrites. *J. Neurosci.* **2012**, *32*, 14033–14049. [CrossRef] [PubMed]
4. Fanara, P.; Husted, K.H.; Selle, K.; Wong, P.Y.A.; Banerjee, J.; Brandt, R.; Hellerstein, M.K. Changes in microtubule turnover accompany synaptic plasticity and memory formation in response to contextual fear conditioning in mice. *Neuroscience* **2010**, *168*, 167–178. [CrossRef] [PubMed]
5. Hoogenraad, C.C.; Bradke, F. Control of neuronal polarity and plasticity—A renaissance for microtubules? *Trends Cell Biol.* **2009**, *19*, 669–676. [CrossRef] [PubMed]
6. Jaworski, J.; Kapitein, L.C.; Gouveia, S.M.; Dortland, B.R.; Wulf, P.S.; Grigoriev, I.; Camera, P.; Spangler, S.A.; Di Stefano, P.; Demmers, J.; et al. Dynamic Microtubules Regulate Dendritic Spine Morphology and Synaptic Plasticity. *Neuron* **2009**, *61*, 85–100. [CrossRef] [PubMed]
7. Khodiyar, V.K.; Maltais, L.J.; Ruef, B.J.; Sneddon, K.M.B.; Smith, J.R.; Shimoyama, M.; Cabral, F.; Dumontet, C.; Dutcher, S.K.; Harvey, R.J.; et al. A revised nomenclature for the human and rodent alpha-tubulin gene family. *Genomics* **2007**, *90*, 285–289. [CrossRef] [PubMed]
8. Findeisen, P.; Mühlhausen, S.; Dempewolf, S.; Hertzog, J.; Zietlow, A.; Carlomagno, T.; Kollmar, M. Six subgroups and extensive recent duplications characterize the evolution of the eukaryotic tubulin protein family. *Genome Biol. Evol.* **2014**, *6*, 2274–2288. [CrossRef] [PubMed]
9. Hall, J.L.; Cowan, N.J. Structural features and restricted expression of a human alpha-tubulin gene. *Nucleic Acids Res.* **1985**, *13*, 207–223. [CrossRef] [PubMed]
10. Lewis, S.A.; Lee, M.G.S.; Cowan, N.J. Five mouse tubulin isotypes and their regulated expression during development. *J. Cell Biol.* **1985**, *101*, 852–861. [CrossRef] [PubMed]
11. Keays, D.A.; Tian, G.; Poirier, K.; Huang, G.-J.; Siebold, C.; Cleak, J.; Oliver, P.L.; Fray, M.; Harvey, R.J.; Molnár, Z.; et al. Mutations in alpha-tubulin cause abnormal neuronal migration in mice and lissencephaly in humans. *Cell* **2007**, *128*, 45–57. [CrossRef] [PubMed]
12. Poirier, K.; Keays, D.A.; Francis, F.; Saillour, Y.; Bahi, N.; Manouvrier, S.; Fallet-Bianco, C.; Pasquier, L.; Toutain, A.; Tuy, F.P.D.; et al. Large spectrum of lissencephaly and pachygyria phenotypes resulting from de novo missense mutations in tubulin alpha 1A (TUBA1A). *Hum. Mutat.* **2007**, *28*, 1055–1064. [CrossRef] [PubMed]

13. Bahi-Buisson, N.; Poirier, K.; Boddaert, N.; Saillour, Y.; Castelnau, L.; Philip, N.; Buyse, G.; Villard, L.; Joriot, S.; Marret, S.; et al. Refinement of cortical dysgeneses spectrum associated with TUBA1A mutations. *J. Med. Genet.* **2008**, *45*, 647–653. [CrossRef] [PubMed]
14. Fallet-Bianco, C.; Loeuillet, L.; Poirier, K.; Loget, P.; Chapon, F.; Pasquier, L.; Saillour, Y.; Beldjord, C.; Chelly, J.; Francis, F. Neuropathological phenotype of a distinct form of lissencephaly associated with mutations in TUBA1A. *Brain* **2008**, *131*, 2304–2320. [CrossRef] [PubMed]
15. Morris-Rosendahl, D.J.; Najm, J.; Lachmeijer, A.M.A.; Sztriha, L.; Martins, M.; Kuechler, A.; Haug, V.; Zeschnigk, C.; Martin, P.; Santos, M.; et al. Refining the phenotype of alpha-1a Tubulin (TUBA1A) mutation in patients with classical lissencephaly. *Clin. Genet.* **2008**, *74*, 425–433. [CrossRef] [PubMed]
16. Kumar, R.A.; Pilz, D.T.; Babatz, T.D.; Cushion, T.D.; Harvey, K.; Topf, M.; Yates, L.; Robb, S.; Uyanik, G.; Mancini, G.M.S.; et al. TUBA1A mutations cause wide spectrum lissencephaly (smooth brain) and suggest that multiple neuronal migration pathways converge on alpha tubulins. *Hum. Mol. Genet.* **2010**, *19*, 2817–2827. [CrossRef] [PubMed]
17. Lecourtois, M.; Poirier, K.; Friocourt, G.; Jaglin, X.; Goldenberg, A.; Saugier-Veber, P.; Chelly, J.; Laquerrière, A. Human lissencephaly with cerebellar hypoplasia due to mutations in TUBA1A: Expansion of the foetal neuropathological phenotype. *Acta Neuropathol.* **2010**, *119*, 779–789. [CrossRef] [PubMed]
18. Poirier, K.; Saillour, Y.; Fourniol, F.; Francis, F.; Souville, I.; Valence, S.; Desguerre, I.; Marie Lepage, J.; Boddaert, N.; Jacquemont, M.L.; et al. Expanding the spectrum of TUBA1A-related cortical dysgenesis to Polymicrogyria. *Eur. J. Hum. Genet.* **2013**, *21*, 381–385. [CrossRef] [PubMed]
19. Cushion, T.D.; Dobyns, W.B.; Mullins, J.G.L.; Stoodley, N.; Chung, S.K.; Fry, A.E.; Hehr, U.; Gunny, R.; Aylsworth, A.S.; Prabhakar, P.; et al. Overlapping cortical malformations and mutations in TUBB2B and TUBA1A. *Brain* **2013**, *136*, 536–548. [CrossRef] [PubMed]
20. Bahi-Buisson, N.; Poirier, K.; Fourniol, F.; Saillour, Y.; Valence, S.; Lebrun, N.; Hully, M.; Bianco, C.F.; Boddaert, N.; Elie, C.; et al. The wide spectrum of tubulinopathies: What are the key features for the diagnosis? *Brain* **2014**, *137*, 1676–1770. [CrossRef] [PubMed]
21. Fallet-Bianco, C.; Laquerrière, A.; Poirier, K.; Razavi, F.; Guimiot, F.; Dias, P.; Loeuillet, L.; Lascelles, K.; Beldjord, C.; Carion, N.; et al. Mutations in tubulin genes are frequent causes of various foetal malformations of cortical development including microlissencephaly. *Acta Neuropathol. Commun.* **2014**, *2*, 69. [CrossRef] [PubMed]
22. Kraemer, J.A.; Erb, M.L.; Waddling, C.A.; Montabana, E.A.; Zehr, E.A.; Wang, H.; Nguyen, K.; Pham, D.S.L.; Agard, D.A.; Pogliano, J. A phage tubulin assembles dynamic filaments by an atypical mechanism to center viral DNA within the host cell. *Cell* **2012**, *149*, 1488–1499. [CrossRef] [PubMed]
23. Amos, L.A.; van den Ent, F.; Löwe, J. Structural/functional homology between the bacterial and eukaryotic cytoskeletons. *Curr. Opin. Cell Biol.* **2004**, *16*, 24–31. [CrossRef] [PubMed]
24. Nogales, E.; Wolf, S.G.; Downing, K.H. Structure of the alpha beta tubulin dimer by electron crystallography. *Nature* **1998**, *391*, 199–203. [CrossRef] [PubMed]
25. Spiegelman, B.M.; Penningroth, S.M.; Kirschner, M.W. Turnover of tubulin and the N site GTP in Chinese hamster ovary cells. *Cell* **1977**, *12*, 587–600. [CrossRef]
26. Carlier, M.F.; Pantaloni, D. Kinetic analysis of guanosine 5′-triphosphate hydrolysis associated with tubulin polymerization. *Biochemistry* **1981**, *20*, 1918–1924. [CrossRef] [PubMed]
27. Alushin, G.M.; Lander, G.C.; Kellogg, E.H.; Zhang, R.; Baker, D.; Nogales, E. High-resolution microtubule structures reveal the structural transitions in αβ-tubulin upon GTP hydrolysis. *Cell* **2014**, *157*, 1117–1129. [CrossRef] [PubMed]
28. Zhang, R.; Alushin, G.M.; Brown, A.; Nogales, E. Mechanistic Origin of Microtubule Dynamic Instability and Its Modulation by EB Proteins. *Cell* **2015**, *162*, 849–859. [CrossRef] [PubMed]
29. Geyer, E.A.; Burns, A.; Lalonde, B.A.; Ye, X.; Piedra, F.-A.; Huffaker, T.C.; Rice, L.M. A mutation uncouples the tubulin conformational and GTPase cycles, revealing allosteric control of microtubule dynamics. *eLife* **2015**, *4*, e10113. [CrossRef] [PubMed]
30. Wang, H.-W.; Nogales, E. Nucleotide-dependent bending flexibility of tubulin regulates microtubule assembly. *Nature* **2005**, *435*, 911–915. [CrossRef] [PubMed]
31. Rice, L.M.; Montabana, E.A.; Agard, D.A. The lattice as allosteric effector: Structural studies of alphabeta- and gamma-tubulin clarify the role of GTP in microtubule assembly. *Proc. Natl. Acad. Sci. USA* **2008**, *105*, 5378–5383. [CrossRef] [PubMed]

32. Nogales, E.; Whittaker, M.; Milligan, R.A.; Downing, K.H. High-resolution model of the microtubule. *Cell* **1999**, *96*, 79–88. [CrossRef]
33. Sept, D.; Baker, N.A.; McCammon, J.A. The physical basis of microtubule structure and stability. *Protein Sci.* **2003**, *12*, 2257–2261. [CrossRef] [PubMed]
34. Kononova, O.; Kholodov, Y.; Theisen, K.E.; Marx, K.A.; Dima, R.I.; Ataullakhanov, F.I.; Grishchuk, E.L.; Barsegov, V. 52 Tubulin bond energies and microtubule biomechanics determined from nanoindentation in silico. *J. Biomol. Struct. Dyn.* **2015**, *33* (Suppl. 1), 35–36. [CrossRef] [PubMed]
35. Mandelkow, E.M.; Mandelkow, E.; Milligan, R.A. Microtubule dynamics and microtubule caps: A time-resolved cryo-electron microscopy study. *J. Cell Biol.* **1991**, *114*, 977–991. [CrossRef] [PubMed]
36. Hirokawa, N.; Takemura, R. Molecular motors and mechanisms of directional transport in neurons. *Nat. Rev. Neurosci.* **2005**, *6*, 201–214. [CrossRef] [PubMed]
37. Conde, C.; Cáceres, A. Microtubule assembly, organization and dynamics in axons and dendrites. *Nat. Rev. Neurosci.* **2009**, *10*, 319–332. [CrossRef] [PubMed]
38. Moon, H.M.; Wynshaw-Boris, A. Cytoskeleton in action: Lissencephaly, a neuronal migration disorder. *Wiley Interdiscip. Rev. Dev. Biol.* **2014**, *2*, 229–245. [CrossRef] [PubMed]
39. Dehmelt, L.; Halpain, S. Actin and Microtubules in Neurite Initiation: Are MAPs the Missing Link? *J. Neurobiol.* **2004**, *58*, 18–33. [CrossRef] [PubMed]
40. Hancock, W.O. The Kinesin-1 Chemomechanical Cycle: Stepping Toward a Consensus. *Biophys. J.* **2016**, *110*, 1216–1225. [CrossRef] [PubMed]
41. Hirokawa, N.; Tanaka, Y. Kinesin superfamily proteins (KIFs): Various functions and their relevance for important phenomena in life and diseases. *Exp. Cell Res.* **2015**, *334*, 16–25. [CrossRef] [PubMed]
42. Tsai, J.-W.; Lian, W.-N.; Kemal, S.; Kriegstein, A.R.; Vallee, R.B. Kinesin 3 and cytoplasmic dynein mediate interkinetic nuclear migration in neural stem cells. *Nat. Neurosci.* **2010**, *13*, 1463–1471. [CrossRef] [PubMed]
43. Homma, N.; Takei, Y.; Tanaka, Y.; Nakata, T.; Terada, S.; Kikkawa, M.; Noda, Y.; Hirokawa, N. Kinesin superfamily protein 2A (KIF2A) functions in suppression of collateral branch extension. *Cell* **2003**, *114*, 229–239. [CrossRef]
44. Schmidt, H.; Carter, A.P. Review: Structure and mechanism of the dynein motor ATPase. *Biopolymers* **2016**, *105*, 557–567. [CrossRef] [PubMed]
45. Schlager, M.A.; Hoang, H.T.; Urnavicius, L.; Bullock, S.L.; Carter, A.P. In vitro reconstitution of a highly processive recombinant human dynein complex. *EMBO J.* **2014**, *33*, 1855–1868. [CrossRef] [PubMed]
46. McKenney, R.J.; Huynh, W.; Tanenbaum, M.E.; Bhabha, G.; Vale, R.D. Activation of cytoplasmic dynein motility by dynactin-cargo adapter complexes. *Science* **2014**, *345*, 337–341. [CrossRef] [PubMed]
47. Ayloo, S.; Lazarus, J.E.; Dodda, A.; Tokito, M.; Ostap, E.M.; Holzbaur, E.L.F. Dynactin functions as both a dynamic tether and brake during dynein-driven motility. *Nat. Commun.* **2014**, *5*, 4807. [CrossRef] [PubMed]
48. McKenney, R.J.; Vershinin, M.; Kunwar, A.; Vallee, R.B.; Gross, S.P. LIS1 and NudE induce a persistent dynein force-producing state. *Cell* **2010**, *141*, 304–314. [CrossRef] [PubMed]
49. Toropova, K.; Zou, S.; Roberts, A.J.; Redwine, W.B.; Goodman, B.S.; Reck-Peterson, S.L.; Leschziner, A.E. Lis1 regulates dynein by sterically blocking its mechanochemical cycle. *eLife* **2014**, *3*. [CrossRef] [PubMed]
50. Gutierrez, P.A.; Ackermann, B.E.; Vershinin, M.; McKenney, R.J. Differential effects of the dynein-regulatory factor Lissencephaly-1 on processive dynein-dynactin motility. *J. Biol. Chem.* **2017**, *292*, 12245–12255. [CrossRef] [PubMed]
51. Yao, X.; Arst, H.N.; Wang, X.; Xiang, X. Discovery of a vezatin-like protein for dynein-mediated early endosome transport. *Mol. Biol. Cell* **2015**, *26*, 3816–3827. [CrossRef] [PubMed]
52. Olenick, M.A.; Tokito, M.; Boczkowska, M.; Dominguez, R.; Holzbaur, E.L.F. Hook Adaptors Induce Unidirectional Processive Motility by Enhancing the Dynein-Dynactin Interaction. *J. Biol. Chem.* **2016**, *291*, 18239–18251. [CrossRef] [PubMed]
53. Schroeder, C.M.; Vale, R.D. Assembly and activation of dynein-dynactin by the cargo adaptor protein Hook3. *J. Cell Biol.* **2016**, *214*, 309–318. [CrossRef] [PubMed]
54. Redwine, W.B.; DeSantis, M.E.; Hollyer, I.; Htet, Z.M.; Tran, P.T.; Swanson, S.K.; Florens, L.; Washburn, M.P.; Reck-Peterson, S.L. The human cytoplasmic dynein interactome reveals novel activators of motility. *eLife* **2017**, *6*. [CrossRef] [PubMed]
55. Yu, I.; Garnham, C.P.; Roll-Mecak, A. Writing and Reading the Tubulin Code. *J. Biol. Chem.* **2015**, *290*, 17163–17172. [CrossRef] [PubMed]

56. Janke, C. The tubulin code: Molecular components, readout mechanisms, and functions. *J. Cell Biol.* **2014**, *206*, 461–472. [CrossRef] [PubMed]
57. Duan, J.; Gorovsky, M.A. Both carboxy-terminal tails of alpha- and beta-tubulin are essential, but either one will suffice. *Curr. Biol.* **2002**, *12*, 313–316. [CrossRef]
58. Peris, L.; Thery, M.; Fauré, J.; Saoudi, Y.; Lafanechère, L.; Chilton, J.K.; Gordon-Weeks, P.; Galjart, N.; Bornens, M.; Wordeman, L.; et al. Tubulin tyrosination is a major factor affecting the recruitment of CAP-Gly proteins at microtubule plus ends. *J. Cell Biol.* **2006**, *174*, 839–849. [CrossRef] [PubMed]
59. Caudron, F.; Andrieux, A.; Job, D.; Boscheron, C. A new role for kinesin-directed transport of Bik1p (CLIP-170) in Saccharomyces cerevisiae. *J. Cell Sci.* **2008**, *121*, 1506–1513. [CrossRef] [PubMed]
60. Aiken, J.; Sept, D.; Costanzo, M.; Boone, C.; Cooper, J.A.; Moore, J.K. Genome-wide analysis reveals novel and discrete functions for tubulin carboxy-terminal tails. *Curr. Biol.* **2014**, *24*, 1295–1303. [CrossRef] [PubMed]
61. Fees, C.P.; Aiken, J.; O'Toole, E.T.; Giddings, T.H.; Moore, J.K. The negatively charged carboxy-terminal tail of β-tubulin promotes proper chromosome segregation. *Mol. Biol. Cell* **2016**, *27*, 1786–1796. [CrossRef] [PubMed]
62. Löwe, J.; Li, H.; Downing, K.H.; Nogales, E. Refined structure of alpha beta-tubulin at 3.5 A resolution. *J. Mol. Biol.* **2001**, *313*, 1045–1057. [CrossRef] [PubMed]
63. Eddé, B.; Rossier, J.; Le Caer, J.P.; Desbruyères, E.; Gros, F.; Denoulet, P. Posttranslational glutamylation of alpha-tubulin. *Science* **1990**, *247*, 83–85. [CrossRef] [PubMed]
64. Cowan, N.J.; Dobner, P.R.; Fuchs, E.V.; Cleveland, D.W. Expression of human alpha-tubulin genes: Interspecies conservation of 3′ untranslated regions. *Mol. Cell. Biol.* **1983**, *3*, 1738–1745. [CrossRef] [PubMed]
65. Cleveland, D.W.; Lopata, M.A.; MacDonald, R.J.; Cowan, N.J.; Rutter, W.J.; Kirschner, M.W. Number and evolutionary conservation of alpha- and beta-tubulin and cytoplasmic beta- and gamma-actin genes using specific cloned cDNA probes. *Cell* **1980**, *20*, 95–105. [CrossRef]
66. Baraban, M.; Anselme, I.; Schneider-Maunoury, S.; Giudicelli, F. Zebrafish Embryonic Neurons Transport Messenger RNA to Axons and Growth Cones In Vivo. *J. Neurosci.* **2013**, *33*, 15726–15734. [CrossRef] [PubMed]
67. Denoulet, P.; Eddé, B.; Gros, F. Differential expression of several neurospecific beta-tubulin mRNAs in the mouse brain during development. *Gene* **1986**, *50*, 289–297. [CrossRef]
68. Leandro-García, L.J.; Leskelä, S.; Landa, I.; Montero-Conde, C.; López-Jiménez, E.; Letón, R.; Cascón, A.; Robledo, M.; Rodríguez-Antona, C. Tumoral and tissue-specific expression of the major human beta-tubulin isotypes. *Cytoskeleton (Hoboken)* **2010**, *67*, 214–223. [CrossRef] [PubMed]
69. Joshi, H.C.; Cleveland, D.W. Differential utilization of beta-tubulin isotypes in differentiating neurites. *J. Cell Biol.* **1989**, *109*, 663–673. [CrossRef] [PubMed]
70. Perry, B.; Jensen-Smith, H.C.; Ludueña, R.F.; Hallworth, R. Selective expression of beta tubulin isotypes in gerbil vestibular sensory epithelia and neurons. *J. Assoc. Res. Otolaryngol.* **2003**, *4*, 329–338. [CrossRef] [PubMed]
71. Hao, L.; Thein, M.; Brust-Mascher, I.; Civelekoglu-Scholey, G.; Lu, Y.; Acar, S.; Prevo, B.; Shaham, S.; Scholey, J.M. Intraflagellar transport delivers tubulin isotypes to sensory cilium middle and distal segments. *Nat. Cell Biol.* **2011**, *13*, 790–798. [CrossRef] [PubMed]
72. Miller, F.D.; Naus, C.C.; Durand, M.; Bloom, F.E.; Milner, R.J. Isotypes of alpha-tubulin are differentially regulated during neuronal maturation. *J. Cell Biol.* **1987**, *105*, 3065–3073. [CrossRef] [PubMed]
73. Hutchens, J.A.; Hoyle, H.D.; Turner, F.R.; Raff, E.C. Structurally similar Drosophila alpha-tubulins are functionally distinct in vivo. *Mol. Biol. Cell* **1997**, *8*, 481–500. [CrossRef] [PubMed]
74. Hurd, D.D.; Miller, R.M.; Núñez, L.; Portman, D.S. Specific alpha- and beta-tubulin isotypes optimize the functions of sensory Cilia in Caenorhabditis elegans. *Genetics* **2010**, *185*, 883–896. [CrossRef] [PubMed]
75. Silva, M.; Morsci, N.; Nguyen, K.C.Q.; Rizvi, A.; Rongo, C.; Hall, D.H.; Barr, M.M. Cell-Specific α-Tubulin Isotype Regulates Ciliary Microtubule Ultrastructure, Intraflagellar Transport, and Extracellular Vesicle Biology. *Curr. Biol.* **2017**, *27*, 968–980. [CrossRef] [PubMed]
76. Burke, D.; Gasdaska, P.; Hartwell, L. Dominant effects of tubulin overexpression in Saccharomyces cerevisiae. *Mol. Cell. Biol.* **1989**, *9*, 1049–1059. [CrossRef] [PubMed]

77. Weinstein, B.; Solomon, F. Phenotypic consequences of tubulin overproduction in Saccharomyces cerevisiae: Differences between alpha-tubulin and beta-tubulin. *Mol. Cell. Biol.* **1990**, *10*, 5295–5304. [CrossRef] [PubMed]
78. Katz, W.; Weinstein, B.; Solomon, F. Regulation of tubulin levels and microtubule assembly in Saccharomyces cerevisiae: Consequences of altered tubulin gene copy number. *Mol. Cell. Biol.* **1990**, *10*, 5286–5294. [CrossRef] [PubMed]
79. Janke, C.; Kneussel, M. Tubulin post-translational modifications: Encoding functions on the neuronal microtubule cytoskeleton. *Trends Neurosci.* **2010**, *33*, 362–372. [CrossRef] [PubMed]
80. Song, Y.; Brady, S.T. Post-translational modifications of tubulin: Pathways to functional diversity of microtubules. *Trends Cell Biol.* **2015**, *25*, 125–136. [CrossRef] [PubMed]
81. Prota, A.E.; Magiera, M.M.; Kuijpers, M.; Bargsten, K.; Frey, D.; Wieser, M.; Jaussi, R.; Hoogenraad, C.C.; Kammerer, R.A.; Janke, C.; et al. Structural basis of tubulin tyrosination by tubulin tyrosine ligase. *J. Cell Biol.* **2013**, *200*, 259–270. [CrossRef] [PubMed]
82. Szyk, A.; Deaconescu, A.M.; Piszczek, G.; Roll-Mecak, A. Tubulin tyrosine ligase structure reveals adaptation of an ancient fold to bind and modify tubulin. *Nat. Struct. Mol. Biol.* **2011**, *18*, 1250–1258. [CrossRef] [PubMed]
83. Brown, A.; Li, Y.; Slaughter, T.; Black, M.M. Composite microtubules of the axon: Quantitative analysis of tyrosinated and acetylated tubulin along individual axonal microtubules. *J. Cell Sci.* **1993**, *352*, 339–352.
84. Baas, P.W.; Black, M.M. Individual microtubules in the axon consist of domains that differ in both composition and stability. *J. Cell Biol.* **1990**, *111*, 495–509. [CrossRef] [PubMed]
85. Gundersen, G.G.; Khawaja, S.; Bulinski, J.C. Postpolymerization detyrosination of alpha-tubulin: A mechanism for subcellular differentiation of microtubules. *J. Cell Biol.* **1987**, *105*, 251–264. [CrossRef] [PubMed]
86. Robson, S.J.; Burgoyne, R.D. Differential localisation of tyrosinated, detyrosinated, and acetylated alpha-tubulins in neurites and growth cones of dorsal root ganglion neurons. *Cell Motil. Cytoskeleton* **1989**, *12*, 273–282. [CrossRef] [PubMed]
87. Brown, A.; Slaughter, T.; Black, M.M. Newly assembled microtubules are concentrated in the proximal and distal regions of growing axons. *J. Cell Biol.* **1992**, *119*, 867–882. [CrossRef] [PubMed]
88. Baas, P.W.; Ahmad, F.J.; Pienkowski, T.P.; Brown, A.; Black, M.M. Sites of microtubule stabilization for the axon. *J. Neurosci.* **1993**, *13*, 2177–2185. [PubMed]
89. Erck, C.; Peris, L.; Andrieux, A.; Meissirel, C.; Gruber, A.D.; Vernet, M.; Schweitzer, A.; Saoudi, Y.; Pointu, H.; Bosc, C.; et al. A vital role of tubulin-tyrosine-ligase for neuronal organization. *Proc. Natl. Acad. Sci. USA* **2005**, *102*, 7853–7858. [CrossRef] [PubMed]
90. Honnappa, S.; Okhrimenko, O.; Jaussi, R.; Jawhari, H.; Jelesarov, I.; Winkler, F.K.; Steinmetz, M.O. Key interaction modes of dynamic +TIP networks. *Mol. Cell* **2006**, *23*, 663–671. [CrossRef] [PubMed]
91. Weisbrich, A.; Honnappa, S.; Jaussi, R.; Okhrimenko, O.; Frey, D.; Jelesarov, I.; Akhmanova, A.; Steinmetz, M.O. Structure-function relationship of CAP-Gly domains. *Nat. Struct. Mol. Biol.* **2007**, *14*, 959–967. [CrossRef] [PubMed]
92. Wang, Q.; Crevenna, A.H.; Kunze, I.; Mizuno, N. Structural basis for the extended CAP-Gly domains of p150(glued) binding to microtubules and the implication for tubulin dynamics. *Proc. Natl. Acad. Sci. USA* **2014**, *111*, 11347–11352. [CrossRef] [PubMed]
93. Nirschl, J.J.; Magiera, M.M.; Lazarus, J.E.; Janke, C.; Holzbaur, E.L.F. α-Tubulin Tyrosination and CLIP-170 Phosphorylation Regulate the Initiation of Dynein-Driven Transport in Neurons. *Cell Rep.* **2016**, *14*, 2637–2652. [CrossRef] [PubMed]
94. McKenney, R.J.; Huynh, W.; Vale, R.D.; Sirajuddin, M. Tyrosination of α-tubulin controls the initiation of processive dynein-dynactin motility. *EMBO J.* **2016**, *35*, 1175–1185. [CrossRef] [PubMed]
95. Janke, C.; Rogowski, K.; Wloga, D.; Regnard, C.; Kajava, A.V.; Strub, J.-M.; Temurak, N.; van Dijk, J.; Boucher, D.; van Dorsselaer, A.; et al. Tubulin polyglutamylase enzymes are members of the TTL domain protein family. *Science* **2005**, *308*, 1758–1762. [CrossRef] [PubMed]
96. Van Dijk, J.; Rogowski, K.; Miro, J.; Lacroix, B.; Eddé, B.; Janke, C. A targeted multienzyme mechanism for selective microtubule polyglutamylation. *Mol. Cell* **2007**, *26*, 437–448. [CrossRef] [PubMed]

97. Natarajan, K.; Gadadhar, S.; Souphron, J.; Magiera, M.M.; Janke, C. Molecular interactions between tubulin tails and glutamylases reveal determinants of glutamylation patterns. *EMBO Rep.* **2017**, *18*, 1013–1026. [CrossRef] [PubMed]
98. Rogowski, K.; van Dijk, J.; Magiera, M.M.; Bosc, C.; Deloulme, J.-C.; Bosson, A.; Peris, L.; Gold, N.D.; Lacroix, B.; Grau, M.B.; et al. A family of protein-deglutamylating enzymes associated with neurodegeneration. *Cell* **2010**, *143*, 564–578. [CrossRef] [PubMed]
99. Valenstein, M.L.; Roll-Mecak, A. Graded Control of Microtubule Severing by Tubulin Glutamylation. *Cell* **2016**, *164*, 911–921. [CrossRef] [PubMed]
100. Lacroix, B.; Van Dijk, J.; Gold, N.D.; Guizetti, J.; Aldrian-Herrada, G.; Rogowski, K.; Gerlich, D.W.; Janke, C. Tubulin polyglutamylation stimulates spastin-mediated microtubule severing. *J. Cell Biol.* **2010**, *189*, 945–954. [CrossRef] [PubMed]
101. Ikegami, K.; Heier, R.L.; Taruishi, M.; Takagi, H.; Mukai, M.; Shimma, S.; Taira, S.; Hatanaka, K.; Morone, N.; Yao, I.; et al. Loss of alpha-tubulin polyglutamylation in ROSA22 mice is associated with abnormal targeting of KIF1A and modulated synaptic function. *Proc. Natl. Acad. Sci. USA* **2007**, *104*, 3213–3218. [CrossRef] [PubMed]
102. Chakrabarti, L.; Eng, J.; Martinez, R.A.; Jackson, S.; Huang, J.; Possin, D.E.; Sopher, B.L.; La Spada, A.R. The zinc-binding domain of Nna1 is required to prevent retinal photoreceptor loss and cerebellar ataxia in Purkinje cell degeneration (pcd) mice. *Vision Res.* **2008**, *48*, 1999–2005. [CrossRef] [PubMed]
103. Wang, T.; Parris, J.; Li, L.; Morgan, J.I. The carboxypeptidase-like substrate-binding site in Nna1 is essential for the rescue of the Purkinje cell degeneration (pcd) phenotype. *Mol. Cell. Neurosci.* **2006**, *33*, 200–213. [CrossRef] [PubMed]
104. Fernandez-Gonzalez, A.; La Spada, A.R.; Treadaway, J.; Higdon, J.C.; Harris, B.S.; Sidman, R.L.; Morgan, J.I.; Zuo, J. Purkinje cell degeneration (pcd) phenotypes caused by mutations in the axotomy-induced gene, Nna1. *Science* **2002**, *295*, 1904–1906. [CrossRef] [PubMed]
105. Brady, S.T.; Tytell, M.; Lasek, R.J. Axonal tubulin and axonal microtubules: Biochemical evidence for cold stability. *J. Cell Biol.* **1984**, *99*, 1716–1724. [CrossRef] [PubMed]
106. Song, Y.; Kirkpatrick, L.L.; Schilling, A.B.; Helseth, D.L.; Chabot, N.; Keillor, J.W.; Johnson, G.V.W.; Brady, S.T. Transglutaminase and polyamination of tubulin: Posttranslational modification for stabilizing axonal microtubules. *Neuron* **2013**, *78*, 109–123. [CrossRef] [PubMed]
107. Xu, Z.; Schaedel, L.; Portran, D.; Aguilar, A.; Gaillard, J.; Marinkovich, M.P.; Théry, M.; Nachury, M.V. Microtubules acquire resistance from mechanical breakage through intralumenal acetylation. *Science* **2017**, *356*, 328–332. [CrossRef] [PubMed]
108. Portran, D.; Schaedel, L.; Xu, Z.; Théry, M.; Nachury, M.V. Tubulin acetylation protects long-lived microtubules against mechanical ageing. *Nat. Cell Biol.* **2017**, *19*, 391–398. [CrossRef] [PubMed]
109. Topalidou, I.; Keller, C.; Kalebic, N.; Nguyen, K.C.Q.; Somhegyi, H.; Politi, K.A.; Heppenstall, P.; Hall, D.H.; Chalfie, M. Genetically separable functions of the MEC-17 tubulin acetyltransferase affect microtubule organization. *Curr. Biol.* **2012**, *22*, 1057–1065. [CrossRef] [PubMed]
110. Cueva, J.G.; Hsin, J.; Huang, K.C.; Goodman, M.B. Posttranslational acetylation of α-tubulin constrains protofilament number in native microtubules. *Curr. Biol.* **2012**, *22*, 1066–1074. [CrossRef] [PubMed]
111. Reed, N.A.; Cai, D.; Blasius, T.L.; Jih, G.T.; Meyhofer, E.; Gaertig, J.; Verhey, K.J. Microtubule Acetylation Promotes Kinesin-1 Binding and Transport. *Curr. Biol.* **2006**, *16*, 2166–2172. [CrossRef] [PubMed]
112. Alper, J.D.; Decker, F.; Agana, B.; Howard, J. The motility of axonemal dynein is regulated by the tubulin code. *Biophys. J.* **2014**, *107*, 2872–2880. [CrossRef] [PubMed]
113. Dehmelt, L.; Smart, F.M.; Ozer, R.S.; Halpain, S. The role of microtubule-associated protein 2c in the reorganization of microtubules and lamellipodia during neurite initiation. *J. Neurosci.* **2003**, *23*, 9479–9490. [PubMed]
114. Witte, H.; Neukirchen, D.; Bradke, F. Microtubule stabilization specifies initial neuronal polarization. *J. Cell Biol.* **2008**, *180*, 619–632. [CrossRef] [PubMed]
115. Knops, J.; Kosik, K.S.; Lee, G.; Pardee, J.D.; Cohen-Gould, L.; McConlogue, L. Overexpression of tau in a nonneuronal cell induces long cellular processes. *J. Cell Biol.* **1991**, *114*, 725–733. [CrossRef] [PubMed]
116. Del Castillo, U.; Winding, M.; Lu, W.; Gelfand, V.I. Interplay between kinesin-1 and cortical dynein during axonal outgrowth and microtubule organization in Drosophila neurons. *eLife* **2015**, *4*, 1–20. [CrossRef] [PubMed]

117. Baas, P.W.; Black, M.M.; Banker, G.A. Changes in microtubule polarity orientation during the development of hippocampal neurons in culture. *J. Cell Biol.* **1989**, *109*, 3085–3094. [CrossRef] [PubMed]
118. Baas, P.W.; Deitch, J.S.; Black, M.M.; Banker, G.A. Polarity orientation of microtubules in hippocampal neurons: Uniformity in the axon and nonuniformity in the dendrite. *Proc. Natl. Acad. Sci. USA* **1988**, *85*, 8335–8339. [CrossRef] [PubMed]
119. Lowery, L.A.; Stout, A.; Faris, A.E.; Ding, L.; Baird, M.A.; Davidson, M.W.; Danuser, G.; Van Vactor, D. Growth cone-specific functions of XMAP215 in restricting microtubule dynamics and promoting axonal outgrowth. *Neural Dev.* **2013**, *8*, 22. [CrossRef] [PubMed]
120. Hirokawa, N.; Noda, Y.; Tanaka, Y.; Niwa, S. Kinesin superfamily motor proteins and intracellular transport. *Nat. Rev. Mol. Cell Biol.* **2009**, *10*, 682–696. [CrossRef] [PubMed]
121. Vale, R.D.; Schnapp, B.J.; Mitchison, T.; Steuer, E.; Reese, T.S.; Sheetz, M.P. Different axoplasmic proteins generate movement in opposite directions along microtubules in vitro. *Cell* **1985**, *43*, 623–632. [CrossRef]
122. Buck, K.B.; Zheng, J.Q. Growth cone turning induced by direct local modification of microtubule dynamics. *J. Neurosci.* **2002**, *22*, 9358–9367. [PubMed]
123. Rivas, R.J.; Hatten, M.E. Motility and cytoskeletal organization of migrating cerebellar granule neurons. *J. Neurosci.* **1995**, *15*, 981–989. [PubMed]
124. Tsai, J.-W.; Bremner, K.H.; Vallee, R.B. Dual subcellular roles for LIS1 and dynein in radial neuronal migration in live brain tissue. *Nat. Neurosci.* **2007**, *10*, 970–979. [CrossRef] [PubMed]
125. Fry, A.E.; Cushion, T.D.; Pilz, D.T. The genetics of lissencephaly. *Am. J. Med. Genet. Part C Semin. Med. Genet.* **2014**, *166*, 198–210. [CrossRef] [PubMed]
126. Stratton, R.F.; Dobyns, W.B.; Airhart, S.D.; Ledbetter, D.H. New chromosomal syndrome: Miller-Dieker syndrome and monosomy 17p13. *Hum. Genet.* **1984**, *67*, 193–200. [CrossRef] [PubMed]
127. Valiente, M.; Marín, O. Neuronal migration mechanisms in development and disease. *Curr. Opin. Neurobiol.* **2010**, *20*, 68–78. [CrossRef] [PubMed]
128. Huang, J.; Roberts, A.J.; Leschziner, A.E.; Reck-Peterson, S.L. Lis1 acts as a "clutch" between the ATPase and microtubule-binding domains of the dynein motor. *Cell* **2012**, *150*, 975–986. [CrossRef] [PubMed]
129. De Rouvroit, C.L.; Goffinet, A.M. Neuronal migration. *Mech. Dev.* **2001**, *105*, 47–56. [CrossRef]
130. Horesh, D.; Sapir, T.; Francis, F.; Wolf, S.G.; Caspi, M.; Elbaum, M.; Chelly, J.; Reiner, O. Doublecortin, a stabilizer of microtubules. *Hum. Mol. Genet.* **1999**, *8*, 1599–1610. [CrossRef] [PubMed]
131. Tanaka, T.; Serneo, F.F.; Higgins, C.; Gambello, M.J.; Wynshaw-Boris, A.; Gleeson, J.G. Lis1 and doublecortin function with dynein to mediate coupling of the nucleus to the centrosome in neuronal migration. *J. Cell Biol.* **2004**, *165*, 709–721. [CrossRef] [PubMed]
132. Jansen, A.C.; Oostra, A.; Desprechins, B.; De Vlaeminck, Y.; Verhelst, H.; Régal, L.; Verloo, P.; Bockaert, N.; Keymolen, K.; Seneca, S.; et al. TUBA1A mutations: From isolated lissencephaly to familial polymicrogyria. *Neurology* **2011**, *76*, 988–992. [CrossRef] [PubMed]
133. Yokoi, S.; Ishihara, N.; Miya, F.; Tsutsumi, M.; Yanagihara, I.; Fujita, N.; Yamamoto, H.; Kato, M.; Okamoto, N.; Tsunoda, T.; et al. TUBA1A mutation can cause a hydranencephaly-like severe form of cortical dysgenesis. *Sci. Rep.* **2015**, *5*, 15165. [CrossRef] [PubMed]
134. Shimojima, K.; Narita, A.; Maegaki, Y.; Saito, A.; Furukawa, T.; Yamamoto, T. Whole-exome sequencing identifies a de novo TUBA1A mutation in a patient with sporadic malformations of cortical development: A case report. *BMC Res. Notes* **2014**, *7*, 465. [CrossRef] [PubMed]
135. McMichael, G.; Bainbridge, M.N.; Haan, E.; Corbett, M.; Gardner, A.; Thompson, S.; van Bon, B.W.M.; van Eyk, C.L.; Broadbent, J.; Reynolds, C.; et al. Whole-exome sequencing points to considerable genetic heterogeneity of cerebral palsy. *Mol. Psychiatry* **2015**, *20*, 176–182. [CrossRef] [PubMed]
136. Hikita, N.; Hattori, H.; Kato, M.; Sakuma, S.; Morotomi, Y.; Ishida, H.; Seto, T.; Tanaka, K.; Shimono, T.; Shintaku, H.; et al. A case of TUBA1A mutation presenting with lissencephaly and Hirschsprung disease. *Brain Dev.* **2014**, *36*, 159–162. [CrossRef] [PubMed]
137. Oegema, R.; Cushion, T.D.; Phelps, I.G.; Chung, S.-K.; Dempsey, J.C.; Collins, S.; Mullins, J.G.L.; Dudding, T.; Gill, H.; Green, A.J.; et al. Recognizable cerebellar dysplasia associated with mutations in multiple tubulin genes. *Hum. Mol. Genet.* **2015**, *24*, 5313–5325. [CrossRef] [PubMed]

138. Bamba, Y.; Shofuda, T.; Kato, M.; Pooh, R.K.; Tateishi, Y.; Takanashi, J.-I.; Utsunomiya, H.; Sumida, M.; Kanematsu, D.; Suemizu, H.; et al. In vitro characterization of neurite extension using induced pluripotent stem cells derived from lissencephaly patients with TUBA1A missense mutations. *Mol. Brain* **2016**, 1–14. [CrossRef] [PubMed]
139. Myers, K.A.; Bello-Espinosa, L.E.; Kherani, A.; Wei, X.-C.; Innes, A.M. TUBA1A Mutation Associated With Eye Abnormalities in Addition to Brain Malformation. *Pediatr. Neurol.* **2015**, *53*, 442–444. [CrossRef] [PubMed]
140. Romaniello, R.; Arrigoni, F.; Cavallini, A.; Tenderini, E.; Baschirotto, C.; Triulzi, F.; Bassi, M.-T.; Borgatti, R. Brain malformations and mutations in α- and β-tubulin genes: A review of the literature and description of two new cases. *Dev. Med. Child Neurol.* **2014**, *56*, 354–360. [CrossRef] [PubMed]
141. Kamiya, K.; Tanaka, F.; Ikeno, M.; Okumura, A.; Aoki, S. DTI tractography of lissencephaly caused by TUBA1A mutation. *Neurol. Sci.* **2014**, *35*, 801–803. [CrossRef] [PubMed]
142. Sohal, A.P.S.; Montgomery, T.; Mitra, D.; Ramesh, V. TUBA1A mutation-associated lissencephaly: Case report and review of the literature. *Pediatr. Neurol.* **2012**, *46*, 127–131. [CrossRef] [PubMed]
143. Judkins, A.R.; Martinez, D.; Ferreira, P.; Dobyns, W.B.; Golden, J.A. Polymicrogyria includes fusion of the molecular layer and decreased neuronal populations but normal cortical laminar organization. *J. Neuropathol. Exp. Neurol.* **2011**, *70*, 438–443. [CrossRef] [PubMed]
144. Stutterd, C.A.; Leventer, R.J. Polymicrogyria: A common and heterogeneous malformation of cortical development. *Am. J. Med. Genet. Part C Semin. Med. Genet.* **2014**, *166*, 227–239. [CrossRef] [PubMed]
145. Gressens, P.; Baes, M.; Leroux, P.; Lombet, A.; Van Veldhoven, P.; Janssen, A.; Vamecq, J.; Marret, S.; Evrard, P. Neuronal migration disorder in Zellweger mice is secondary to glutamate receptor dysfunction. *Ann. Neurol.* **2000**, *48*, 336–343. [CrossRef]
146. Barkovich, A.J.; Guerrini, R.; Kuzniecky, R.I.; Jackson, G.D.; Dobyns, W.B. A developmental and genetic classification for malformations of cortical development: Update 2012. *Brain* **2012**, *135*, 1348–1369. [CrossRef] [PubMed]
147. Zanni, G.; Colafati, G.S.; Barresi, S.; Randisi, F.; Talamanca, L.F.; Genovese, E.; Bellacchio, E.; Bartuli, A.; Bernardi, B.; Bertini, E. Description of a novel TUBA1A mutation in Arg-390 associated with asymmetrical polymicrogyria and mid-hindbrain dysgenesis. *Eur. J. Paediatr. Neurol.* **2013**, *17*, 361–365. [CrossRef] [PubMed]
148. Faheem, M.; Naseer, M.I.; Rasool, M.; Chaudhary, A.G.; Kumosani, T.A.; Ilyas, A.M.; Pushparaj, P.; Ahmed, F.; Algahtani, H.A.; Al-Qahtani, M.H.; et al. Molecular genetics of human primary microcephaly: An overview. *BMC Med. Genom.* **2015**, *8* (Suppl. 1), S4. [CrossRef] [PubMed]
149. Harris, S.R. Measuring head circumference: Update on infant microcephaly. *Can. Fam. Physician* **2015**, *61*, 680–684. [PubMed]
150. Woods, C.G.; Parker, A. Investigating microcephaly. *Arch. Dis. Child.* **2013**, *98*, 707–713. [CrossRef] [PubMed]
151. Tian, G.; Jaglin, X.H.; Keays, D.A.; Francis, F.; Chelly, J.; Cowan, N.J. Disease-associated mutations in TUBA1A result in a spectrum of defects in the tubulin folding and heterodimer assembly pathway. *Hum. Mol. Genet.* **2010**, *19*, 3599–3613. [CrossRef] [PubMed]
152. Tian, G.; Kong, X.-P.; Jaglin, X.H.; Chelly, J.; Keays, D.; Cowan, N.J. A pachygyria-causing alpha-tubulin mutation results in inefficient cycling with CCT and a deficient interaction with TBCB. *Mol. Biol. Cell* **2008**, *19*, 1152–1161. [CrossRef] [PubMed]
153. Belvindrah, R.; Natarajan, K.; Shabajee, P.; Bruel-Jungerman, E.; Bernard, J.; Goutierre, M.; Moutkine, I.; Jaglin, X.H.; Savariradjane, M.; Irinopoulou, T.; et al. Mutation of the α-tubulin Tuba1a leads to straighter microtubules and perturbs neuronal migration. *J. Cell Biol.* **2017**, *216*, 2443–2461. [CrossRef] [PubMed]
154. Zhang, Y.; Chen, K.; Sloan, S.A.; Bennett, M.L.; Scholze, A.R.; O'Keeffe, S.; Phatnani, H.P.; Guarnieri, P.; Caneda, C.; Ruderisch, N.; et al. An RNA-Sequencing Transcriptome and Splicing Database of Glia, Neurons, and Vascular Cells of the Cerebral Cortex. *J. Neurosci.* **2014**, *34*, 11929–11947. [CrossRef] [PubMed]
155. Gartz Hanson, M.; Aiken, J.; Sietsema, D.V.; Sept, D.; Bates, E.A.; Niswander, L.; Moore, J.K. Novel α-tubulin mutation disrupts neural development and tubulin proteostasis. *Dev. Biol.* **2016**, *409*, 406–419. [CrossRef] [PubMed]
156. Gloster, A.; Wu, W.; Speelman, A.; Weiss, S.; Causing, C.; Pozniak, C.; Reynolds, B.; Chang, E.; Toma, J.G.; Miller, F.D. The T alpha 1 alpha-tubulin promoter specifies gene expression as a function of neuronal growth and regeneration in transgenic mice. *J. Neurosci.* **1994**, *14*, 7319–7330. [PubMed]

157. Bamji, S.X.; Miller, F.D. Comparison of the expression of a Tα1:nlacZ transgene and Tα1 α-tubulin mRNA in the mature central nervous system. *J. Comp. Neurol.* **1996**, *374*, 52–69. [CrossRef]
158. Gloster, A.; El-Bizri, H.; Bamji, S.X.; Rogers, D.; Miller, F.D. Early induction of Talpha1 alpha-tubulin transcription in neurons of the developing nervous system. *J. Comp. Neurol.* **1999**, *405*, 45–60. [CrossRef]
159. Coksaygan, T.; Magnus, T.; Cai, J.; Mughal, M.; Lepore, A.; Xue, H.; Fischer, I.; Rao, M.S. Neurogenesis in Talpha-1 tubulin transgenic mice during development and after injury. *Exp. Neurol.* **2006**, *197*, 475–485. [CrossRef] [PubMed]
160. Keays, D.A.; Cleak, J.; Huang, G.J.; Edwards, A.; Braun, A.; Treiber, C.D.; Pidsley, R.; Flint, J. The role of Tuba1a in adult hippocampal neurogenesis and the formation of the dentate gyrus. *Dev. Neurosci.* **2010**, *32*, 268–277. [CrossRef] [PubMed]
161. Goldman, D.; Hankin, M.; Li, Z.; Dai, X.; Ding, J. Transgenic zebrafish for studying nervous system development and regeneration. *Transgenic Res.* **2001**, *10*, 21–33. [CrossRef] [PubMed]
162. Veldman, M.B.; Bemben, M.A.; Goldman, D. Tuba1a gene expression is regulated by KLF6/7 and is necessary for CNS development and regeneration in zebrafish. *Mol. Cell. Neurosci.* **2010**, *43*, 370–383. [CrossRef] [PubMed]
163. Ramachandran, R.; Reifler, A.; Parent, J.M.; Goldman, D. Conditional gene expression and lineage tracing of tuba1a expressing cells during zebrafish development and retina regeneration. *J. Comp. Neurol.* **2010**, *518*, 4196–4212. [CrossRef] [PubMed]
164. Mathew, T.C.; Miller, F.D. Increased expression of Tα1 α-tubulin mRNA during collateral and NGF-induced sprouting of sympathetic neurons. *Dev. Biol.* **1990**, *141*, 84–92. [CrossRef]
165. Mohiuddin, L.; Fernyhough, P.; Tomlinson, D.R. Acidic fibroblast growth factor enhances neurite outgrowth and stimulates expression of GAP-43 and T alpha 1 alpha-tubulin in cultured neurones from adult rat dorsal root ganglia. *Neurosci. Lett.* **1996**, *215*, 111–114. [CrossRef]
166. Ménard, C.; Hein, P.; Paquin, A.; Savelson, A.; Yang, X.M.; Lederfein, D.; Barnabé-Heider, F.; Mir, A.A.; Sterneck, E.; Peterson, A.C.; et al. An essential role for a MEK-C/EBP pathway during growth factor-regulated cortical neurogenesis. *Neuron* **2002**, *36*, 597–610. [CrossRef]
167. Miller, F.D.; Tetzlaff, W.; Bisby, M.A.; Fawcett, J.W.; Milner, R.J. Rapid induction of the major embryonic alpha-tubulin mRNA, T alpha 1, during nerve regeneration in adult rats. *J. Neurosci.* **1989**, *9*, 1452–1463. [PubMed]
168. Kost, S.A.; Oblinger, M.M. Immature corticospinal neurons respond to axotomy with changes in tubulin gene expression. *Brain Res. Bull.* **1993**, *30*, 469–475. [CrossRef]
169. Mikucki, S.A.; Oblinger, M.M. Corticospinal neurons exhibit a novel pattern of cytoskeletal gene expression after injury. *J. Neurosci. Res.* **1991**, *30*, 213–225. [CrossRef] [PubMed]
170. Tetzlaff, W.; Alexander, S.W.; Miller, F.D.; Bisby, M.A. Response of facial and rubrospinal neurons to axotomy: Changes in mRNA expression for cytoskeletal proteins and GAP-43. *J. Neurosci.* **1991**, *11*, 2528–2544. [PubMed]
171. Theriault, E.; Tetzlaff, W.; Tator, C.H. Elevated gene expression in the red nucleus after spinal cord compression injury. *Neuroreport* **1992**, *3*, 559–562. [CrossRef] [PubMed]
172. Wu, W.; Mathew, T.C.; Miller, F.D. Evidence that the loss of homeostatic signals induces regeneration-associated alterations in neuronal gene expression. *Dev. Biol.* **1993**, *158*, 456–466. [CrossRef] [PubMed]
173. Wu, W.; Toma, J.G.; Chan, H.; Smith, R.; Miller, F.D. Disruption of fast axonal transport in vivo leads to alterations in Schwann cell gene expression. *Dev. Biol.* **1994**, *163*, 423–439. [CrossRef] [PubMed]
174. Mathew, T.C.; Miller, F.D. Induction of T alpha 1 alpha-tubulin mRNA during neuronal regeneration is a function of the amount of axon lost. *Dev. Biol.* **1993**, *158*, 467–474. [CrossRef] [PubMed]
175. Ramón y Cajal, S. *Degeneration and Regeneration of the Nervous System*; Oxford University Press: New York, NY, USA, 1959.
176. Richardson, P.M.; Issa, V.M.; Aguayo, A.J. Regeneration of long spinal axons in the rat. *J. Neurocytol.* **1984**, *13*, 165–182. [CrossRef] [PubMed]
177. Wu, W.D.; Gloster, A.; Miller, F.D. Transcriptional repression of the growth-associated T alpha 1 alpha-tubulin gene by target contact. *J. Neurosci. Res.* **1997**, *48*, 477–487. [CrossRef]

3

Ttc21b Is Required in Bergmann Glia for Proper Granule Cell Radial Migration

Ashley M. Driver [1], Christopher Shumrick [1] and Rolf W. Stottmann [1,2,*]

[1] Division of Human Genetics, Department of Pediatrics, Cincinnati Children's Hospital Medical Center, Cincinnati, OH 45229, USA; Ashley.Driver@uwsp.edu (A.M.D.); shumricm@mail.uc.edu (C.S.)

[2] Division of Developmental Biology, Department of Pediatrics, Cincinnati Children's Hospital Medical Center, Cincinnati, OH 45229, USA

* Correspondence: rolf.stottmann@cchmc.org

Abstract: Proper cerebellar development is dependent on tightly regulated proliferation, migration, and differentiation events. Disruptions in any of these leads to a range of cerebellar phenotypes from ataxia to childhood tumors. Animal models have shown that proper regulation of *sonic hedgehog* (*Shh*) signaling is crucial for normal cerebellar architecture, and increased signaling leads to cerebellar tumor formation. Primary cilia are known to be required for the proper regulation of multiple developmental signaling pathways, including *Shh*. *Tetratricopeptide Repeat Domain 21B* (*Ttc21b*) is required for proper primary cilia form and function, and is primarily thought to restrict *Shh* signaling. Here we investigated a role for *Ttc21b* in cerebellar development. Surprisingly, *Ttc21b* ablation in Bergmann glia resulted in the accumulation of ectopic granule cells in the lower/posterior lobes of the cerebellum and a reduction in *Shh* signaling. *Ttc21b* ablation in just Purkinje cells resulted in a similar phenotype seen in fewer cells, but across the entire extent of the cerebellum. These results suggest that *Ttc21b* expression is required for Bergmann glia structure and signaling in the developing cerebellum, and in some contexts, augments rather than attenuates *Shh* signaling.

Keywords: *Ttc21b*; cilia; *Shh*; cerebellum; postnatal; migration; Purkinje cell; Bergmann glia; Cre; LoxP

---

1. Introduction

Development of the cerebellum includes significant postnatal growth and morphological change. The most numerous cell type in the mammalian brain is the cerebellar granule cell, with estimates of approximately 100 billion cells in humans [1]. In mouse and human, cerebellar granule cells are produced postnatally in the external granule layer (EGL). The post-mitotic granule cells migrate radially on Bergmann glial fibers to the interior of the cerebellum and ultimately form the inner granule layer (IGL). In humans, disrupted cerebellar development can result in a wide variety of structural malformations, including cerebellar hypoplasia, failure of proper foliation, and tumor formation [2].

Sonic hedgehog (*Shh*) signaling has become a well-recognized pathway involved in early cerebellar development. Granule cell proliferation and proper development of the Bergmann glial migratory tracts are both known to require sonic hedgehog (SHH) secreted from Purkinje cells [3–6]. The downstream GLI transcription factors have dynamic expression patterns during early cerebellar development [7].

Altered levels of *Shh* signaling have severe cerebellar developmental consequences, and can cause tumor formation. Loss of *Shh* signaling results in a reduction of granule cell proliferation and cerebellar hypoplasia [8–10]. In addition to hypoplasia, conditional ablation of *Gli2* with an *En1-Cre* led to abnormal Purkinje cell and Bergmann glia morphology [8]. Deletion of *Gli1 and Gli2* leads to further disruptions in foliation and patterning [8]. In humans, impaired SHH signaling has been implicated in the cerebellar anomalies seen in both Joubert and Meckel Syndrome [11]. Conversely, activation of SHH signaling via loss of *Patched*-1 or constitutively activated Smoothened mutations results in medulloblastoma formation [12–16].

The primary cilium is a non-motile sensory organelle that is now well established to be critical for proper modulation of the SHH pathway [17]. Many proteins are trafficked through the primary cilium via intraflagellar transport (IFT). GLI proteins in particular require IFT through the primary cilium for proper processing into transcriptional repressor/activator forms [18,19]. Loss of cilia and/or IFT leads to multiple defects affecting SHH signal transduction, including perturbation of GLI processing patterns. Consistent with this role in SHH signal transduction, loss of primary cilia causes cerebellar hypoplasia [10] and modulates the effects of other mutations on tumor formation [14].

Among the IFT proteins, *Tetratricopeptide Repeat Domain 21B* (*Ttc21b*; *Ift139*) is required for normal ciliary retrograde transport [19] and early forebrain patterning and development [20,21]. Multiple studies have shown that the TTC21B protein (or the IFT139 orthologue) is found within the primary cilium [19,22,23]. Loss of *Ttc21b* has been shown to result in abnormal GLI3 processing, and thus increased *Shh* signaling [19,21]. In humans, *TTC21B* is one of the most commonly mutated genes in ciliopathy patients, and variants are present in disorders of the cerebellum such as Joubert Syndrome [22]. We hypothesized loss of *Ttc21b* during postnatal development may affect SHH signaling and cerebellar development. To investigate this and circumvent the perinatal lethality of *Ttc21b* mutants [19], we used the Cre-loxP system to separately ablate *Ttc21b* in Bergmann glia and Purkinje cells. Surprisingly, loss of *Ttc21b* resulted in ectopic granule cells in the posterior lobes of the cerebellum. Analysis of the affected lobes showed abnormal morphology of both Bergmann glia and Purkinje cells, as well as reduced SHH signaling. These results demonstrate a unique requirement for *Ttc21b* in cerebellar astrocytes and the first demonstration of reduced SHH signaling upon loss of *Ttc21b*.

2. Materials and Methods

2.1. Mouse Husbandry

All animals were maintained through a protocol approved by the Cincinnati Children's Hospital Medical Center IACUC committee (IACUC2016-0098). Mice were housed in a vivarium with a 12-h light cycle with food and water ad libitum. All mouse alleles used in this study were previously published: $Ttc21b^{alien}$ is a null allele of *Ttc21b* [19], $Ttc21b^{tm1a(KOMP)Wtsi\text{-}lacZ}$ ($Ttc21b^{lacZ}$) for analysis of *Ttc21b* expression via the LacZ expression cassette [24], $Ttc21b^{flox}$ [20,24]); a Cre recombinase reporter allele B6.129(Cg)-$Gt(ROSA)26Sor^{tm4(ACTB\text{-}tdTomato,\text{-}EGFP)Luo}/J$ ($ROSA^{dTom/EGFP}$) [25]; FVB-Tg(GFAP-cre)$^{25Mes/J}$ (GFAP-Cre) [26] and B6.129-Tg(Pcp2-cre)$^{2Mpin/J}$ (Pcp2-Cre) [27]. $Gli1^{tm2AlJ}/J$ ($Gli1^{lacZ}$) mice were used to measure *Gli1* expression [28]. For adult analysis, 4-month-old animals were sedated and transcardial perfusion with paraformaldehyde was performed using standard procedures. Brains were subsequently harvested and fixed for 48 h in 4% paraformaldehyde at room temperature (RT) and embedded in paraffin. For postnatal brain collection, pups were sedated with isoflurane to minimize discomfort followed by decapitation. Brains were collected and fixed for 24 h at 4 °C and embedded in O.C.T. medium (Tissue Tek) for cryosectioning or in paraffin.

2.2. Histology and Immunohistochemistry

For histological analysis, brain sections were collected on a Leica microtome at a thickness of 5 μM. Sections were de-paraffinized, hematoxylin and eosin stained, and sealed using Cytoseal (Thermo Scientific, Waltham, MA, USA). Images were obtained with a Zeiss Stereoscope (all paired images are shown at the same magnification). For 3,3′-Diaminobenzidine (DAB) immunohistochemistry, sections were de-paraffinized and bleached with hydrogen peroxide. Sections were incubated overnight in primary antibody (NeuN, 1:500; Ki67 1:500) at RT. Sections were then washed in phosphate-buffered saline (PBS) and incubated for two hours in biotinylated swine-anti rabbit secondary antibody (1:500; Vector Laboratories, Burlingame, CA, USA) for 2 h at RT. Sections were then washed in PBS and incubated in avidin-biotin-complex (ABC) solution (Vector Laboratories) for 1 h at RT. Sections were then washed and visualized using DAB for 3 min followed by a PBS wash, ethanol dehydration, xylene incubation, and mounted with Cytoseal (Thermo). For fluorescent immunohistochemistry, sections were collected on a Leica cryostat at 12 μM thickness and washed in PBS prior to antigen retrieval with sodium citrate buffer (as appropriate for each antibody). Sections were then washed with PBS and incubated in blocking buffer (PBS + 4% normal goat serum and 0.1% Triton-X) for 1 h at RT. Sections were incubated overnight at 4 °C in primary antibody diluted in blocking buffer: Calbindin 1:1000 (Abcam, ab25085), glial fibrillary acidic protein (GFAP) 1:500 (Abcam, ab7260), and phospho-histone H3 (pHH3) 1:500 (SIGMA, H0412). Sections were then washed and incubated in Alexa Fluor-conjugated secondary antibodies diluted in blocking buffer (1:500, Thermo Scientific, Waltham, MA, USA) followed by PBS washes and DAPI counterstain (1:1000) for 10 min. To visualize the GFP expressed by the $ROSA^{dTom/EGFP}$ Cre reporter allele, sections were only incubated for 1 h at RT with anti-GFP antibody (1:100, Life Technologies, A21311) followed by washes and DAPI counterstain. Sections were sealed with ProLong Gold (Life Technologies) and images were obtained on a Nikon C2 confocal microscope. All paired images were taken at the same magnification.

2.3. Section In Situ Hybridization and LacZ Staining

For RNA section in situ hybridization (SISH) and LacZ staining, sections were collected on a Leica cryostat at a thickness of 20 μM. The in situ hybridization was done according to standard protocols. For LacZ staining, sections were washed in PBS followed by fixation in cold 0.5% PFA on ice followed by PBS wash and incubation in LacZ buffer on ice. Sections were then incubated in X-Gal solution at 4 °C for 1–3 days, after which they were washed in PBS, counterstained in eosin, and dehydrated. Sections were sealed with Cytoseal (Thermo) and images were taken on a Zeiss Stereoscope with all paired images taken at the same magnification.

3. Results

3.1. Ttc21b Is Enriched in the Purkinje Cell Layer of the Developing Cerebellum and GFAP-Cre Ablation of Ttc21b Results in Ectopic Granule Cells in the Cerebellum

We used a $Ttc21b^{lacZ}$ allele to determine the pattern of *Ttc21b* expression in the developing cerebellum. LacZ staining from P1–P10 showed expression in the Purkinje cell layer (PCL) and the deep cerebellar nuclei (DCN; Figure 1). The Purkinje cell layer at these stages is comprised of both Purkinje cells and the cell bodies of Bergmann glia [29].

Given the known roles of SHH signaling in early postnatal cerebellar development and the role of *Ttc21b* in regulating *Shh* signaling, we sought to determine if *Ttc21b* was required in the developing cerebellum. To circumvent the perinatal lethality of *Ttc21b* null mutations [19], we used a conditional genetic approach and performed a genetic ablation using GFAP-Cre [26] (Figure S1A–C). Consistent with previous reports, GFAP-Cre-stimulated recombination was observed in the telencephalic hemispheres and in the cerebellum. To confirm the cell types exhibiting Cre activity, we used a Cre recombinase reporter allele, $ROSA^{dTom/EGFP}$, in combination with immunohistochemistry. We saw significant co-localization of GFP immunoreactivity indicating Cre activity with GLAST marking immature astrocytes (Figure S1G–I), but no such co-localization in the Calbindin-positive Purkinje cells (Figure S1D–F).

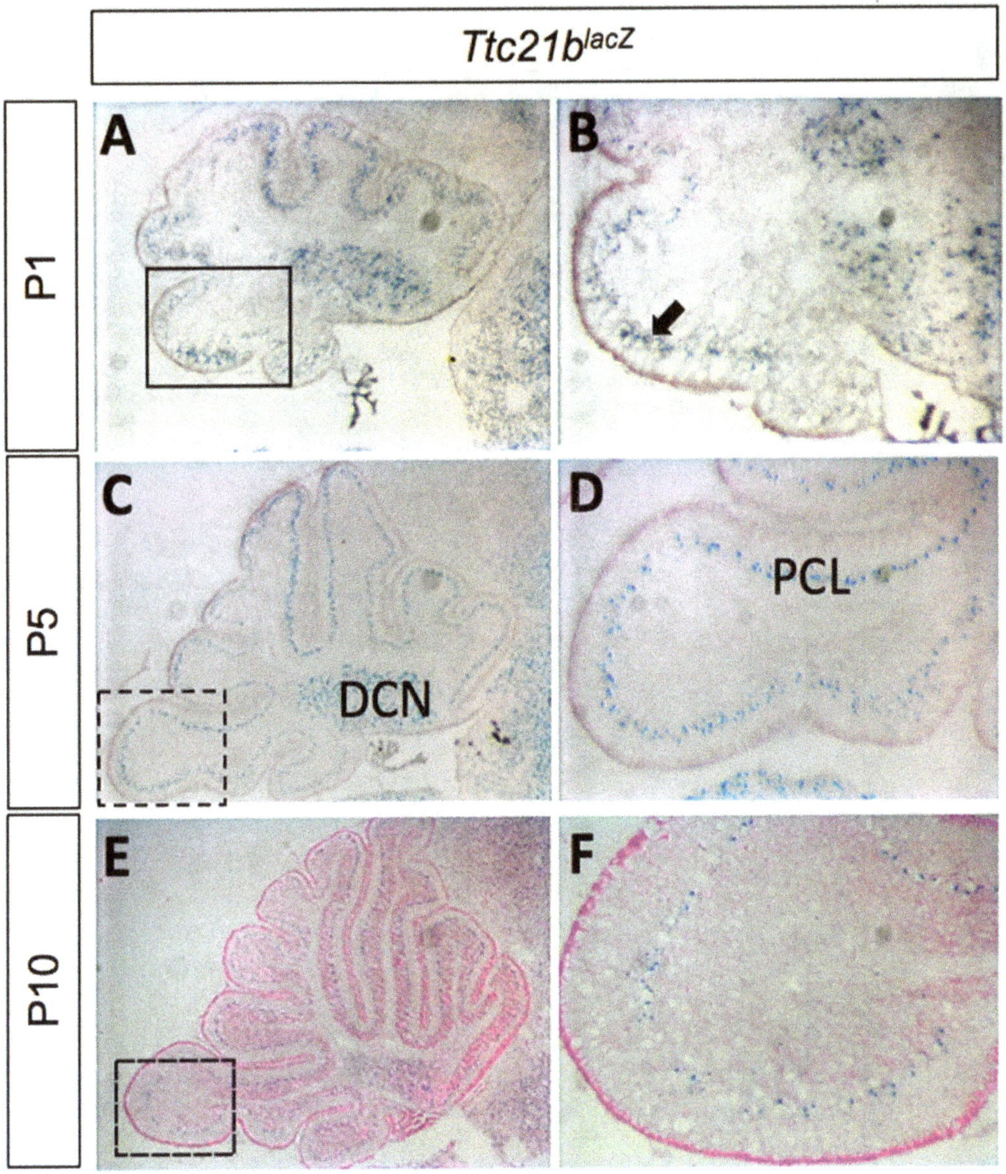

Figure 1. *Ttc21b* is expressed in the postnatal mouse cerebellum. (**A–F**) LacZ expression from the *Ttc21b*^{lacZ} reporter is evident in the Purkinje cell layer of the cerebellum from P1–P10. Boxes in **A**,**C**,**E** indicate areas of focus in **B**,**D**,**F**, respectively. $n \geq 3$ for LacZ staining. DCN: deep cerebellar nuclei; PCL: Purkinje cell layer.

We genetically ablated *Ttc21b* from Bergmann glia by crossing *GFAP-Cre;Ttc21b*$^{aln/wt}$ mice with *Ttc21b*$^{flox/flox}$ animals to generate *GFAP-Cre;Ttc21b*$^{flox/aln}$ mice lacking functional *Ttc21b* in the entire GFAP-expressing lineage. Histological analysis of *GFAP-Cre;Ttc21b*$^{flox/aln}$ animals at P7 showed dense areas of granule cells in the nodular region of the cerebellum which were not seen in control animals (Figure 2A,B, arrowhead; [30]. This phenotype was also observed in 4-month-old adult cerebella. Interestingly, the ectopic cells were found in the posterior cerebellum (lobes IX and X; Figure 2C–F). We demonstrated that these cells are differentiated granule cells with immunohistochemistry for NeuN as a marker for differentiated neurons. Ki67 immunostaining for proliferating cells indicated that these are not mitotically active (Figure 2G–J). This suggests that the ectopic cells, while displaced, are not forming tumors as has been previously shown upon manipulation of SHH signaling in the cerebellum.

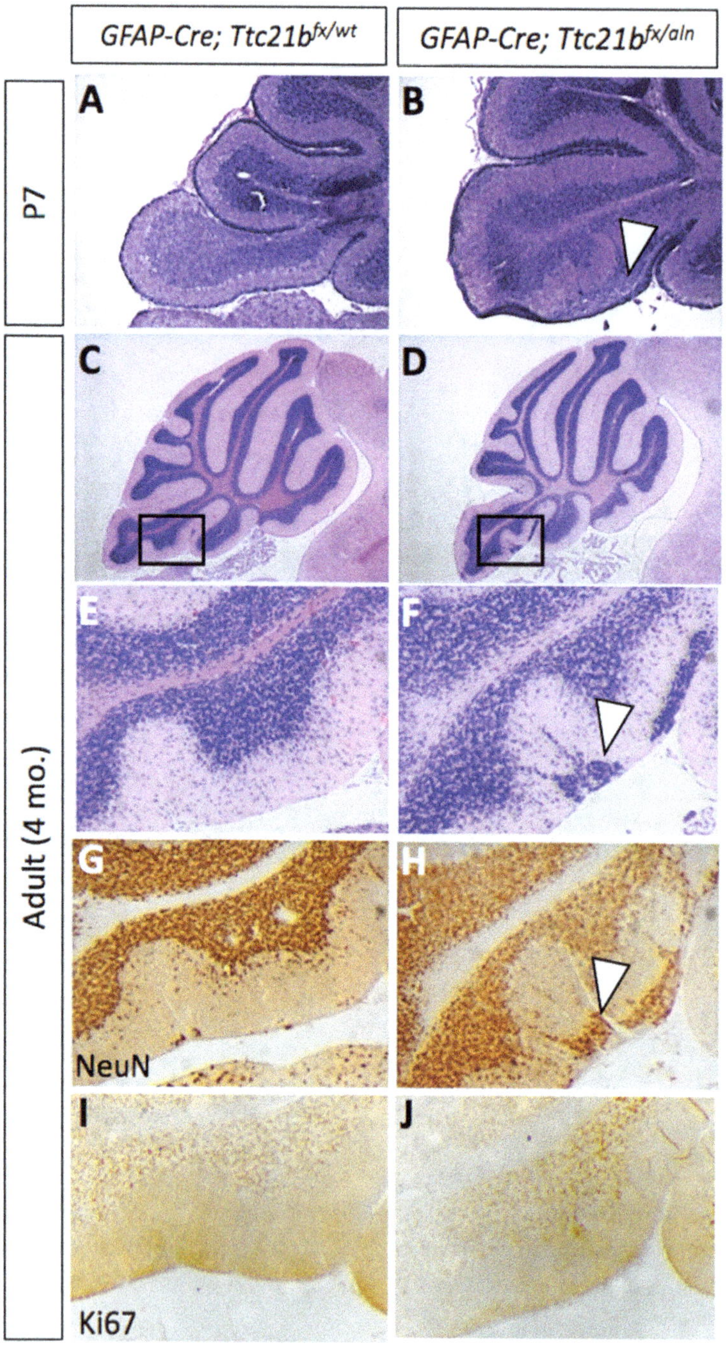

Figure 2. Ablation of *Ttc21b* from astrocytes results in ectopic granule cells in the cerebellum. H&E staining at P7 (**A**,**B**) and 4 months of age (**C**–**F**) show granule cells present in the outer layers of the cerebellum of *GFAP-Cre;Ttc21b*$^{fx/aln}$ animals which are not found in controls (n > 4). NeuN immunoreactivity indicates that these ectopic cells are differentiated granule cells (**G**,**H**, arrowheads). Ki67 immunostaining for proliferative cells is negative in adults (**I**,**J**), suggesting that these are not proliferating tumors. n ≥ 4 for adult histology, n ≥ 3 for adult DAB staining and P7 histology.

3.2. Purkinje Cell and Bergmann Glial Cell Development Are Disrupted upon Loss of Ttc21b

We next sought to determine why the ectopic cells may be present in the *GFAP-Cre;Ttc21b*$^{flox/aln}$ animals. We initially reasoned that they may be granule cells that did not properly migrate from the site of formation in the external granule layer. This migration occurs along the radially oriented Bergmann glial fibers as part of normal granule cell development [6]. To assess this hypothesis, we performed an immunohistochemical analysis of Bergmann glial cell development.

Immature Bergmann glia were specified in approximately the appropriate numbers and appeared to have established radial fibers by P1, as shown by immunoreactivity for both Nestin and GLAST (Figure 3A–D). Purkinje cells were specified and migrated to the appropriate position in the *GFAP-Cre;Ttc21b*$^{flox/aln}$ cerebella at P1 (Figure 3E,F). However, by P11 the *GFAP-Cre;Ttc21b*$^{flox/aln}$ animals showed disrupted Bergmann glia fibers in the lower lobes of the cerebellum failing to form appropriate scaffolds for migration (Figure 3G,H). Interestingly, the anterior lobes where we did not note ectopic cells appeared to have normal Bergmann glia in the *GFAP-Cre;Ttc21b*$^{flox/aln}$ animals (Figure 3K,L). At P11, the Purkinje cell bodies were irregularly aligned, and the dendritic arbors were more extensive in the *GFAP-Cre;Ttc21b*$^{flox/aln}$ animals as compared to control (Figure 3K,L).

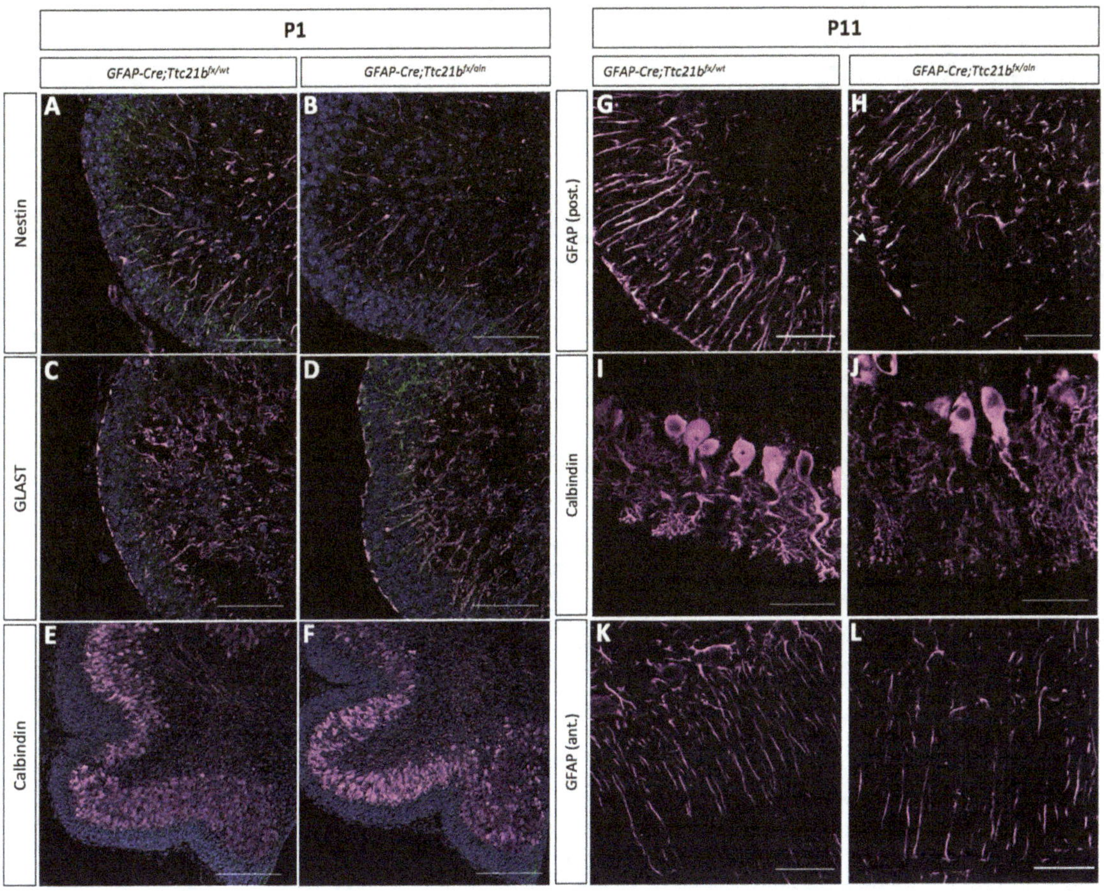

Figure 3. Purkinje cells and Bergmann glia development is disrupted in the *GFAP-Cre;Ttc21b*$^{fx/aln}$ cerebella. (**A**–**D**) Immunostaining for immature astrocytes at P1 with Nestin (**A**,**B**) and GLAST (**C**,**D**) appears similar between control and *GFAP-Cre;Ttc21b*$^{fx/aln}$ animals. (**E**,**F**) Calbindin immunostaining highlights the presence of Purkinje cells at P1 in both control and *GFAP-Cre;Ttc21b*$^{fx/aln}$ animals. (**G**–**J**) P11 GFAP-positive mature astrocytes (**G**,**H**) and Calbindin-positive Purkinje cells (**I**–**L**) appear abnormal in the lower lobes of the *GFAP-Cre;Ttc21b*$^{fx/aln}$ cerebella. (**K**,**L**) Anterior lobes still show normal Bergmann glia morphology in all animals. n = 3 for each genotype. Scale bars 50 μm.

3.3. SHH Signaling Is Reduced in GFAP-Cre;Ttc21b$^{flox/aln}$ *Cerebella*

As the primary cilium is a major hub for *Shh* signaling and loss of *Ttc21b* has been shown to affect SHH signaling, we then investigated SHH pathway activity in the *GFAP-Cre;Ttc21b*$^{flox/aln}$ animals with the *Gli1*^{lacZ} allele [31]. *Gli1*^{lacZ} expression was reduced in the Purkinje cell layer of *Gli1*^{lacZ}*;GFAP-Cre;Ttc21b*$^{flox/aln}$ animals (Figure 4A–D, arrow). Since *Gli1* is not expressed by Purkinje cells, these results suggest a loss of SHH signaling activity in Bergmann glia [7]. RNA section in situ hybridization for another *Shh* transcriptional target—*Patched-1* (*Ptch1*)—also showed reduced expression in the Purkinje cell layer of the *GFAP-Cre;Ttc21b$^{flox/alien}$* animals as compared to controls at P11 (Figure 4E,F, arrowhead). However, expression of the *Shh* itself was at similar levels between *GFAP-Cre;Ttc21b$^{flox/aln}$* and control animals, suggesting that Purkinje cell expression of the secreted SHH ligand is not affected (Figure 5G,H). Taken together, these data show that *Shh* signaling is reduced in the adult *GFAP-Cre;Ttc21b$^{flox/aln}$* cerebellum—a striking difference to previous observations of *Ttc21b* function in the nervous system.

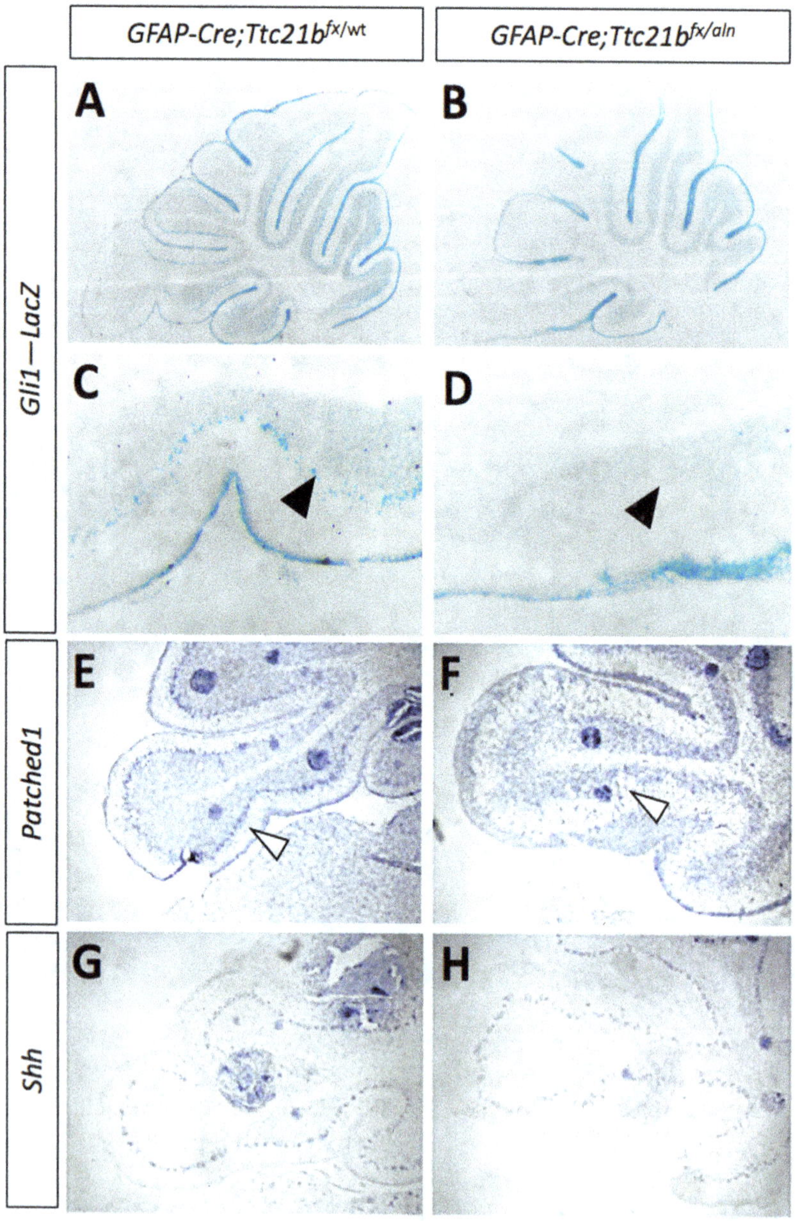

Figure 4. *Shh* signaling is reduced in *GFAP-Cre;Ttc21b$^{fx/aln}$* cerebella. (**A–D**) P11 *Gli1*-LacZ activity is reduced in in the Purkinje cell layer (arrows) of the *GFAP-Cre;Ttc21b$^{fx/aln}$* animals (**B**,**D**) as compared to control (**A**,**C**). (**E**,**F**) *Patched1* expression is also reduced in the *GFAP-Cre;Ttc21b$^{fx/aln}$* cerebella at P11 (**F**) compared to control (**E**), especially in the Purkinje cell layer (open arrowhead). (**G**,**H**) *Shh* expression appears similar between control (**G**) and *GFAP-Cre;Ttc21b$^{fx/aln}$* (**H**), but the organization of *Shh*-expressing cells again appears disorganized in the lower lobe of the cerebellum in ablated animals.

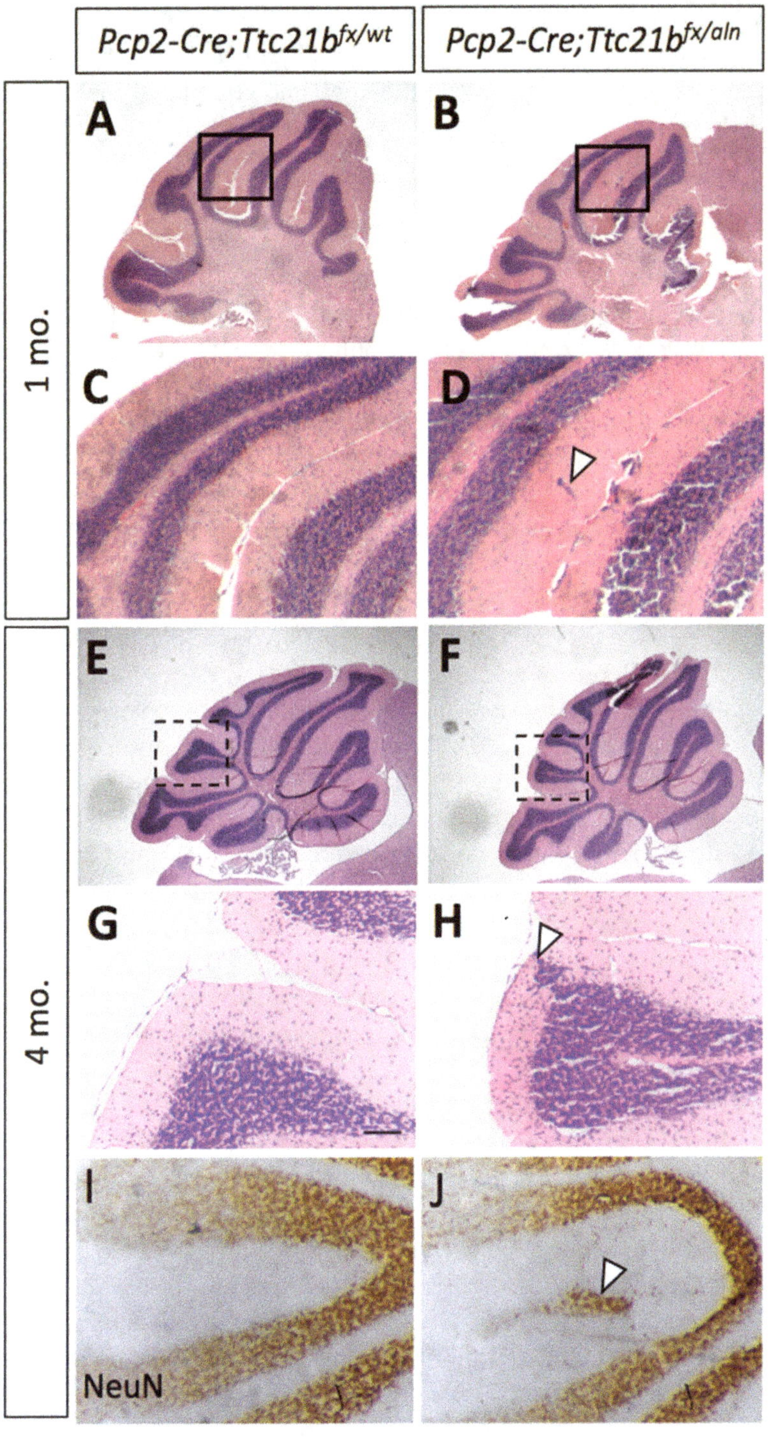

Figure 5. Ablation of *Ttc21b* with *Pcp2-Cre* results in similar, but milder, ectopic granule cell phenotype. (**A**,**B**) Small areas of displaced granule cells (arrowhead) are present in *Pcp2-Cre;Ttc21b*$^{fx/aln}$ brains at 1 month of age. (**B**,**D**: boxes in **A**,**B** show areas highlighted in **C**,**D**). (**E–I**) These cells are still visible at 4 months of age in the *Pcp2-Cre;Ttc21b*$^{fx/aln}$ animals (**H**, arrowhead) and are immuno-positive for NeuN, indicating that these cells are differentiated granule cells (**J**, arrowhead). $n \geq 3$ for each genotype.

3.4. Purkinje Cell Ablation of Ttc21b Results in Milder Ectopic Granule Cell Phenotype

Since we observed *Ttc21b* expression in the Purkinje cell layers which have both Purkinje cells and Bergmann glia, we performed a similar genetic ablation of *Ttc21b* in the Purkinje cells with the *Pcp2-Cre* [27] (Figure S2). At 1 month of age, histological analysis again revealed small areas of ectopic cells in the outer layer of the *Pcp2-Cre;Ttc21b^flox/aln^* cerebella (Figure 5A–D, arrowhead). This was also periodically seen in 4-month-old *Pcp2-Cre;Ttc21b^flox/aln^* (Figure 5E–H, n = 4). Immunohistochemistry for NeuN again suggested that these cells were differentiated granule cells which failed to migrate to the IGL (Figure 5I,J). Interestingly, this phenotype was less penetrant than that in *GFAP-Cre;Ttc21b^flox/aln^* animals, but was not solely restricted to the posterior lobes.

4. Discussion

Here we have demonstrated a role for *Ttc21b* in the both the Bergmann glia astrocytes and Purkinje cells of the developing mouse cerebellum. Using a *GFAP-Cre* genetic ablation of *Ttc21b*, we showed that a few granule cells are retained in the outer layer of the cerebellum, specifically in lobes IX and X. This appears to be a failure of granule cell migration, as the Bergmann glia—substrates for granule cell migration—have reduced radial projections in the lower lobes of the ablated animals. In contrast, a *Pcp2-Cre* Purkinje cell-specific ablation has a similar—but much less severe—phenotype, with only sparse areas of ectopic granule cells. Additionally, while GFAP-Cre recombination occurs in Bergmann glia, we noted effects on Purkinje cell development, supporting previous suggestions that a critical interaction between these two cells types is required for proper Purkinje cell dendritic development. We also observed that ablation of *Ttc21b* resulted in a reduction in SHH signaling in the Purkinje cell layer, which is the first time we have a reduction on the *Shh* pathway upon loss of *Ttc21b*.

The radial fibers of the Bergmann glia serve as primary migratory tracts for granule cells in the developing cerebellum [6], and disruptions of these radial fibers have been previously shown to be associated with ectopic granule cells [32,33]. We only observed disruption of mature Bergmann glia in the affected lobes of the mutant, suggesting that this is a primary cause of the failed granule cell migration. Additionally, we noted abnormalities in Purkinje cell morphology in *GFAP-Cre;Ttc21b^flox/aln^* animals consistent with numerous interactions between Bergmann glia and Purkinje neurons [34–36]. The ectopic granule cell phenotype was much less severe in the *Pcp2-Cre; Ttc21b^flox/aln^* animals. This could suggest differing cell-specific requirements for *Ttc21b*. Indeed, patients with *TTC21B* variants have shown cytoskeletal abnormalities in human podocytes in addition to abnormal cilia, suggesting multiple roles for *TTC21B* [37]. Additionally, *Pcp2-Cre* recombination does occur later in development than GFAP-Cre (P5) and may not completely ablate *Ttc21b* until after many cells have already completed migration.

Prior reports of *Ttc21b^aln/aln^* mutants have shown increased SHH signaling in addition to abnormal GLI processing. Embryonically, loss of *Ttc21b* led to increased translation and abnormal proteolytic processing of GLI3 [19]. Inappropriate SHH activity has also been demonstrated in embryonic forebrain of *Ttc21b^aln/aln^* mutants with abnormal GLI3 processing [19,21]. In a postnatal deletion of *Ttc21b*, increased GLI2 activity led to polycystic kidneys [24]. In contrast to these findings, we report a reduction in downstream *Patched* and *Gli1*, consistent with reduced *Shh* signaling.

While *GFAP-Cre* should lead to loss of *Ttc21b* throughout the cerebellum, we only observed the ectopic granule cell phenotype in the posterior lobes. GLI2 is enriched in the Purkinje cell layer in the posterior lobes of the developing cerebellum around P5, and has been shown to be the major effector of positive SHH signaling [7]. While all three GLI transcription factors are expressed in the developing cerebellum, *Gli1* becomes restricted to the Bergmann glia in the Purkinje cell layer [7]. However, it is unlikely that the phenotype reported here is solely due to an effect on *Gli1*, as *Gli1* homozygous null mutants have a phenotypically normal cerebellum [38]. Both *Gli2* and *Gli3* null mice have abnormal cerebellar development, with primary defects in foliation [39]. Additionally, loss of *Gli2* has been shown to result in abnormal Bergmann glia morphology and Purkinje cell alignment [8]. Primary cilia are required for the proper processing of GLI2 protein into a transcriptional activator and GLI3 protein into a transcriptional repressor form. The loss of normal cilia function may have different effects on SHH signaling in different tissues, depending on which GLI protein has a more prominent role at that particular time in development. Given the enriched expression of *Gli2* in the same area as the ectopic cells and its role in stimulating SHH signaling, the reduced SHH noted here in the postnatal cerebellum may be more reflective of the loss of the GLI2 activator form than the GLI3 repressor.

The role of both ciliary genes and SHH signaling is a rapidly expanding field of study in human cerebellar development. SHH signaling is dysregulated in cells from Joubert syndrome and Meckel syndrome patients [11]. Interestingly, *TTC21B* variants are associated with Joubert syndrome patients [40]. Our study shows *Ttc21b* has a role in the glial cerebellar cells and loss of *Ttc21b* in different tissues can lead to increased and decreased SHH signaling.

Supplementary Materials: The following are available online at www.mdpi.com/2221-3759/5/4/18/s1. Figure S1: GFAP-Cre recombination occurs in the Bergmann glia of the cerebellum, Figure S2: Pcp2-Cre recombination occurs in the Purkinje cells of the cerebellum.

Acknowledgments: This work is supported by the National Institutes of Health (R01 NS085023).

Author Contributions: A.M.D. and R.W.S. conceived and designed the experiments; A.M.D. and C.S. performed the experiments; A.M.D., C.S., and R.W.S. analyzed the data; A.M.D. and R.W.S. wrote the paper.

Conflicts of Interest: The authors declare no conflict of interest. The funding sponsors had no role in the design of the study; in the collection, analyses, or interpretation of data; in the writing of the manuscript, and in the decision to publish the results.

References

1. Andersen, B.B.; Korbo, L.; Pakkenberg, B. A quantitative study of the human cerebellum with unbiased stereological techniques. *J. Comp. Neurol.* **1992**, *326*, 549–560. [CrossRef] [PubMed]
2. Ten Donkelaar, H.J.; Lammens, M. Development of the human cerebellum and its disorders. *Clin. Perinatol.* **2009**, *36*, 513–530. [CrossRef] [PubMed]
3. Dahmane, N.; Ruiz i Altaba, A. Sonic hedgehog regulates the growth and patterning of the cerebellum. *Development* **1999**, *126*, 3089–3100. [PubMed]
4. Wallace, V.A. Purkinje-cell-derived Sonic hedgehog regulates granule neuron precursor cell proliferation in the developing mouse cerebellum. *Curr. Biol.* **1999**, *9*, 445–448. [CrossRef]
5. Wechsler-Reya, R.J.; Scott, M.P. Control of neuronal precursor proliferation in the cerebellum by Sonic Hedgehog. *Neuron* **1999**, *22*, 103–114. [CrossRef]
6. Xu, H.; Yang, Y.; Tang, X.; Zhao, M.; Liang, F.; Xu, P.; Hou, B.; Xing, Y.; Bao, X.; Fan, X. Bergmann glia function in granule cell migration during cerebellum development. *Mol. Neurobiol.* **2013**, *47*, 833–844. [CrossRef] [PubMed]
7. Corrales, J.D.; Rocco, G.L.; Blaess, S.; Guo, Q.; Joyner, A.L. Spatial pattern of sonic hedgehog signaling through Gli genes during cerebellum development. *Development* **2004**, *131*, 5581–5590. [CrossRef] [PubMed]
8. Corrales, J.D.; Blaess, S.; Mahoney, E.M.; Joyner, A.L. The level of sonic hedgehog signaling regulates the complexity of cerebellar foliation. *Development* **2006**, *133*, 1811–1821. [CrossRef] [PubMed]
9. Lewis, P.M.; Gritli-Linde, A.; Smeyne, R.; Kottmann, A.; McMahon, A.P. Sonic hedgehog signaling is required for expansion of granule neuron precursors and patterning of the mouse cerebellum. *Dev. Biol.* **2004**, *270*, 393–410. [CrossRef] [PubMed]
10. Spassky, N.; Han, Y.G.; Aguilar, A.; Strehl, L.; Besse, L.; Laclef, C.; Ros, M.R.; Garcia-Verdugo, J.M.; Alvarez-Buylla, A. Primary cilia are required for cerebellar development and Shh-dependent expansion of progenitor pool. *Dev. Biol.* **2008**, *317*, 246–259. [CrossRef] [PubMed]
11. Aguilar, A.; Meunier, A.; Strehl, L.; Martinovic, J.; Bonniere, M.; Attie-Bitach, T.; Encha-Razavi, F.; Spassky, N. Analysis of human samples reveals impaired SHH-dependent cerebellar development in Joubert syndrome/Meckel syndrome. *Proc. Natl. Acad. Sci. USA* **2012**, *109*, 16951–16956. [CrossRef] [PubMed]
12. Goodrich, L.V.; Scott, M.P. Hedgehog and patched in neural development and disease. *Neuron* **1998**, *21*, 1243–1257. [CrossRef]
13. Hallahan, A.R.; Pritchard, J.I.; Hansen, S.; Benson, M.; Stoeck, J.; Hatton, B.A.; Russell, T.L.; Ellenbogen, R.G.; Bernstein, I.D.; Beachy, P.A.; et al. The SmoA1 mouse model reveals that notch signaling is critical for the growth and survival of sonic hedgehog-induced medulloblastomas. *Cancer Res.* **2004**, *64*, 7794–7800. [CrossRef] [PubMed]

14. Han, Y.G.; Kim, H.J.; Dlugosz, A.A.; Ellison, D.W.; Gilbertson, R.J.; Alvarez-Buylla, A. Dual and opposing roles of primary cilia in medulloblastoma development. *Nat. Med.* **2009**, *15*, 1062–1065. [CrossRef] [PubMed]
15. Oliver, T.G.; Read, T.A.; Kessler, J.D.; Mehmeti, A.; Wells, J.F.; Huynh, T.T.; Lin, S.M.; Wechsler-Reya, R.J. Loss of patched and disruption of granule cell development in a pre-neoplastic stage of medulloblastoma. *Development* **2005**, *132*, 2425–2439. [CrossRef] [PubMed]
16. Yang, Z.J.; Ellis, T.; Markant, S.L.; Read, T.A.; Kessler, J.D.; Bourboulas, M.; Schuller, U.; Machold, R.; Fishell, G.; Rowitch, D.H.; et al. Medulloblastoma can be initiated by deletion of Patched in lineage-restricted progenitors or stem cells. *Cancer Cell* **2008**, *14*, 135–145. [CrossRef] [PubMed]
17. Bangs, F.; Anderson, K.V. Primary Cilia and Mammalian Hedgehog Signaling. *Cold Spring Harb. Perspect. Biol.* **2017**, *9*, a028175. [CrossRef] [PubMed]
18. Liu, A.; Wang, B.; Niswander, L.A. Mouse intraflagellar transport proteins regulate both the activator and repressor functions of Gli transcription factors. *Development* **2005**, *132*, 3103–3111. [CrossRef] [PubMed]
19. Tran, P.V.; Haycraft, C.J.; Besschetnova, T.Y.; Turbe-Doan, A.; Stottmann, R.W.; Herron, B.J.; Chesebro, A.L.; Qiu, H.; Scherz, P.J.; Shah, J.V.; et al. THM1 negatively modulates mouse sonic hedgehog signal transduction and affects retrograde intraflagellar transport in cilia. *Nat. Genet.* **2008**, *40*, 403–410. [CrossRef] [PubMed]
20. Snedeker, J.; Schock, E.N.; Struve, J.N.; Chang, C.F.; Cionni, M.; Tran, P.V.; Brugmann, S.A.; Stottmann, R.W. Unique spatiotemporal requirements for intraflagellar transport genes during forebrain development. *PLoS ONE* **2017**, *12*, e0173258. [CrossRef] [PubMed]
21. Stottmann, R.W.; Tran, P.V.; Turbe-Doan, A.; Beier, D.R. Ttc21b is required to restrict sonic hedgehog activity in the developing mouse forebrain. *Dev. Biol.* **2009**, *335*, 166–178. [CrossRef] [PubMed]
22. Davis, E.E.; Zhang, Q.; Liu, Q.; Diplas, B.H.; Davey, L.M.; Hartley, J.; Stoetzel, C.; Szymanska, K.; Ramaswami, G.; Logan, C.V.; et al. TTC21B contributes both causal and modifying alleles across the ciliopathy spectrum. *Nat. Genet.* **2011**, *43*, 189–196. [CrossRef] [PubMed]
23. Niwa, S. The nephronophthisis-related gene ift-139 is required for ciliogenesis in Caenorhabditis elegans. *Sci. Rep.* **2016**, *6*, 31544. [CrossRef] [PubMed]
24. Tran, P.V.; Talbott, G.C.; Turbe-Doan, A.; Jacobs, D.T.; Schonfeld, M.P.; Silva, L.M.; Chatterjee, A.; Prysak, M.; Allard, B.A.; Beier, D.R. Downregulating hedgehog signaling reduces renal cystogenic potential of mouse models. *J. Am. Soc. Nephrol.* **2014**, *25*, 2201–2212. [CrossRef] [PubMed]
25. Muzumdar, M.D.; Tasic, B.; Miyamichi, K.; Li, L.; Luo, L. A global double-fluorescent Cre reporter mouse. *Genesis* **2007**, *45*, 593–605. [CrossRef] [PubMed]
26. Zhuo, L.; Theis, M.; Alvarez-Maya, I.; Brenner, M.; Willecke, K.; Messing, A. hGFAP-cre transgenic mice for manipulation of glial and neuronal function in vivo. *Genesis* **2001**, *31*, 85–94. [CrossRef] [PubMed]
27. Barski, J.J.; Dethleffsen, K.; Meyer, M. Cre recombinase expression in cerebellar Purkinje cells. *Genesis* **2000**, *28*, 93–98. [CrossRef]
28. Bai, C.B.; Auerbach, W.; Lee, J.S.; Stephen, D.; Joyner, A.L. Gli2, but not Gli1, is required for initial Shh signaling and ectopic activation of the Shh pathway. *Development* **2002**, *129*, 4753–4761. [PubMed]
29. Yamada, K.; Watanabe, M. Cytodifferentiation of Bergmann glia and its relationship with Purkinje cells. *Anat. Sci. Int.* **2002**, *77*, 94–108. [CrossRef] [PubMed]
30. White, J.J.; Sillitoe, R.V. Development of the cerebellum: From gene expression patterns to circuit maps. *Wiley Interdiscip. Rev. Dev. Biol.* **2013**, *2*, 149–164. [CrossRef] [PubMed]
31. Lee, J.; Platt, K.A.; Censullo, P.; Ruiz i Altaba, A. Gli1 is a target of Sonic hedgehog that induces ventral neural tube development. *Development* **1997**, *124*, 2537–2552. [PubMed]
32. Frick, A.; Grammel, D.; Schmidt, F.; Poschl, J.; Priller, M.; Pagella, P.; von Bueren, A.O.; Peraud, A.; Tonn, J.C.; Herms, J.; et al. Proper cerebellar development requires expression of beta1-integrin in Bergmann glia, but not in granule neurons. *Glia* **2012**, *60*, 820–832. [CrossRef] [PubMed]
33. Inouye, M.; Hayasaka, S.; Funahashi, A.; Yamamura, H. Gamma-radiation produces abnormal Bergmann fibers and ectopic granule cells in mouse cerebellar cortex. *J. Radiat. Res.* **1992**, *33*, 275–281. [CrossRef] [PubMed]
34. Bellamy, T.C. Interactions between Purkinje neurones and Bergmann glia. *Cerebellum* **2006**, *5*, 116–126. [CrossRef] [PubMed]
35. Lordkipanidze, T.; Dunaevsky, A. Purkinje cell dendrites grow in alignment with Bergmann glia. *Glia* **2005**, *51*, 229–234. [CrossRef] [PubMed]

36. Wang, F.; Xu, Q.; Wang, W.; Takano, T.; Nedergaard, M. Bergmann glia modulate cerebellar Purkinje cell bistability via Ca^{2+}-dependent K^+ uptake. *Proc. Natl. Acad. Sci. USA* **2012**, *109*, 7911–7916. [CrossRef] [PubMed]
37. Huynh Cong, E.; Bizet, A.A.; Boyer, O.; Woerner, S.; Gribouval, O.; Filhol, E.; Arrondel, C.; Thomas, S.; Silbermann, F.; Canaud, G.; et al. A homozygous missense mutation in the ciliary gene TTC21B causes familial FSGS. *J. Am. Soc. Nephrol.* **2014**, *25*, 2435–2443. [CrossRef] [PubMed]
38. Park, C.; Longo, C.M.; Ackerman, S.L. Genetic and physical mapping of the cerebellar deficient folia (cdf) locus on mouse chromosome 6. *Genomics* **2000**, *69*, 135–138. [CrossRef] [PubMed]
39. Blaess, S.; Stephen, D.; Joyner, A.L. Gli3 coordinates three-dimensional patterning and growth of the tectum and cerebellum by integrating Shh and Fgf8 signaling. *Development* **2008**, *135*, 2093–2103. [CrossRef] [PubMed]
40. Romani, M.; Micalizzi, A.; Valente, E.M. Joubert syndrome: Congenital cerebellar ataxia with the molar tooth. *Lancet Neurol.* **2013**, *12*, 894–905. [CrossRef]

4

Flying the RNA Nest: *Drosophila* Reveals Novel Insights into the Transcriptome Dynamics of Early Development

Fabio Alexis Lefebvre [1,2] and Éric Lécuyer [1,2,3,4,*]

[1] Institut de Recherches Cliniques de Montréal (IRCM), Montréal, QC H2W 1R7, Canada; fabioalexis.lefebvre@ircm.qc.ca
[2] Département de Biochimie, Université de Montréal, Montréal, QC H3T 1J4, Canada
[3] Division of Experimental Medicine, McGill University, Montréal, QC H3A 0G4, Canada
[4] IRCM, RNA Biology Laboratory, 110 Avenue des Pins, Ouest, Montréal, QC H2W 1R7, Canada
* Correspondence: Eric.Lecuyer@ircm.qc.ca

Abstract: Early development is punctuated by a series of pervasive and fast paced transitions. These events reshape a differentiated oocyte into a totipotent embryo and allow it to gradually mount a genetic program of its own, thereby framing a new organism. Specifically, developmental transitions that ensure the maternal to embryonic control of developmental events entail a deep remodeling of transcriptional and transcriptomic landscapes. *Drosophila* provides an elegant and genetically tractable system to investigate these conserved changes at a dazzling developmental pace. Here, we review recent studies applying emerging technologies such as ribosome profiling, in situ Hi-C chromatin probing and live embryo RNA imaging to investigate the transcriptional dynamics at play during *Drosophila* embryogenesis. In light of this new literature, we revisit the main models of zygotic genome activation (ZGA). We also review the contributions played by zygotic transcription in shaping embryogenesis and explore emerging concepts of processes such as transcriptional bursting and transcriptional memory.

Keywords: embryogenesis; *Drosophila*; development; RNA; transcriptomics; RNA; zygotic genome activation; posttranscriptional regulation; maternal-to-zygotic transition; epigenetics

---

1. Introduction

Early development unfolds through exquisitely coordinated and deeply conserved biological transitions. Fertilization entails the remodeling of a differentiated oocyte into a totipotent embryo, which involves profound genomic, transcriptomic and proteomic re-organization [1]. Early embryos execute a dazzling proliferative push driven by maternally provided gene products to increase cell number. The rapid pace of these early divisions, termed cleavage cycles, is achieved through copious supplies of maternal cyclins, an abbreviated DNA replication execution and the absence of active growth and mitotic checkpoints [2]. In most systems, zygotic nuclei remain transcriptionally silent during this period. As embryogenesis proceeds, cell cycle duration progressively lengthens, reflecting the gradual emergence of the DNA replication checkpoint and the increasing destabilization of maternal cyclins [3] (Figure 1). Interphase lengthening broadly coincides with progressive zygotic genome activation (ZGA), i.e., de novo expression of robust populations of functional transcripts [4–6]. New zygotic products gradually take over the pool of maternal RNAs, which undergo progressive clearance through regulated degradation mechanisms [7–9] (Figure 1). The juxtaposition of ZGA and maternal clearance gradually remodels the transcriptome, a process termed the maternal-to-zygotic transition (MZT). The MZT ends at a key developmental time point called the midblastula transition

(MBT), which typically involves a dramatic cascade of anatomical and physiological changes that are dependent on zygotic transcription.

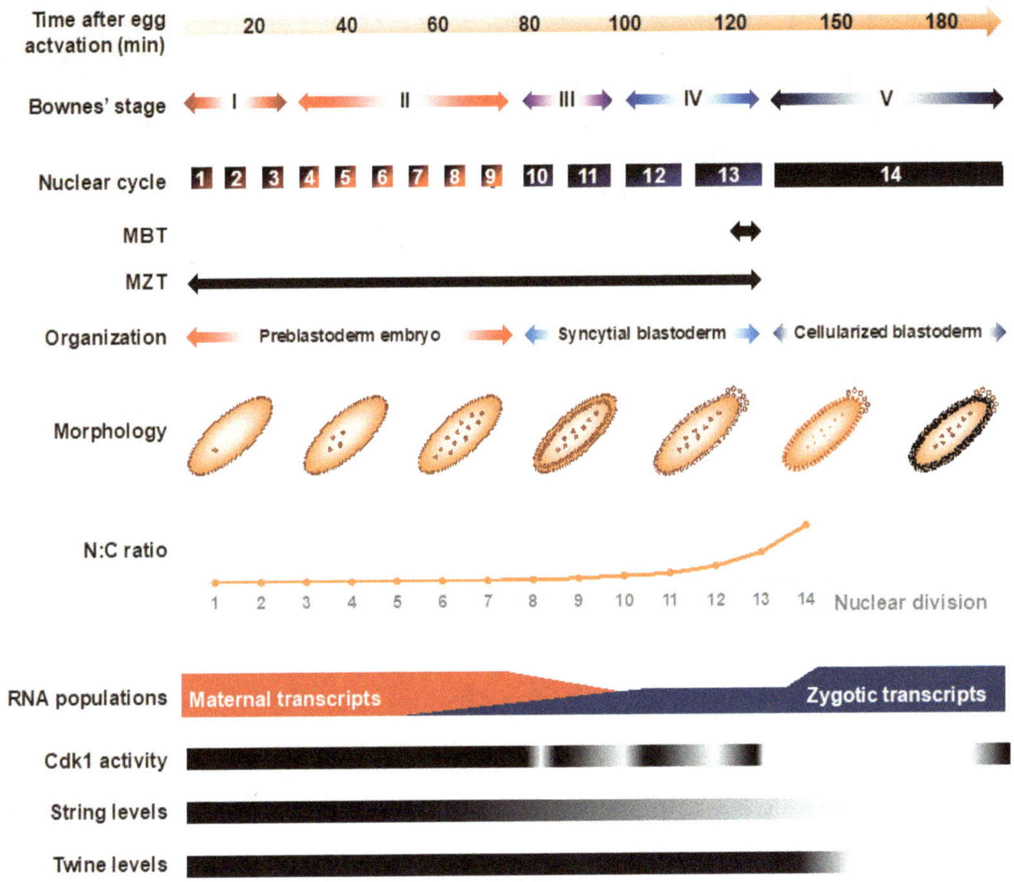

Figure 1. Overview of morphologic and transcriptomic features of early embryonic development. Comparative timescale of Bownes' stages, nuclear divisions and histological organization as of function of time after egg activation. The nucleocytoplasmic (N:C) ratio increases rapidly during the cleavage cycles, driven by fast-paced mitotic cycles in absence of cytokinesis and growth of the syncytium. The transcriptomic landscape that prevails during the first embryonic cycles reflects maternally-deposited transcripts encoded in ovarian nurse cells and deposited in the oocyte before egg activation. Zygotic genome activation (ZGA) begins in late preblastoderm embryos and leads to the progressive accumulation of zygotic transcripts. Concomitantly, a large fraction of maternally deposited RNAs undergo targeted degradation.

Drosophila and many arthropods display a facultative parthenogenetic mode of reproduction, meaning that egg activation can take place independently of sperm entry, although this rarely occurs. The egg-to-embryo transition is triggered by changes in pressure and osmotic balance as the mature oocyte exits through the uterus [10,11]. Egg activation involves the completion of meiosis and the initiation of fast-paced mitotic divisions, thereby setting the onset of embryogenesis. In *Drosophila*, the first rounds of nuclear divisions arise every 9 min, leading to the formation of 6000 nuclei in only 2 h [2] (Foe and Alberts, 1983). To facilitate the rapid pace of these divisions and the synchronization of mitotic entry, nuclear cycles (NC) take place within a single large syncytial cell [12,13] (Figure 1). The first syncytial divisions are metasynchronous, and proceed under maternal control, while zygotic nuclei remain largely transcriptionally quiescent [14,15].

During the interphase of NC8 and NC9, nuclei start migrating to the cortical periphery, forming a syncytial blastoderm embryo [16]. This process coincides with gradual interphase lengthening, reflecting increasing long periods of Cyclin-dependent kinase 1 (Cdk1) inhibition. With its cyclin partners, Cdk1 acts as the chief regulator of cell cycle progression through the phosphorylation of a wide range of protein targets, which notably mediate S phase initiation, spindle assembly and sister chromatid alignment [2]. Cdk1 inhibition expands as the cytoplasmic pool of maternal cyclins, notably String and Twine, is progressively depleted and as the DNA replication checkpoints emerge (Figure 1). In hiatus during the first cycles, the checkpoint safeguards genome integrity by preventing mitotic entry when single-stranded (ss)DNA is sensed, underlying incomplete replication or extensive damage [17,18]. When the MBT takes place at NC13, interphase duration suddenly triples as mitotic synchrony is lost and cortical nuclei secede from the syncytium to form a well-defined primordial epithelium, a process termed cellularization [2]. En masse zygotic transcription ensues in cells now endowed with motility and a susceptibility for apoptosis—new-found attributes that will play crucial roles in gastrulation, neurulation and organogenesis [19,20].

Evolutionarily conserved features of the MZT have been well reviewed by Tadros and Lipsitz (2009) and more recently by Langley and colleagues (2014). This exciting area of developmental biology has further expanded over the last few years, along with our understanding of the complex cross-talks resulting in the interrelated emergence of ZGA, checkpoint activation and cellularization. Indeed, the deployment of disruptive technologies to track translation, probe chromosome conformation and image single RNA molecules in *Drosophila* have revealed new insights into the organization of the transcriptome during embryogenesis. Here, we focus on key findings established in recent years that paint an increasingly resolved picture of transcriptome dynamics during early development. We first review the changes in transcriptomic landscapes that take place as maternally-deposited transcripts are selectively localized, translated and degraded. We discuss the different models proposed to regulate the onset of zygotic transcription and focus on the remodeling of chromatin during early embryogenesis. We then review the roles played by zygotic transcription in promoting cellularization, maternal clearance and replication checkpoint activation. Finally, we outline novel properties of zygotic transcription revealed by real-time, in vivo imaging, including transcriptional bursting and transcriptional memory.

2. Deposition, Localization, Translation and Clearance of Maternal Transcripts

Early *Drosophila* development is driven by maternal proteins and RNA transcripts synthesized by multiploid nurse cells and deposited in the egg during oogenesis [21,22]. Long before fertilization, structural and biosynthetic factors such as ribosomal and glycolytic proteins are endowed in the oocyte along with their corresponding mRNAs [23]. These factors will direct rapid mitotic cycling and sustain DNA replication in early embryogenesis, while zygotic nuclei remain transcriptionally quiescent. Maternal deposition is widespread: up to 65% of all annotated *Drosophila* mRNAs can be detected during the first syncytial cycles [24–26]. Among these products, a set of maternal mRNAs acquire spatially-resolved localization in the oocyte through a series of symmetry breaking events during ovarian development. After egg activation, the asymmetric distributions of these transcripts, notably *bicoid*, *oskar* and *gurken*, defines anteroposterior and dorsoventral axes, which will later determine the body plan of the developing fly [27–29]. Aside from these classically-defined examples, recent large-scale *fluorescent* in situ *hybridization* (FISH) surveys have demonstrated that subellular localization is a pervasive feature of the *Drosophila* transcriptome. Indeed, detailed profiling of RNA expression/localization features in fly oocytes and embryos, as tabulated in the Fly-FISH (http://fly-fish.ccbr.utoronto.ca/) and Dresden Ovary Table/DOT (http://tomancak-srv1.mpi-cbg.de/DOT/main.html) database repositories, revealed that the vast majority of mRNAs and long non-coding RNAs are localized in a striking array of distribution patterns [24,30]. While these studies underline the dynamic localization properties of both maternal and zygotic RNA populations, the underlying regulatory mechanisms and functions for the most part remain to be characterized.

The transcriptomic landscape is highly dynamic during the cleavage cycles, as several RNA degradation pathways operate to selectively remove large sets of deposited transcripts. Indeed, a study using chromosomal deletions to track the post-transcriptional dynamics of maternal and zygotic transcripts has shown that approximately 35% of maternally-deposited mRNAs are cleared out by the MBT [31]. Comparisons of degradation dynamics in unfertilized and fertilized embryos have revealed the prevalence of at least two pathways. The early onset maternal pathway operates independently of ZGA and accounts for the destabilization of approximately 20% of all mRNAs [8] (Figure 2). This maternal degradation program reflects the coordinated activity of RNA-binding proteins (RBPs) that associate with specific subsets of maternal transcripts and recruit the CCR4-NOT deadenylase complex to initiate their degradation [32]. Maternal clearance is a highly-coordinated process determined by the interplay of *cis*-acting motifs, generally found in the 3′UTR of the target transcripts, and RBPs that adopt tightly regulated spatial and temporal distributions. Indeed, mRNAs encoding the RBP Smaug (SMG) form an anterior-to-posterior gradient in the oocyte and are translationally regulated at egg activation [33]. SMG activity peaks at NC10, enacting an elegant spatial and temporal regulation of maternal clearance [32–34]. In addition, three RBPs, Pumilio (PUM), Brain tumor (BRAT) and ME31B have more recently been shown to have non-overlapping roles in mediating the decay of hundreds of maternal transcripts [35,36].

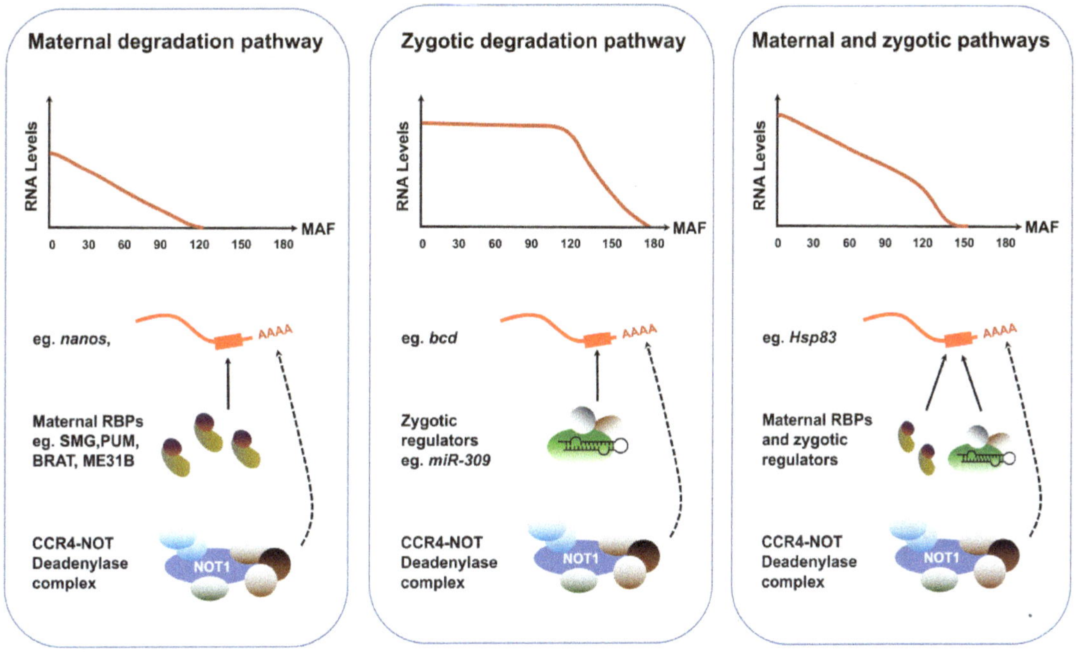

Figure 2. Alternative degradation profiles of maternally-deposited transcripts. The clearance of maternally-deposited transcripts can proceed through a strictly maternal pathway, a strictly zygotic pathway or a combination of both maternal and zygotic effectors. Transcripts strictly targeted by the maternal pathway, such as *nanos* (*nos*) display identical dynamics in fertilized and activated eggs. The maternal RNA-binding proteins Smaug, Pumilio, Brat and/or ME31B selectively interact with these RNAs through a consensus motif and recruit the CCR4/POP2/NOT deadenylase complex to initiate their degradation. Transcripts targeted through the zygotic pathway include *bicoid* (*bcd*) and their degradation depends on the ZGA. Transcripts targeted through both maternal and zygotic degradation pathways include *Hsp83* and their clearance relies on the activity of both maternally deposited and zygotically encoded factors. MAF: Minutes after fertilization.

Strong evidence in zebrafish, *Xenopus* and mouse has long suggested that some of the zygotic actors involved in maternal clearance are microRNAs (miRNAs) [37]. Small-interfering RNAs are versatile regulators that can destabilize vast pools of mRNA targets [37,38]. In support of this model, SMG is required for the zygotic expression of the miR-309 family [39]. Among the miRNAs expressed during the MZT, over 70 exhibit compromised levels in *smg* mutants. In addition, Argonaute-1, a key component of the miRNA-induced silencing complex, showed decreased levels in *smg* mutants. Furthermore, the clearance of predicted maternal targets of SMG-dependent miRNAs is hampered in *smg* mutants, suggesting that these small RNAs contribute to the zygotic component of maternal mRNA degradation [39,40]. Because of the specificity and diversity of clearance pathways effective during the MZT, the degradation kinetics of maternal transcripts exhibit striking diversity (Figure 3). The abundance of transcripts strictly targeted by maternal degradation RBPs, such as *nanos* (*nos*), starts declining linearly after egg activation. By contrast, targets of zygotic degradation effectors such as *bicoid* (*bcd*) are untouched during the first cycles and their removal begins after ZGA. Finally, transcripts targeted through both maternal and zygotic pathways include *Hsp83* and adopt highly specific degradation kinetics that reflect the contributions and levels of both maternal and zygotic factors.

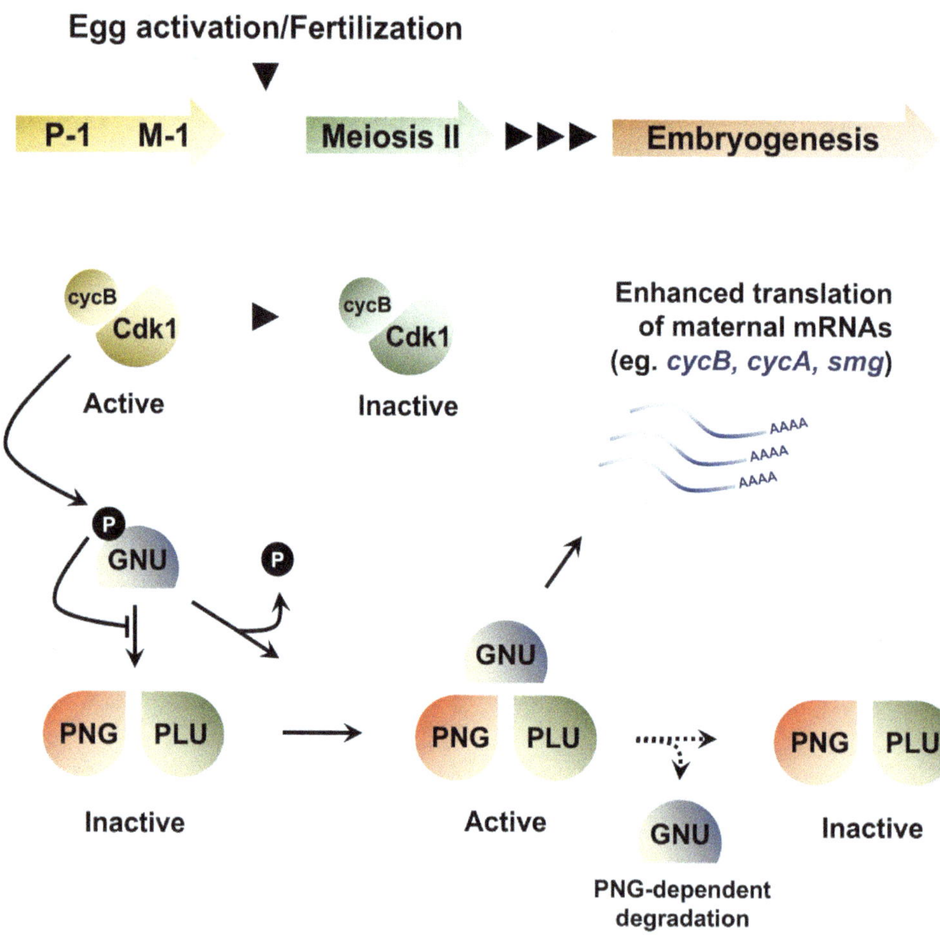

Figure 3. Developmental regulation of the PNG kinase coupled to cell cycle progression. In mature oocytes arrested at metaphase of meiosis I, cyclinB-CDK1 dependent phospohorylation of GNU exerts an inhibition of PNG complex assembly and activation. The completion of meiosis that follows egg activation results in CDK1 inactivation, prompting the dephosphorylation of GNU. In meiosis II, the accumulation of dephosphorylated GNU proteins leads to the spontaneous assembly of an active PNG kinase complex, consisting of GNU, PNG and PLU. The PNG kinase regulates the translation of hundreds of maternal mRNAs, including *cycB* and and *smg*. In addition, GNU protein degradation is promoted by PNG activity, enacting a negative feedback loop that restricts the activity of the complex to the temporal context of early embryonic development.

The execution of maternal clearance is tightly coupled to translational regulation of RBPs that mediate mRNA degradation. Indeed, the levels of SMG depend on the PAN GU (PNG) Ser/Thr kinase complex, a key translational regulator activated at the oocyte-to-embryo transition. The PNG complex is involved in reprogramming the proteome as the egg becomes an embryo by regulating the translation of hundreds of maternal mRNAs (Figure 3). Its notable targets include *cyclin B* (*cycB*), which encodes a key regulator of the embryonic cell cycle [41–43]. The complex consists of three proteins, PNG, GNU and PLU. Mutations in any of these genes leads to a profound decrease in Cyclin B protein levels, without affecting the corresponding mRNA levels. A recent study revealed a feedback loop involved in the regulation of PNG activity by the Cyclin B/CDK1 complex at the oocyte-to-embryo transition [44]. In mature oocytes, PNG activity is kept in check through inhibitory CyclinB/CDK1-dependent phosphorylation of the GNU activating subunit, preventing its association with PNG. Meiosis completion coincides with a decrease in CyclinB/CDK1 activity, leading to GNU dephosphorylation, which can then activate the PNG kinase, unleashing its widespread translational activation. However, sustained PNG kinase activity leads to a decrease in GNU levels, providing a mechanism to end PNG kinase function after egg activation and restrict its activity to the oocyte-to-embryo transition period (Figure 3).

In another study, ribosome profiling performed on mature oocytes and activated eggs derived from *png* mutant mothers revealed compromised translational efficiencies in nearly 1000 mRNAs [45]. Surprisingly, it was found that translational upregulation poorly reflects on protein levels in fertilized eggs, leading the authors to propose the existence of a "resetting" process in which enhanced translation is counterbalanced by proteasomal degradation, perhaps enabling the removal of proteins bearing oocyte-specific posttranslational modifications. Interestingly, this model is reminiscent of the well-characterized MZT phenomenon that takes place later in embryogenesis, as many maternal transcripts are degraded and subsequently expressed de novo from the zygotic genome [22].

Overall, spatiotemporal regulation enacted through mRNA localization and translation control are key features of embryonic transitions. In the absence of large-scale zygotic transcription, maternally deposited mRNAs and their regulated translation drive the developmental program of early embryogenesis. These maternal transcripts are then selectively removed, a process relying on the recognition of *cis*-acting motifs by maternal RBPs that are tightly regulated in time and space. As zygotic transcription emerges, a second phase of maternal clearance unfolds through the activity of zygotically expressed determinants (e.g., miRNAs and RBPs).

3. Models of Zygotic Genome Activation

As maternal clearance takes place during the cleavage cycles, the transcriptomic landscape is remodeled and new populations of transcripts arise upon activation of the zygotic genome. The mechanisms behind ZGA onset are poorly understood and remain an outstanding question in developmental biology, although several models have been proposed [7] (Figure 4). Seminal work involving the injection of plasmids in *Xenopus* embryos showed that early transcription is possible prior to ZGA [46,47]. However, the expression of plasmid DNA is rapidly silenced and resumes at the normal timing of transcription initiation. This observation suggests that the early zygotic genome is transcriptionally competent and actively repressed during the cleavage cycles. Moreover, ZGA is a gradual process and delineating its onset has remained challenging. Indeed, the emergence of the first zygotic transcripts has long been associated with the acquisition of the syncytial blastoderm morphology, at NC8. However, the detection of a small subset of zygotic genes before NC7, including the transcription factor *engrailed* (*en*), has recently been reported in preblastoderm embryos, with key implications in establishing the synchrony of early mitotic cycles [48]. Regardless of the exact onset of their expression, the first zygotic products display a conserved tendency to encode few exons and their length is shorter than maternally provided mRNAs [5]. In line with this observation, one hypothesis is that transcription is systematically attempted during the first cleavage cycles but that nascent transcripts are largely aborted, due to excessively fast cycling [7,15,46,49–51]. Hence, zygotic genes may be shorter than populations contributed maternally because Pol II can complete their transcription prior to an intervening mitotic entry [5,6].

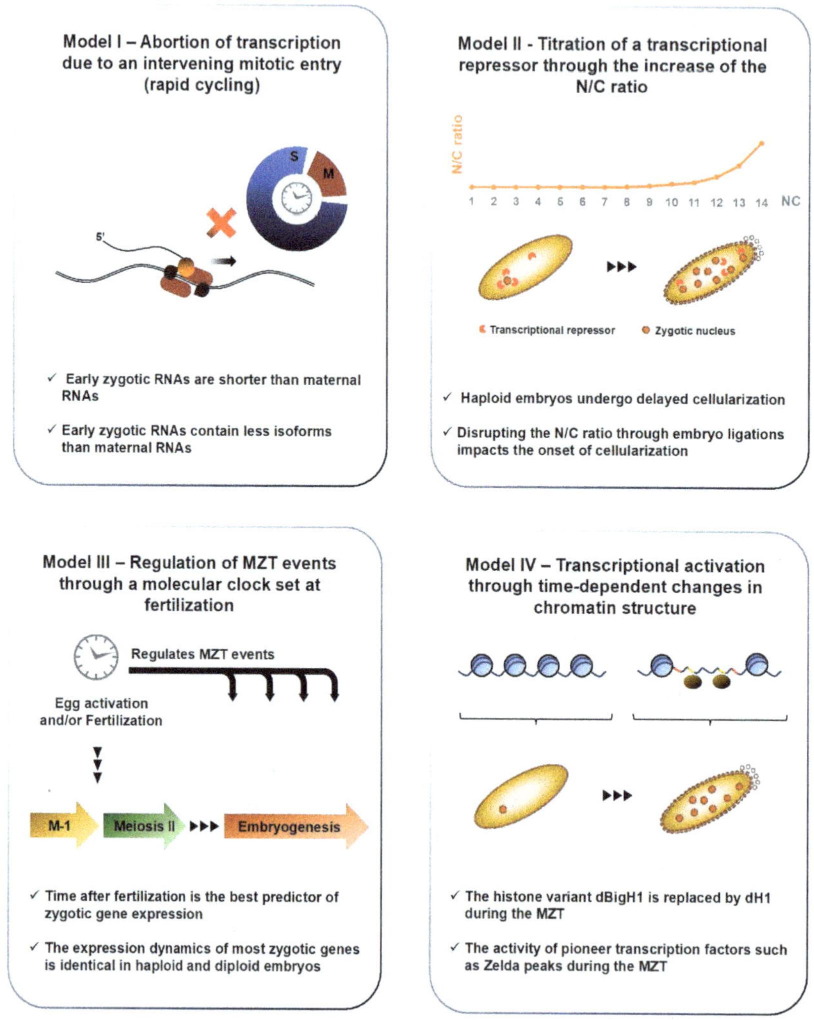

Figure 4. Models of zygotic genome activation. Several models have been proposed to contribute to ZGA. Each is supported by a set of empirical evidence (outlined under each cartoon) and these different propositions likely contribute synergistically to the emergence of zygotic transcription. Model I posits that early zygotic transcription is restricted due to the short duration of interphases during early embryogenesis, effectively preventing the complete transcription of long genes. Model II stipulates that zygotic transcription is prohibited during early embryogenesis due to the abundance of a maternally-inherited transcriptional repressor. The titration of this repressor against the increasing mass of zygotic nuclei would progressively lead to transcriptional competence. Model III proposes that egg activation sets in a molecular clock, which times key events of the MZT, including ZGA. Model IV postulates that chromatin is kept in a state that precludes transcription during early embryogenesis and is progressively remodeled through active changes in its composition to promote gene expression.

One long-standing model posits that ZGA is triggered once a critical value of nucleocytoplasmic (N/C) ratio is breached in the syncytium [2,4,15,52,53]. Indeed, the ratio of nuclei to cytoplasm increases rapidly during early embryogenesis, as the nuclear count expands exponentially and the volume of syncytial cytoplasm is kept constant due to growth inhibition. Tampering with this ratio by constricting embryos and compounding or reducing chromosome size impacts the number of syncytial divisions and the onset of cellularization [53]. It is thought that the increasing DNA mass titrates a maternal factor required to sustain fast-paced proliferation, until its concentration has decreased sufficiently to terminate the early cell cycle program. Similarly, a transcriptional repressor inherited maternally could be embedded in chromatin and diluted as the mass of zygotic DNA increases.

After a certain N/C threshold is breached, the concentration of this repressor would have sufficiently decreased to allow zygotic transcription to take place. One enduring rival of the N/C proposition has been called the "molecular clock" model; it proposes that egg activation sets a chronological countdown that times MZT events, including ZGA [2,15]. A study comparing the onset of transcription in haploid and diploid embryos, which differ markedly in N/C ratio, found little difference in the expression dynamics of most zygotic genes, suggesting that the molecular clock model is the best overall predictor of ZGA onset in *Drosophila*. Interestingly, the authors identified a small subset of genes regulated in a N/C-dependent manner, suggesting that both models prevail, albeit at distinct loci and frequencies [54]. Nevertheless, the molecular identity that senses the clock to activate zygotic transcription independently of the N/C ratio remains elusive.

Hence, diverse mechanisms have been proposed to account for ZGA onset, each of which is supported by empirical evidence [7]. How these different propositions converge to enact a precise execution of genome activation remains elusive. The key to ZGA regulation might lie in the reorganization of chromatin. Interestingly, a rapidly expanding literature is helping to define how chromatin states relate to transcription in early embryos (Figures 4 and 5).

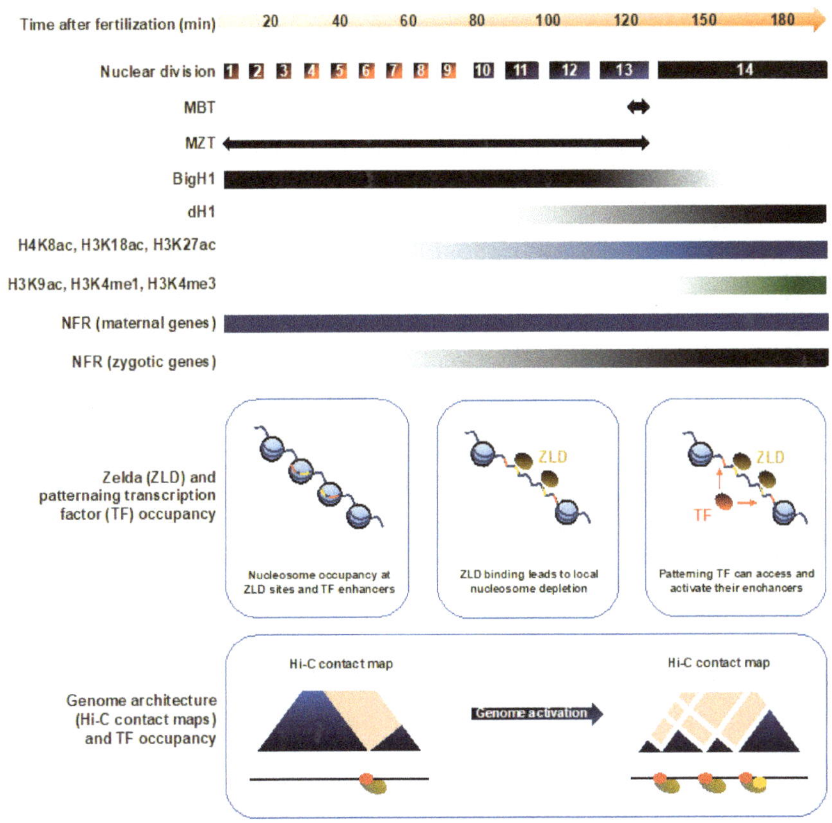

Figure 5. Developmental regulation of chromatin landscapes and genomic architecture. The histone H1 variant BigH1 is a constitutive chromatin component in the germline and in fertilized embryos until the MBT. The acetylation marks H4K8ac, H3K18ac and H3K27ac appear at the ZGA and scale up with the prevalence of zygotic transcription in syncytial embryos. By contrast, H3K9ac and H3K4me1/3 emerge around the MBT. Nucleosome-free regions (NFR) are found upstream of maternally-deposited genes throughout embryonic development but their appearance upstream of zygotic genes is concomitant with their transcription. Endowed with pioneer transcription factor activity, the binding of Zelda (ZLD) to its consensus sequence leads to local nucleosome depletion around NC10, exposing surrounding enhancers to promote the recruitment of patterning transcription factors by NC14. Hi-C data shows that chromatin is poorly organized prior to NC10. The emergence of intricate long-range interactions emerges after ZGA, by NC14.

4. Chromatin Rearrangements and Zygotic Genome Activation

Histone proteins are fundamental components of chromatin and developmentally-regulated changes in their expression could have profound impacts on genome activation. Indeed, most metazoans express tissue-specific variants of the linker histone H1 and the somatic H1 is often replaced by a developmental variant during early embryogenesis [55,56]. In a recent article, Pérez-Montero et al. identified the first H1 variant in *Drosophila*, called dBigH1, and demonstrate its involvement in ZGA regulation [57]. Ubiquitous in preblastoderm and syncytial blastoderm embryos, dBigH1 is progressively replaced by dH1 in somatic cells at the MBT, except in primordial germ cells (PGCs), in which it is retained well after gastrulation (Figure 5). *BigH1*^{00} mutants exhibit high embryonic lethality and a range of developmental defects, including altered nuclear distributions and highly asynchronous divisions. Interestingly, ChIP showed that Pol II is recruited to chromatin earlier in *BigH1*^{00} mutants than in *wt* embryos. In addition, zygotic mRNAs were more abundant in *BigH1*^{00} mutants than in *wt* embryos 2h after fertilization. Together, these results show that BigH1, a novel *Drosophila* variant of the linker histone, regulates ZGA onset and is removed from chromatin prior to the MBT.

The histone code refers to a set of posttranslational modifications that modulate chromatin compaction and the accessibility of DNA elements [58]. Changes in this epigenetic landscape likely contribute to the emergence of zygotic transcription. In a recent study, Li et al. investigated the genome-wide distribution of nine histone marks using staged embryo collections at NC8, NC12 and NC14. Prior to ZGA (NC8), chromatin exists in a relatively simple state, lacking histone methylation (me) and displaying low levels of histone acetylation (ac) or nucleosome free regions (NFRs), a hallmark of transcriptional activity [59]. The acetylation marks H4K8ac, H3K18ac and H3K27ac appeared along with transcription by NC12. By contrast, H3K9ac and the methylation marks H3K4me1, H3K4me3, H3K27me3 and H3K36me3 are only apparent after the MBT, at NC14 (Figure 5). As reported by previous studies, NFRs are prevalent upstream of maternally deposited genes even in absence of zygotic transcription, suggesting that nucleosome depletion is stable across development [60]. To identify putative blastoderm enhancers, Li et al. calculated the cumulative binding of 16 early transcription factors and examined the sites showing the highest cumulative occupation, excluding known promoters and coding regions. They found that putative enhancers display relatively high nucleosome density at NC8, with the appearance of acetylation marks by NC12 and H3K4me1 by early NC14, whereas the repressive mark H3K27me3 only spreads in surrounding regions by late NC14 (Figure 5).

The factors that act between NC8 and NC14 to deplete enhancer-associated nucleosomes represent major instigators of ZGA. Transcription factors (TFs) that can recognize their binding sites in a closed chromatin context to promote chromatin remodeling are termed "pioneer TFs". Several studies have identified such pioneer activity for the Zn-finger transcription factor Zelda/Vielfaltig (ZLD/VFL), a master regulator of early zygotic gene expression [61,62]. Indeed, ZLD is detected by NC2 in syncytial embryos and its binding displays a striking correlation with the timing and magnitude of early zygotic transcription [63]. ZLD has been shown to prime enhancers by lowering the nucleosome barrier sufficiently to promote the accession of specific binding motifs by their associated TFs [64]. These observations suggest that ZLD may act as a global genome activator in *Drosophila*, like Nanog, Pouf5f3 and SoxB1 in vertebrates [65,66]. Indeed, Li et al. found that nearly all the putative blastoderm enhancers identified through cumulative TF binding at NC14 are already bound by ZLD at NC8. Moreover, the ZLD consensus motif CAGGTAG was the single most enriched sequence associated to the early enhancer marks H3K27ac, H3K18ac and H4K8ac. Finally, H3K4me1 was lost and H3K18ac strongly compromised at ZLD-bound regions in embryos obtained from *zld*$^{-}$ germline clones. Overall, Li et al. show that histone marks are depleted during the first cleavage cycles and emerge between NC8 and NC14. Importantly, ZLD is a pioneer TF of the MZT: it can bind its genomic sites in condensed chromatin at NC8 and promote the recruitment of other factors that carry out profound chromatin remodeling at the MBT [67,68].

In addition to histone modifications, the three-dimensional folding of chromosomes can bring distant genomic loci in close physical proximity, with profound impacts on gene expression [69]. Topologically associated domains (TADs) are regions of high contact probabilities that display significant insulation from neighboring loci, enabling enhancer-promoter contacts and the coordination of gene expression programs [70,71]. Chromosome conformation capture (3C), its adaptations (4C and 5C) and the recent genome-wide variant Hi-C can reveal TADs with increasing resolution. They have been optimized in *Drosophila* embryos, enabling investigations into the developmental implications of genome architecture [72–75]. Recently, *Hug* et al. performed Hi-C at time-points surrounding the ZGA to determine when chromatin architecture is established during development and how its emergence relates to the onset of zygotic transcription [76]. NC8 embryos display poorly organized chromatin, exhibiting broadly uniform contact probabilities through large genomic distances. By contrast, NC13, NC14 and gastrula embryos revealed increasingly strong enrichments of chromatin associations within TADs and sharply declining contact frequencies with the loci surrounding TADs. This picture suggests that chromatin architecture is rapidly remodeled from an unordered state in preblastoderm embryos to a structured organization by NC14 (Figure 5). These boundaries are tightly maintained in later-stage embryos and in Kc167 cells, consistent with highly stable TAD boundaries described in other models (Dixon et al., 2012).

ChIP-seq revealed a strong dose-dependent correlation between Pol II occupancy and TAD boundary-like regions, especially at housekeeping genes and across developmental stages. Analysis of the early zygotic *Bsg25/Elba3* locus, which is switched off before gastrulation, showed that loss of Pol II occupancy at NC14 coincides with the loss of its boundary-like structure [77]. Together, these results suggest that Pol II binding contributes to chromatin conformation reorganization. To test the role of transcription, Hug et al. injected NC8 embryos with the Pol II inhibitors α-amanitin and triptolide before performing Hi-C to examine chromatin architecture at the MBT. Inter-TAD insulation was compromised at NC14, as well as the co-localization of housekeeping gene boundaries, although extensive long-distance contacts were still prevalent independent of transcription. ZLD occupancy showed striking correlations with TAD boundaries by NC12, hinting at a potential role in their establishment. To explore this hypothesis, Hug et al. performed in situ Hi-C on NC14 *zld*$^-$ embryos, which revealed a loss of insulation of TAD-boundaries at strong ZLD sites, especially at boundaries established in early cycles. Collectively, Hug et al. provide strong evidence that the establishment of long-range interactions broadly coincide with ZGA. Although transcription is not required for the emergence of chromatin conformation, loci transcribed early act as nucleation sites and contribute to the establishment of TAD boundaries. Similarly, ZLD binding contribute significantly to TAD boundary insulation, consistent with ZLD roles as global activator of the zygotic genome.

In brief, ZGA coincides with a profound reorganization of chromatin. Prior to NC8, chromatin exhibits a simple and disorganized state, with few histone modifications, NFRs or TADs. The germline-specific histone variant dBigH1 is embedded in chromatin, possibly contributing to its transcriptional silencing. Through pioneer TF activity, factors such as ZLD disrupt the nucleosome barrier between NC8 and NC12 to expose zygotic enhancers and promote the transcription of their target genes. Concomitantly, the activating histone marks H4K8ac, H3K18ac and H3K27ac appear and TADs emerge. By the MBT, dBigH1 has been replaced by histone H1, long-range interactions have gained complexity and stability and the histone marks H3K4me1, H3K4me3, H3K27me3 and H3K36me3 are established.

5. ZGA as a Driver of Embryonic Development

After its emergence, zygotic transcription becomes a major driver of embryonic development. Its contribution is twofold: zygotic products directly enact important functions, notably transcription factors that reshape the developmental program and miRNAs that contribute to maternal clearance. In addition, the process of transcription itself seems to mediate changes in the biology of the embryo. Indeed, active transcription can expose ssDNA and may cause replication stalling when facing a replication fork. These processes have recently been linked to the activation of the DNA replication checkpoint before the MBT. Indeed, studies taking advantage of mutants with impaired ZGA have revealed that transcription contributes to maternal clearance and determines the onset of cellularization and replication checkpoint activation.

Sung et al. characterized a fly model exhibiting a point mutation in the 3′ untranslated region (3′UTR) of the *RNPII215* gene, which encodes the large subunit of Pol II [78]. These mutants, termed X161 embryos, undergo premature zygotic transcription onset, providing an appealing model to investigate the complex relationships between zygotic transcription and other key events of the MBT (Figure 6). Interestingly, X161 embryos terminate the syncytial stage after the completion of 12 NC rather than 13 NC, suggesting that interfering with ZGA onset impacts the timing of cellularization. To confirm this observation, Sung et al. considered mutants for the master transcription factor Zelda (ZLD), which fail to transcribe a broad set of early zygotic genes. They found that X161 *zld* double mutants all undergo 13 syncytial NCs, like *wt* embryos. Since ZLD loss-of-function rescues the premature transcription phenotype of X161 mutants, a normal number of syncytial cycles in X161 *zld* double mutants suggests that early transcription causes the precocious cellularization phenotype of X161 single mutants. In addition, Sung et al. used the X161 model to test the contribution of the nucleocytoplasmic ratio on cellularization. Haploid X161 embryos, which present a lowered N/C ratio, underwent only 12 NC, the same number as diploid X161 mutants. This result suggests that the N/C ratio acts independently of ZGA and does not directly regulate the onset of cellularization in *Drosophila*. Together, these experiments provide strong evidence that the onset of zygotic transcription times key events of early embryonic development.

As discussed earlier, maternal clearance is a complex process relying on factors contributed maternally and on the expression of zygotic products. Sung et al. surveyed the levels of three canonical targets of maternal clearance, *string*, *twine* and *smaug* in X161 mutants to monitor the impact of ZGA onset on maternal clearance. They found that the degradation of these maternal transcripts, which starts during the 14th interphase in *wt* embryos, is already well advanced by NC13 in X161 embryos, in agreement with reports of a zygotic contribution to maternal clearance. The authors also found that premature ZGA leads to a precocious requirement for a functional replication checkpoint. Checkpoint activation pauses M phase entry until the completion of DNA replication to safeguard genome integrity. Its emergence is a gradual process completed at the MBT. The Ser/Thr kinase Chk1, encoded by the *Drosophila* gene *grapes* (*grp*), is a key component of the DNA damage response (DDR), signal transduction cascades that sense DNA lesions to halt mitotic entry [17,79]. Chk1 activity is required for progression through the MBT and its loss leads to genomic instability exemplified by chromatin defects and embryonic lethality. Chk1 activity is dispensable prior to NC13, but necessary around the MBT, when its loss leads to the apparition of genotoxic lesions. Therefore, the requirement for Chk1 activity can be used as a proxy to score the onset of checkpoint activation. Sung et al. found that X161 *grp* double mutants display nuclear envelope and chromatin condensation defects by the 13th interphase, one cycle earlier than *grp* single mutants, supporting a role of transcription in the onset of the DNA replication checkpoint activation.

Interestingly, the nuclear retention of zygotic transcripts has been identified as a new facet of the DDR during early embryogenesis. Indeed, Iampietro et al. showed that syncytial-stage embryos challenged with genotoxic stress undergo extensive nuclear fallout at the MBT, a mechanism of programmed elimination [18]. The authors showed that fallout nuclei display widespread nuclear retention of diverse zygotic transcripts, including histone mRNAs. The nuclear retention of histone mRNAs is linked to a Chk2-mediated phosphorylation of the stem loop binding protein (SLBP), which orchestrates the posttranscriptional processing and nuclear export of histone mRNAs. In turn, the nuclear retention of essential mRNAs such as histones leads to a local depletion of their corresponding proteins in the vicinity of damaged nuclei, promoting their fallout and elimination from the somatic pool. Prior to the establishment of a robust DNA replication checkpoint, the propensity of syncytial embryos to the accumulation of DNA lesion is thus mitigated through a Chk2-mediated nuclear fallout process that relies on the nuclear retention of essential mRNAs. These results reveal a novel role of posttranscriptional transport routes in ensuring genome integrity surveillance during embryogenesis.

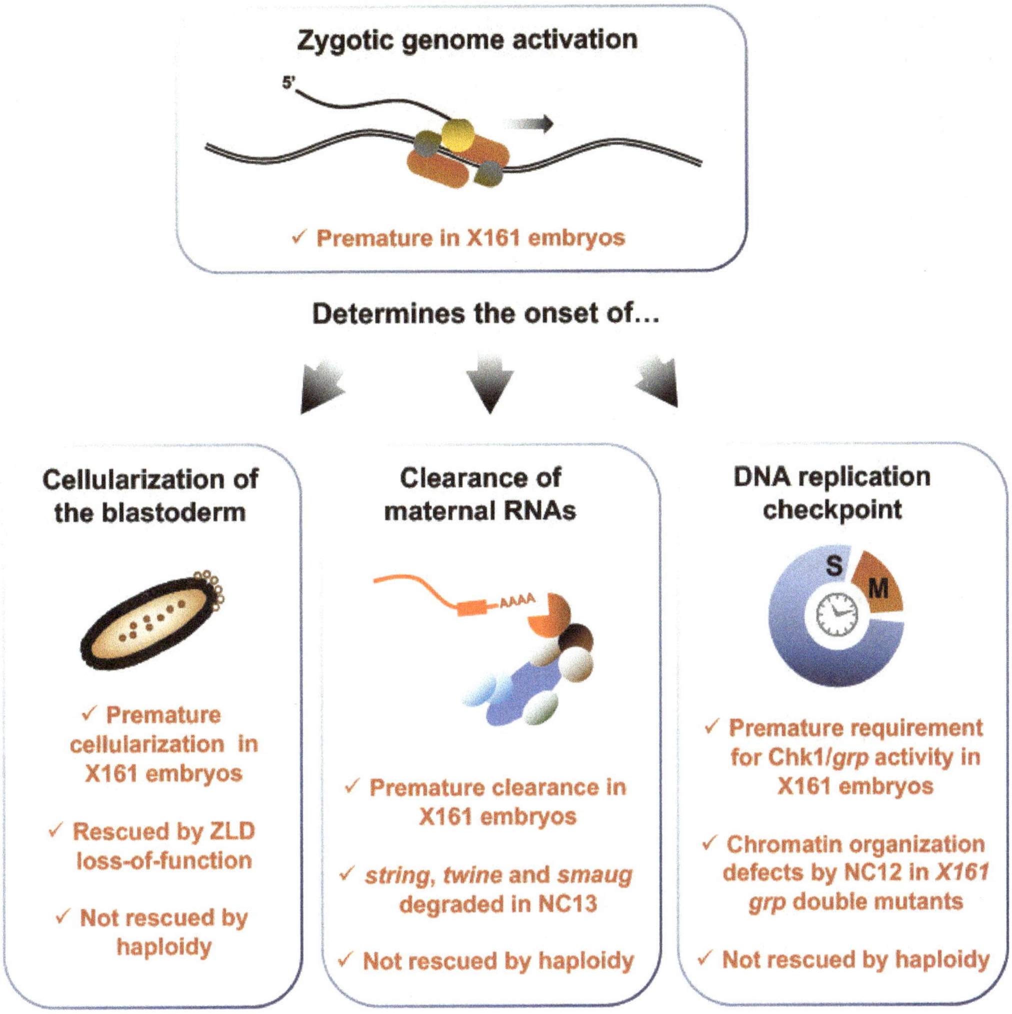

Figure 6. Zygotic genome activation times the onset of the DNA replication checkpoint, maternal clearance and cellularization. X161 mutants display premature zygotic transcription due to a point mutation in the *RNPII215* gene, which encodes a subunit of Pol II. This disruption leads to premature cellularization, which is rescued by altering transcription in X161 *zld* double mutants. Precocious transcription also leads to an early deployment of maternal clearance, exemplified by the premature degradation of the maternal transcripts *string*, *twine* and *smaug*. In addition, early ZGA leads to a premature activation of the replication checkpoint, as inferred from a precipitate requirement for Chk1/grp activity. None of these phenotypes are rescued in haploid X161 embryos, which exhibit a decreased N/C ratio, meaning that the N/C ratio does not act upstream of transcription activation in the regulation of cellularization, maternal clearance and checkpoint activation.

Blythe and Wieschaus provided further evidence of the interplay between zygotic transcription and replication checkpoint activation. These authors found that checkpoint activation correlates with the amount of DNA engaged by Pol II, independently of the N/C ratio [80,81]. Through ChIP-seq analyses, then found that Pol II distributions are not severely impaired in *grp* mutants, with widespread genomic occupancy at NC12, NC13 and NC14. This result suggests that the transcriptional machinery is in place independently of the functionality of the replication checkpoint. To investigate the links between checkpoint activation and ZGA at the molecular level, the authors took advantage of RPA70, an important effector of the DDR. RPA70 binds stress-induced ssDNA produced when replication is stalled, leading to ATR (*mei-41*) recruitment and checkpoint activation. Assessment of RPA70 occupancy through fluorescent microscopy and ChIP-seq revealed a strong correlation with Pol II binding sites, consistent with the hypothesis that Pol II engagement activates the checkpoint at the MBT. In *zld*⁻ mutants, ChIP-seq analyses revealed altered RPA70 occupancy at *zld*-dependent promoters. This result suggests that transcription contributes to checkpoint activation.

To test this hypothesis, the authors attempted to rescue the lethality phenotype associated to mutations in the DDR factor ATR by altering ZGA through different approaches. They showed that most embryos from double *zld mei-41* (ATR) mutants complete cleavage cycles and that many escape the mitotic catastrophe that characterizes *mei-41* mutants. In addition to the *zld*⁻ model, they used a heterozygous deficiency in the transcriptional activator Trithorax-like/GAGA (*Trl*), associated to defects in the genomic recruitment of poised Pol II, to interfere with ZGA. They found that *mei-41; Df(3L)ED4545/+ (Df(trl)/+)* embryos complete cleavage cycles without a mitotic catastrophe after a slightly lengthened NC13 and eventually yield hatching larvae. In addition, heterozygosity of the *cyclin B* gene (*Df(cycB)/+*), which lengthens NC13 time, effectively suppressed the mitotic catastrophe of *mei-41* mutants. Together, these rescue experiments show that reducing the source of replication stalling by interfering with transcription (in *zld* and *Trl* mutants) and providing more time to allow DNA replication (in *cycB* mutants) can bypass the MBT requirement for a functional replication checkpoint. In conjunction with evidence of RPA70 colocalization with the transcriptional machinery, these results strongly suggest that the replication checkpoint is activated in response to ZGA.

In brief, recent studies have used loss-of-function analyses to reveal the contributions of ZGA to key facets of embryonic development. Mutants exhibiting a premature ZGA undergo early cellularization, promptly enact maternal clearance and acquire a precocious requirement for effectors of the DDR. These effects are independent of the N/C ratio and can be rescued by modulating zygotic transcription. Moreover, effectors of the DDR are recruited to chromatin at Pol II occupied loci after ZGA and interfering with the scope of zygotic transcription can bypass the requirement for a functional checkpoint. In addition, DNA damage elicits a Chk2-dependent clearance of damaged nuclei in syncytial embryos through the nuclear retention of essential mRNAs, providing an elegant mechanism to safeguard genome integrity prior to the establishment of a robust DNA replication checkpoint.

6. Emerging Properties of Zygotic Transcription

While zygotic transcription drives key events of embryonic development, the properties and dynamics of the emerging transcriptional process itself have been challenging to study. The advent of approaches enabling RNA labeling in vivo and in real time has provided a clearer picture of zygotic transcription. Originally developed by Singer and colleagues in yeast, the MS2 system takes advantage of the strong affinity of bacteriophage coat proteins (e.g., MS2, PP7) for specific RNA stem-loops [82]. For imaging purposes, MS2 phage coat protein fused to a fluorescent reporter (e.g., GFP, mCherry) are co-expressed in a transgenic organism along with an RNA fusion that encompasses the target transcript and MS2 stem-loop repeats (Figure 7). Stable tethering of the coat fusion protein allows for durable tracking of the target RNA, which can be expressed in its endogenous regulatory context to recapitulate physiological properties [83]. Over the last decade, several groups have harnessed the power of the MS2 imaging system to study RNA dynamics and localization during *Drosophila* development [84–89] This system has notably been used to investigate the dynamics of transcriptional bursting, calculate Pol II elongation rate at the MBT and monitor post-mitotic transcriptional reactivation.

Quantitative RNA detection methods suggest that transcriptional bursting is a key property of gene expression in diverse systems [90,91]. The term "bursting" refers to the episodic, discontinuous emergence of nascent transcripts at Pol II-bound loci. To investigate the links between enhancer control and transcriptional bursting at the MBT, *Fukaya* et al. placed well characterized enhancers upstream and downstream of reporter genes flanked by *MS2* and *PP7* stem loops [92]. They performed live-embryo imaging of the *MS2-yellow* reporter containing different *snail* (*sna*) enhancers of varying strength, along with its proximal promoter in different configurations. They observed major differences in bursting frequencies produced by the *sna* primary enhancer and by its shadow enhancer, a redundant regulatory sequence, which were correlated to the discrepancy in total RNA outputs. This analysis was extended to the *rhomboid* (*rho*), *Krüppel*; (*Kr*) and *Abdominal-B* (*Abd-B*) enhancers. By testing a set of conditions, Fukaya et al. showed that differential core promoter, distal enhancer and anteroposterior gradient positioning all affect bursting frequency, in line with discrepancies in total RNA outputs. Thus, Fukaya et al. identified the regulation of transcriptional bursting frequency as a key determinant of developmental gene activity at the MBT.

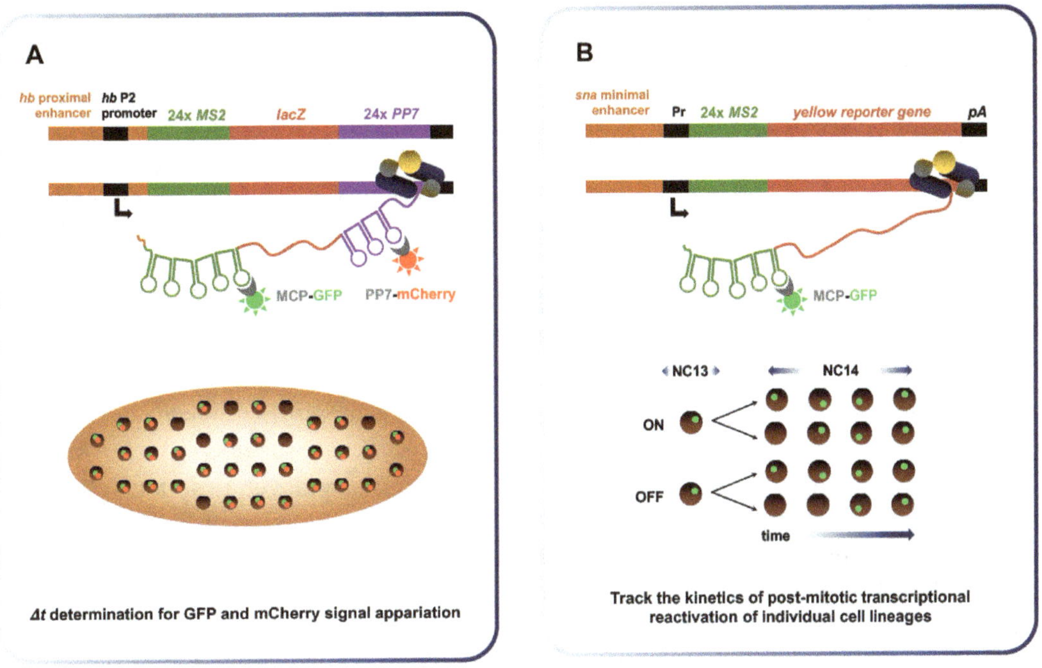

Figure 7. Real-time imaging of zygotic transcription enables the determination of Pol II elongation rate and demonstrates transcriptional memory at NC14. (**A**) Fukaya et al. used a dual fluorescence approach involving the MS2 system to measure Pol II elongation rates at NC13. They integrated a construct encompassing a *lacZ* reporter flanked by 24 *MS2* repeats at its 5′ end and 24 *PP7* repeats at the 3′ end. Its expression in conjunction with the MCP-GFP and mCherry-PP7 coat proteins leads to the emission of dual green and red fluorescence. By measuring the delay between the emission of the green and red signals at the single molecule level, the time required to transcribe the intervening *lacZ* sequence can be determined. This value is then used to calculate the elongation rate of Pol II. (**B**) Ferraro et al. provided evidence of transcriptional memory by monitoring post-mitotic reactivation. Sensitized transgenes were used to obtain sporadic expression of the *yellow* reporter downstream of 24 *MS2* repeats. They tracked expression during NC13 and NC14 in a lineage-specific manner. After mitosis, the authors found that the daughter of nuclei having expressed the reporter during NC13 where four times more likely to re-express it rapidly. This result indicates that transcription prior to mitosis increases the chance and rapidity of re-expression across cell generations, a phenomenon termed transcriptional memory.

Puzzling disparities have long prevailed between reported rates of Pol II elongation (1.1–1.5 kb/min) and robust detection of several long de novo transcripts before the MBT [14,93]. Indeed, established elongation rates cannot account for the zygotic transcription of the 22 kb-long unit of Short gastrulation (*sog*) in NC13, when the time window permissive to transcription is narrowly restricted by a hasty interphase (10–12 min). Fukaya et al. solved this long-lasting paradox by revisiting Pol II elongation rates in early embryogenesis using dual-fluorescence through the MS2 imaging system [94] (Figure 7A). They measured an elongation rate of 2.4 kb/min, nearly twice that of previous estimates.

This figure is compatible with endogenous *sog* transcription during the 13th interphase. In addition, they found that replacing the promoter or introducing a reporter containing an intron had little impact on elongation rate measurements, suggesting that elongation is not the rate-limiting step in transcription.

The inheritance of transcriptional states from mother to daughter cells, termed transcriptional memory, has been documented in the amoeba *Dictyostelium* [95,96]. In a recent study, Ferraro et al. monitored post-mitotic transcriptional reactivation of stochastically expressed transgenes using the MS2 imaging system [97]. This work provided the first evidence that transcriptional memory prevails at the massive wave of zygotic expression between the 13th and the 14th division. The authors used sensitized transgenes exhibiting patterns of sporadic expression to individually image the behavior of single cell lineages (Figure 7B). Daughter cells derived from nuclei that expressed the transgene during NC13, called memory mothers, were four times more likely to show early reactivation during NC14 interphase than daughters arising from non-memory mothers. Quantitative analyses of average fluorescence intensities revealed that memory nuclei produce, on average, two-fold more total mRNA than non-memory nuclei during NC14. These results provide strong evidence that transcriptional memory prevails during *Drosophila* MBT and impacts total RNA output, likely through modifications incurred at the level of nucleosomes, bound TFs or histone modifications following a first round of transcription. Ferraro et al. envision this emerging property of early transcription as a mechanism of developmental homeostasis, which could help ensure that cells retain the properties of their progenitors.

In brief, the recent deployment of in vivo imaging to document the transcriptional process in real time has revealed new insights into the dynamics of zygotic transcription. In vivo imaging has established the notion of transcriptional bursting, and shown that total gene-specific outputs at ZGA reflect the frequency of transcriptional bursts. It has enabled the revision of Pol II elongation rate at the MBT, reconciling the expression of lengthy genes such as *sog* with a short NC13 interphase. Quantitative imaging also showed that transcriptional memory prevails in *Drosophila* embryogenesis, promoting the rapid post-mitotic re-activation of sequences expressed during NC13, which likely contributes to developmental homeostasis.

7. Conclusions

In this review, we aimed to provide an overview of recent findings relevant to the transcriptome dynamics of early *Drosophila* development. Maternal control is essential in syncytial-stage embryos to sustain fast-paced proliferation in absence of a sizeable transcriptional output. Thus, we discussed the processes of RNA maternal deposition, localization and targeted clearance. Indeed, *Drosophila* embryos host a wealth of complex posttranscriptional regulatory processes. The spatiotemporal dosage of RBPs such as SMG reflect translational fine-tuning which, in turn, modulates the dynamics of hundreds of maternal mRNAs. RNA localization is highly prevalent in the large syncytial embryo and likely plays key roles in orchestrating the developmental program. Indeed, protein-coding transcripts adopt a large diversity of spatial distributions in early embryos, including subembryonic and exclusionary patterns, asymmetric anteroposterior localization and more resolved patterns such as membrane, microtubule or mitotic apparatus associations [24]. In many cases, the functional relevance of these mRNA localization events remains untapped, and awaits further characterization. Indeed, the fly model represents a powerful system to further dissect the *cis*- and *trans*-determinants regulating RNA localization events.

Maternal control is gradually met with an increasing contribution of the zygotic genome, as it progressively acquires transcriptional competence. The mechanisms accounting for ZGA onset are multifaceted and their underpinnings remain unclear. As the N/C ratio increases, a maternal factor responsible for transcriptional quiescence could be diluted against the mass of DNA, loosening the efficiency of the repression. Studies in *Xenopus* have identified maternal histone as putative transcriptional repressors [98]. In line with this finding, a recently identified linker H1 variant, dBigH1, has been associated to transcriptional quiescence. Indeed, dBigH1 is cleared out prior to the MBT and *big1* mutants show signs of disorganized chromatin and early zygotic transcription. Nevertheless, we know that the repressor titration model is not sufficient to account for the transcriptional silence of the zygotic genome in *Drosophila*. Indeed, most zygotic transcripts display similar expression kinetics in haploid and diploid embryos, which present very different N/C ratios [54]. This observation points to the molecular clock model, which proposes that egg activation sets a timer in motion to eventually trigger ZGA. One possible interpretation is that maternally contributed transcripts encoding pioneer TFs such as ZLD require time to be translated and accumulate sufficiently before pioneer activity has reached a level amenable to widespread ZGA. Together, a molecular clock set at egg activation and the rapid increase in nucleocytoplasmic ratio likely converge to exert changes in the structure of zygotic chromatin. Specific loci may exhibit enhanced sensitivity to pioneer TF activity. In addition, titration of the maternal repressor may not occur at a homogeneous rate across the genome. Such effects could account for the gradual nature of ZGA and explain why only a subset of zygotic genes display N/C-dependent expression dynamics.

Independently of its underlying mechanisms, once it has been triggered, zygotic transcription contributes to shaping a complex genome topology before the MBT, which largely remains in place throughout the life of the fly. After NC10, the growing population of transcripts produced by cortical nuclei play key roles in driving the course of development through the MBT. The transcriptional process itself exposes ssDNA and triggers the activation of the DNA replication checkpoint, possibly through the formation of stalled replication forks. Cellularization onset is linked to ZGA timing and it coincides with the loss of mitotic synchrony and the expression of additional zygotic genes. Real-time imaging has revealed that zygotic transcription proceeds as bursts, with the frequency of bursting events linked to its total RNA output. An example of transcriptional memory in *Drosophila*, the preferential post-mitotic reactivation of loci transcribed at NC13 has been demonstrated through this approach. In addition, embryonic Pol II elongation rates have been revised via the development of a dual fluorescence system. The deployment of real-time, in vivo RNA labeling to study transcription in *Drosophila* is still very recent. Future applications will likely contribute to clarify how ZGA is triggered. Indeed, important insights could be revealed by monitoring transcription in mutants of specific chromatin components, such as *bigH1*, or pioneer TFs, such as *zld*. This technology can notably reveal transcriptional dynamics at the single cell level, a sizeable advantage when investigating heterogeneous and multifaceted responses such as ZGA.

Acknowledgments: This work was supported by grants from the Canadian Institutes of Health Research (CIHR; MOP-111161) and Natural Sciences and Engineering Research Council of Canada (NSERC; Discovery Grant #386644), as well as a Junior 2 research scholar of the Fonds de Recherche du Québec-Santé (FRQS) to E.L., F.A.L. is funded by a Frederick Banting and Charles Best Canada Graduate Scholarship from CIHR.

References

1. Von Stetina, J.R.; Orr-Weaver, T.L. Developmental control of oocyte maturation and egg activation in metazoan models. *Cold Spring Harb. Perspect. Biol.* **2011**, *3*, a005553. [CrossRef] [PubMed]
2. Farrell, J.A.; O'Farrell, P.H. From egg to gastrula: how the cell cycle is remodeled during the Drosophila mid-blastula transition. *Annu. Rev. Genet.* **2014**, *48*, 269–294. [CrossRef] [PubMed]
3. Shermoen, A.W.; McCleland, M.L.; O'Farrell, P.H. Developmental control of late replication and S phase length. *Curr. Biol.* **2010**, *20*, 2067–2077. [CrossRef] [PubMed]
4. Pritchard, D.K.; Schubiger, G. Activation of transcription in Drosophila embryos is a gradual process mediated by the nucleocytoplasmic ratio. *Genes Dev.* **1996**, *10*, 1131–1142. [CrossRef] [PubMed]
5. Heyn, P.; Kircher, M.; Dahl, A.; Kelso, J.; Tomancak, P.; Kalinka, A.T.; Neugebauer, K.M. The earliest transcribed zygotic genes are short, newly evolved, and different across species. *Cell Rep.* **2014**, *6*, 285–292. [CrossRef] [PubMed]
6. Heyn, P.; Neugebauer, K.M. Purification of Zygotically Transcribed RNA through Metabolic Labeling of Early Zebrafish Embryos. *Methods Mol. Biol.* **2017**, *1605*, 121–131. [PubMed]

7. Palfy, M.; Joseph, S.R.; Vastenhouw, N.L. The timing of zygotic genome activation. *Curr. Opin. Genet. Dev.* **2017**, *43*, 53–60. [CrossRef] [PubMed]
8. Bashirullah, A.; Halsell, S.R.; Cooperstock, R.L.; Kloc, M.; Karaiskakis, A.; Fisher, W.W.; Fu, W.; Hamilton, J.K.; Etkin, L.D.; Lipshitz, H.D. Joint action of two RNA degradation pathways controls the timing of maternal transcript elimination at the midblastula transition in Drosophila melanogaster. *EMBO J.* **1999**, *18*, 2610–2620. [CrossRef] [PubMed]
9. Tadros, W.; Lipshitz, H.D. The maternal-to-zygotic transition: a play in two acts. *Development* **2009**, *136*, 3033–3042. [CrossRef] [PubMed]
10. Riparbelli, M.G.; Gottardo, M.; Callaini, G. Parthenogenesis in Insects: The Centriole Renaissance. *Results Probl. Cell Differ.* **2017**, *63*, 435–479. [PubMed]
11. Van der Kooi, C.J.; Schwander, T. Parthenogenesis: Birth of a new lineage or reproductive accident? *Curr. Biol.* **2015**, *25*, R659–R661. [CrossRef] [PubMed]
12. Rabinowitz, M. Studies on the Cytology and Early Embryology of the Egg of Drosophila Melanogaster. *J. Morphol.* **1941**. [CrossRef]
13. Deneke, V.E.; Melbinger, A.; Vergassola, M.; Di Talia, S. Waves of Cdk1 Activity in S Phase Synchronize the Cell Cycle in Drosophila Embryos. *Dev. Cell* **2016**, *38*, 399–412. [CrossRef] [PubMed]
14. Shermoen, A.W.; O'Farrell, P.H. Progression of the cell cycle through mitosis leads to abortion of nascent transcripts. *Cell* **1991**, *67*, 303–310. [CrossRef]
15. Langley, A.R.; Smith, J.C.; Stemple, D.L.; Harvey, S.A. New insights into the maternal to zygotic transition. *Development* **2014**, *141*, 3834–3841. [CrossRef] [PubMed]
16. Fogarty, P.; Campbell, S.D.; Abu-Shumays, R.; Phalle, B.S.; Yu, K.R.; Uy, G.L.; Goldberg, M.L.; Sullivan, W. The Drosophila grapes gene is related to checkpoint gene chk1/rad27 and is required for late syncytial division fidelity. *Curr. Biol.* **1997**, *7*, 418–426. [CrossRef]
17. Sibon, O.C.; Stevenson, V.A.; Theurkauf, W.E. DNA-replication checkpoint control at the Drosophila midblastula transition. *Nature* **1997**, *388*, 93–97. [CrossRef] [PubMed]
18. Iampietro, C.; Bergalet, J.; Wang, X.; Cody, N.A.; Chin, A.; Lefebvre, F.A.; Douziech, M.; Krause, H.M.; Lecuyer, E. Developmentally regulated elimination of damaged nuclei involves a Chk2-dependent mechanism of mRNA nuclear retention. *Dev. Cell* **2014**, *29*, 468–481. [CrossRef] [PubMed]
19. O'Farrell, P.H.; Stumpff, J.; Su, T.T. Embryonic cleavage cycles: How is a mouse like a fly? *Curr. Biol.* **2004**, *14*, R35–R45. [CrossRef] [PubMed]
20. Stack, J.H.; Newport, J.W. Developmentally regulated activation of apoptosis early in Xenopus gastrulation results in cyclin A degradation during interphase of the cell cycle. *Development* **1997**, *124*, 3185–3195. [PubMed]
21. Marlow, F.L. *Maternal Control of Development in Vertebrates: My Mother Made Me Do It!* Morgan & Claypool Life Sciences: San Rafael, CA, USA, 2010.
22. Tadros, W.; Lipshitz, H.D. Setting the stage for development: mRNA translation and stability during oocyte maturation and egg activation in Drosophila. *Dev. Dyn.* **2005**, *232*, 593–608. [CrossRef] [PubMed]
23. Ding, D.; Lipshitz, H.D. A molecular screen for polar-localised maternal RNAs in the early embryo of Drosophila. *Zygote* **1993**, *1*, 257–271. [CrossRef] [PubMed]
24. Lecuyer, E.; Yoshida, H.; Parthasarathy, N.; Alm, C.; Babak, T.; Cerovina, T.; Hughes, T.R.; Tomancak, P.; Krause, H.M. Global analysis of mRNA localization reveals a prominent role in organizing cellular architecture and function. *Cell* **2007**, *131*, 174–187. [CrossRef] [PubMed]
25. Lefebvre, F.A.; Benoit Bouvrette, L.P.; Bergalet, J.; Lecuyer, E. Biochemical Fractionation of Time-Resolved Drosophila Embryos Reveals Similar Transcriptomic Alterations in Replication Checkpoint and Histone mRNA Processing Mutants. *J. Mol. Biol.* **2017**, *21*, 3264–3279. [CrossRef] [PubMed]
26. Lefebvre, F.A.; Bouvrette, L.P.B.; Bergalet, J.; Lecuyer, E. Data for the generation of RNA spatiotemporal distributions and interpretation of Chk1 and SLBP protein depletion phenotypes during Drosophila embryogenesis. *Data Brief* **2017**, *13*, 28–31. [CrossRef] [PubMed]
27. Johnstone, O.; Lasko, P. Translational regulation and RNA localization in Drosophila oocytes and embryos. *Annu. Rev. Genet.* **2001**, *35*, 365–406. [CrossRef] [PubMed]
28. Martin, K.C.; Ephrussi, A. mRNA localization: Gene expression in the spatial dimension. *Cell* **2009**, *136*, 719–730. [CrossRef] [PubMed]

29. Roth, S.; Lynch, J.A. Symmetry breaking during Drosophila oogenesis. *Cold Spring Harb. Perspect. Biol.* **2009**, *1*, a001891. [CrossRef] [PubMed]
30. Jambor, H.; Surendranath, V.; Kalinka, A.T.; Mejstrik, P.; Saalfeld, S.; Tomancak, P. Systematic imaging reveals features and changing localization of mRNAs in Drosophila development. *eLife* **2015**, *4*. [CrossRef] [PubMed]
31. De Renzis, S.; Elemento, O.; Tavazoie, S.; Wieschaus, E.F. Unmasking Activation of the Zygotic Genome Using Chromosomal Deletions in the Drosophila Embryo. *PLoS Biol.* **2007**, *5*, e117. [CrossRef]
32. Semotok, J.L.; Cooperstock, R.L.; Pinder, B.D.; Vari, H.K.; Lipshitz, H.D.; Smibert, C.A. Smaug recruits the CCR4/POP2/NOT deadenylase complex to trigger maternal transcript localization in the early Drosophila embryo. *Curr. Biol.* **2005**, *15*, 284–294. [CrossRef] [PubMed]
33. Benoit, B.; He, C.H.; Zhang, F.; Votruba, S.M.; Tadros, W.; Westwood, J.T.; Smibert, C.A.; Lipshitz, H.D.; Theurkauf, W.E. An essential role for the RNA-binding protein Smaug during the Drosophila maternal-to-zygotic transition. *Development* **2009**, *136*, 923–932. [CrossRef] [PubMed]
34. Semotok, J.L.; Luo, H.; Cooperstock, R.L.; Karaiskakis, A.; Vari, H.K.; Smibert, C.A.; Lipshitz, H.D. Drosophila maternal Hsp83 mRNA destabilization is directed by multiple SMAUG recognition elements in the open reading frame. *Mol. Cell. Biol.* **2008**, *28*, 6757–6772. [CrossRef] [PubMed]
35. Laver, J.D.; Li, X.; Ray, D.; Cook, K.B.; Hahn, N.A.; Nabeel-Shah, S.; Kekis, M.; Luo, H.; Marsolais, A.J.; Fung, K.Y.; et al. Brain tumor is a sequence-specific RNA-binding protein that directs maternal mRNA clearance during the Drosophila maternal-to-zygotic transition. *Genome Biol.* **2015**, *16*, 94. [CrossRef] [PubMed]
36. Wang, M.; Ly, M.; Lugowski, A.; Laver, J.D.; Lipshitz, H.D.; Smibert, C.A.; Rissland, O.S. ME31B globally represses maternal mRNAs by two distinct mechanisms during the Drosophila maternal-to-zygotic transition. *eLife* **2017**, *6*. [CrossRef] [PubMed]
37. Giraldez, A.J. microRNAs, the cell's Nepenthe: clearing the past during the maternal-to-zygotic transition and cellular reprogramming. *Curr. Opin. Genet. Dev.* **2010**, *20*, 369–375. [CrossRef] [PubMed]
38. Giraldez, A.J.; Mishima, Y.; Rihel, J.; Grocock, R.J.; Van Dongen, S.; Inoue, K.; Enright, A.J.; Schier, A.F. Zebrafish MiR-430 promotes deadenylation and clearance of maternal mRNAs. *Science* **2006**, *312*, 75–79. [CrossRef] [PubMed]
39. Luo, H.; Li, X.; Claycomb, J.M.; Lipshitz, H.D. The Smaug RNA-Binding Protein Is Essential for microRNA Synthesis During the Drosophila Maternal-to-zygotic Transition. *G3 (Bethesda)* **2016**, *6*, 3541–3551. [CrossRef] [PubMed]
40. Bushati, N.; Stark, A.; Brennecke, J.; Cohen, S.M. Temporal reciprocity of miRNAs and their targets during the maternal-to-zygotic transition in Drosophila. *Curr. Biol.* **2008**, *18*, 501–506. [CrossRef] [PubMed]
41. Tadros, W.; Goldman, A.L.; Babak, T.; Menzies, F.; Vardy, L.; Orr-Weaver, T.; Hughes, T.R.; Westwood, J.T.; Smibert, C.A.; Lipshitz, H.D. SMAUG is a major regulator of maternal mRNA destabilization in Drosophila and its translation is activated by the PAN GU kinase. *Dev. Cell* **2007**, *12*, 143–155. [CrossRef] [PubMed]
42. Tay, J.; Hodgman, R.; Richter, J.D. The control of cyclin B1 mRNA translation during mouse oocyte maturation. *Dev. Biol.* **2000**, *221*, 1–9. [CrossRef] [PubMed]
43. Vardy, L.; Orr-Weaver, T.L. The Drosophila PNG kinase complex regulates the translation of cyclin B. *Dev. Cell* **2007**, *12*, 157–166. [CrossRef] [PubMed]
44. Hara, M.; Petrova, B.; Orr-Weaver, T.L. Control of PNG kinase, a key regulator of mRNA translation, is coupled to meiosis completion at egg activation. *eLife* **2017**, *6*. [CrossRef] [PubMed]
45. Kronja, I.; Yuan, B.; Eichhorn, S.W.; Dzeyk, K.; Krijgsveld, J.; Bartel, D.P.; Orr-Weaver, T.L. Widespread changes in the posttranscriptional landscape at the Drosophila oocyte-to-embryo transition. *Cell Rep.* **2014**, *7*, 1495–1508. [CrossRef] [PubMed]
46. Almouzni, G.; Wolffe, A.P. Constraints on transcriptional activator function contribute to transcriptional quiescence during early Xenopus embryogenesis. *EMBO J.* **1995**, *14*, 1752–1765. [PubMed]
47. Gilbert, S.F. *Developmental Biology*, 10th ed.; Sinauer Associates: Sunderland, MA, USA, 2014.
48. Ali-Murthy, Z.; Lott, S.E.; Eisen, M.B.; Kornberg, T.B. An essential role for zygotic expression in the pre-cellular Drosophila embryo. *PLoS Genet.* **2013**, *9*, e1003428. [CrossRef] [PubMed]
49. Kimelman, D.; Kirschner, M.; Scherson, T. The events of the midblastula transition in Xenopus are regulated by changes in the cell cycle. *Cell* **1987**, *48*, 399–407. [CrossRef]

50. Veenstra, G.J.; Destree, O.H.; Wolffe, A.P. Translation of maternal TATA-binding protein mRNA potentiates basal but not activated transcription in Xenopus embryos at the midblastula transition. *Mol. Cell. Biol.* **1999**, *19*, 7972–7982. [CrossRef] [PubMed]
51. Collart, C.; Allen, G.E.; Bradshaw, C.R.; Smith, J.C.; Zegerman, P. Titration of four replication factors is essential for the Xenopus laevis midblastula transition. *Science* **2013**, *341*, 893–896. [CrossRef] [PubMed]
52. Newport, J.; Kirschner, M. A major developmental transition in early Xenopus embryos: I. Characterization and timing of cellular changes at the midblastula stage. *Cell* **1982**, *30*, 675–686. [CrossRef]
53. Edgar, B.A.; Kiehle, C.P.; Schubiger, G. Cell cycle control by the nucleo-cytoplasmic ratio in early Drosophila development. *Cell* **1986**, *44*, 365–372. [CrossRef]
54. Lu, X.; Li, J.M.; Elemento, O.; Tavazoie, S.; Wieschaus, E.F. Coupling of zygotic transcription to mitotic control at the Drosophila mid-blastula transition. *Development* **2009**, *136*, 2101–2110. [CrossRef] [PubMed]
55. Clarke, H.J.; McLay, D.W.; Mohamed, O.A. Linker histone transitions during mammalian oogenesis and embryogenesis. *Dev. Genet.* **1998**, *22*, 17–30. [CrossRef]
56. Saeki, H.; Ohsumi, K.; Aihara, H.; Ito, T.; Hirose, S.; Ura, K.; Kaneda, Y. Linker histone variants control chromatin dynamics during early embryogenesis. *Proc. Natl. Acad. Sci. USA* **2005**, *102*, 5697–5702. [CrossRef] [PubMed]
57. Perez-Montero, S.; Carbonell, A.; Moran, T.; Vaquero, A.; Azorin, F. The embryonic linker histone H1 variant of Drosophila, dBigH1, regulates zygotic genome activation. *Dev. Cell* **2013**, *26*, 578–590. [CrossRef] [PubMed]
58. Jenuwein, T.; Allis, C.D. Translating the histone code. *Science* **2001**, *293*, 1074–1080. [CrossRef] [PubMed]
59. Li, X.Y.; Harrison, M.M.; Villalta, J.E.; Kaplan, T.; Eisen, M.B. Establishment of regions of genomic activity during the Drosophila maternal to zygotic transition. *eLife* **2014**, *3*. [CrossRef] [PubMed]
60. Gaertner, B.; Johnston, J.; Chen, K.; Wallaschek, N.; Paulson, A.; Garruss, A.S.; Gaudenz, K.; De Kumar, B.; Krumlauf, R.; Zeitlinger, J. Poised RNA polymerase II changes over developmental time and prepares genes for future expression. *Cell Rep.* **2012**, *2*, 1670–1683. [CrossRef] [PubMed]
61. Nien, C.Y.; Liang, H.L.; Butcher, S.; Sun, Y.; Fu, S.; Gocha, T.; Kirov, N.; Manak, J.R.; Rushlow, C. Temporal coordination of gene networks by Zelda in the early Drosophila embryo. *PLoS Genet.* **2011**, *7*, e1002339. [CrossRef] [PubMed]
62. Harrison, M.M.; Li, X.Y.; Kaplan, T.; Botchan, M.R.; Eisen, M.B. Zelda binding in the early Drosophila melanogaster embryo marks regions subsequently activated at the maternal-to-zygotic transition. *PLoS Genet.* **2011**, *7*, e1002266. [CrossRef] [PubMed]
63. Liang, H.L.; Nien, C.Y.; Liu, H.Y.; Metzstein, M.M.; Kirov, N.; Rushlow, C. The zinc-finger protein Zelda is a key activator of the early zygotic genome in Drosophila. *Nature* **2008**, *456*, 400–403. [CrossRef] [PubMed]
64. Sun, Y.; Nien, C.Y.; Chen, K.; Liu, H.Y.; Johnston, J.; Zeitlinger, J.; Rushlow, C. Zelda overcomes the high intrinsic nucleosome barrier at enhancers during Drosophila zygotic genome activation. *Genome Res.* **2015**, *25*, 1703–1714. [CrossRef] [PubMed]
65. Lee, M.T.; Bonneau, A.R.; Takacs, C.M.; Bazzini, A.A.; DiVito, K.R.; Fleming, E.S.; Giraldez, A.J. Nanog, Pou5f1 and SoxB1 activate zygotic gene expression during the maternal-to-zygotic transition. *Nature* **2013**, *503*, 360–364. [CrossRef] [PubMed]
66. Leichsenring, M.; Maes, J.; Mossner, R.; Driever, W.; Onichtchouk, D. Pou5f1 transcription factor controls zygotic gene activation in vertebrates. *Science* **2013**, *341*, 1005–1009. [CrossRef] [PubMed]
67. Iwafuchi-Doi, M.; Zaret, K.S. Cell fate control by pioneer transcription factors. *Development* **2016**, *143*, 1833–1837. [CrossRef] [PubMed]
68. Zaret, K.S.; Mango, S.E. Pioneer transcription factors, chromatin dynamics, and cell fate control. *Curr. Opin. Genet. Dev.* **2016**, *37*, 76–81. [CrossRef] [PubMed]
69. Misteli, T. Beyond the sequence: Cellular organization of genome function. *Cell* **2007**, *128*, 787–800. [CrossRef] [PubMed]
70. Dixon, J.R.; Selvaraj, S.; Yue, F.; Kim, A.; Li, Y.; Shen, Y.; Hu, M.; Liu, J.S.; Ren, B. Topological domains in mammalian genomes identified by analysis of chromatin interactions. *Nature* **2012**, *485*, 376–380. [CrossRef] [PubMed]
71. Lieberman-Aiden, E.; van Berkum, N.L.; Williams, L.; Imakaev, M.; Ragoczy, T.; Telling, A.; Amit, I.; Lajoie, B.R.; Sabo, P.J.; Dorschner, M.O.; et al. Comprehensive mapping of long-range interactions reveals folding principles of the human genome. *Science* **2009**, *326*, 289–293. [CrossRef] [PubMed]

72. Van Berkum, N.L.; Lieberman-Aiden, E.; Williams, L.; Imakaev, M.; Gnirke, A.; Mirny, L.A.; Dekker, J.; Lander, E.S. Hi-C: A method to study the three-dimensional architecture of genomes. *J. Vis. Exp.* **2010**. [CrossRef] [PubMed]
73. Sexton, T.; Yaffe, E.; Kenigsberg, E.; Bantignies, F.; Leblanc, B.; Hoichman, M.; Parrinello, H.; Tanay, A.; Cavalli, G. Three-dimensional folding and functional organization principles of the Drosophila genome. *Cell* **2012**, *148*, 458–472. [CrossRef] [PubMed]
74. Tolhuis, B.; Blom, M.; van Lohuizen, M. Chromosome conformation capture on chip in single Drosophila melanogaster tissues. *Methods* **2012**, *58*, 231–242. [CrossRef] [PubMed]
75. Li, H.B. Chromosome Conformation Capture in Drosophila. *Methods Mol. Biol.* **2016**, *1480*, 207–212. [PubMed]
76. Hug, C.B.; Grimaldi, A.G.; Kruse, K.; Vaquerizas, J.M. Chromatin Architecture Emerges during Zygotic Genome Activation Independent of Transcription. *Cell* **2017**, *169*, 216–228. [CrossRef] [PubMed]
77. Dai, Q.; Ren, A.; Westholm, J.O.; Duan, H.; Patel, D.J.; Lai, E.C. Common and distinct DNA-binding and regulatory activities of the BEN-solo transcription factor family. *Genes Dev.* **2015**, *29*, 48–62. [CrossRef] [PubMed]
78. Sung, H.W.; Spangenberg, S.; Vogt, N.; Grosshans, J. Number of nuclear divisions in the Drosophila blastoderm controlled by onset of zygotic transcription. *Curr. Biol.* **2013**, *23*, 133–138. [CrossRef] [PubMed]
79. Fogarty, P.; Kalpin, R.F.; Sullivan, W. The Drosophila maternal-effect mutation grapes causes a metaphase arrest at nuclear cycle 13. *Development* **1994**, *120*, 2131–2142. [PubMed]
80. Blythe, S.A.; Wieschaus, E.F. Zygotic genome activation triggers the DNA replication checkpoint at the midblastula transition. *Cell* **2015**, *160*, 1169–1181. [CrossRef] [PubMed]
81. Laver, J.D.; Lipshitz, H.D. Transcription gets to the checkpoint. *Cell* **2015**, *160*, 1043–1044. [CrossRef] [PubMed]
82. Bertrand, E.; Chartrand, P.; Schaefer, M.; Shenoy, S.M.; Singer, R.H.; Long, R.M. Localization of ASH1 mRNA particles in living yeast. *Mol. Cell* **1998**, *2*, 437–445. [CrossRef]
83. Weil, T.T.; Parton, R.M.; Davis, I. Making the message clear: visualizing mRNA localization. *Trends Cell Biol.* **2010**, *20*, 380–390. [CrossRef] [PubMed]
84. Forrest, K.M.; Gavis, E.R. Live imaging of endogenous RNA reveals a diffusion and entrapment mechanism for nanos mRNA localization in Drosophila. *Curr. Biol.* **2003**, *13*, 1159–1168. [CrossRef]
85. Belaya, K.; St Johnston, D. Using the mRNA-MS2/MS2CP-FP system to study mRNA transport during Drosophila oogenesis. *Methods Mol. Biol.* **2011**, *714*, 265–283. [PubMed]
86. Trovisco, V.; Belaya, K.; Nashchekin, D.; Irion, U.; Sirinakis, G.; Butler, R.; Lee, J.J.; Gavis, E.R.; St Johnston, D. bicoid mRNA localises to the Drosophila oocyte anterior by random Dynein-mediated transport and anchoring. *eLife* **2016**, *5*. [CrossRef] [PubMed]
87. Halstead, J.M.; Lionnet, T.; Wilbertz, J.H.; Wippich, F.; Ephrussi, A.; Singer, R.H.; Chao, J.A. Translation. An RNA biosensor for imaging the first round of translation from single cells to living animals. *Science* **2015**, *347*, 1367–1671. [CrossRef] [PubMed]
88. Halstead, J.M.; Wilbertz, J.H.; Wippich, F.; Lionnet, T.; Ephrussi, A.; Chao, J.A. TRICK: A Single-Molecule Method for Imaging the First Round of Translation in Living Cells and Animals. *Methods Enzymol.* **2016**, *572*, 123–157. [PubMed]
89. Bothma, J.P.; Garcia, H.G.; Ng, S.; Perry, M.W.; Gregor, T.; Levine, M. Enhancer additivity and non-additivity are determined by enhancer strength in the Drosophila embryo. *eLife* **2015**, *4*. [CrossRef] [PubMed]
90. Sanchez, A.; Golding, I. Genetic determinants and cellular constraints in noisy gene expression. *Science* **2013**, *342*, 1188–1193. [CrossRef] [PubMed]
91. Chong, S.; Chen, C.; Ge, H.; Xie, X.S. Mechanism of transcriptional bursting in bacteria. *Cell* **2014**, *158*, 314–326. [CrossRef] [PubMed]
92. Fukaya, T.; Lim, B.; Levine, M. Enhancer Control of Transcriptional Bursting. *Cell* **2016**, *166*, 358–368. [CrossRef] [PubMed]
93. O'Brien, T.; Lis, J.T. Rapid changes in Drosophila transcription after an instantaneous heat shock. *Mol. Cell. Biol.* **1993**, *13*, 3456–3463. [CrossRef] [PubMed]
94. Fukaya, T.; Lim, B.; Levine, M. Rapid Rates of Pol II Elongation in the Drosophila Embryo. *Curr. Biol.* **2017**, *27*, 1387–1391. [CrossRef] [PubMed]

95. Muramoto, T.; Muller, I.; Thomas, G.; Melvin, A.; Chubb, J.R. Methylation of H3K4 Is required for inheritance of active transcriptional states. *Curr. Biol.* **2010**, *20*, 397–406. [CrossRef] [PubMed]
96. Zhao, R.; Nakamura, T.; Fu, Y.; Lazar, Z.; Spector, D.L. Gene bookmarking accelerates the kinetics of post-mitotic transcriptional re-activation. *Nat. Cell Biol.* **2011**, *13*, 1295–1304. [CrossRef] [PubMed]
97. Ferraro, T.; Esposito, E.; Mancini, L.; Ng, S.; Lucas, T.; Coppey, M.; Dostatni, N.; Walczak, A.M.; Levine, M.; Lagha, M. Transcriptional Memory in the Drosophila Embryo. *Curr. Biol.* **2016**, *26*, 212–218. [CrossRef] [PubMed]
98. Amodeo, A.A.; Jukam, D.; Straight, A.F.; Skotheim, J.M. Histone titration against the genome sets the DNA-to-cytoplasm threshold for the Xenopus midblastula transition. *Proc. Natl. Acad. Sci. USA* **2015**, *112*, E1086–E1095. [CrossRef] [PubMed]

5

Ethanol Exposure Causes Muscle Degeneration in Zebrafish

Elizabeth C. Coffey [1], Maggie E. Pasquarella [1], Michelle F. Goody [1] and Clarissa A. Henry [1,2,*]

1 School of Biology and Ecology, University of Maine, Orono, ME 04469, USA; elizabeth.c.mason@maine.edu (E.C.C.); margaret.pasquarella@maine.edu (M.E.P.); Michelle.Goody@maine.edu (M.F.G.)
2 Graduate School of Biomedical Sciences and Engineering, University of Maine, Orono, ME 04469, USA
* Correspondence: clarissa.henry@maine.edu

Abstract: Alcoholic myopathies are characterized by neuromusculoskeletal symptoms such as compromised movement and weakness. Although these symptoms have been attributed to neurological damage, EtOH may also target skeletal muscle. EtOH exposure during zebrafish primary muscle development or adulthood results in smaller muscle fibers. However, the effects of EtOH exposure on skeletal muscle during the growth period that follows primary muscle development are not well understood. We determined the effects of EtOH exposure on muscle during this phase of development. Strikingly, muscle fibers at this stage are acutely sensitive to EtOH treatment: EtOH induces muscle degeneration. The severity of EtOH-induced muscle damage varies but muscle becomes more refractory to EtOH as muscle develops. NF-kB induction in muscle indicates that EtOH triggers a pro-inflammatory response. EtOH-induced muscle damage is p53-independent. Uptake of Evans blue dye shows that EtOH treatment causes sarcolemmal instability before muscle fiber detachment. Dystrophin-null *sapje* mutant zebrafish also exhibit sarcolemmal instability. We tested whether Trichostatin A (TSA), which reduces muscle degeneration in *sapje* mutants, would affect EtOH-treated zebrafish. We found that TSA and EtOH are a lethal combination. EtOH does, however, exacerbate muscle degeneration in *sapje* mutants. EtOH also disrupts adhesion of muscle fibers to their extracellular matrix at the myotendinous junction: some detached muscle fibers retain beta-Dystroglycan indicating failure of muscle end attachments. Overexpression of Paxillin, which reduces muscle degeneration in zebrafish deficient for beta-Dystroglycan, is not sufficient to rescue degeneration. Taken together, our results suggest that EtOH exposure has pleiotropic deleterious effects on skeletal muscle.

Keywords: zebrafish; muscle; ethanol; alcohol; myopathy; fetal alcohol spectrum disorder; muscular dystrophy; Paxillin

---

1. Introduction

Ethanol (EtOH) is a teratogen that can cause multisystemic diseases at any stage of life. The rate of lifetime alcohol use disorders is approximately 12% in the US [1]. Chronic exposure in adults (alcoholism) can impair motor skills and cause fatigue, pain, and cramping [2]. These symptoms are similar to neuropathy [3]. Children with fetal alcohol spectrum disorders (FASD), which arise from maternal consumption of alcohol during pregnancy, exhibit compromised motor skills and full body weakness [4]. These symptoms are also attributed to neurological defects. FASD occurs in about 1 out of every 1000 live births [5]. In addition to the effect of EtOH on the nervous system, data indicate that skeletal muscle may also be affected by EtOH. Alcoholism in adults can cause a 30% reduction in muscle mass [3]. Early life stage exposure to EtOH causes skeletal muscle necrosis in chick embryos [6], decreases both fiber number and fiber size in rat pups born to EtOH-treated mothers [7], and decreases fiber size in zebrafish embryos exposed to EtOH during primary muscle development [8]. Adult zebrafish exposed to EtOH have decreased muscle fiber size and exhibit abnormal behavior that varies between strains [9,10]. The effects of EtOH exposure between primary muscle development and adulthood are not understood.

Zebrafish are an ideal model for studying the effects of EtOH exposure on skeletal muscle after primary muscle development. Zebrafish skeletal muscle is structurally similar to that of other vertebrates, with individual muscle fibers and segmentally reiterated groups of muscle fibers surrounded by distinct and dynamic regions of extracellular matrix (ECM) [11–17]. An important substructure in the muscle ECM is the myotendinous junction (MTJ). The MTJ serves as the site of force transmission from muscle to bone and is conserved in fish and mammals [18]. The Dystrophin-glycoprotein complex (DGC) is one adhesion complex that anchors muscle fibers to laminin at the MTJ [19]. Dystrophin is the DGC component that provides a critical link between the muscle cell membrane (sarcolemma) and the cytoskeleton. Duchenne/Becker muscular dystrophy (DMD), results from mutations in *dystrophin* [20]. Mutations in the *laminin a2* gene cause congenital muscular dystrophy with merosin deficiency [21]. Despite the fact that these two muscular dystrophies result from mutations in components within the same adhesion complex, zebrafish and mammalian models for these two diseases showed that the cellular pathology underlying their muscle degeneration is different [22–24]. In the zebrafish model of DMD, damage to the sarcolemma is observed before muscle fibers detach from the ECM at the MTJ [24]. In contrast, muscle fibers in the zebrafish model of merosin-deficient congenital muscular dystrophy detach from the MTJ in the absence of sarcolemmal damage [24]. Thus, the primary cause of muscle degeneration in dystrophin-null *sapje* mutant zebrafish is sarcolemmal damage. In contrast, the primary cause of muscle damage in *laminin a2* mutant zebrafish is failure of adhesion of muscle fibers to the MTJ. We found that human influenza A virus causes both sarcolemmal damage and disrupts muscle fiber adhesion to the MTJ [25], indicating that these modes of damage are not mutually exclusive.

In this study, we show that EtOH treatment after primary muscle development disrupts muscle homeostasis and causes muscle degeneration. Similar to previous reports [8,9], the penetrance and severity of EtOH-induced muscle damage was somewhat variable. Despite this variability, the severity of muscle damage caused by EtOH exposure significantly decreased by 48 hpf, suggesting that muscle becomes more refractory to EtOH exposure as development proceeds. EtOH treatment caused sarcolemmal damage and disrupted adhesion of muscle fibers to the MTJ. Interestingly, neither a treatment that improves pathology in zebrafish with sarcolemmal damage (Trichostatin A (TSA) treatment) nor a treatment that improves pathology in zebrafish with disrupted muscle-MTJ adhesion (Paxillin overexpression) was beneficial in EtOH-exposed zebrafish. EtOH treatment triggered a robust, pro-inflammatory cellular response as indicated by aberrant activation of NF-kB muscle. Taken together, these data suggest that EtOH exposure induces cellular stress, sarcolemmal damage, and disrupts muscle fiber-MTJ adhesion.

2. Materials and Methods

2.1. Zebrafish Husbandry, Transgenic Lines, and Mutant Lines

Zebrafish embryos were retrieved from natural spawns of adult zebrafish maintained on a 14-h light/10-h dark cycle. Strains used in this study were AB, *Tg(actb2:pxn-EGFP)*^{mai1} [26], *Tg(hsp70l:pxn-EGFP)*, *Tg(Nf-kB:EGFP)*^{nc1} [27], *tp53*^{M214K} [28], and *sapje*^{ta222a} [29]. Embryos were grown in embryo rearing media (1X ERM) with methylene blue at 28.5 or 33 degrees Celsius and staged according to [30]. All protocols conformed to the University of Maine Institutional Animal Care and Usage Committee's Guidelines that do not require permission for the use of zebrafish 3 days old and younger.

2.2. Ethanol (EtOH) and Trichostatin A (TSA) Treatment

EtOH dilutions were chosen based on [8]. Solutions of 2% EtOH diluted in 1X ERM from either 95% or 100% EtOH stocks were continuously administered beginning at different stages of development (24, 30, 36, or 48 hpf) through 72 hpf. Treatments with the histone deacetylase (HDAC) inhibitor TSA were carried out according to [31]. TSA was dissolved in either DMSO or EtOH. TSA (200 nM) was diluted in 1X ERM and administered continuously from 24 to 72 hpf alone or from 24 to 30 hpf alone and then concurrently with 2% EtOH from 30 to 72 hpf. Embryos were dechorionated prior to EtOH or TSA administration. All solutions were changed daily.

2.3. *Heat Shock Treatment*

Heat shock was done at 6 or 24 hpf in a 37–38 degrees Celsius incubator for 1–2 h. All heat shock-treated embryos were screened using a GFP 470 filter on a Zeiss Discovery.V12 microscope to identify fluorescent transgenic embryos.

2.4. *Evans Blue Dye (EBD) Injection*

Zebrafish were anesthetized in 0.16 mg/mL tricaine in 1X ERM and side mounted on 1% agarose-lined Petri dishes in a minimal volume of liquid. EBD (1% w/v) was dissolved in a 0.9% saline solution. Injection needles were pulled from glass capillary tubes containing a filament on a Sutter Flaming/Brown Micropipette Puller. EBD solution was loaded into injection needles and injected into the circulation of anesthetized ~72 hpf zebrafish using a MPPI-3 pressure injector from ASI. EBD injected zebrafish were maintained in 1X ERM for 6 h before live imaging.

2.5. *Immunohistochemistry*

Zebrafish embryo fixation, phalloidin staining, and beta-Dystroglycan immunohistochemistry was done according to [32].

2.6. *Imaging*

Images were taken on a Zeiss Axio Imager Z1 microscope with a Zeiss ApoTome attachment or an Olympus Fluoview IX-81 inverted microscope with FV1000 confocal system. Embryos were side mounted in 80% glycerol/20% PBS and viewed with the 20× objective. Exposure times were kept constant throughout the imaging of an experiment allowing for the comparison of fluorescent levels between images. Live imaging for EBD experiments involved side mounting ~78 hpf zebrafish embryos in 0.4% low melt agarose containing 0.16 mg/mL tricaine and imaging via confocal microscopy. Linear adjustments were made to images in Adobe Photoshop and figures were collated in Adobe Illustrator.

2.7. *Analysis and Statistics*

Statistical comparisons of two groups were conducted with student t-tests. Statistical comparisons of multiple groups were determined using ANOVA followed by Tukey's honest significant difference multiple comparisons test. For ANOVA details, please see the supplementary materials. Statistical analysis was done using Prism 7 or Microsoft Excel.

3. Results

3.1. *Late Embryonic Exposure to EtOH Causes Muscle Damage in Zebrafish*

Zebrafish primary muscle development entails the elongation of short muscle precursor cells into long myofibers that attach to the ECM at the MTJ and undergo fusion to generate myotubes [33]. This process occurs by 24 post fertilization (hpf) (Figure 1A–C). Exposure to 2 or 2.5% EtOH during primary muscle development (8–24 hpf) results in smaller muscle fibers and abnormally wide MTJs [8]. We asked if muscle integrity would be disrupted in zebrafish that were treated with EtOH after primary muscle development during a period of muscle growth (Figure 1A). We first assessed survival and found that treatment with 0–2% EtOH did not drastically affect survival at 72 hpf (Figure 1D). In contrast, most zebrafish treated with 2.5% EtOH died by 72 hpf (Figure 1D). Exposure to lower concentrations of EtOH (0.5–1%) did not cause gross morphological phenotypes

(Figure 1E–G1). Exposure to 1.5% EtOH or higher resulted in fully penetrant pericardial edema (Figure 1H–I1 $n = 52/52$ and 53/53 for 1.5 and 2% EtOH, respectively). EtOH exposure also caused muscle damage. Control zebrafish had linear, organized muscle fibers at 24 hpf (Figure 1C, white arrow) and at 3 dpf (Figure 1J–J1). EtOH treatment caused muscle fiber degeneration (Figure 1K–K1, white arrows) although there was not a clear dose-dependent effect (Figure S1L,M, supplementary materials). These data suggest that EtOH negatively affects skeletal muscle homeostasis in zebrafish.

Although treatment with 2% EtOH diluted from a 95% (190 proof) stock resulted in fiber detachment, the penetrance of this phenotype varied across experimental replicates (Figure 1N,O). We also treated zebrafish with EtOH diluted from a 100% (200 proof) stock and observed the same fiber detachment phenotype with similar variability in phenotypic frequency between experimental replicates (Figure 1N,O). Therefore, the variable penetrance of the EtOH-induced skeletal muscle phenotype is independent of pericardial edema and the stock solution used. All subsequent EtOH dilutions were made from the 95% stock. These data indicate that EtOH treatment after primary muscle development results in muscle degeneration.

3.2. Zebrafish Muscle Is Less Sensitive to EtOH That Is Administered Later in Development

Zebrafish primary muscle development is followed by a period of growth [34,35]. Less obvious, but significant changes also occur in the ECM at the MTJ during this time (reviewed in [36]). Thus, we asked whether there were any critical time windows in which EtOH administration would result in increased muscle damage. We treated embryos continuously with 2% EtOH beginning at 24, 30, 36, or 48 hpf and fixed them at 3 dpf. We found a spectrum of muscle damage severity following EtOH administration at each time point. However, muscle phenotypes were significantly less severe when zebrafish were EtOH-treated at 48 hpf (Figure 2A). To better analyze the muscle damage spectrum, we binned fiber detachment phenotypes into one of three groups: none (0 muscle segments with fiber detachments), mild (1–10 segments with fiber detachments), or severe (11+ segments with fiber detachments). Zebrafish treated with EtOH beginning at 24, 30, or 36 hpf displayed the full spectrum of muscle damage severity (Figure 2B–K1). In contrast, treatment at 48 hpf resulted in either no or only mild muscle damage (Figure 2B,L–M1). These data indicate that zebrafish are more refractory to EtOH-induced muscle damage later in development.

3.3. p53-Dependent Apoptosis Is Not Required for EtOH-Induced Fiber Detachments

To begin characterizing the muscle damage caused by EtOH treatment, we first investigated whether inflammation or apoptosis may be involved. Activation of the transcription factor NF-kB is part of a pro-inflammatory, cellular stress response that can result in tissue damage. We used the *Tg(NF-kB:EGFP)* [27] zebrafish line to ask whether NF-kB-dependent signaling is altered upon EtOH exposure. *Tg(NF-kB:EGFP)* zebrafish were continuously exposed to 0 or 2% EtOH from 30 hpf on and then fixed at 3 dpf. Muscle fibers in control *Tg(NF-kB:EGFP)* zebrafish were long and linear (Figure 3A1,A2) and did not express GFP at 72 hpf (Figure 3A3–A6, [25]). EtOH treatment increased NF-kB activity as evidenced by increased GFP expression in muscle fibers (Figure 3B3–B6). Therefore, muscle cells respond to EtOH exposure by activating NF-kB-dependent transcription, which may itself contribute to muscle damage.

The transcription factor p53 induces apoptosis in various cell types in response to stress and is important for muscle homeostasis [37,38]. We used homozygous *tp53*^{M214K} mutant zebrafish [28] to ask whether EtOH causes fiber detachment via p53-dependent apoptosis. There were no significant differences in the proportion of treated embryos that displayed fiber detachments (Figure 3C, $p = 0.19$) or the average number of segments per embryo with fiber detachments (Figure 3D, $p = 0.29$) in EtOH-treated, homozygous *tp53* mutants and ABs. Therefore, EtOH-induced muscle degeneration could involve pro-inflammatory NF-kB-dependent signaling, but occurs independently of p53-mediated apoptosis.

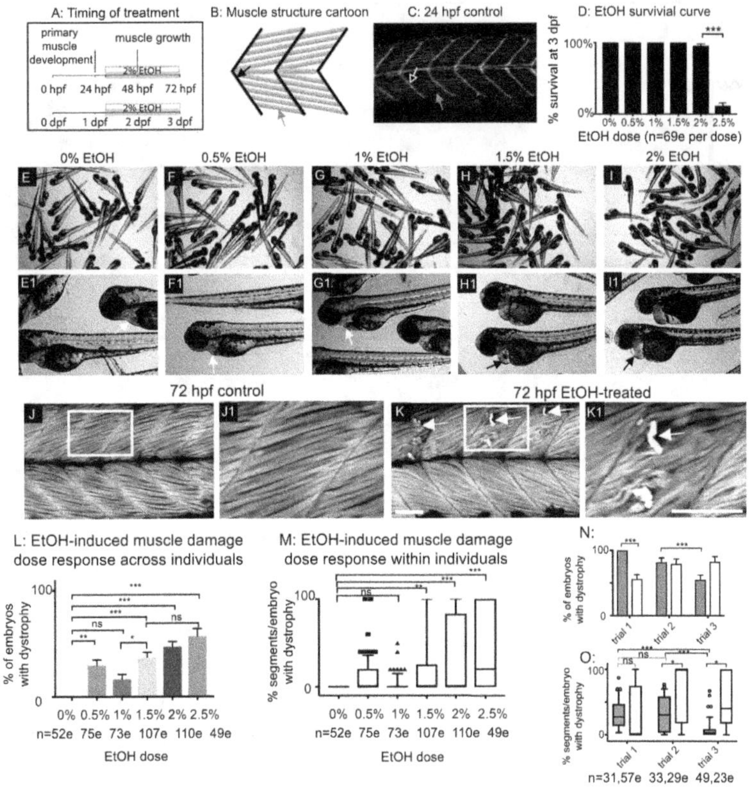

Figure 1. Late embryonic exposure to EtOH causes muscle damage in zebrafish. (**A**) Zebrafish muscle development timeline showing that 2% EtOH was administered after primary muscle development is complete. (**B**) Cartoon showing zebrafish muscle structure. Elongated myofibers (gray arrow) attach at the myotendinous junction (MTJ) (black arrow). (**C**) Control embryo. Muscle fibers have elongated (gray arrow) and attached to the MTJ (black arrow outlined in white) by 24 hpf. (**D**) Quantification of survival at 72 hpf following continuous treatment with one of six different doses of EtOH from 30 to 72 hpf. (**E–I1**) Brightfield images of 72 hpf zebrafish treated with one of six different doses of EtOH. Numbered panels are higher magnification images of zebrafish treated with the same EtOH dose as the corresponding lettered panels. (**E–E1**) 0% EtOH. Note normal heart morphology (**E1**, white arrow). (**F–F1**) 0.5% EtOH. Note normal heart morphology (**F1**, white arrow). (**G–G1**) 1% EtOH. Note normal heart morphology (**G1**, white arrow). (**H–H1**) 1.5% EtOH. Note pericardial edema (**H1**, black arrow). (**I–I1**) 2% EtOH. Note pericardial edema (**I1**, black arrow). (**J–K1**) Anterior left, dorsal-top, side-mounted, 72 hpf embryos stained with phalloidin (white) to visualize actin. White boxes in **J** and **K** correspond to zoomed-in, numbered panels **J1** and **K1**, respectively. (**J–J1**) Control embryo. Muscle fibers remain attached to the MTJ. (**K–K1**) Embryo treated continuously with 2% EtOH from 30 to 72 hpf. White arrows indicate muscle fibers that have detached from the MTJ. (**L**) Quantification of the percent of EtOH-treated embryos at each of the EtOH treatment doses that displayed one or more fiber detachments. (**M**) Quantification of the percent of imaged segments with fiber detachments per embryo across the six different EtOH treatment doses. (**N–O**) Quantification of the variability in EtOH-induced fiber detachments across experimental replicates. (**N**) Percent of EtOH-treated embryos with at least one fiber detachment across three trials with either 2% EtOH diluted from 95% stock (gray bars) or 2% EtOH diluted from 100% stock (white bars). (**O**) Percent of imaged segments per embryo with fiber detachments across three trials with either 2% EtOH diluted from 95% stock (gray boxes) or 2% EtOH diluted from 100% stock (white boxes). Note the persistence of variability in muscle damage across three trials regardless of EtOH stock solution used. Scale bars are 50 micrometers. "e" = embryo. Error bars are standard error of the mean and whiskers are 10th–90th percentiles. * $p < 0.05$, ** $p < 0.01$, *** $p < 0.001$.

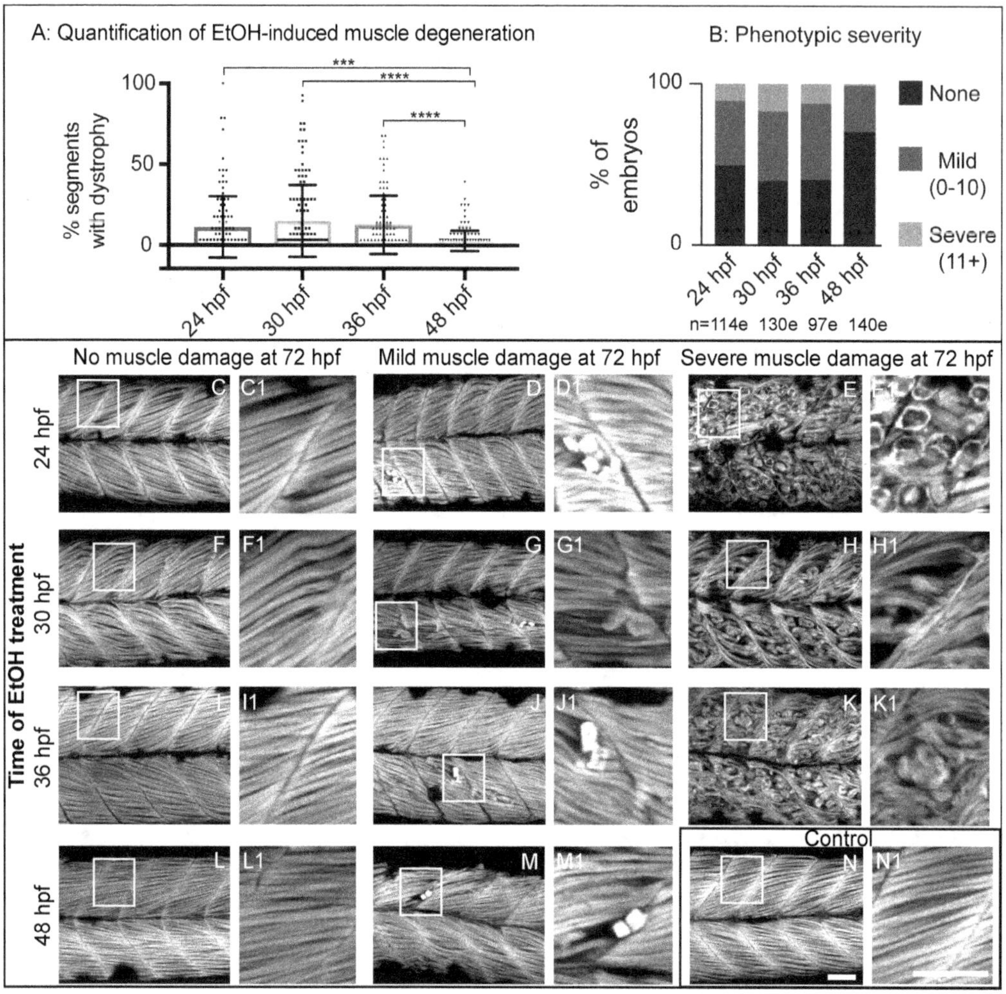

Figure 2. Zebrafish muscle is less sensitive to EtOH administered later in development. (**A**) Quantification of the percent of imaged segments with fiber detachments per embryo. 2% EtOH treatment from 48 to 72 hpf caused significantly fewer segments with fiber detachments per embryo compared to EtOH treatment beginning at 24, 30, or 36 hpf. (**B**) Quantification of the percent of embryos at each of the four EtOH treatment time points that fell into each of the three phenotype bins. (**C–N**) Anterior left, dorsal top, side-mounted, 72 hpf embryos stained with phalloidin (white) to visualize actin. Embryos treated continuously with 2% EtOH from 24 hpf, 30 hpf, 36 hpf, or 48 hpf through 72 hpf showed a spectrum of muscle damage severity that was binned into one of three groups: None (0 segments per embryo with dystrophy), Mild (1–10 segments per embryo with dystrophy), or Severe (11+ segments per embryo with dystrophy). White boxes correspond to zoomed-in, numbered panels. (**C–E1**) Embryos continuously treated with 2% EtOH between 24 and 72 hpf showed no muscle damage (**C–C1**), mild damage (**D–D1**), or severe damage (**E–E1**). (**F–H1**) Embryos continuously treated with 2% EtOH between 30 and 72 hpf showed no muscle damage (**F–F1**), mild damage (**G–G1**), or severe damage (**H–H1**). (**I–K1**) Embryos continuously treated with 2% EtOH between 36 and 72 hpf showed no muscle damage (**I–I1**), mild damage (**J–J1**), or severe damage (**K–K1**). (**L–M1**) Embryos continuously treated with 2% EtOH between 48 and 72 hpf showed no muscle damage (**L–L1**) or mild damage (**M–M1**). (**N–N1**) Control embryo. Scale bars are 50 micrometers. "e" = embryo. Error bars are standard error of the mean. *** $p < 0.001$, **** $p < 0.0001$.

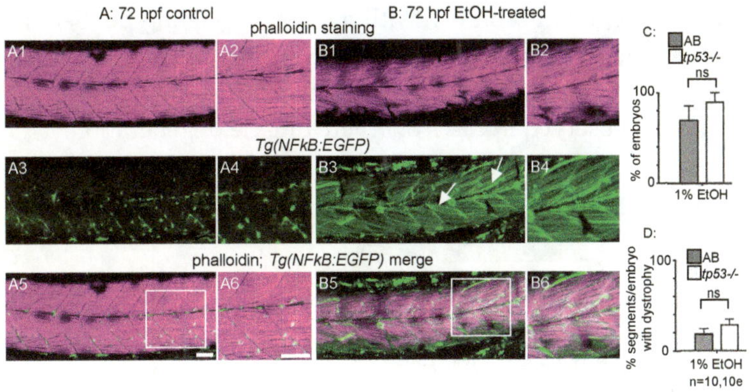

Figure 3. NF-kB-dependent inflammation is present, but p53-dependent apoptosis is not required for EtOH-induced fiber detachments. (**A1–B6**) Anterior left, dorsal top, side-mounted 72 hpf *Tg(NF-kB:EGFP)* embryos stained with phalloidin (pseudo-colored fuchsia) to visualize actin. (**A1–6**) 72 hpf *Tg(NF-kB:EGFP)* control embryo. (**A1–2**) Phalloidin channel only. (**A2**) Zoomed-in view corresponds to white box in A5. (**A3–4**) GFP channel only. (**A4**) Zoomed-in panel corresponds to white box in A5. Note low NF-kB activity in muscle fibers. (**A5**) Phalloidin and NF-kB-EGFP channels merged. (A6) Zoomed-in panel corresponds to white box in A5. (**B1–6**) 72 hpf *Tg(NF-kB:EGFP)* embryo treated continuously with 2% EtOH from 30 to 72 hpf. (**B1–2**) Phalloidin channel only. (**B2**) Zoomed-in view corresponds to white box in B5. (**B3–4**) GFP channel only. (**B4**) Zoomed-in view corresponds to white box in B5. Note increased NF-kB activity in muscle fibers due to EtOH exposure. White arrows points to muscle fibers expressing GFP. (**B5–6**) Phalloidin and NF-kB-EGFP channels merged. (**B6**) Zoomed-in view corresponds to white box in B5. (**C–D**) Quantification of fiber detachment frequency within and between AB or *tp53* mutant embryos treated with EtOH from 30 to 72 hpf. (**C**) Percent of EtOH-treated embryos displaying one or more fiber detachments. There was no significant difference in fiber detachment frequency between embryos depending on AB (gray bars) vs. *tp53* mutant (white bars) genetic background at 1% EtOH. (**D**) Percent of imaged segments with fiber detachments per embryo. There was no significant difference in fiber detachment frequency within embryos depending on AB (gray bars) vs. *tp53* mutant (white bars) genetic background at 1% EtOH. Scale bars are 50 micrometers. "e" = embryo. Error bars are standard error of the mean.

3.4. EtOH-Induced Fiber Detachments Result from Sarcolemmal Instability and Failure of Muscle End Attachments

Muscle damage in zebrafish models studied thus far can result from two distinct cellular etiologies: failure to maintain sarcolemmal integrity or failure to maintain adhesion at the MTJ external to the sarcolemma [24]. Use of cell impermeable Evans blue dye (EBD), which is only taken up by muscle when sarcolemmal integrity is compromised, and beta-Dystroglycan (beta-DG) immunohistochemistry can be used to distinguish between these two etiologies [24]. Retention of the membrane-integral protein beta-DG at the unanchored ends of retracted fibers indicates MTJ adhesion failure [39]. As shown previously [39], EBD was excluded from muscle fibers in controls (Figure 4A). EBD was observed in retracted myofibers as well as long, attached myofibers in EtOH-treated zebrafish (Figure 4B–D), suggesting that EtOH disrupts sarcolemmal integrity. Every zebrafish with muscle damage showed EBD uptake regardless of whether the phenotype was mild or severe (n = 30 zebrafish over 3 trials). Beta-DG robustly localized to MTJs in untreated controls (Figure 4E–E3). Beta-DG also localized to MTJs in EtOH-treated embryos that did not have muscle degeneration (not shown). Beta-DG remained at the MTJ and did not accompany retracted fibers in 19/21 EtOH-treated zebrafish with the mild phenotype (Figure 4F–G3, white arrows indicate retracted fibers unaccompanied by beta-DG). In 2/21 EtOH-treated zebrafish with the mild muscle phenotype, beta-DG remained with at least one detached muscle fiber in each embryo (not shown). Thus, the vast majority of muscle fibers in EtOH-treated zebrafish with the mild phenotype do not retain beta-DG at their unanchored ends. The opposite was observed in zebrafish with the severe phenotype. Beta-DG was observed at the unanchored ends of retracted muscle fibers in 5/6 severe zebrafish (Figure 4H–I3). Taken together, these data suggest that EtOH treatment disrupts both sarcolemmal integrity and adhesion to the MTJ, the latter is more frequently observed in cases of severe muscle degeneration.

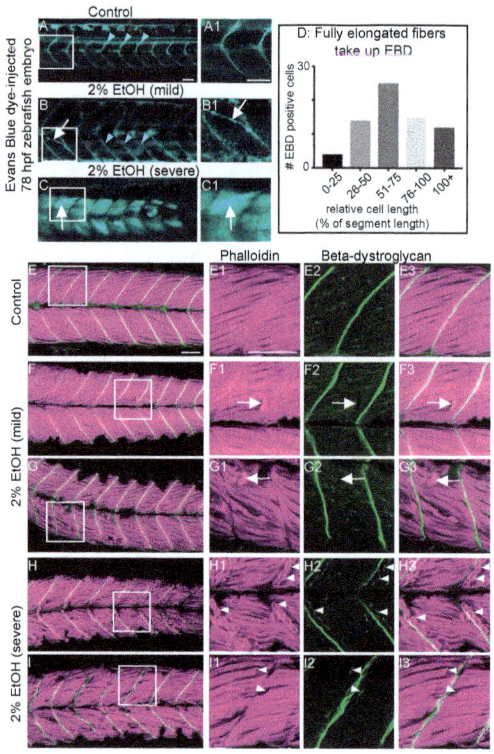

Figure 4. EtOH-induced fiber detachments result from sarcolemmal instability and failure of muscle end attachments. (**A–C1**) Anterior left, dorsal top, side-mounted, live, 78 hpf embryos treated continuously with 2% EtOH from 30 to 72 hpf, injected with 1% Evans Blue Dye (EBD) (pseudocolored blue) at 72 hpf. White boxes in A, B, and C correspond to zoomed-in, numbered panels A1, B1, and C1, respectively. (**A–A1**) Control embryo. EBD is in the blood stream (blue arrowheads). (**B–B1**) Embryo with mild fiber detachment phenotype that was induced by continuous treatment of 2% EtOH from 30 to 72 hpf. EBD is in the blood stream (blue arrowheads). EBD permeates some long muscle fibers (white arrows in B and B1). (**C–C1**) Embryo with severe fiber detachment phenotype that was induced by continuous treatment of 2% EtOH from 30 to 72 hpf. EBD is in the blood stream (blue arrowheads). EBD permeates long (white arrows in C and C1) and short muscle fibers. (**D**) Quantification of the relative length of EBD-penetrated myofibers. EBD penetrated myofibers ranging from short, retracted fibers to fully elongated, diagonally arrayed, fast-twitch myofibers in EtOH-treated embryos. (**E–I3**) Anterior left, dorsal top, side-mounted, 72 hpf embryos stained with phalloidin (pseudo-colored fuchsia) to visualize actin and anti-beta-DG antibodies (green). Lettered panels show phalloidin and beta-DG channels merged. White boxes correspond to numbered panels on the right. Panels numbered 1 show zoomed-in images of phalloidin staining only. Panels numbered 2 show zoomed-in images of beta-DG staining only. Panels numbered 3 show zoomed-in images of phalloidin and beta-DG channels merged. (**E–E3**) Control embryo. Fibers remained long and beta-DG is localized to the MTJ. (**F–G3**) Embryos treated continuously with 2% EtOH from 30 to 72 hpf showing mild fiber detachment. White arrowheads in F1–F3 and G1–G3 point to retracted fibers unaccompanied by beta-DG. (**H–I3**) Embryos treated continuously with 2% EtOH from 30 to 72 hpf with severe fiber detachment. White arrowheads in H1–H3 and I1–I3 point to retracted fibers that retained beta-DG at their detached ends. Scale bar is 50 micrometers.

3.5. Overexpression of Paxillin Does Not Rescue Fiber Detachments Caused by Loss of Sarcolemmal Integrity

Muscle fibers in dystrophin-null *sapje* mutant zebrafish uptake EBD prior to overt muscle degeneration [24,40]. Previous work from our lab has shown that overexpression of the cytoplasmic, Integrin adaptor protein Paxillin reduces fiber detachments caused by failure of MTJ adhesion external to the sarcolemma [26]. However, it is unknown whether Paxillin overexpression can reduce muscle degeneration that is preceded by sarcolemmal failure or whether Paxillin overexpression will benefit severe EtOH-induced muscle degeneration given that MTJ adhesion failure was observed in these cases. We thus asked whether Paxillin overexpression would reduce muscle degeneration in the contexts of EtOH exposure or in *sapje* mutant zebrafish. Transgenic zebrafish that constitutively express Paxillin (*Tg(actb2:pxn-EGFP)*) have normal muscle (Figure 5A–A1, [26]). Surprisingly, Paxillin overexpression significantly increased both the prevalence and severity of EtOH-induced muscle damage (Figure 5B–B1,C,D, $p < 0.03$ and 0.02 respectively). In contrast, Paxillin overexpression did not significantly change the phenotype of *sapje* mutants (Figure 5E, $p = 0.11$).

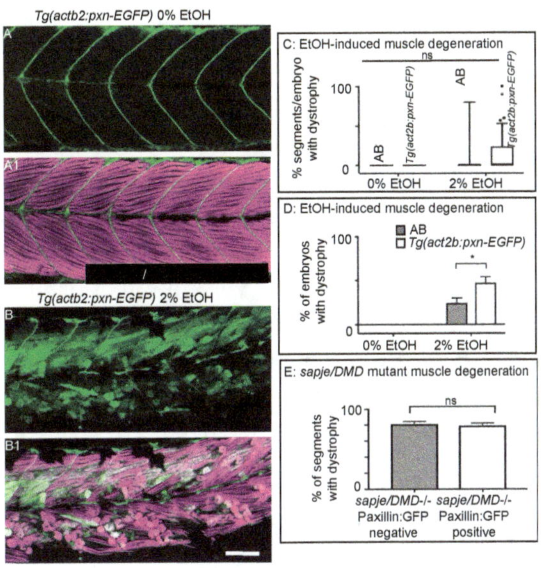

Figure 5. Paxillin overexpression does not rescue fiber detachments caused by loss of sarcolemmal integrity. (**A–B1**) Anterior left, dorsal top, side mounted, 72 hpf *Tg(act2b:pxn-EGFP)* embryos overexpressing Paxillin-EGFP (green) stained with phalloidin (pseudocolored fuchsia) to visualize actin. (**A–A1**) *Tg(act2b:pxn-EGFP)* control embryo. Fibers remain long and Paxillin-EGFP localizes to the MTJ. (**A**) GFP channel only. (**A1**) Phalloidin and GFP channels merged. (B–B1) *Tg(act2b:pxn-EGFP)* embryo treated continuously with 2% EtOH from 30 to 72 hpf. Fibers detach from the MTJ. (B) GFP channel only. (B1) Phalloidin and GFP channels merged. (**C**) Quantification of the percentage of imaged segments/embryo that had EtOH-induced dystrophy. Embryos that constitutively overexpressed Paxillin-EGFP (n = 53e) had more segments with fiber detachments compared to AB embryos (n = 54e) after continuous treatment with 2% EtOH. (**D**) Quantification of percent of embryos with EtOH-induced fiber detachments. Fiber detachment frequency was significantly greater in embryos that constitutively overexpressed Paxillin-EGFP compared to AB embryos after continuous treatment with 2% EtOH. (**E**) Quantification of the proportion of imaged segments/embryo with dystrophy. *sapje* mutants that constitutively overexpressed Paxillin-EGFP (white bars) (n = 28e) showed no significant difference in the percent of segments that contained fiber detachments compared to those that did not constitutively overexpress Paxillin-EGFP (gray bars) (n = 23e) ($p = 0.11$). These data suggest that Paxillin overexpression does not reduce fiber detachments due to loss of sarcolemmal integrity. Scale bars are 50 micrometers. Error bars are standard error of the mean. * $p < 0.05$.

3.6. Combining the HDAC Inhibitor TSA with EtOH Is Lethal for Zebrafish Embryos

The fact that Paxillin overexpression exacerbates muscle damage in EtOH-treated zebrafish but not *sapje* mutant zebrafish suggests that the underlying mechanisms of muscle damage are distinct despite EBD uptake in both contexts. We asked whether the HDAC-inhibitor TSA, which significantly reduces muscle degeneration in *sapje* mutant zebrafish [31],, would alter the penetrance or severity of EtOH-induced muscle damage. We found that the combination of EtOH and TSA was lethal in zebrafish (Table 1) before EtOH-induced fiber detachments occurred. Interestingly, this combination of small molecules was lethal regardless of whether the TSA was dissolved in DMSO or EtOH (data not shown).

Table 1. EtOH and Trichostatin A (TSA) is a Lethal Combination for Zebrafish Embryos.

Experimental Group	Total Embryos	Embryos Alive at 72 hpf	%Survival
0.66% DMSO	48	48	100
TSA (200 nM) with 0.66% DMSO	52	52	100
EtOH (2%)	50	36	72
EtOH (2%) with 0.66% DMSO	27	20	74
EtOH (2%) + TSA (200 nM) with 0.66% DMSO	54	2	3.7

3.7. EtOH Exposure Exacerbates Muscle Degeneration in Sapje Mutant Zebrafish

Given the differences between *sapje* mutant and EtOH-treated zebrafish, we asked whether EtOH treatment would additively worsen muscle damage in *sapje* mutant zebrafish. Wildtype zebrafish not exposed to EtOH displayed normal muscle (Figure 6A). *sapje* mutants not exposed to EtOH or wildtype embryos exposed to 2% EtOH had relatively few foci of muscle damage (Figure 6B–C, white arrows). *sapje* mutants treated with 2% EtOH displayed more segments with muscle damage per embryo than either unexposed *sapje* mutants or exposed wildtypes alone (Figure 6D, white arrows). Although EtOH-treated *sapje* mutant zebrafish trended towards having more muscle degeneration, the interaction was not quite significant (Figure 6E, $p = 0.06$).

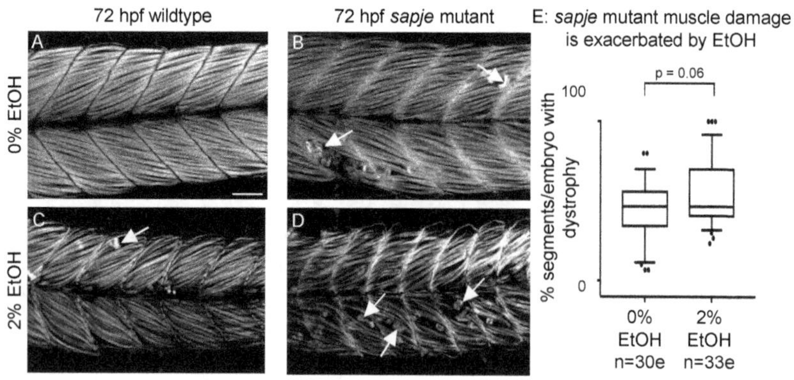

Figure 6. EtOH exacerbates dystrophy in *sapje* mutants. (**A–D**) Anterior left, dorsal top, side-mounted, 72 hpf embryos stained with phalloidin (white) to visualize actin. (**A**) Wildtype control embryo. Muscle fibers are attached to the MTJ. (**B**) *sapje* mutant control embryo. A few muscle fibers detach from the MTJ (white arrows). (**C**) Wildtype embryo treated continuously with 2% EtOH from 30 to 72 hpf. A few muscle fibers detach from the MTJ (white arrow). (**D**) *sapje* mutant embryo treated continuously with 2% EtOH from 30 to 72 hpf. Note the increased number of muscle fibers detached from the MTJ (white arrows). (**E**) Quantification of fiber detachment frequency in *sapje* mutant embryos without and with EtOH exposure. *sapje* mutant embryos that were treated continuously with 2% EtOH showed an average increase in fiber detachments per embryo ($p = 0.06$). "e" = embryo. Whiskers are 10th–90th percentiles.

4. Discussion

EtOH exposure can cause a wide range of symptoms in both fetal and adult organisms. Most of what is known about these diseases, such as in FASD, centers on the neurological defects [41]. Our data support a growing body of evidence suggesting that EtOH exposure damages skeletal muscle. We show for the first time that EtOH exposure after primary muscle development (after 24 hpf) causes muscle fiber detachment and degeneration in zebrafish (Figures 1 and 2). This result contrasts with the effects of EtOH exposure *during* muscle development. EtOH treatment from 8 to 24 hpf causes abnormally wide MTJs and smaller slow-twitch muscle fibers [8] but does not cause muscle degeneration. The mechanisms underlying muscle degeneration in zebrafish exposed to EtOH after 24 hpf are not known. We show that EtOH treatment results in a pro-inflammatory stress response, as indicated by NF-kB induction in skeletal muscle (Figure 3). EtOH treatment also causes sarcolemmal damage, which is visualized by uptake of EBD (Figure 4). In severely affected embryos, EtOH treatment results in failure of muscle end attachments to the ECM at the MTJ (Figure 4). We tested whether two approaches that reduce muscle damage in contexts where there is sarcolemmal damage or disrupted muscle-MTJ adhesion could reduce degeneration upon EtOH exposure. Neither of these approaches worked: TSA interacts with EtOH and is lethal, and Paxillin overexpression exacerbated muscle degeneration. Taken together, these data indicate that EtOH likely has pleiotropic effects on zebrafish skeletal muscle.

4.1. Variability in the Penetrance and Severity of Muscle Damage Caused by EtOH

We report here that there is some variation in the penetrance and severity of skeletal muscle damage accompanying EtOH treatment between trials. We do not know if this variation is specific to this time point because previous reports have compiled trials. However, it has been shown that EtOH affects neurobehavioral responses differently in different strains of zebrafish [8]. Although we use the AB strain, we have not characterized the genetic variation in our zebrafish facility. Thus, one future avenue of investigation would be to determine the potential genetic modifiers that provide more resilience to EtOH treatment in some zebrafish.

The decreasing muscle damage associated with increased age at initial EtOH exposure is likely due to either less EtOH ingestion and/or enhanced EtOH metabolism in older embryos. It was observed that the concentration of EtOH within zebrafish embryos was 2.7–4.2 fold lower in 24 hpf zebrafish and 5.8–6.2 fold less in 48 hpf zebrafish than the waterborne EtOH concentration [42]. Therefore, zebrafish may develop a barrier to EtOH as they age [42]. However, another study found the percentage of the waterborne EtOH concentration detected within embryos steadily rose from about 15–30% when embryos were exposed to the same concentration of EtOH for increasing lengths of time (i.e., 10–48 h of continuous EtOH exposure) [43]. Furthermore, phenotypes of interest are usually elicited upon exposing adult zebrafish to 0.5–1% EtOH (v/v) whereas the EtOH dose used on zebrafish embryos is usually 1.5–2.5% EtOH. Together, these observations potentially contradict the idea that zebrafish develop a barrier to EtOH as they age.

4.2. EtOH Treatment Damages the Muscle Sarcolemma and Disrupts Muscle Fiber-MTJ Adhesion

Muscle atrophy is a well-known symptom of many chemical or environmental insults and genetic perturbations. Previous studies showed that muscle fibers can detach before undergoing cell death via at least two distinct cellular etiologies in zebrafish models of genetic muscle diseases [24]. Here, we show that EtOH exposure can result in mild (10 or fewer segments with muscle degeneration) or severe (> 10 segments with muscle degeneration) phenotypes. The fact that muscle fibers in both phenotypic categories uptake EBD indicates that EtOH causes damage to the muscle sarcolemma regardless of phenotypic severity. In contrast, disruption of muscle fiber-MTJ adhesion correlated with phenotypic severity. Out of 21 EtOH exposed zebrafish with mild muscle damage, only two zebrafish had one or more muscle fibers each that retained beta-Dystroglycan at their terminal ends. In contrast, the vast majority of muscle fibers retained beta-Dystroglycan at their terminal ends in zebrafish with the severe muscle phenotype (5/6 zebrafish, when damage is severe it is impossible to count how many fibers have detached). It will be interesting in the future to determine the mechanisms underlying disruption of muscle cell-MTJ adhesion in zebrafish with the severe phenotype.

4.3. Interaction of Environment and Muscle Disease

One hallmark of many congenital muscular dystrophies is their phenotypic variation but the underlying causes of variation are not known. We previously showed that the human influenza A virus (IAV) damages muscle in zebrafish embryos. IAV infection significantly increases muscle degeneration and decreases survival in the zebrafish model of Duchenne Muscular Dystrophy (dystrophin-deficient *sapje* mutant zebrafish) [25]. Here we asked whether EtOH treatment similarly exacerbated muscle degeneration in *sapje* mutant zebrafish. Although we did not observe a significant interaction, there was a clear trend towards worsening of the *sapje* phenotype (Figure 6, $p = 0.06$). Taken together these data highlight the utility of the zebrafish model to integrate gene environment interactions and suggest that environmental insults may have more pleiotropic deleterious effects on muscle tissue than genetic perturbations.

Supplementary Materials: The supplementary materials are available online www.mdpi.com/2221-3759/6/1/7/s1.

Acknowledgments: The authors would like to acknowledge members of the Henry lab, especially Mary Astumian for technical assistance. We would also like to thank Mark Nilan for excellent fish care. This work was supported by the March of Dimes Award number #1-FY14-284 to C.A.H. and the National Institute of Child Health and Human Development Award number R15HD088217 to C.A.H. The funders had no role in study design, data collection, data analysis, decision to publish, or preparation of the manuscript.

Author Contributions: E.C., M.P., and M.G. performed the experiments; E.C., M.P., M.G., and C.H. conceived and designed the experiments, analyzed the data, and wrote the paper.

Conflicts of Interest: The authors declare no conflict of interest.

References

1. Haberstick, B.C.; Young, S.E.; Zeiger, J.S.; Lessem, J.M.; Hewitt, J.K.; Hopfer, C.J. Prevalence and correlates of alcohol and cannabis use disorders in the United States: Results from the national longitudinal study of adolescent health. *Drug Alcohol. Depend.* **2014**, *136*, 158–161. [CrossRef] [PubMed]
2. Rehm, J. The Risks Associated With Alcohol Use and Alcoholism. *Alcohol. Res. Health* **2011**, *34*, 135–143. [PubMed]
3. Preedy, V.R.; Adachi, J.; Ueno, Y.; Ahmed, S.; Mantle, D.; Mullatti, N.; Rajendram, R.; Peters, T.J. Alcoholic skeletal muscle myopathy: Definitions, features, contribution of neuropathy, impact and diagnosis. *Eur. J. Neurol.* **2001**, *8*, 677–687. [CrossRef] [PubMed]
4. Memo, L.; Gnoato, E.; Caminiti, S.; Pichini, S.; Tarani, L. Fetal alcohol spectrum disorders and fetal alcohol syndrome: The state of the art and new diagnostic tools. *Early Hum. Dev.* **2013**, *89* (Suppl. S1), S40–S43. [CrossRef]
5. Centers for Disease Control and Prevention (CDC). Fetal alcohol syndrome—Alaska, Arizona, Colorado, and New York, 1995–1997. *Morb. Mortal. Wkly. Rep.* **2002**, *51*, 433–435.
6. Chaudhuri, J.D. Effect of a single dose of ethanol on developing skeletal muscle of chick embryos. *Alcohol* **2004**, *34*, 279–283. [CrossRef] [PubMed]
7. David, P.; Subramaniam, K. Prenatal alcohol exposure and early postnatal changes in the developing nerve-muscle system. *Birth Defects Res. A Clin. Mol. Teratol.* **2005**, *73*, 897–903. [CrossRef] [PubMed]
8. Sylvain, N.J.; Brewster, D.L.; Ali, D.W. Zebrafish embryos exposed to alcohol undergo abnormal development of motor neurons and muscle fibers. *Neurotoxicol. Teratol.* **2010**, *32*, 472–480. [CrossRef] [PubMed]
9. Dlugos, C.A.; Rabin, R.A. Ethanol effects on three strains of zebrafish: Model system for genetic investigations. *Pharmacol. Biochem. Behav.* **2003**, *74*, 471–480. [CrossRef]

10. Khayrullin, A.; Smith, L.; Mistry, D.; Dukes, A.; Pan, Y.A.; Hamrick, M.W. Chronic alcohol exposure induces muscle atrophy (myopathy) in zebrafish and alters the expression of microRNAs targeting the Notch pathway in skeletal muscle. *Biochem. Biophys. Res. Commun.* **2016**, *479*, 590–595. [CrossRef] [PubMed]
11. Gillies, A.R.; Lieber, R.L. Structure and Function of the Skeletal Muscle Extracellular Matrix. *Muscle Nerve* **2011**, *44*, 318–331. [CrossRef] [PubMed]
12. Bajanca, F.; Luz, M.; Duxson, M.J.; Thorsteinsdóttir, S. Integrins in the mouse myotome: Developmental changes and differences between the epaxial and hypaxial lineage. *Dev. Dyn.* **2004**, *231*, 402–415. [CrossRef] [PubMed]
13. Bentzinger, C.F.; Wang, Y.X.; von Maltzahn, J.; Soleimani, V.D.; Yin, H.; Rudnicki, M.A. Fibronectin regulates Wnt7a signaling and satellite cell expansion. *Cell Stem Cell* **2013**, *12*, 75–87. [CrossRef] [PubMed]
14. Costa, M.L.; Escaleira, R.C.; Jazenko, F.; Mermelstein, C.S. Cell adhesion in zebrafish myogenesis: Distribution of intermediate filaments, microfilaments, intracellular adhesion structures and extracellular matrix. *Cell Motil. Cytoskeleton* **2008**, *65*, 801–815. [CrossRef] [PubMed]
15. Deries, M.; Gonçalves, A.B.; Vaz, R.; Martins, G.G.; Rodrigues, G.; Thorsteinsdóttir, S. Extracellular matrix remodeling accompanies axial muscle development and morphogenesis in the mouse. *Dev. Dyn.* **2012**, *241*, 350–364. [CrossRef] [PubMed]
16. Marques, L.; Thorsteinsdóttir, S. Dynamics of Akt activation during mouse embryo development: Distinct subcellular patterns distinguish proliferating versus differentiating cells. *Differ. Res. Biol. Divers.* **2013**, *86*, 48–56. [CrossRef] [PubMed]
17. Snow, C.J.; Henry, C.A. Dynamic formation of microenvironments at the myotendinous junction correlates with muscle fiber morphogenesis in zebrafish. *Gene Expr. Patterns* **2009**, *9*, 37–42. [CrossRef] [PubMed]
18. Gemballa, S.; Vogel, F. Spatial arrangement of white muscle fibers and myoseptal tendons in fishes. *Comp. Biochem. Physiol. A Mol. Integr. Physiol.* **2002**, *133*, 1013–1037. [CrossRef]
19. Han, R.; Kanagawa, M.; Yoshida-Moriguchi, T.; Rader, E.P.; Ng, R.A.; Michele, D.E.; Muirhead, D.E.; Kunz, S.; Moore, S.A.; Iannaccone, S.T.; et al. Basal lamina strengthens cell membrane integrity via the laminin G domain-binding motif of alpha-dystroglycan. *Proc. Natl. Acad. Sci. USA.* **2009**, *106*, 12573–12579. [CrossRef] [PubMed]
20. Hoffman, E.P.; Brown, R.H.; Kunkel, L.M. Dystrophin: The protein product of the Duchenne muscular dystrophy locus. *Cell* **1987**, *51*, 919–928. [CrossRef]
21. Tomé, F.M.; Evangelista, T.; Leclerc, A.; Sunada, Y.; Manole, E.; Estournet, B.; Barois, A.; Campbell, K.P.; Fardeau, M. Congenital muscular dystrophy with merosin deficiency. *C. R. Acad. Sci. III* **1994**, *317*, 351–357. [PubMed]
22. Matsuda, R.; Nishikawa, A.; Tanaka, H. Visualization of dystrophic muscle fibers in mdx mouse by vital staining with Evans blue: Evidence of apoptosis in dystrophin-deficient muscle. *J. Biochem. (Tokyo)* **1995**, *118*, 959–964. [CrossRef] [PubMed]
23. Straub, V.; Rafael, J.A.; Chamberlain, J.S.; Campbell, K.P. Animal models for muscular dystrophy show different patterns of sarcolemmal disruption. *J. Cell Biol.* **1997**, *139*, 375–385. [CrossRef] [PubMed]
24. Hall, T.E.; Bryson-Richardson, R.J.; Berger, S.; Jacoby, A.S.; Cole, N.J.; Hollway, G.E.; Berger, J.; Currie, P.D. The zebrafish candyfloss mutant implicates extracellular matrix adhesion failure in laminin alpha2-deficient congenital muscular dystrophy. *Proc. Natl. Acad. Sci. USA* **2007**, *104*, 7092–7097. [CrossRef] [PubMed]
25. Goody, M.; Jurczyszak, D.; Kim, C.; Henry, C. Influenza A Virus Infection Damages Zebrafish Skeletal Muscle and Exacerbates Disease in Zebrafish Modeling Duchenne Muscular Dystrophy. *PLoS Curr.* **2017**, *9*. [CrossRef] [PubMed]
26. Goody, M.F.; Kelly, M.W.; Reynolds, C.J.; Khalil, A.; Crawford, B.D.; Henry, C.A. NAD+ biosynthesis ameliorates a zebrafish model of muscular dystrophy. *PLoS Biol.* **2012**, *10*, e1001409. [CrossRef] [PubMed]
27. Kanther, M.; Sun, X.; Mühlbauer, M.; Mackey, L.C.; Flynn, E.J.; Bagnat, M.; Jobin, C.; Rawls, J.F. Microbial colonization induces dynamic temporal and spatial patterns of NF-κB activation in the zebrafish digestive tract. *Gastroenterology* **2011**, *141*, 197–207. [CrossRef] [PubMed]
28. Berghmans, S.; Murphey, R.D.; Wienholds, E.; Neuberg, D.; Kutok, J.L.; Fletcher, C.D.M.; Morris, J.P.; Liu, T.X.; Schulte-Merker, S.; Kanki, J.P.; et al. tp53 mutant zebrafish develop malignant peripheral nerve sheath tumors. *Proc. Natl. Acad. Sci. USA.* **2005**, *102*, 407–412. [CrossRef] [PubMed]
29. Bassett, D.; Currie, P.D. Identification of a zebrafish model of muscular dystrophy. *Clin. Exp. Pharmacol. Physiol.* **2004**, *31*, 537–540. [CrossRef] [PubMed]

30. Kimmel, C.B.; Ballard, W.W.; Kimmel, S.R.; Ullmann, B.; Schilling, T.F. Stages of embryonic development of the zebrafish. *Dev. Dyn.* **1995**, *203*, 253–310. [CrossRef] [PubMed]
31. Johnson, N.M.; Farr, G.H.; Maves, L. The HDAC Inhibitor TSA Ameliorates a Zebrafish Model of Duchenne Muscular Dystrophy. *PLoS Curr.* **2013**, *5*. [CrossRef] [PubMed]
32. Goody, M.F.; Kelly, M.W.; Lessard, K.N.; Khalil, A.; Henry, C.A. Nrk2b-mediated NAD+ production regulates cell adhesion and is required for muscle morphogenesis in vivo: Nrk2b and NAD+ in muscle morphogenesis. *Dev. Biol.* **2010**, *344*, 809–826. [CrossRef] [PubMed]
33. Snow, C.J.; Goody, M.; Kelly, M.W.; Oster, E.C.; Jones, R.; Khalil, A.; Henry, C.A. Time-lapse analysis and mathematical characterization elucidate novel mechanisms underlying muscle morphogenesis. *PLoS Genet.* **2008**, *4*, e1000219. [CrossRef] [PubMed]
34. Hollway, G.E.; Bryson-Richardson, R.J.; Berger, S.; Cole, N.J.; Hall, T.E.; Currie, P.D. Whole-somite rotation generates muscle progenitor cell compartments in the developing zebrafish embryo. *Dev. Cell* **2007**, *12*, 207–219. [CrossRef] [PubMed]
35. Stellabotte, F.; Dobbs-McAuliffe, B.; Fernández, D.A.; Feng, X.; Devoto, S.H. Dynamic somite cell rearrangements lead to distinct waves of myotome growth. *Dev. Camb. Engl.* **2007**, *134*, 1253–1257. [CrossRef] [PubMed]
36. Goody, M.F.; Sher, R.B.; Henry, C.A. Hanging on for the ride: Adhesion to the extracellular matrix mediates cellular responses in skeletal muscle morphogenesis and disease. *Dev. Biol.* **2015**, *401*, 75–91. [CrossRef] [PubMed]
37. Baldelli, S.; Ciriolo, M.R. Altered S-nitrosylation of p53 is responsible for impaired antioxidant response in skeletal muscle during aging. *Aging* **2016**, *8*, 3450–3467. [CrossRef] [PubMed]
38. Schwarzkopf, M.; Coletti, D.; Sassoon, D.; Marazzi, G. Muscle cachexia is regulated by a p53-PW1/Peg3 -dependent pathway. *Genes Dev.* **2006**, *20*, 3440–3452. [CrossRef] [PubMed]
39. Bassett, D.I.; Bryson-Richardson, R.J.; Daggett, D.F.; Gautier, P.; Keenan, D.G.; Currie, P.D. Dystrophin is required for the formation of stable muscle attachments in the zebrafish embryo. *Dev. Camb. Engl.* **2003**, *130*, 5851–5860. [CrossRef] [PubMed]
40. Roostalu, U.; Strähle, U. In vivo imaging of molecular interactions at damaged sarcolemma. *Dev. Cell* **2012**, *22*, 515–529. [CrossRef] [PubMed]
41. Doney, R.; Lucas, B.R.; Watkins, R.E.; Tsang, T.W.; Sauer, K.; Howat, P.; Latimer, J.; Fitzpatrick, J.P.; Oscar, J.; Carter, M.; et al. Visual-motor integration, visual perception, and fine motor coordination in a population of children with high levels of Fetal Alcohol Spectrum Disorder. *Res. Dev. Disabil.* **2016**, *55*, 346–357. [CrossRef] [PubMed]
42. Lovely, C.B.; Nobles, R.D.; Eberhart, J.K. Developmental age strengthens barriers to ethanol accumulation in zebrafish. *Alcohol. Fayettev. N* **2014**, *48*, 595–602. [CrossRef] [PubMed]
43. Reimers, M.J.; Flockton, A.R.; Tanguay, R.L. Ethanol- and acetaldehyde-mediated developmental toxicity in zebrafish. *Neurotoxicol. Teratol.* **2004**, *26*, 769–781. [CrossRef] [PubMed]

6

Requirement of the Dynein-Adaptor Spindly for Mitotic and Post-Mitotic Functions in *Drosophila*

Giuliana D. Clemente [1], Matthew R. Hannaford [1], Hamze Beati [1,2], Katja Kapp [2], Jens Januschke [1], Eric R. Griffis [3,†] and Hans-Arno J. Müller [1,2,*]

1 Division of Cell and Developmental Biology, School of Life Sciences, University of Dundee, Dundee DD1 5EH, UK; giuliana.clemente@manchester.ac.uk (G.D.C.); m.r.hannaford@dundee.ac.uk (M.R.H.); hamze.beati@uni-kassel.de (H.B.); j.januschke@dundee.ac.uk (J.J.)

2 Institut für Biologie, Fachgebiet Entwicklungsgenetik, Universität Kassel, Heinrich-Plett Str. 40, 34321 Kassel, Germany; k.kapp@uni-kassel.de

3 Centre of Gene Regulation and Expression, School of Life Sciences, University of Dundee, Dundee DD1 5EH, UK; ergriffis@gmail.com

* Correspondence: h.a.muller@uni-kassel.de

† Present Address: Nikon Instruments Incorporated, 1300 Walt Whitman Rd., Melville, NY 11747, USA.

Abstract: Spindly was originally identified as a specific regulator of Dynein activity at the kinetochore. In early prometaphase, Spindly recruits the Dynein/Dynactin complex, promoting the establishment of stable kinetochore-microtubule interactions and progression into anaphase. While details of Spindly function in mitosis have been worked out in cultured human cells and in the *C. elegans* zygote, the function of Spindly within the context of an organism has not yet been addressed. Here, we present loss- and gain-of-function studies of Spindly using transgenic RNAi in *Drosophila*. Knock-down of Spindly in the female germ line results in mitotic arrest during embryonic cleavage divisions. We investigated the requirements of Spindly protein domains for its localisation and function, and found that the carboxy-terminal region controls Spindly localisation in a cell-type specific manner. Overexpression of Spindly in the female germ line is embryonic lethal and results in altered egg morphology. To determine whether Spindly plays a role in post-mitotic cells, we altered Spindly protein levels in migrating cells and found that ovarian border cell migration is sensitive to the levels of Spindly protein. Our study uncovers novel functions of Spindly and a differential, functional requirement for its carboxy-terminal region in *Drosophila*.

Keywords: *Drosophila*; mitosis; cell migration; mitotic spindle; Dynein

---

1. Introduction

In early *Drosophila* embryogenesis, nuclei divide in a common cytoplasm without intervening cytokinesis. The nuclei divide synchronously in rapid mitotic cycles oscillating between S- and M-phases that are supported by maternal-derived gene products [1,2]. As for an archetypal cell cycle, entry into mitosis is driven by the activation of Cdk1-Cyclin-B complexes, whereas mitotic exit is accompanied by the timely inactivation of these complexes in a wave that propagates from the spindle poles towards the metaphase plate [3,4]. In most cell types, mitotic exit is under the strict control of the Spindle Assembly Checkpoint (SAC), which halts the progression into anaphase by inhibiting the activity of the E3 ubiquitin ligase APC/C until each chromosome is stably attached to the fibres of the mitotic spindle [5–7]. Once all of the chromosomes have achieved bi-orientation, the SAC is inactivated by the Dynein-dependent transport of key SAC components, such as Mad1 and Mad2, from the outer kinetochores [8–13]. Mitotic control and SAC regulation in *Drosophila* syncytial embryos are unique. In this system, mitosis is very robust and it does not need the SAC to ensure faithful chromosome segregation [14,15]. Indeed, *mad-2*-null embryos undergo mitosis, normally with very little rate of aneuploidy and develop into viable, fertile adults [14].

Despite the lack of the requirement of a SAC in normal *Drosophila* syncytial divisions, SAC activity is still crucial when microtubule attachment and chromosome alignment are compromised. Mutations in genes encoding for the checkpoint proteins Rough deal (Rod), ZW-10, Mps1, Bubr1 and Bub3 are lethal and accompanied by a high rate of aneuploidy [16–20]. These severe mitotic phenotypes were attributed not only to defects in the mitotic checkpoint, but also to impaired kinetochore activity. A good example is offered by the evolutionary conserved ROD-Zwilch-ZW10 (RZZ) complex [21]; the RZZ complex has a dual activity in controlling the "on-off" state of the mitotic checkpoint. First, it promotes the stable localisation of Mad1-Mad2 at kinetochores of unattached chromosomes [12,22]. Secondly, by recruiting Dynein to kinetochores, the RZZ complex regulates the establishment of stable end-on kinetochore-microtubule attachments, thereby contributing to the bilateral orientation of chromosomes at the spindle equator. Finally, it contributes to SAC silencing through the recruitment of Dynein upon microtubule-kinetochore attachment [23,24].

How does the RZZ complex localise Dynein at kinetochores? The RZZ complex not only directly interacts with the Dynactin complex and the Dynein intermediate chain, but it localises Dynein also by interaction with the Dynein co-factor Spindly [9,25–29]. *Drosophila* Spindly has a predicted rod-like structure that is organised in four coiled-coil domains in the amino-terminal (N-terminal) half and four positively-charged repeats in the carboxy-terminal (C-terminal) half of the protein [9]. Spindly localises to the outer kinetochore by directly interacting with the RZZ complex through its C-terminal domain [30–33]. Once bound to RZZ, Spindly extends its N-terminal half out from the kinetochore to bind Dynein [27,30,34]. In this respect, Spindly shares similarities to other Dynein-adaptor proteins, such as BicD-2 [35,36], which rely on their C-terminal domain to interact with cargoes (i.e., the kinetochore) while the N-terminal portion mediates interaction with the Dynein-Dynactin motor complex [9]. Thus, Spindly controls cell-cycle progression by linking the RZZ complex with the Dynein motor, thereby acting as a molecular bridge between chromosome bi-orientation and Dynein-dependent checkpoint silencing [27]. In addition to its role in SAC function, the phenotype of Spindly depleted S2 cells during interphase suggested a role for Spindly in the control of cell shape and the organisation of the cytoskeleton [9]. However, a function for Spindly in a tissue or an organismal context has thus far remained elusive.

Here, we analysed Spindly function in *Drosophila* using loss-of-function and gain-of-function experiments in a tissue-specific fashion. In neuroblasts, Green Fluorescent Protein (GFP)-tagged Spindly localises dynamically during mitosis and like the human protein, its localisation to the kinetochore depends on the C-terminal region. Conversely, deletion of the C-terminal region only mildly affects Spindly function in the early embryo, and it is still able to localise to the kinetochore in a dynamic fashion. Thus, *Drosophila* Spindly exhibits a differential requirement for its C-terminal region in different cell types. In addition to the mitotic defects due to the loss of function, overexpression of Spindly in the germ line resulted in eggs with altered shape, a mild ventralisation, and severely compromised embryogenesis. Finally, we demonstrate that normal levels of Spindly protein are required for migration of the follicular border cells in oogenesis. We conclude that, in addition to its functions in mitotic cells, *Drosophila* Spindly is important for the movement of post-mitotic cells.

2. Results

2.1. Requirements for Dynamic Subcellular Localisation of Spindly

Spindly localises dynamically in dividing cultured cells [9,27,30]. During prophase and metaphase, Spindly associates with the kinetochore and moves poleward in early anaphase. While details of the requirement of Spindly protein domains for its localisation are available for the human homologue [30], this information is still missing for *Drosophila* Spindly. Therefore, to perform a domain-function analysis, we generated a similar range of GFP-tagged Spindly protein constructs, as studied by Barisic et al. [30] and expressed them in larval neuroblasts, using the Gal4/UAS system [37] (Figure 1A). Larval neuroblasts (NBs) are rather large cells, which can be cultured in vitro and are readily identified and imaged during mitosis at high spatial resolution [38]. For these reasons, larval NBs provide an excellent model to study the dynamic localisation of proteins during mitosis. Overexpression of GFP::Spindly in neuroblasts revealed that the protein localises to the kinetochores in a dynamic fashion during mitosis (Figuer 1B, Supplementary Materials Movie 1). A construct that lacks the C-terminal 380 amino acids did not localise to kinetochores during neuroblast division, suggesting that, like in humans and *C. elegans,* the C-terminal half is essential for normal Spindly localisation to kinetochores (Figure 1C; Supplementary Materials Movie 2). All the other constructs localised to the kinetochores during mitosis (Figure 1D–F; Supplementary Materials Movies 3–5). Another interesting observation was that in 78% of cases (n = 40), the protein lacking the N-terminal 228 amino acids localised to cytoplasmic punctae in interphase, which were never observed with any of the other constructs (Figure 1E; Supplementary Materials Movie 4). The speckles do not seem to be polarised in any way. With the caveat that overexpression of GFP-conjugated fusion proteins may cause the abnormal localisation of proteins in a wild type background, our results suggest that the N-terminal domain is required for the normal distribution of the protein in interphase. Furthermore, our experiments confirm that similar to human Spindly, the C-terminal domain is important for binding to the kinetochore.

2.2. Requirement of Spindly for Initiation of Syncytial Cleavage Division

Since Spindly knock-down by RNAi worked efficiently in *Drosophila* S2 cells [9], we selected shRNAs that were directed against *spindly* mRNA that can be expressed under temporal and spatial control by the Gal4/UAS system [26,37,39]. A maternal Gal4 driver (alpha-*tubulin67>Gal4*) was used to express UAS-controlled shRNA targeting *spindly* mRNA in oocytes and embryos (called *spindly*[mat67>RNAi] hereafter). The efficacy of the transgenic RNAi knockdown was assayed by immunoblotting of embryo lysates with an affinity-purified Spindly antibody [9], and a marked reduction in the levels of Spindly protein was observed at all of the temperatures tested (Supplementary Materials Figure S1).

spindly[mat67>RNAi] had severe effects on embryogenesis causing a remarkable embryonic lethality (Figure 2A). The well-established function of Spindly for mitosis suggested that *spindly*[mat67>RNAi] should affect the syncytial mitotic divisions in the early embryo explaining the embryonic lethality upon its depletion (Figure 2A). Indeed, the immunostaining of mitotic spindles in early embryos revealed that 83% (n = 100) of the *spindly*[mat67>RNAi] embryos showed abnormal syncytial cleavage divisions (Figure 2B–F). The majority of embryos arrested during metaphase of the first 1–3 mitotic divisions (Figure 2C–F). In 40% of the embryos (n = 30), only a single metaphase-arrested mitotic spindle was detected within the egg (Figure 2E,E′), whereas only in 13% of the cases a variable number of two to four mitotic spindles were observed. We also detected in many cases (46.6%) characteristic polar body rosettes [40], suggesting that egg activation occurs and meiosis is able to resume in *spindly*[mat67>RNAi] embryos (Figure 2F,F′). We conclude that Spindly is essential for syncytial cleavage division to commence.

2.3. Requirement of Spindly for Syncytial Cleavage Divisions

The alpha-*tubulin67>Gal4* driver exhibits a high level of Gal4 expression in the female germ line, which blocks syncytial cleavages when driving shRNAs directed against *spindly* mRNA. The expression of alpha-*tubulin67>Gal4* was first detected when egg chambers leave the germarium (Supplementary Materials Figure S2A). To analyse the requirement of Spindly during syncytial divisions, we expressed *spindly* shRNA using the weaker *nanos (nos)*>Gal4 driver (hereafter *spindly*[*nos*>RNAi]). According to the *nos* expression pattern, nos>Gal4 should be already expressed in the germarium [41,42]. After 3–5 days, *spindly*[*nos*>RNAi] females stopped laying eggs, presumably because the knock-down of Spindly blocked the mitotic division in the germarium, including stem cell divisions. *spindly*[*nos*>RNAi] females produced embryos in which Spindly was efficiently knocked down in a temperature-dependent fashion (Figure 3A–C). At 21 °C to 25 °C, *spindly*[*nos*>RNAi] efficiently depleted Spindly protein in early embryos, being the most efficient at 23 °C to 25 °C (Figure 3A). *spindly*[*nos*>RNAi] embryos exhibited temperature-dependent rates of survival: at 23 °C only 16% (n = 100) of embryos hatched, while at 21 °C, about half of the *spindly*[*nos*>RNAi] embryos survived to larval stages (n = 100) (Figure 3B). Immunostaining of *spindly*[*nos*>RNAi] embryos was performed to further analyse the cause of the embryonic lethality and revealed a requirement of Spindly for normal syncytial divisions (Figure 3C).

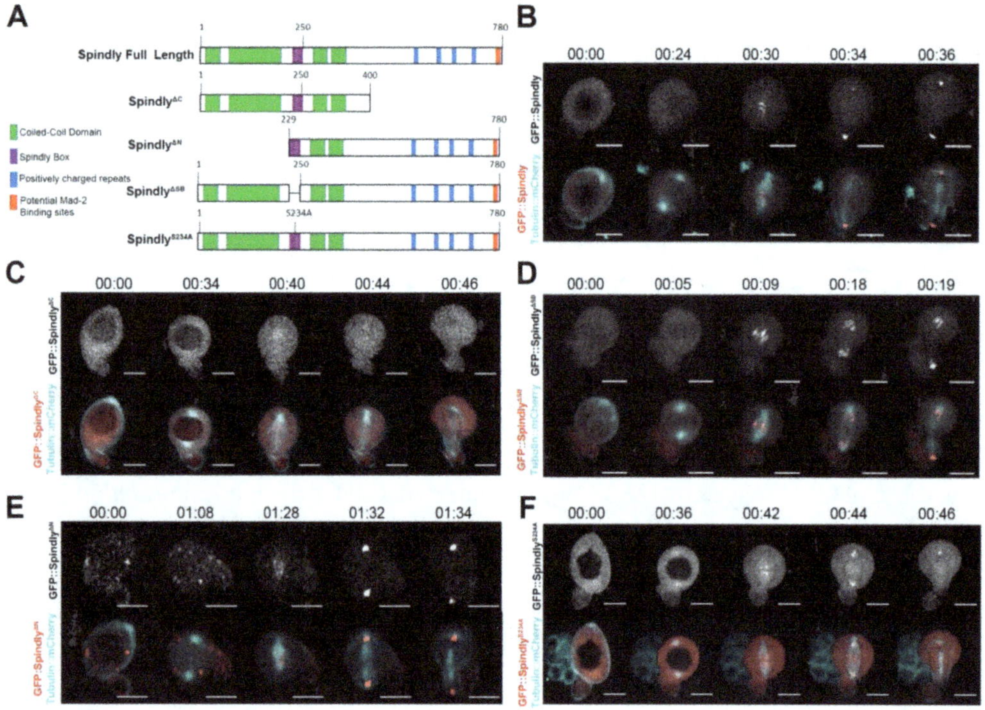

Figure 1. Identification of Spindly protein domain(s) required for dynamic localisation in mitosis. (**A**) Schematic drawing showing the full-length *Drosophila* Spindly protein and mutant constructs generated in this study. Spindly is a predicted coiled-coil domain containing protein (N-terminal half, in green) characterised by a positively charged C-terminal domain (in blue). The highly conserved repeat motif is depicted (Spindly box, in purple). Spindly localises to the outer kinetochore plate by interacting with the RZZ complex through its C-terminal domain [30–33]. In addition to the importance of the C-terminal half, other regions of the protein contribute to kinetochore targeting. Human Spindly requires the first coil-coiled domain, the Spindly box and a region between residues 440–564 for kinetochore binding [30]; (**B**–**F**) Representative examples of dividing neuroblasts overexpressing the indicated *UAS*>>[Green Fluorescent Protein (GFP)]::Spindly variants under the control of the *worniu*>Gal4 driver; (**B**) GFP::Spindly full length gets recruited to kinetochores at early pro-metaphase (00:26) and can be seen moving from KT to poles at anaphase (00:32). In all cases (n = 21) the protein localised this way; (**C**) GFP::Spindly$^{\Delta Ct}$ never localised at kinetochores throughout mitosis (n = 43). Mutant constructs deleted of the Spindly box (GFP::Spindly$^{\Delta SB}$) (n = 21) (**D**); N-terminal half (n = 40) (**E**) or a point mutant S234A (GFP::Spindly^{S234A}) (n = 33) (**F**) showed a localisation dynamics similar to the full-length protein. Note the speckled pattern of GFP::Spindly$^{\Delta Nt}$ in interphase, which was observed in 31/40 cases. The minor differences between the distribution of the full length GFP::Spindly and the GFP::Spindly$^{\Delta SB}$ (**D**) or GFP::Spindly^{S234A} (**F**) proteins are due to small variations in the expression of the *wor*>*Gal4* driver and slight differences in the time resolution of the still images; for a full comparison, we refer to the respective movies. Time stamp: hh:min. Scale bars: 10 μm.

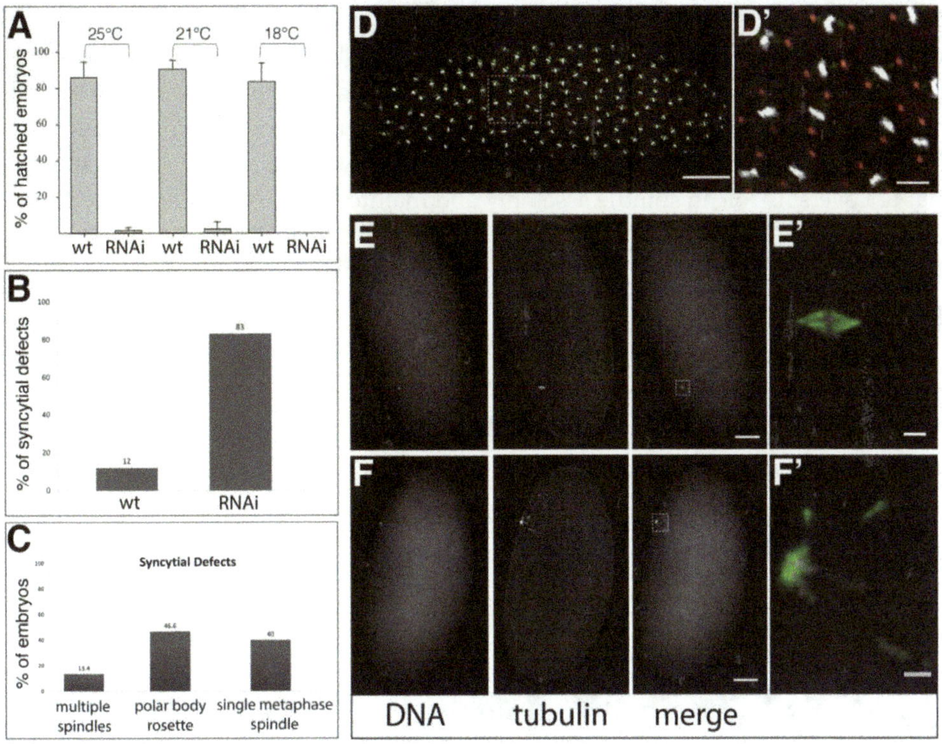

Figure 2. Phenotype of *mat-tub*>>*Gal4* driven Spindly knock-down. (**A**) Larval hatching rates of *spindly*[*mat67*>>RNAi] embryos at temperatures indicated ($n = 100$); for protein levels see Supplementary Materials Figure S1; (**B**) Penetrance of mitotic arrest during syncytial divisions in *spindly mat67*>>RNAi embryos ($n = 100$). (**C**) Expressivity of syncytial cleavage defects in *spindly mat67*>>RNAi embryos ($n = 30$); (**D**,**D′**) Mitotic domains in wild-type syncytial blastoderm embryo. Nuclei are stained with DAPI (grey), anti-beta-tubulin (green) and anti-Centrosomin (red). Scale bars represent 50 μm (**D**) and 10 μm (**D′**); (**E**,**F**) Syncytial division defects in *spindly mat67*>>RNAi embryos. The images are maximum-intensity projections of 10 μm z-stack (at spacing of 1 μm); (**E**,**E′**) Embryos often contained a single metaphase-arrested nucleus; (**F**,**F′**) In some cases condensed polar body rosettes were visible. (DAPI (grey) and beta-tubulin (green). Scale bars: 50 μm (**E**,**F**) and 5 μm (**E′**,**F′**).

Fluorescence microscopy was used to assess the mitotic defects in *spindly*[*nos*>RNAi] embryos by analysing centrosome positioning and mitotic spindle morphology. At 23 °C, the majority of *spindly*[*nos*>RNAi] embryos had arrested before the end of syncytial divisions (76% $n = 50$; Figure 3C). We further divided the phenotypes of these embryos into different classes (Figure 3D). 22% of all embryos did not show any DNA staining, while 20% arrested development after a few rounds of syncytial mitotic cycles. Of this latter class, 10% of the embryos were characterised by patches of multiple centrosomes (Figure 3E,E′). Another class exhibited the abnormal organisation of the mitotic spindle, where condensed chromatin was found associated with patches of tubulin (12%, $n = 50$; Figure 3F,F′). Unlike wild-type embryos, the nuclei of *spindly*[*nos*>RNAi] embryos were unequally distributed in the cortical cytoplasm of the embryos and the regular spacing of the mitotic domains was lost (Figure 3G). In summary, at the highest temperature tested, *spindly*[*nos*>RNAi] resulted in severe disorganisation of centrosome distribution, chromatin structure, and nuclear distribution in syncytial stages.

At 21 °C, only just under 50% of *spindly*[*nos*>RNAi] embryos failed to hatch and exhibited defects in the synchrony of syncytial divisions and various spindle defects (Figures 3C and 4A). The most severe spindle arrests might result from the abnormal spacing of the mitotic domains in the cortex of the syncytial embryo. Often mitotic spindles appeared to be stacked together, fusing into highly abnormal structures with accumulations of tubulin (Figure 4B). About 20% of the abnormal embryos exhibited various centrosome defects, including centrosomes that are detaching from one or both spindle poles or a pair of centrosomes that are associated with one or both poles (Figure 4C). The phenotypes suggest that centrosomes might undergo abnormal replication or they detach and move away from the spindle pole or that free centrosomes are left behind by nuclei dropping out of the cortex. The results suggest that the lower levels of *spindly*[*nos*>RNAi] mediated knock-down produced a threshold of Spindly activity that uncovers its role in spindle assembly and maintenance.

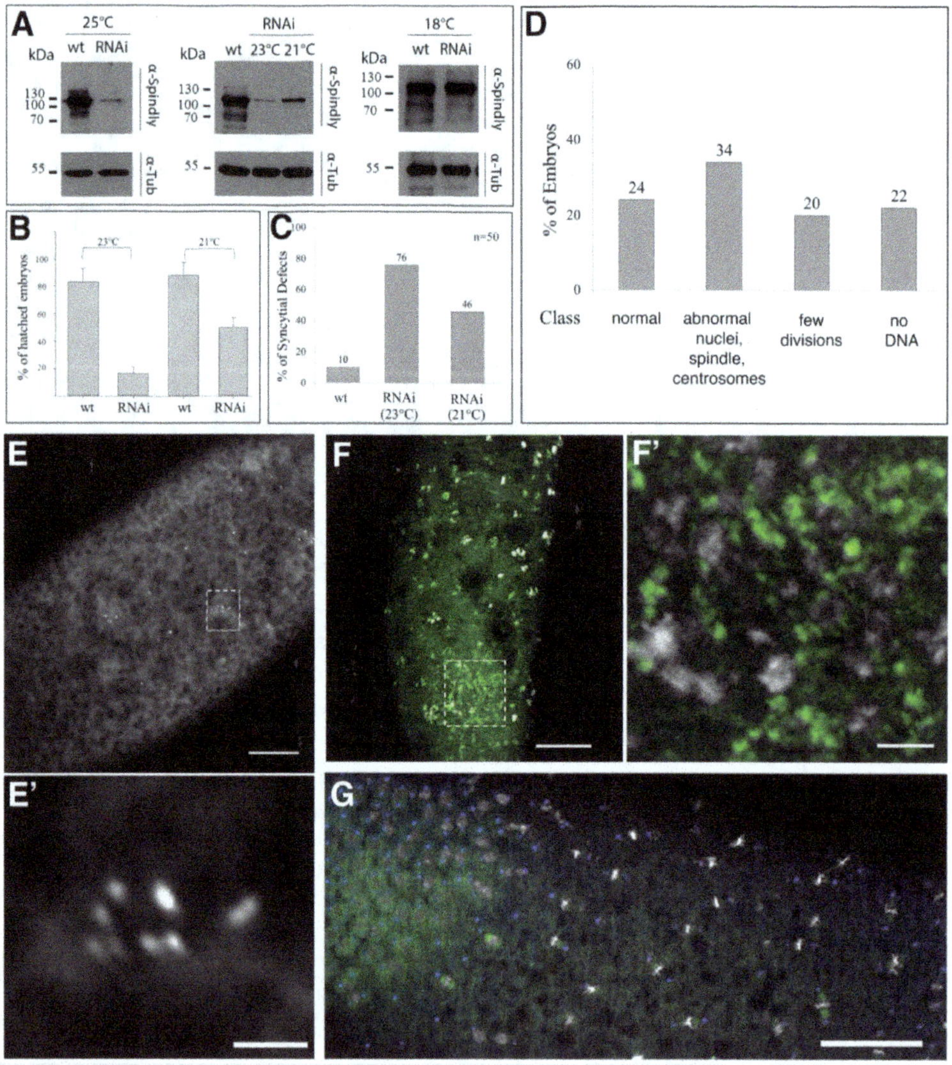

Figure 3. Phenotype of *nos*>>Gal4 driven Spindly knock-down. (**A**) Immunoblot analysis of protein lysates of 0–3 h old w^{1118} embryos (wt) or *spindly*[*nos*>RNAi] (RNAi) embryos at different temperatures (loading control—α-Tubulin (Tub); molecular weight standards in kDa); (**B**) Viability of *spindly*[*nos*>RNAi] scored by larval hatching rate (n = 100; data represent 3 independent replicates); (**C**) Penetrance of mitotic arrests in syncytial divisions of *spindly*[*nos*>RNAi] embryos featuring phenotypes described in (**D–G**); (**D**) Expressivity of syncytial cleavage defects of *spindly*[*nos*>RNAi] embryos at 23 °C. Occurrence of phenotypic classes of *spindly*[*nos*>RNAi] embryos (n = 50); (**E**) Syncytial stages with patches of multiple centrosomes (anti-Centrosomin staining); (**E′**) higher magnification of boxed area in (**E**); scale bars: 50 μm (**E**) and 10 μm (**E′**)); (**F**) Condensed chromatin associated with patches of tubulin (detail indicated by yellow box; DAPI (grey), beta-tubulin (green), scale bars represent 50 μm (**F**) and 10 μm (**F′**)); (**G**) Asynchrony of mitotic divisions, as indicated by an area of mitotic divisions and areas enriched in interphase nuclei (scale bar: 50 μm).

In early embryos, the observed fusion of mitotic spindles suggests a function of Spindly in the separation of mitotic domains during cortical syncytial divisions. A characteristic feature of *spindly*[*nos*>RNAi] syncytial embryos was a defect in synchrony of syncytial divisions. *spindly*[*nos*>RNAi] contained areas with different rates of mitotic activity, i.e., part of the embryo contained interphase nuclei, while in adjacent areas nuclei were still in mitotic phases (Figure 4D). This asynchrony in mitotic divisions and the uneven spacing of nuclei is quite common in mutants affecting mitotic progression and may suggest that Spindly protein levels are important for the wave-like progression of mitotic divisions in the syncytial embryo [43].

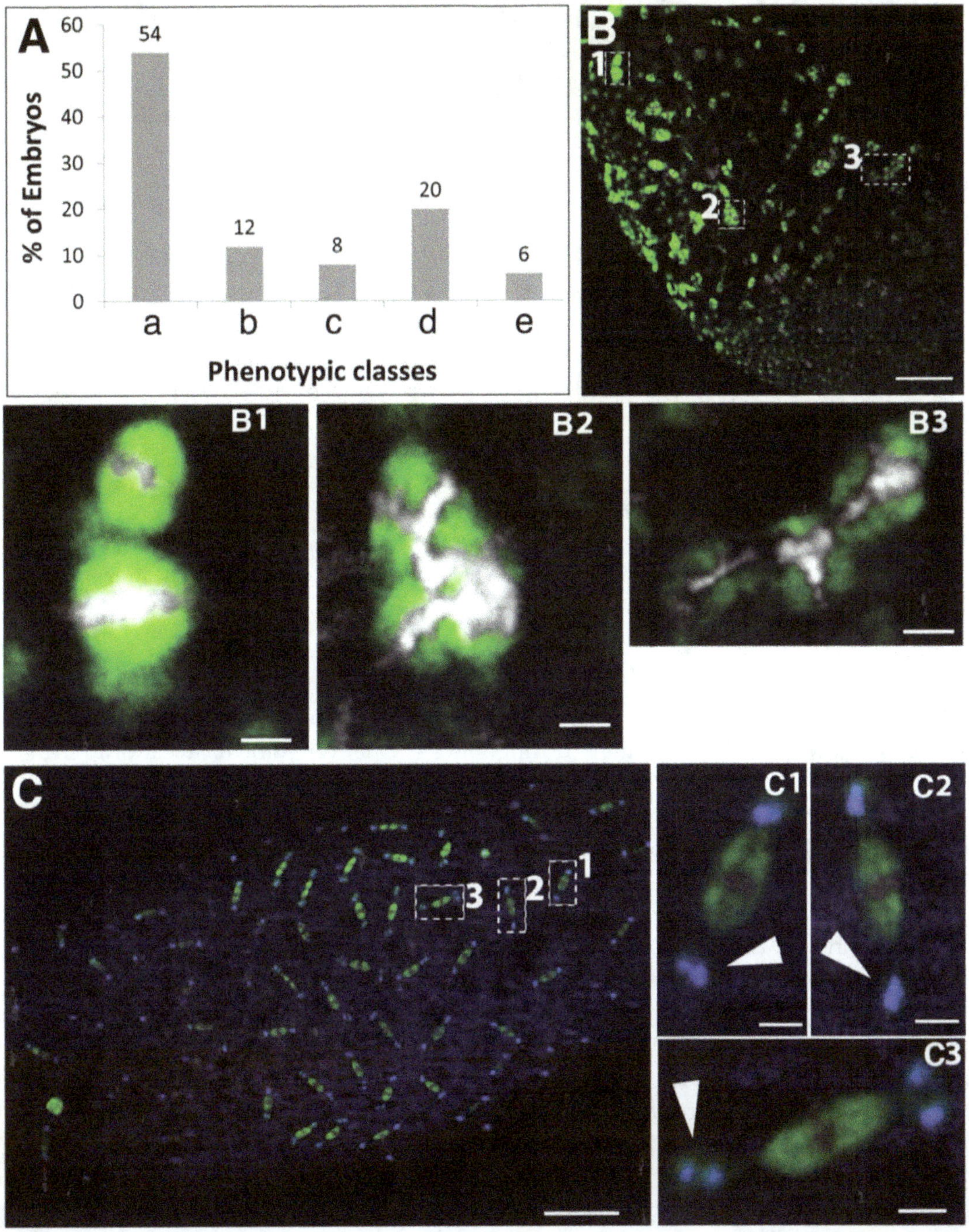

Figure 4. *Cont.*

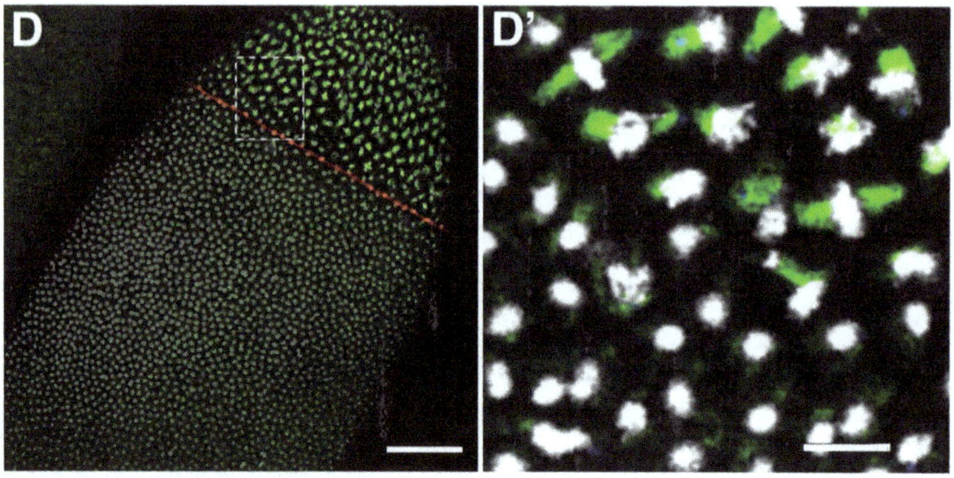

Figure 4. Expressivity of syncytial cleavage defects in *spindly*[*nos*>RNAi] embryos. (**A**) *spindly*[*nos*>RNAi] embryos raised at 21 °C displayed a range of mitotic defects of the following classes ($n = 50$): (a) normal development; (b) abnormal mitotic figures, examples in (**B**); (c) centrosome abnormalities (detachment, monopolar spindles, centrosome clustering; examples in (**C**)); (d) uneven spacing of the mitotic domains and asynchrony of nuclear divisions (examples in (**D**)); (e) dead embryos at later stages; (**B**) Abnormal mitotic figures; irregular spacing between adjacent mitotic domains. Details are shown from the dashed yellow boxes in merged panel. (**B1**) Mitotic spindles collapse and fuse; (**B2**) abnormal tubulin-rich structures associated with chromatin; (**B3**) tri-polar spindle (Scale bars represent 50 μm (**B**) and 5 μm (**B1–B3**); (**C**) Centrosome abnormalities: Defects in cohesion between centrosomes and spindle poles. Details from merged panel marked with a dashed yellow box: (**C1**) a pair of centrosomes associated to a single spindle pole (marked by arrowhead); (**C2,C3**) centrosomes detach from one (**C2**) or both poles (**C3**); arrowheads point to detaching centrosomes. Scale bar represents 50 μm (**C**) and 5 μm (**C1–C3**); (**D**) embryo with domains of distinct mitotic rate; red line marks boundary between a region of interphase nuclei and a region with mitotic divisions. (**D′**) higher magnification of boxed area in (**D**). Scale bars represent 50 μm (**D**) and 10 μm (**D′**).

2.4. Requirements of Spindly Protein Domains in Early Development

RNAi mediated knock-down of Spindly produced phenotypes that are consistent with a previously known function of Spindly during mitosis. One well-known caveat of RNAi is that associated phenotypes might be produced not (only) by the depletion of the cognate mRNA, but by the depletion of other mRNAs that share sequence similarity. To exclude that the effects of the knock-down were caused by such off-target effects we generated an RNAi-resistant construct encoding an N-terminally GFP-tagged full-length version of Spindly, in which the shRNA target site was mutated without changing the polypeptide sequence. The overexpression of the RNAi-resistant GFP-tagged Spindly transgene restored the viability of *spindly*[mat67>RNAi] embryos to wild-type levels (Figure 5A). The rescue also provided us with a basis for investigating the requirements for different protein domains of Spindly for its function and localisation (see Figure 1A). We found that the C-terminal region was largely dispensable for the rescue of *spindly*[mat67>RNAi] embryos, while the N-terminal region and the highly conserved Serine 234 residue within the Spindly box were essential for the Spindly function (Figure 5A). As a control for the expression of the protein constructs in the *spindly*[mat67>RNAi] background, we analysed the protein levels by Western blotting (Figure 5B). The GFP::Spindly[S234A] and the GFP::Spindly[Δ-N-term] proteins were expressed at levels that are similar to the GFP-tagged full length protein, indicating that the lack of rescue was not caused by poor protein stability or expression.

Our results indicated that the C-terminal half of Spindly was dispensable for its function in the embryo. However, in neuroblasts, the C-terminal region was required for the localisation of the protein to kinetochore (Figure 1C). This discrepancy raised the possibilities that either Spindly has kinetochore-independent functions that are sufficient to complement its loss in the embryo, or that Spindly binds to the kinetochore in a cell-type specific manner. To discriminate between these possibilities, we imaged GFP::Spindly and the GFP::Spindly[Δ-C-term] protein in *spindly*[mat67>RNAi] embryos (Figure 5C,D). We found that the GFP::Spindly accumulated on punctate structures as consistent with a localisation to the kinetochore in a dynamic fashion (Figure 5C,E; Supplementary Materials, Movie 6). In prometaphase, GFP::Spindly associated in a dot-like fashion with areas of condensed chromatin and during metaphase these punctae were lined up in a series of dots at the metaphase plate (Figure 5C). As anaphase proceeded, the punctae moved towards the spindle poles and eventually accumulated at one side of the reforming nucleus (Supplementary Materials Movie 6). The GFP::Spindly[Δ-C-term] protein behaved similarly to the full-length GFP::Spindly protein, however the accumulation into punctae on the condensed chromosomes were less pronounced compared to the full-length protein (Figure 5D,F). Nevertheless, the dynamics of the GFP::Spindly[Δ-C-term] punctae during mitosis was comparable to the full-length protein (Supplementary Materials, Movie 7). We conclude that, despite the requirement of the C-terminal region in neuroblasts, Spindly is able to localise without the C-terminal region in a similar fashion to the full-length protein, albeit in a somewhat reduced level.

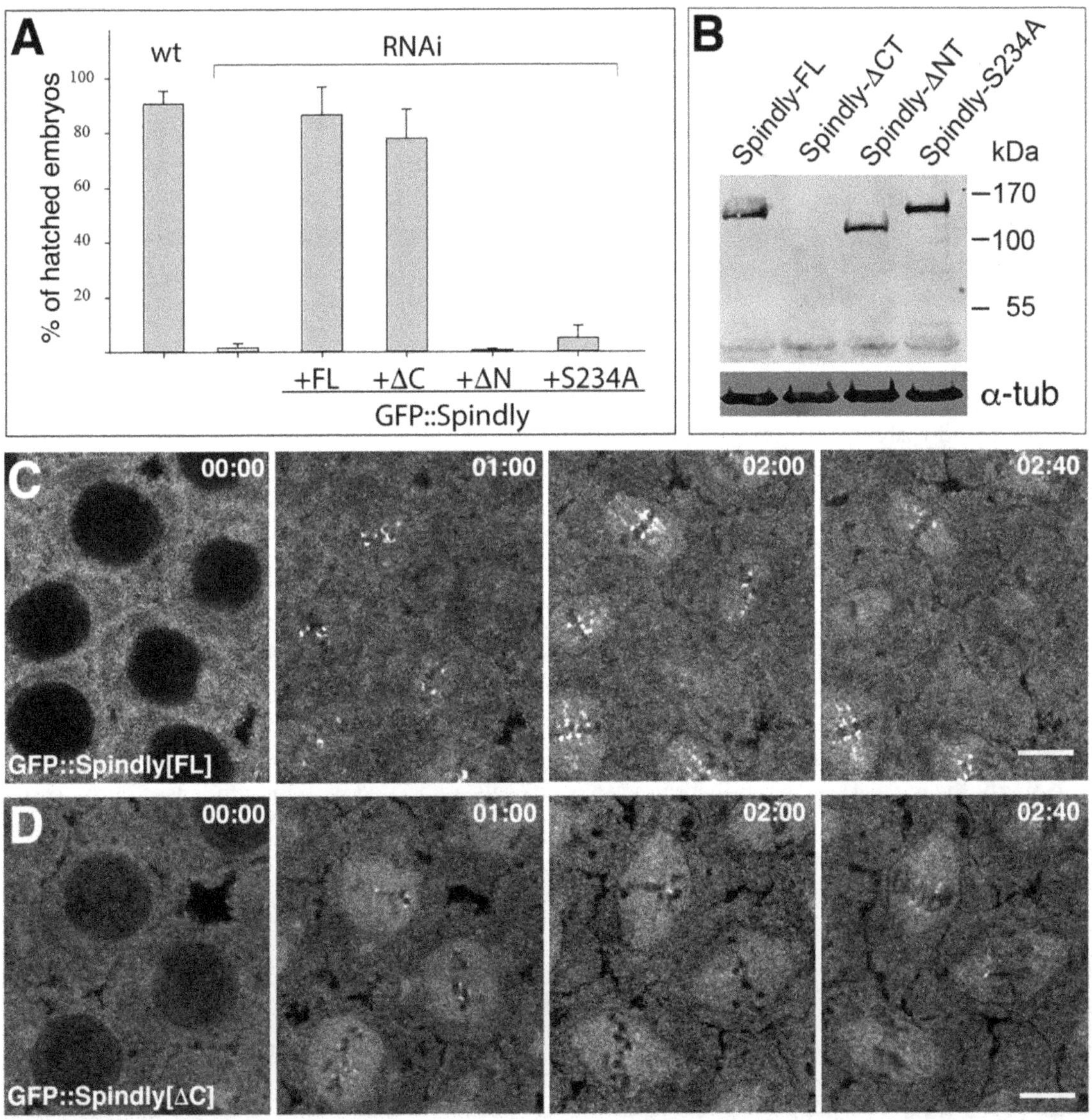

Figure 5. *Cont.*

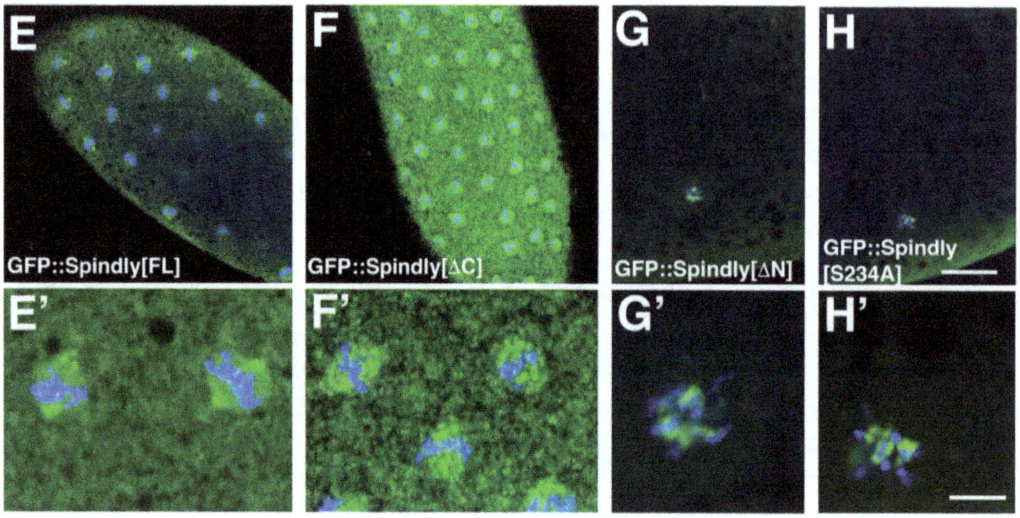

Figure 5. Domain-function analyses of Spindly in the embryo. (**A**) Larval hatching rates in wild type (wt; w^{1118}), *spindly*[*mat67*>>RNAi] (RNAi) and *spindly*[*mat67*>>RNAi] embryos expressing either GFP::Spindly full-length (+FL), GFP::Spindly$^{\Delta Ct}$ (+ΔCt), GFP::Spindly$^{\Delta Nt}$ (+ΔNt) or GFP::Spindly^{S234A} (+S234A) (n = 100); (**B**) Western blot analysis of lysates obtained from *spindly*[*mat67*>>RNAi] embryos expressing full length GFP::Spindly (Spindly-FL), GFP::Spindly[Δ-Cterm] (Spindly-ΔC), GFP::Spindly[Δ-Nterm] (Spindly-ΔN), GFP::Spindly[S234A] (Spindly-S234A). Affinity-purified anti Spindly antibodies were used for the analysis, which recognise the C-terminal domain of the protein. Therefore, no signal was detected from the GFP::Spindly[Δ-Cterm] construct; (**C**,**D**) Single frames from time-lapse movies (Supplementary Materials Movies 6,7, respectively) depicting syncytial cleavage division 12 of *spindly*[*mat67*>>RNAi] embryos expressing either full length GFP::Spindly protein (**C**) or GFP::Spindly lacking the C-terminal domain (Spindly-ΔC) (**D**); Note dynamic localisation of GFP::Spindly accumulating in punctate labelling associated with chromosomes, negatively stained darker areas. Scale bars: 5 µm; (**E**–**H**) *spindly*[*mat67*>>RNAi] embryos expressing either GFP::Spindly full-length (GFP::Spindly[FL]) (**E**,**E'**), GFP::Spindly$^{\Delta Ct}$ (GFP::Spindly[ΔC]) (**F**,**F'**), GFP::Spindly$^{\Delta N}$ (GFP::Spindly[ΔN]) (**G**,**G'**) or GFP::Spindly^{S234A} (GFP::Spindly[S234A], (**H**,**H'**) fixed and stained with anti-GFP (green) antibodies and DAPI (blue); (**E'**–**H'**) show corresponding higher power micrographs. Note accumulation of GFP::Spindly constructs to condensed chromosomes in (**E'**,**F'**); (**G**,**G'**,**H**,**H'**) show polar body rosettes; note the accumulation of GFP::Spindly constructs with the condensed chromosomes. Scale bars represent 50 µm for (**E**–**H**), and 5 µm for (**E'**–**H'**), respectively.

We also examined the distribution of the GFP::Spindly[S234A] and the GFP::Spindly[Δ-N-term] proteins, which were not able to rescue the *spindly*[mat67>RNAi] lethality (Figure 5G,H). The phenotype of such embryos was largely equal to the phenotype that was described for *spindly*[mat67>RNAi] (Figure 2), with most embryos exhibiting arrested meiosis stages with polar body rosettes (Figure 5G',H'). Interestingly, the mutant GFP::Spindly proteins were closely associated with the condensed chromosomes of the rosettes, suggesting that, similar to the neuroblasts, their chromosome binding was unimpaired, but that these proteins are otherwise non-functional. Furthermore, the result with the C-terminally deleted proteins demonstrated that *Drosophila* Spindly exhibits a differential cell-type specific requirement for its C-terminal region for binding to the kinetochore.

2.5. Spindly Overexpression Causes Defects in Egg Shape and Embryogenesis

The loss-of-function studies reported here demonstrated that maternally supplied Spindly is essential for sustaining the syncytial cell cycles in the embryo. Importantly, the phenotypes were specific for Spindly RNAi knockdown, as a full rescue of the defects was observed by the expression of an RNAi-resistant Spindly protein. In the course of these experiments, we noted that the expression of GFP::Spindly using the maternal *alpha-tubulin67c>Gal4,* in a wild-type background that is caused defects in embryogenesis. In *mat67>>GFP::Spindly* embryos, levels of Spindly-GFP were higher than endogenous protein levels in lysates from 0–3 h-old embryos at 25 °C (Figure 6A).

The majority (70%, $n = 100$) of *mat67>>GFP::Spindly* embryos did not hatch, but the co-expression of Spindly[shRNA] was able to fully rescue this effect, indicating that this phenotype was caused by an altered level of Spindly protein, rather than by an abnormal effect of the GFP::Spindly fusion protein (Figure 6B). The embryonic lethality of *mat67>>GFP::Spindly* resulted in three classes of larval cuticle phenotypes (Figure 6C). Within the largest class, embryos did not form any cuticle, suggesting that Spindly is either involved in cuticle formation or the embryos died early in development. The remaining embryos exhibited cuticle phenotypes, suggesting a range of morphogenetic defects (Figure 6C). In addition to embryonic defects, the normal egg shape was impaired by *mat67>>GFP::Spindly* expression, resulting in eggs that were shorter along their anterior-posterior axis (Figure 6D,E). In addition, the morphology of the dorsal-anterior appendages of the chorion suggested a mild ventralisation phenotype as the distance between the attachments of the appendages was significantly shorter in *mat67>>GFP::Spindly* eggs (Figure 6F,G). These data indicate that the overexpression of Spindly in the female germ line affects egg formation and patterning in oogenesis and embryonic survival.

To shed some light on the way how Spindly overexpression affected oogenesis, we examined the distribution of GFP::Spindly in the egg chambers (Figure 6H,H'). We find that in stage 10, egg chambers the overexpressed GFP::Spindly in the germ line is equally distributed in the nurse cells, but in the oocyte the protein accumulates in the posterior pole of the oocyte cortex. The distribution of the GFP::Spindly protein was similar in *spindly*[mat67>RNAi] ovaries, indicating that the excess overexpressed protein localised like the endogenous protein (Supplementary Materials Figure S2B). The localisation of Spindly protein in stage 10 egg chambers is reminiscent of the localisation of components that are transported along microtubules towards their plus ends, which become localised to the posterior cortex of the oocyte. These results are consistent with the localisation of overexpressed Spindly protein in cultured cells [9], and suggest that the overexpressed GFP::Spindly protein might interfere with the function of microtubule plus ends or proteins that are associated with these.

2.6. Requirement of Spindly in Ovarian Border Cell Migration

Knockdown of Spindly drastically alters S2 cell morphology in interphase: Spindly-depleted cells have malformed actin lamellae and abnormal microtubules bundles that protrude from the cell body [9]. Furthermore, the overexpression of a GFP-tagged Spindly revealed its accumulation at the plus ends of microtubules in S2 cells and in the oocyte (Figure 6H; [9]). We therefore addressed whether Spindly has a role in regulating the motile behaviour of cells.

The border cells (BCs) in the *Drosophila* ovary provide an excellent model system to study genetic requirements of collective cell migration [44] The BCs form a cluster of 6–10 post-mitotic cells at the anterior domain of somatic follicular epithelium of the egg chamber around stage 8 of egg chamber development (Figure 7A). Subsequently, BCs undergo an epithelial-mesenchymal transition and migrate posteriorly as a coherent cluster, reaching the nurse cell-oocyte boundary by stage 10 of egg-chamber development (Figure 7A) [44]. Spindly expression was knocked down by driving shRNAi with the BC-specific *slow border cells* (*slbo*) Gal4 driver (*slbo>Gal4*) [45]. In wild-type egg chambers the majority of BC clusters had migrated halfway through the egg chamber at late stage 9 (69%, $n = 42$), while a small percentage (14.3%) had already completed their migration at this stage of egg-chamber development (Figure 7B,D,E). In ovaries from *slbo*>Spindly[RNAi] females, 50% ($n = 46$) of the BC clusters had completed migration at mid-stage 9 of oogenesis (Figure 7C–E). To compare the BC cluster migratory phenotypes upon Spindly-RNAi in a quantitative fashion, the normalised migration (MI) and the completion (CI) indices were calculated [46]. Values of both indices suggested that the down-regulation of Spindly protein level confers a migratory advantage to BC (Figure 7D). Thus, we concluded that knockdown of Spindly caused the BC cluster to migrate either earlier or faster along the anterior-posterior axis towards the oocyte, suggesting that Spindly negatively regulates border cell migration.

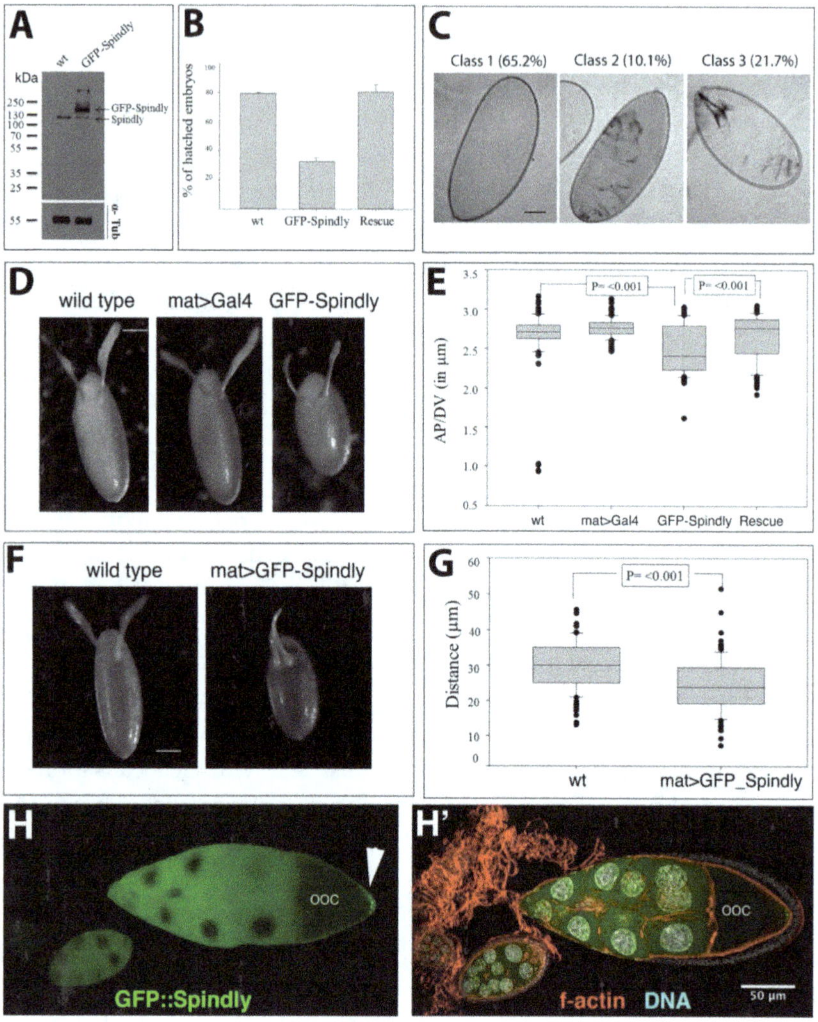

Figure 6. Overexpression of Spindly affects oogenesis and embryogenesis. (**A**) Immunoblot analysis of Spindly in lysates from wild-type (*wt*) and *nos>>Gal4::UAS>>GFP::Spindly* (GFP::Spindly) embryos. Endogenous Spindly and GFP::Spindly bands are indicated (α-tubulin—loading control; molecular weight markers in kDa); (**B**) Survival rate of embryos upon Spindly overexpression scored by larval hatching (n = 100). Viability was restored by knock-down of endogenous Spindly with UAS>>Spindly[RNAi] (Rescue); (**C**) Larval cuticle defects in *mat67>>Gal4:UAS>>GFP::Spindly.* The majority of embryos died before cuticle deposition (65.2%, class 1). The remaining 31.8% of embryos exhibited dorsal holes (class 2) or segmentation defects (class 3) (n = 138); (**D**) Eggs laid by *mat67>>Gal4::UAS>>GFP::Spindly* females exhibited abnormal rounded shape (48%, n = 100). Scale bar represents 50 µm; (**E**) Quantification of egg size by calculating the ratio between the length (A/P axis) and the width (D/V axis) for a total of 100 eggs (Mann-Whitney Test ($p \leq 0.001$). This phenotype is suppressed by UAS>>Spindly[RNAi] knock-down of endogenous Spindly (Rescue); (**F**) Moderate ventralisation defect in chorions of *mat67>>Gal4::UAS>>GFP::Spindly* eggs. The dorsal appendages were shorter and apposed closer to each other along the dorsal midline compared to controls. Scale bar represents 50 µm; (**G**) Quantification of the distance between dorsal appendages shown in (**F**) (Mann-Whitney Test $p \leq 0.001$; n = 100); (**H**) Part of an ovariole from an ovary of a *mat67>>Gal4::UAS>>GFP::Spindly* female fixed and stained with antibodies against GFP (green), f-actin (red) and DAPI (blue). Note the accumulation of GFP::Spindly (arrowhead) at the posterior pole of the oocyte (ooc) in the egg chamber (stage 10). Scale bar: 50 µm.

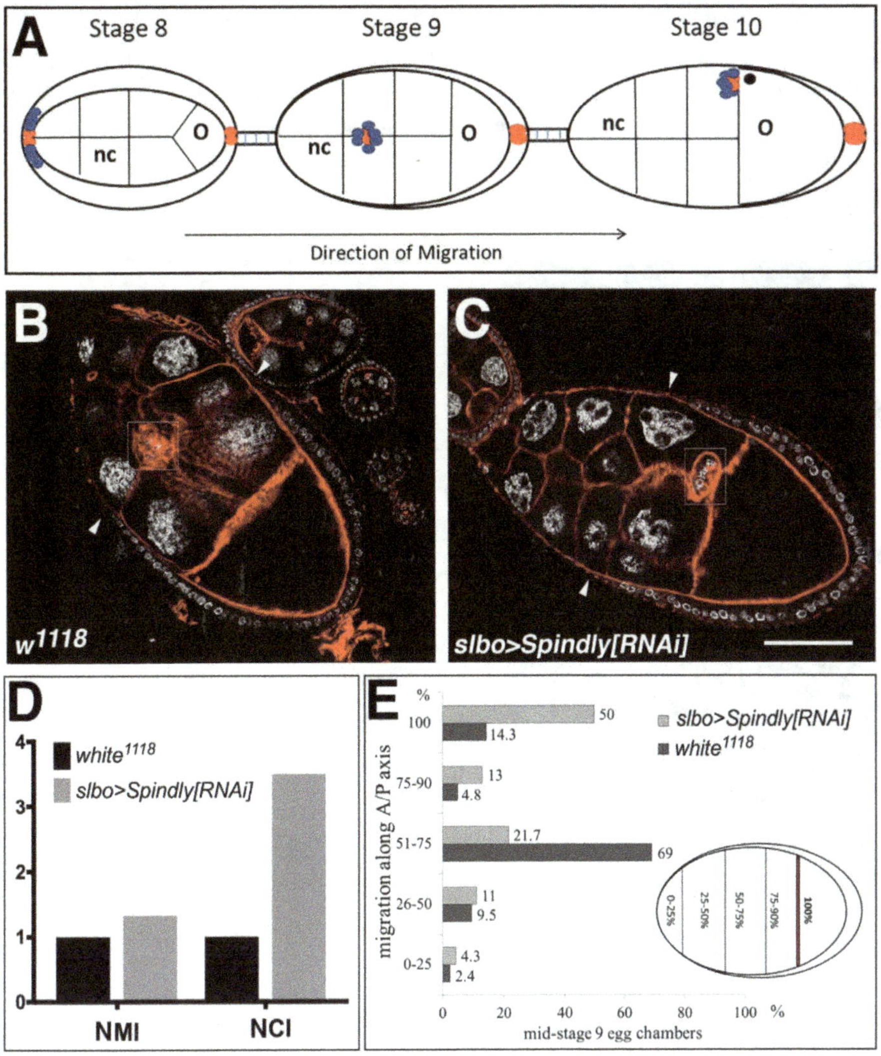

Figure 7. Spindly knockdown in migrating ovarian border cells. (**A**) Schematic drawing of border cell migration and outer follicle cells rearrangement. Egg chambers are depicted at stage 8–10 of oogenesis, where each egg chamber consists of 16 germ line cells (15 nurse cells (nc) and one oocyte (o)) and are surrounded by a layer of somatic follicle cells. At stage 8, polar cells (in red) at the anterior of the follicular epithelium recruit neighbouring cells (in blue) to form the migrating border cell cluster. At stage 9, the cluster delaminates from the epithelium and starts migrating posteriorly towards the oocyte and arriving at the nurse-cell-oocyte boundary at stage 10; (**B**,**C**) Immunostaining of stage 9 egg chambers from wild-type (w^{1118}) (**B**) and *slbo>Gal4;UAS>SpindlyRNAi* females (**C**); (DAPI (grey), f-actin (red). Arrowheads indicate the border between squamous and columnar follicle epithelium; scale bar: 50 µm; (**D**) Normalised migration index (NMI) and Normalised completion index (NCI) are indicated; for calculation see Materials and Methods; (**E**) Quantification of the phenotype described in (**B**,**C**). Premature border cell migration was observed in 50% of stage 9 egg-chambers (n = 42).

The sensitivity of BCs to Spindly levels prompted us to test the effect of Spindly overexpression on their migration. Specifically, we asked whether overexpression of Spindly could change BC migration at stage 10 of egg chamber development (Figure 8E,F). Indeed, when analysing the migration of the border cells, we found that in 46% (n = 113) of BC clusters their migration towards the oocyte was delayed at stage 10 of egg-chamber development (Figure 8A,B,E,F). While *slbo*>Gal4 females on their own also exhibited a mild incomplete migration phenotype at stage 10 (12.6%; n = 103), the penetrance of the phenotype was significantly lower when compared to *slbo>>GFP::Spindly* overexpression (Figure 8C,E,F). As additional control, the overexpression of GFP-tagged histone

H3 did not alter border cell migration (n = 100), supporting the result that the increased Spindly expression causes incomplete border cell migration (Figure 8D–F). When comparing the normalised migration and completion indices, overexpression of GFP::Spindly reduced the completion of BC migration to about 50% (Figure 8E). We conclude that the overexpression of GFP::Spindly produces the opposite phenotype as the knockdown of Spindly supporting our conclusion that Spindly is involved in timely migration of the BC in oogenesis.

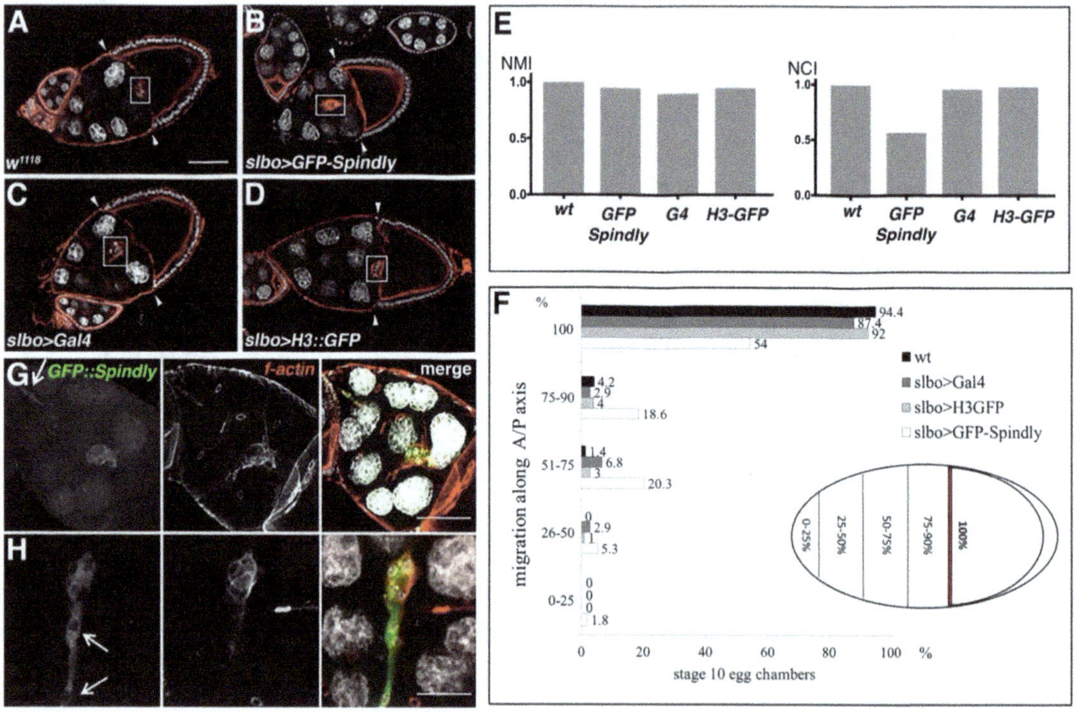

Figure 8. Spindly overexpression in migrating ovarian border cells. (**A–D**) Confocal optical sections of fixed and immunolabeled egg-chambers at stage 10 of oogenesis of the indicated genotypes. Migration of border cells was delayed by GFP::Spindly overexpression (**B**) compared to w^{1118} (**A**), *slbo>Gal4* (**C**) and *slbo>Gal4;H3::GFP* (**D**) (*slbo>H3::GFP*) as controls. Arrowheads indicate the border between squamous and columnar follicle epithelium DAPI (grey); f-actin (red); scale bars: 50 μm; (**E**) Comparison of normalised migration indices (NMI) and normalised completion index (NCI); (**F**) Quantification of the phenotype described in (**E**). 46% of the clusters overexpressing Spindly did not complete migration by stage 10 of egg-chamber development (n = 113); (**G**) Morphology of border cell clusters in egg chambers of *slbo>Gal4; UAS>GFP::Spindly* females. Confocal optical section of stage 10 egg-chambers (maximum-intensity projection of 10 μm z-stack at a spacing of 1 μm). Loss of border cell cluster integrity is marked by white arrows. In 11,5% of cases the border cell cluster had lost integrity or in most severe cases single cells had detached from the main cluster (**H**) and migration was delayed (n = 113) (DAPI (grey), f-actin (red), GFP (green); scale bars represent 20 μm).

Overexpression of Spindly in BCs also affected their morphology in a way suggesting abnormal adhesive properties and cytoskeletal architecture. Many of the BC clusters lost their integrity and the cell-cell contacts between the BCs were compromised to the extent that individual cells detached from the BC cluster, and thus were apparently migrating as individual cells (9.7%, n = 113) (Figure 8G,H). These results suggested that excess Spindly negatively affects the migration and the integrity of the border cell cluster during migration. We conclude that correct Spindly levels are required in post-mitotic follicle BC to control or modulate their migration possibly by controlling the integrity and adhesion of the cluster.

3. Discussion

3.1. *Spindly Function in Early Drosophila Embryogenesis*

In this work, we present the first investigation of the mitotic and post-mitotic activities of *Drosophila* Spindly in a variety of cellular and developmental processes. The temperature sensitive Gal4/UAS system in combination with shRNA-mediated knockdown provided a means to generate embryos with different maternally derived levels of Spindly protein; these combinations produced phenotypes as expected from a series of hypomorphic to amorphic mutant alleles. Consistent with its central role in the mitotic checkpoint, *spindly*[RNAi] embryos exhibit mild to severe defects in mitosis. We showed that Spindly has an important role in the transition from a mature oocyte into cleavage stages of embryonic development. Analysis of nuclear proliferation in *spindly*[RNAi] embryos revealed abnormal mitotic progression, which results in an early arrest of embryo development.

The phenotype of Spindly[RNAi] embryos in mitosis are reminiscent of defects that were observed in *Drosophila* embryos injected with Dynein inhibitors [47]. It is well known that the maintenance of the spacing and architecture of the spindle machinery is achieved by the coordinated and complementary activity of different spindle-associated Dynein and Kinesin motor proteins [48,49]. Loss of Spindly function might alter the balance between the opposing forces that govern spindle stability, resulting in defective spindle morphology and chromosome segregation. In *Drosophila*, Dynein plays a major role in maintaining a tight association between the centrosomes and the spindle poles [50,51]. Therefore, in Spindly[RNAi] embryos, the replication of detached centrosomes and the association of these free centrosomes to neighbouring spindles likely causes further aberrant mitoses. Although a direct interaction between Spindly and Dynein in *Drosophila* has yet to be demonstrated, our results support the idea that the knock-down of Spindly reflects a loss of Dynein function in establishing and maintaining stable kinetochore-microtubule interactions [27].

Human Spindly forms a ternary complex with Dynein and Dynactin in order to fulfil its functions [9]. The first coiled-coiled domain in the N-terminal portion of Spindly is responsible for the interaction with Dynein LIC-1 [52]. The Spindly box contains highly conserved residues, S256, F258 in human, and F199 in *C. elegans*, which are essential for the interaction with the Dynactin complex, and for the binding of dynein to kinetochores [29,34,52]. In our study, we tested the requirement of these domains for the mitotic function of Spindly. We found that Spindly protein constructs lacking the N-terminal domain or containing a mutation in the conserved S234 residue (S256 in human Spindly) are still able to localise to kinetochores in neuroblasts, but were unable to rescue the lethality that was caused by the knockdown of maternal Spindly in the embryo. We conclude that the requirement of the N-terminal half and the Spindly box for the interaction of Spindly with Dynein may be evolutionarily conserved.

The C-terminal portion of human Spindly is crucial for the localisation of the protein at kinetochores [30]. Here, we show for the first time that in *Drosophila* the C-terminal region exhibits a cell-type specific requirement for kinetochore binding. The C-terminal domain is essential for kinetochore binding in larval neuroblasts, but it is dispensable for kinetochore binding in the early embryo. In the embryo, Spindly protein lacking the C-terminal domain is still functional as it is able to rescue the lethality of the RNAi knockdown of Spindly. Other protein domains are likely to contribute to Spindly localisation making the C-terminal region no longer essential for kinetochore binding in early embryos. One possibility would be that the composition of the outer kinetochore differs between the early embryos and larval neuroblasts, and therefore, Spindly may achieve kinetochore binding by interacting with different binding partners. Similar cell-type specific differences between neuroblasts and syncytial embryos have been reported for the assembly of the kinetochore component *rough deal* (*rod*) [53]. A mutant form of Rod, Rod^{Z3}, was unable to localise to kinetochores in syncytial embryos, but is localised, albeit in a reduced fashion, to kinetochores in larval neuroblasts. Therefore, it is likely that there is a profound difference for the requirement of stable kinetochore-microtubule interactions in syncytial vs. post-embryonic cell divisions. This difference may also be reflected by the robustness of mitotic divisions in *Drosophila* upon inactivation of the SAC. The important SAC component Mad2 is dispensable for viability and fertility in the fly, while other components, e.g., the RZZ complex, are essential for the progression through mitosis [14,53]. Our study adds Spindly to the latter group of SAC components and indicates that like the RZZ complex Spindly might have SAC independent functions that are essential for the progression through mitosis in *Drosophila* embryos.

3.2. Spindly Overexpression Affects Dynein-Dependent Processes in Egg Patterning

Spindly overexpression led to embryonic lethality, altered egg morphology and a mild ventralisation phenotype. Within the ovary, spherical egg chambers elongate dramatically until they acquire the typical ellipsoid shape of the mature egg. The contribution of the somatic follicular epithelium to the process has been well characterised. Egg elongation relies on the interaction between a polarised actin network in the follicle cells and a polarised arrangement of the basal membrane components [54–56]. Maternal effect mutations also often result in a round egg phenotype [57]. One of the best studied class is the so-called dumpless phenotype. This phenotype is caused by reduced cortical, radial, and/or ring canal associated actin bundles or defects in the ring canal morphogenesis required for the fast transport of maternal products from the nurse cells into the oocyte [58]. The slow transport of nurse cell products into the oocyte transport is Dynein-based [59] and mutations in the Dynein light intermediate chain also cause a round egg phenotype and ventralisation (Hüttinger and Müller, unpublished data). We therefore propose that Spindly overexpression might interfere with Dynein-based transport, resulting in altered egg morphology.

The egg shape defect upon Spindly overexpression was associated with a mild ventralisation phenotype. The establishment of dorsal/ventral polarity occurs during mid-oogenesis and ultimately relies upon a dorsalising signal that is generated by the *gurken* gene [60,61]. Transport of *gurken* mRNA to an anterior-dorsal position is dependent on Dynein and requires anchoring by microtubules surrounding the oocyte nucleus [62–64]. Defects in Dynein and its co-factors cause the displacement of the oocyte nucleus, and consequently, of Gurken away from its anterior-dorsal position, resulting in the ventralisation of the eggs [59,65–68]. Since we did not observe any egg shape and ventralisation defects in eggs derived from mat67>Spindly RNAi females, it is unlikely that Spindly acts in these processes. We therefore conclude that Spindly interferes with Dynein dependent process by either directly binding to Dynein itself or another of its cofactors, thereby compromising their function. Loss of the Dynein cofactor Asunder in the germ line results in a similar egg shape and ventralisation defects, as we observed upon the overexpression of Spindly, and therefore represents one candidate target affected by Spindly overexpression [67]. The localization of the excess Spindly protein at the posterior pole of the oocyte also suggests that the protein interferes with Dynein function as Dhc and Dlic have also been reported to accumulate in this area [10,67].

3.3. A Role for Spindly in Cell Migration

The loss- and gain-of-function experiments performed on follicular border cells have opened an ideal system to test the role of Spindly in the regulation of cell migration in vivo. We found that the downregulation of Spindly positively regulates the migration of border cells towards the oocyte. Conversely overexpression of GFP-tagged Spindly resulted in an incomplete migration phenotype and abnormal cluster morphology. The Dynein light chains, Lis-1, and Nud-E have been implicated in border cell migration; however, in contrast to Spindly-RNAi, knock-down of these proteins delayed border cell migration [69]. In addition, Dynein chains are required for the normal organisation of the cluster, proper localisation of adhesion molecules and epithelial cell polarity [69,70]. These data suggest that altering the activity of dynein might cause defects in border cell cluster morphology. We hypothesise that Spindly gain-of-function might have similar consequences to the overexpression of Dynamitin [66,71,72], resulting in the inhibition of dynein activity and altered distribution of polarity markers and/or adhesion molecules. This interpretation would fit with the mild-ventralisation of the egg upon Spindly overexpression, a phenotype that could arise from reduced Dynein activity, and therefore the detachment of the oocyte nucleus from the anterior-dorsal cortex.

4. Materials and Methods

4.1. Drosophila Strains and Culture

Flies were kept on standard medium and grown accordingly to standard laboratory procedures. *white*^{1118} serves as wild-type control strain (Bloomington Stock Centre, Bloomington, IN, USA). To express genes ectopically and for the RNAi experiments, the Gal4-UAS system was used [37]. The Gal4 lines used were: *nanos*>Gal4 VP16 (Ulrike Gaul), *maternal tubulin 67c*>Gal4, *slbo*>Gal4 (Denise Montell). The UAS lines used were: p*UASp*>GFP::Spindly constructs (this work), *UAS*p>Spindly RNAi (FBst0034933, 34933 *Bloomington Stock Centre*), *UAS*>*H_3-GFP*, *worniu*-GAL4 [73], and UAS-mCherry::tubulin [74]. Transgenic flies harbouring the p*UASp*>GFP::Spindly constructs were generated by PhiC31–mediated sequence-directed transformation using an attP site at the genomic location 68A4.

4.2. Embryo Collection

Embryos were collected on yeasted apple juice agar plates. Gal4>Spindly^{RNAi} females were crossed to homozygous UAS::Spindly^{RNAi} males at indicated temperatures. To determine the hatching rates, embryos from an overnight collection were counted and aged at 25 °C, 23 °C, 21 °C, 18 °C for 48 h. The number of hatched embryos was determined by counting the number of intact, unhatched embryos, and subtracting this value from the total number of embryos collected. For larval cuticle preparation, terminally differentiated embryos were collected and dechorionated in bleach. After extensive washes in PBS-0.1% Triton X-100, embryos were mounted in Hoyer's mounting solution/lactic acid (1:1). Samples were incubated overnight at 65 °C. Imaging was performed on an Olympus BX-61, TRF microscope system.

4.3. Culture and Live Imaging of Larval Neuroblasts and Embryos

Live imaging of isolated neuroblasts was performed using published methods [38]. Brains were dissected from third instar larvae and then incubated in collagenase for 20 min. Brains were then manually dissociated with needles in fibrinogen (Sigma-Aldrich F-3879, Gillingham, UK), and dissolved in Schneider's medium (SLS-04-351Q) on a 25 mm glass bottom dish (WPI). Fibrinogen was then clotted by the addition of Thrombin (Sigma T7513). Schneider's medium was then added and a 5 µM slice at the centre of the cell was imaged every 2 min. Imaging was performed using a 100× objective (Oil, NA1.45) on a spinning disk confocal microscope. For embryo live imaging, embryos were dechorionated and mounted on a petri dish (fluorodish, WPI) under halocarbon oil 27 (Sigma). Images were taken at 20 s intervals on a Zeiss LSM880 Airyscan confocal microscope using parallel oversampling with the airyscan module. All data was processed using FIJI [75].

4.4. Site-Directed Mutagenesis

An RNAi-resistant GFP-tagged Spindly transgene was generated by mutagenesis PCR targeting the shRNA-binding site (1611-caggacgcggttgatatcaaa-1632) at every third position of each codon, leaving the polypeptide sequence codified unaltered. The shRNA-resistant nucleotide sequence originated was 5′-caagatgccgtggacattaag-3′. A GFP-tagged variant of the full-length *spindly* cDNA cloned in a Gateway vector was kindly provided by Dr E. Griffis (School of Life Sciences, University of Dundee). This construct was used as a template in a site-directed mutagenesis PCR using the following primers (Table 1) for mutagenesis:

Table 1. Primer.

Primer	5′→3′
Forward	caagatgccgtggacattaagacggagttggaagctccagaattaattcc
Reverse	cttaatgtccacggcatcttgctgttccaagtttaagtcgatttctcg

The DpnI-digested PCR products were used to transform *E. coli* and positive clones were sequenced to verify that only the appropriate mutation was present. The cDNA codifying for the RNAi-resistant GFP::Spindly was then subcloned in a pUASp K10 vector using KpnI-BamHI sites and was used to generate transgenic animals. The same plasmid vector was used to generate Spindly mutant variants. Primers are available upon request.

4.5. Protein Extraction and Western Blotting

0–3 h-old embryos were collected on apple juice plates at different temperatures, as indicated in the text. After dechorionisation, embryos were lysed in RIPA Buffer (50 mM Tris-HCl pH 8.0, 150 mM NaCl, 1% NP-40, 0.5% Sodiumdeoxycholate, 0.1% SDS) containing a mixture of protease inhibitors (Mini-cOmplete EDTA-free-Roche, 1 tablet/10 mL). The homogenised tissue was incubated on ice for 30 min. Subsequently, the samples were centrifuged twice at 13,000 rpm for 30 min at 4 °C to remove insoluble debris. Protein concentration was estimated by reading the absorbance at 280 nm using a photometer (Biophotometer, Eppendorf, Hamburg, Germany). For Western Blotting, 10 μg of proteins for each sample were heat-denatured, resolved on SDS-PAGE, and transferred to PVDF membrane (Whatman, GE Healthcare, Little Chalfont, UK) for immunoblotting. The membrane was immunoblotted with the following primary antibodies: rabbit-anti-Spindly 1:2000 [9], rabbit-anti-actin 1:3000 (Sigma-Aldrich A-2066, Gillingham, UK), and mouse-anti-tubulin 1:2000 (Developmental Studies Hybridoma Bank (DSHB), 12G10; Iowa City, IA, USA). After extensive washes, membrane was incubated with HRP-conjugated antibodies: anti-rabbit-HRP 1:5000 (Thermo Scientific, #31460, Loughborough, UK) and anti-mouse-HRP 1:2000 (Roche, Burgess Hill, UK). Signals were detected using chemi luminescence on X-ray film.

4.6. Embryo Immunostaining

Embryos for immunostaining were collected every three hours at different temperatures, as indicated and dechorionated. Embryos were transferred in a 1:1 solution of heptane and 4% formaldehyde in PBS for 20 min. Alternatively, embryos were incubated in a 1:1 solution of PEM Buffer containing 1 μM Taxol (Tocris Bioscience, Bristol, UK) and heptane to which 1 mL of 20% formaldehyde was added for 10 min. Taxol was added to stabilise microtubules and improve the staining of this cytoskeletal component [76]. After fixation, formaldehyde was replaced with methanol and vigorously shaken. After two washes in methanol, embryos were washed in PBS-0.1% Tween-20 (3 times, 15 min each) and incubated in blocking solution for 1 h at room temperature. Primary antibodies were incubated overnight at 4 °C, rotating. Primary antibodies were used at the following dilutions: rabbit anti-Centrosomin 1:250 (kindly provided by Professor J. Raff, Oxford, UK), and mouse anti-β-tubulin 1:50 (DSHB, E7; Iowa City, IW, USA). Fluorophore-conjugated secondary antibodies (Invitrogen by Thermo Fisher Scientific, Paisley, UK) were used at 1:250 and DAPI at 1 μg/mL. Specimens were mounted in MOWIOL/DABCO and imaged on Olympus BX-61, TRF fluorescence microscope. Images were processed using Volocity (Improvision, Perkin Elmer, Seer Green, UK) and Adobe Photoshop.

4.7. Ovary Dissection and Immunolabeling

1–3 old mated flies were placed on standard medium containing dry yeast for 24 h at 25 °C prior to dissection. Ovaries were dissected in phosphate buffer (PBS) and were fixed in cold 4% formaldehyde in PBS for 15 min. Ovaries were washed repetitively in PBS- 0.1% Tween-20 and rotated for 2 h in a solution of PBS- 0.1% Triton X-100 supplemented with 5% Foetal Calf Serum (FCS). F-actin staining was performed by adding fluorochrome–conjugated Phalloidin (1:50) (Phalloidin-Alexa594, Invitrogen #A12381). Subsequently, the samples were washed three times in PBS-0.1% Tween-20, 15 min each time on a rocking platform. Nuclei were stained with DAPI (1:1000) for 10 min. Ovaries were embedded in Mowiol/DABCO (Sigma Aldrich) and mounted on coverslips. Microscopy was performed using confocal laser-scanning microscopes (Leica SP2, Milton Keynes, UK), Leica TCS SP8 and Zeiss LSM880, Cambridge, UK). Images were processed using Volocity (Improvision, Perkin Elmer, Seer Green, UK) and Adobe Photoshop. Image J [77] was used for image analysis.

4.8. Egg Chamber Staging

Precise identification of egg-chamber stages was based on egg chamber morphology and stage-specific markers. At the transition between stage 8 and 9 of egg-chamber development, concomitantly to the initiation of border cell cluster migration, the anterior follicle cells start to migrate away from each other in an anterior to posterior direction to cover the oocyte. To unequivocally identify mid-stage 9 egg chambers, we took advantage of the morphogenetic rearrangements of the follicular epithelium. Particularly, a line was traced to connect the dorsal and ventral anterior-most cells in the follicular epithelium, and the distance between the position of the follicle cells and the nurse cells (NC)-oocyte boundary was measured. For mid-stage 9 egg chambers, an average distance value of 46 μm was calculated. Image analysis and distance measurements were performed on Image J.

4.9. Quantification of the Migration Phenotypes

For scoring the migration of the border cell cluster, the egg chamber was divided in five regions: 0–25% f migration, 26–50% of migration, 51–75% of migration, 75–90% of migration, and 100% complete migration. The distance travelled by the cluster was measured as the distance between the posterior-most follicle cells and the leading border cell. This value was then normalised to the total distance, that is the distance between the posterior-most follicle cells and the NC-oocyte boundary and expressed as percentage. On the base of these measurements, the number of clusters falling in each region was counted and results represented in histograms. Image analysis and distance measurements were performed on Image J. Migration (MI) and Completion (CI) indices were calculated, as previously described [46]. The MI is the measure of the mean distance that is travelled by the cluster. This index was calculated with the following formula:

$$\mathrm{MI} = \frac{1\times \mathrm{n}\ (100\%) + 0.75\times \mathrm{n}\ (75\%) + 0.5\times \mathrm{n}\ (50\%) + 0.25\times \mathrm{n}\ (25\%) + 0\times \mathrm{n}\ (0\%) + 0.5\times \mathrm{n}\ (\mathrm{splitcluster})}{\mathrm{N}\ (\mathrm{total})}$$

In the formula, n (100%) represents the number of clusters that have completed migration, reaching the NC-oocyte border, n (75%), n (50%), and n (25%) represent the number of clusters that have migrated 75%, 50% and 25% of the total distance, respectively. The formula includes also the number of dissociated clusters (n splitcluster). The CI index represents the number of clusters that have completed migration. CI is calculated as the ratio between the number of clusters that have migrated the total distance [n(100%)] and the total number of egg chambers examined [n(total)].

$$\mathrm{CI} = \frac{\mathrm{n}\ (100\%)}{\mathrm{n}\ (\mathrm{total})}$$

The indices were normalised to the wild-type control.

Supplementary Materials: The following are available online at http://www.mdpi.com/2221-3759/6/2/9/s1.

Author Contributions: G.D.C., E.R.G., J.J., M.R.H. and H.-A.J.M. planned and designed the experiments. G.D.C., H.-A.J.M., M.R.H. H.B., and K.K. performed all experiments. G.D.C., J.J., M.R.H. and H.-A.J.M. evaluated and interpreted the data. G.D.C., M.R.H. and H.-A.J.M. prepared the figures. G.D.C. and H.-A.J.M. wrote the manuscript and J.J., M.R.H., E.R.G. and H.-A.J.M edited the manuscript.

Conflicts of Interest: The authors declare no conflict of interest.

References

1. Foe, V.E.; Odell, G.M.; Edgar, B. Mitosis and morphogenesis in the *Drosophila* embryo: Point and counterpoint. In *The Development of Drosophila Melanogaster*; Bate, M., Martinez-Arias, A., Eds.; Cold Spring Harbor Laboratory Press: Cold Spring Harbor, NY, USA, 1993; pp. 149–300.
2. Lee, L.A.; Orr-Weaver, T.L. Regulation of cell cycles in *Drosophila* development: Intrinsic and extrinsic cues. *Annu. Rev. Genet.* **2003**, *37*, 545–578. [CrossRef] [PubMed]
3. Huang, J.; Raff, J.W. The disappearance of cyclin B at the end of mitosis is regulated spatially in *Drosophila* cells. *EMBO J.* **1999**, *18*, 2184–2195. [CrossRef] [PubMed]
4. Su, T.T.; Sprenger, F.; DiGregorio, P.J.; Campbell, S.D.; O'Farrell, P.H. Exit from mitosis in *Drosophila* syncytial embryos requires proteolysis and cyclin degradation, and is associated with localized dephosphorylation. *Genes Dev.* **1998**, *12*, 1495–1503. [CrossRef] [PubMed]
5. Musacchio, A. Spindle assembly checkpoint: The third decade. *Philos. Trans. R. Soc. Lond. B Biol. Sci.* **2011**, *366*, 3595–3604. [CrossRef] [PubMed]
6. Lara-Gonzalez, P.; Westhorpe, F.G.; Taylor, S.S. The spindle assembly checkpoint. *Curr. Biol.* **2012**, *22*, R966–R980. [CrossRef]
7. Jia, L.; Kim, S.; Yu, H. Tracking spindle checkpoint signals from kinetochores to APC/C. *Trends Biochem. Sci.* **2013**, *38*, 302–311. [CrossRef] [PubMed]
8. Howell, B.J.; McEwen, B.F.; Canman, J.C.; Hoffman, D.B.; Farrar, E.M.; Rieder, C.L.; Salmon, E.D. Cytoplasmic dynein/dynactin drives kinetochore protein transport to the spindle poles and has a role in mitotic spindle checkpoint inactivation. *J. Cell Biol.* **2001**, *155*, 1159–1172. [CrossRef] [PubMed]
9. Griffis, E.R.; Stuurman, N.; Vale, R.D. Spindly, a novel protein essential for silencing the spindle assembly checkpoint, recruits dynein to the kinetochore. *J. Cell Biol.* **2007**, *177*, 1005–1015. [CrossRef] [PubMed]
10. Mische, S.; He, Y.; Ma, L.; Li, M.; Serr, M.; Hays, T.S. Dynein light intermediate chain: An essential subunit that contributes to spindle checkpoint inactivation. *Mol. Biol. Cell* **2008**, *19*, 4918–4929. [CrossRef] [PubMed]
11. Basto, R.; Scaerou, F.; Mische, S.; Wojcik, E.; Lefebvre, C.; Gomes, R.; Hays, T.; Karess, R. In vivo dynamics of the rough deal checkpoint protein during *Drosophila* mitosis. *Curr. Biol.* **2004**, *14*, 56–61. [CrossRef] [PubMed]
12. Buffin, E.; Lefebvre, C.; Huang, J.; Gagou, M.E.; Karess, R.E. Recruitment of Mad2 to the kinetochore requires the Rod/Zw10 complex. *Curr. Biol.* **2005**, *15*, 856–861. [CrossRef] [PubMed]
13. Wojcik, E.; Basto, R.; Serr, M.; Scaerou, F.; Karess, R.; Hays, T. Kinetochore dynein: Its dynamics and role in the transport of the Rough deal checkpoint protein. *Nat. Cell Biol.* **2001**, *3*, 1001–1007. [CrossRef] [PubMed]
14. Buffin, E.; Emre, D.; Karess, R.E. Flies without a spindle checkpoint. *Nat. Cell Biol.* **2007**, *9*, 565–572. [CrossRef] [PubMed]
15. Rahmani, Z.; Gagou, M.E.; Lefebvre, C.; Emre, D.; Karess, R.E. Separating the spindle, checkpoint, and timer functions of BubR1. *J. Cell Biol.* **2009**, *187*, 597–605. [CrossRef] [PubMed]
16. Basu, J.; Bousbaa, H.; Logarinho, E.; Li, Z.; Williams, B.C.; Lopes, C.; Sunkel, C.E.; Goldberg, M.L. Mutations in the essential spindle checkpoint gene bub1 cause chromosome missegregation and fail to block apoptosis in *Drosophila*. *J. Cell Biol.* **1999**, *146*, 13–28. [CrossRef] [PubMed]
17. Williams, B.C.; Karr, T.L.; Montgomery, J.M.; Goldberg, M.L. The *Drosophila* l(1)zw10 gene product, required for accurate mitotic chromosome segregation, is redistributed at anaphase onset. *J. Cell Biol.* **1992**, *118*, 759–773. [CrossRef] [PubMed]
18. Lopes, C.S.; Sampaio, P.; Williams, B.; Goldberg, M.; Sunkel, C.E. The *Drosophila* Bub3 protein is required for the mitotic checkpoint and for normal accumulation of cyclins during G2 and early stages of mitosis. *J. Cell Sci.* **2005**, *118*, 187–198. [CrossRef] [PubMed]
19. Karess, R.E.; Glover, D.M. rough deal: A gene required for proper mitotic segregation in *Drosophila*. *J. Cell Biol.* **1989**, *109*, 2951–2961. [CrossRef] [PubMed]
20. Althoff, F.; Karess, R.E.; Lehner, C.F. Spindle checkpoint-independent inhibition of mitotic chromosome segregation by *Drosophila* Mps1. *Mol. Biol. Cell* **2012**, *23*, 2275–2291. [CrossRef] [PubMed]
21. Karess, R. Rod-Zw10-Zwilch: A key player in the spindle checkpoint. *Trends Cell. Biol.* **2005**, *15*, 386–392. [CrossRef] [PubMed]
22. Kops, G.J.; Kim, Y.; Weaver, B.A.; Mao, Y.; McLeod, I.; Yates, J.R., 3rd; Tagaya, M.; Cleveland, D.W. ZW10 links mitotic checkpoint signaling to the structural kinetochore. *J. Cell Biol.* **2005**, *169*, 49–60. [CrossRef] [PubMed]

23. Starr, D.A.; Williams, B.C.; Hays, T.S.; Goldberg, M.L. ZW10 helps recruit dynactin and dynein to the kinetochore. *J. Cell Biol.* **1998**, *142*, 763–774. [CrossRef] [PubMed]
24. Williams, B.C.; Li, Z.; Liu, S.; Williams, E.V.; Leung, G.; Yen, T.J.; Goldberg, M.L. Zwilch, a new component of the ZW10/ROD complex required for kinetochore functions. *Mol. Biol. Cell* **2003**, *14*, 1379–1391. [CrossRef] [PubMed]
25. Bader, J.R.; Kasuboski, J.M.; Winding, M.; Vaughan, P.S.; Hinchcliffe, E.H.; Vaughan, K.T. Polo-like kinase1 is required for recruitment of dynein to kinetochores during mitosis. *J. Biol. Chem.* **2011**, *286*, 20769–20777. [CrossRef] [PubMed]
26. Whyte, J.; Bader, J.R.; Tauhata, S.B.; Raycroft, M.; Hornick, J.; Pfister, K.K.; Lane, W.S.; Chan, G.K.; Hinchcliffe, E.H.; Vaughan, P.S.; et al. Phosphorylation regulates targeting of cytoplasmic dynein to kinetochores during mitosis. *J. Cell Biol.* **2008**, *183*, 819–834. [CrossRef] [PubMed]
27. Gassmann, R.; Essex, A.; Hu, J.S.; Maddox, P.S.; Motegi, F.; Sugimoto, A.; O'Rourke, S.M.; Bowerman, B.; McLeod, I.; Yates, J.R., 3rd.; et al. A new mechanism controlling kinetochore-microtubule interactions revealed by comparison of two dynein-targeting components: SPDL-1 and the Rod/Zwilch/Zw10 complex. *Genes Dev.* **2008**, *22*, 2385–2399. [CrossRef] [PubMed]
28. Chan, Y.W.; Fava, L.L.; Uldschmid, A.; Schmitz, M.H.; Gerlich, D.W.; Nigg, E.A.; Santamaria, A. Mitotic control of kinetochore-associated dynein and spindle orientation by human Spindly. *J. Cell Biol.* **2009**, *185*, 859–874. [CrossRef] [PubMed]
29. Gassmann, R.; Holland, A.J.; Varma, D.; Wan, X.; Civril, F.; Cleveland, D.W.; Oegema, K.; Salmon, E.D.; Desai, A. Removal of Spindly from microtubule-attached kinetochores controls spindle checkpoint silencing in human cells. *Genes Dev.* **2010**, *24*, 957–971. [CrossRef] [PubMed]
30. Barisic, M.; Sohm, B.; Mikolcevic, P.; Wandke, C.; Rauch, V.; Ringer, T.; Hess, M.; Bonn, G.; Geley, S. Spindly/CCDC99 is required for efficient chromosome congression and mitotic checkpoint regulation. *Mol. Biol. Cell* **2010**, *21*, 1968–1981. [CrossRef] [PubMed]
31. Holland, A.J.; Reis, R.M.; Niessen, S.; Pereira, C.; Andres, D.A.; Spielmann, H.P.; Cleveland, D.W.; Desai, A.; Gassmann, R. Preventing farnesylation of the dynein adaptor Spindly contributes to the mitotic defects caused by farnesyltransferase inhibitors. *Mol. Biol. Cell* **2015**, *26*, 1845–1856. [CrossRef] [PubMed]
32. Moudgil, D.K.; Westcott, N.; Famulski, J.K.; Patel, K.; Macdonald, D.; Hang, H.; Chan, G.K. A novel role of farnesylation in targeting a mitotic checkpoint protein, human Spindly, to kinetochores. *J. Cell Biol.* **2015**, *208*, 881–896. [CrossRef] [PubMed]
33. Mosalaganti, S.; Keller, J.; Altenfeld, A.; Winzker, M.; Rombaut, P.; Saur, M.; Petrovic, A.; Wehenkel, A.; Wohlgemuth, S.; Muller, F.; et al. Structure of the RZZ complex and molecular basis of its interaction with Spindly. *J. Cell Biol.* **2017**, *216*, 961–981. [CrossRef] [PubMed]
34. Varma, D.; Wan, X.; Cheerambathur, D.; Gassmann, R.; Suzuki, A.; Lawrimore, J.; Desai, A.; Salmon, E.D. Spindle assembly checkpoint proteins are positioned close to core microtubule attachment sites at kinetochores. *J. Cell Biol.* **2013**, *202*, 735–746. [CrossRef] [PubMed]
35. Schlager, M.A.; Hoang, H.T.; Urnavicius, L.; Bullock, S.L.; Carter, A.P. In vitro reconstitution of a highly processive recombinant human dynein complex. *EMBO J.* **2014**, *33*, 1855–1868. [CrossRef] [PubMed]
36. Hoogenraad, C.C.; Akhmanova, A. Bicaudal D Family of Motor Adaptors: Linking Dynein Motility to Cargo Binding. *Trends Cell Biol.* **2016**, *26*, 327–340. [CrossRef] [PubMed]
37. Brand, A.H.; Perrimon, N. Targeted gene expression as a means of altering cell fates and generating dominant phenotypes. *Development* **1993**, *118*, 401–415. [PubMed]
38. Pampalona, J.; Januschke, J.; Sampaio, P.; Gonzalez, C. Time-lapse recording of centrosomes and other organelles in *Drosophila* neuroblasts. *Methods Cell Biol.* **2015**, *129*, 301–315. [PubMed]
39. Ni, J.Q.; Liu, L.P.; Binari, R.; Hardy, R.; Shim, H.S.; Cavallaro, A.; Booker, M.; Pfeiffer, B.D.; Markstein, M.; Wang, H.; et al. A *Drosophila* resource of transgenic RNAi lines for neurogenetics. *Genetics* **2009**, *182*, 1089–1100. [CrossRef] [PubMed]
40. Page, A.W.; Orr-Weaver, T.L. Activation of the meiotic divisions in *Drosophila* oocytes. *Dev. Biol.* **1997**, *183*, 195–207. [CrossRef] [PubMed]
41. Tastan, O.Y.; Maines, J.Z.; Li, Y.; McKearin, D.M.; Buszczak, M. *Drosophila* ataxin 2-binding protein 1 marks an intermediate step in the molecular differentiation of female germline cysts. *Development* **2010**, *137*, 3167–3176. [CrossRef] [PubMed]

42. Wang, C.; Dickinson, L.K.; Lehmann, R. Genetics of nanos localization in *Drosophila*. *Dev. Dyn.* **1994**, *199*, 103–115. [CrossRef] [PubMed]
43. Perez-Mongiovi, D.; Malmanche, N.; Bousbaa, H.; Sunkel, C. Maternal expression of the checkpoint protein BubR1 is required for synchrony of syncytial nuclear divisions and polar body arrest in *Drosophila* melanogaster. *Development* **2005**, *132*, 4509–4520. [CrossRef] [PubMed]
44. Montell, D.J. Border-cell migration: The race is on. *Nat. Rev. Mol. Cell Biol.* **2003**, *4*, 13–24. [CrossRef] [PubMed]
45. Rorth, P. Specification of C/EBP function during *Drosophila* development by the bZIP basic region. *Science* **1994**, *266*, 1878–1881. [CrossRef] [PubMed]
46. Assaker, G.; Ramel, D.; Wculek, S.K.; Gonzalez-Gaitan, M.; Emery, G. Spatial restriction of receptor tyrosine kinase activity through a polarized endocytic cycle controls border cell migration. *Proc. Natl. Acad. Sci. USA* **2010**, *107*, 22558–22563. [CrossRef] [PubMed]
47. Sharp, D.J.; Rogers, G.C.; Scholey, J.M. Cytoplasmic dynein is required for poleward chromosome movement during mitosis in *Drosophila* embryos. *Nat. Cell Biol.* **2000**, *2*, 922–930. [CrossRef] [PubMed]
48. Sharp, D.J.; Brown, H.M.; Kwon, M.; Rogers, G.C.; Holland, G.; Scholey, J.M. Functional coordination of three mitotic motors in *Drosophila* embryos. *Mol. Biol. Cell* **2000**, *11*, 241–253. [CrossRef] [PubMed]
49. Wang, H.; Brust-Mascher, I.; Scholey, J.M. Sliding filaments and mitotic spindle organization. *Nat. Cell Biol.* **2014**, *16*, 737–739. [CrossRef] [PubMed]
50. Robinson, J.T.; Wojcik, E.J.; Sanders, M.A.; McGrail, M.; Hays, T.S. Cytoplasmic dynein is required for the nuclear attachment and migration of centrosomes during mitosis in *Drosophila*. *J. Cell Biol.* **1999**, *146*, 597–608. [CrossRef] [PubMed]
51. Morales-Mulia, S.; Scholey, J.M. Spindle pole organization in *Drosophila* S2 cells by dynein, abnormal spindle protein (Asp), and KLP10A. *Mol. Biol. Cell* **2005**, *16*, 3176–3186. [CrossRef] [PubMed]
52. Gama, J.B.; Pereira, C.; Simoes, P.A.; Celestino, R.; Reis, R.M.; Barbosa, D.J.; Pires, H.R.; Carvalho, C.; Amorim, J.; Carvalho, A.X.; et al. Molecular mechanism of dynein recruitment to kinetochores by the Rod-Zw10-Zwilch complex and Spindly. *J. Cell Biol.* **2017**, *216*, 943–960. [CrossRef] [PubMed]
53. Defachelles, L.; Hainline, S.G.; Menant, A.; Lee, L.A.; Karess, R.E. A maternal effect rough deal mutation suggests that multiple pathways regulate *Drosophila* RZZ kinetochore recruitment. *J. Cell Sci.* **2015**, *128*, 1204–1216. [CrossRef] [PubMed]
54. Horne-Badovinac, S. The *Drosophila* egg chamber-a new spin on how tissues elongate. *Integr. Comp. Biol.* **2014**, *54*, 667–676. [CrossRef] [PubMed]
55. Haigo, S.L.; Bilder, D. Global tissue revolutions in a morphogenetic movement controlling elongation. *Science* **2011**, *331*, 1071–1074. [CrossRef] [PubMed]
56. Crest, J.; Diz-Munoz, A.; Chen, D.Y.; Fletcher, D.A.; Bilder, D. Organ sculpting by patterned extracellular matrix stiffness. *Elife* **2017**, *6*. [CrossRef] [PubMed]
57. Luschnig, S.; Moussian, B.; Krauss, J.; Desjeux, I.; Perkovic, J.; Nusslein-Volhard, C. An F1 genetic screen for maternal-effect mutations affecting embryonic pattern formation in *Drosophila* melanogaster. *Genetics* **2004**, *167*, 325–342. [CrossRef] [PubMed]
58. Hudson, A.M.; Cooley, L. Understanding the function of actin-binding proteins through genetic analysis of *Drosophila* oogenesis. *Annu. Rev. Genet.* **2002**, *36*, 455–488. [CrossRef] [PubMed]
59. Januschke, J.; Gervais, L.; Dass, S.; Kaltschmidt, J.A.; Lopez-Schier, H.; St Johnston, D.; Brand, A.H.; Roth, S.; Guichet, A. Polar transport in the *Drosophila* oocyte requires Dynein and Kinesin I cooperation. *Curr. Biol.* **2002**, *12*, 1971–1981. [CrossRef]
60. Forlani, S.; Ferrandon, D.; Saget, O.; Mohier, E. A regulatory function for K10 in the establishment of dorsoventral polarity in the *Drosophila* egg and embryo. *Mech. Dev.* **1993**, *41*, 109–120. [CrossRef]
61. Neuman-Silberberg, F.S.; Schupbach, T. The *Drosophila* TGF-alpha-like protein Gurken: Expression and cellular localization during *Drosophila* oogenesis. *Mech. Dev.* **1996**, *59*, 105–113. [CrossRef]
62. MacDougall, N.; Clark, A.; MacDougall, E.; Davis, I. *Drosophila* gurken (TGFalpha) mRNA localizes as particles that move within the oocyte in two dynein-dependent steps. *Dev. Cell* **2003**, *4*, 307–319. [CrossRef]
63. Rom, I.; Faicevici, A.; Almog, O.; Neuman-Silberberg, F.S. *Drosophila* Dynein light chain (DDLC1) binds to gurken mRNA and is required for its localization. *Biochim. Biophys. Acta* **2007**, *1773*, 1526–1533. [CrossRef] [PubMed]

64. Swan, A.; Nguyen, T.; Suter, B. *Drosophila* Lissencephaly-1 functions with Bic-D and dynein in oocyte determination and nuclear positioning. *Nat. Cell Biol.* **1999**, *1*, 444–449. [CrossRef] [PubMed]
65. Lei, Y.; Warrior, R. The *Drosophila* Lissencephaly1 (DLis1) gene is required for nuclear migration. *Dev. Biol.* **2000**, *226*, 57–72. [CrossRef] [PubMed]
66. Duncan, J.E.; Warrior, R. The cytoplasmic dynein and kinesin motors have interdependent roles in patterning the *Drosophila* oocyte. *Curr. Biol.* **2002**, *12*, 1982–1991. [CrossRef]
67. Sitaram, P.; Merkle, J.A.; Lee, E.; Lee, L.A. Asunder is required for dynein localization and dorsal fate determination during *Drosophila* oogenesis. *Dev. Biol.* **2014**, *386*, 42–52. [CrossRef] [PubMed]
68. Swan, A.; Suter, B. Role of Bicaudal-D in patterning the *Drosophila* egg chamber in mid-oogenesis. *Development* **1996**, *122*, 3577–3586. [PubMed]
69. Yang, N.; Inaki, M.; Cliffe, A.; Rorth, P. Microtubules and Lis-1/NudE/dynein regulate invasive cell-on-cell migration in *Drosophila*. *PLoS ONE* **2012**, *7*, e40632. [CrossRef] [PubMed]
70. Horne-Badovinac, S.; Bilder, D. Dynein regulates epithelial polarity and the apical localization of stardust A mRNA. *PLoS Genet.* **2008**, *4*, e8. [CrossRef] [PubMed]
71. Quintyne, N.J.; Gill, S.R.; Eckley, D.M.; Crego, C.L.; Compton, D.A.; Schroer, T.A. Dynactin is required for microtubule anchoring at centrosomes. *J. Cell Biol.* **1999**, *147*, 321–334. [CrossRef] [PubMed]
72. Schroer, T.A. Motors, clutches and brakes for membrane traffic: A commemorative review in honor of Thomas Kreis. *Traffic* **2000**, *1*, 3–10. [CrossRef] [PubMed]
73. Albertson, R.; Chabu, C.; Sheehan, A.; Doe, C.Q. Scribble protein domain mapping reveals a multistep localization mechanism and domains necessary for establishing cortical polarity. *J. Cell Sci.* **2004**, *117*, 6061–6070. [CrossRef] [PubMed]
74. Rusan, N.M.; Peifer, M. A role for a novel centrosome cycle in asymmetric cell division. *J. Cell Biol.* **2007**, *177*, 13–20. [CrossRef] [PubMed]
75. Schindelin, J.; Arganda-Carreras, I.; Frise, E.; Kaynig, V.; Longair, M.; Pietzsch, T.; Preibisch, S.; Rueden, C.; Saalfeld, S.; Schmid, B.; et al. Fiji: An open-source platform for biological-image analysis. *Nat. Methods* **2012**, *9*, 676–682. [CrossRef] [PubMed]
76. Karr, T.L.; Alberts, B.M. Organization of the cytoskeleton in early *Drosophila* embryos. *J. Cell Biol.* **1986**, *102*, 1494–1509. [CrossRef] [PubMed]
77. Abramoff, M. Image processing with Image J. *Biophotonic Int.* **2004**, *11*, 36–42.

7

Forced Expression of Foxg1 in the Cortical Hem Leads to the Transformation of Cajal-Retzius Cells into Dentate Granule Neurons

Bin Liu, Hongmei Xiao and Chunjie Zhao *

Key Laboratory of Developmental Genes and Human Diseases, MOE, School of Medicine, Southeast University, Nanjing 210009, China; myqj2006@126.com (B.L.); xhmforever@hotmail.com (H.X.)
* Correspondence: zhaocj@seu.edu.cn

Abstract: The Wnt- and BMP-rich cortical hem has been demonstrated to be critical for the pattern formation of the telencephalon, and it is particularly important for the induction of the hippocampus. Meanwhile, the cortical hem is one of the sources of Cajal-Retzius cells. Many Cajal-Retzius cells are produced in the hem and populated to the media-caudal surface of the telencephalon. However, the mechanism of the maintenance of the hem remain unclear. In this study, we generated a transgenic mouse line *CAG-loxp-stop-loxp-Foxg1-IRES-EGFP*. By crossing *Fzd10CreER*^{TM} with this line, combined with tamoxifen induction, Foxg1 was ectopically expressed in the hem from embryonic day 10.5 (E10.5) onwards. We have found the hem-derived Cajal-Retzius cells were transformed into dentate granule neurons accompanied with ectopic expression of Lhx2. However, the morphology of the hem displayed no obvious changes. The hem specific markers, Wnt3a and Wnt2b, were slightly downregulated. Our results indicate that Foxg1 is sufficient to induce the expression of Lhx2 in the dorsal part of the hem. The ectopic Lhx2 and decreased Wnt signals may both contribute to the cell fate switch. Our study provides new insight into the mechanism underlying the maintenance of the hem.

Keywords: cortical hem; Cajal-Retzius cell; cortical patterning; Foxg1; Lhx2; hippocampus; granule cell; Wnt

---

1. Introduction

During the cortical development, regionalization of the cortex is controlled by morphogens secreted from signal centres located at the perimeter of the telencephalon and transcription factors expressed in a gradient pattern along the coordinate axes [1,2]. Regionalization along the medio-lateral axis leads to the progressive subdivision of the telencephalic neuroepithelium into the medial pallium, which gives rise to the hippocampus, including the dentate gyrus (DG), the adjacent dorsal pallium that gives rise to the neocortex, respectively. The Wnt- and BMP-rich cortical hem, located between the telencephalic choroid plexus and hippocampal primordium, has been identified to be one of the cortical organizers to regulate the patterning of the telencephalon [3,4]. A severely-reduced neocortex, particularly dorsomedial neocortex was reported after ablation of the hem [5]. Disruption of *Wnt3a*, a gene specifically expressed in the hem, leads to loss of the whole hippocampus [6]. When *Lef1*, a downstream mediator of Wnt signalling, is deleted, the development of the hippocampus is significantly affected [7]. Previous study also shows that an ectopic hem is capable to induce an ectopic hippocampus [8]. Thus, the cortical hem is necessary and sufficient to induce the adjacent hippocampal primordium. Meanwhile, cell linage mapping has revealed the hem is one of the sources of Cajal-Retzius (CR) cells, the cell population which is crucial for the cortical lamination [9–12].

Despite its critical roles, the mechanism of regulating the hem during the telencephalic development is still unclear.

The transcription factor *Foxg1*, formerly called Brain Factor1, is strongly expressed in the telencephalon, but specifically excluded from the hem. Its expression can be detected as early as E8.5 [13]. Previously, *Foxg1* and the transcription factor *Lhx2* have been shown to play important roles to suppress the hem. Both *Foxg1* and *Lhx2* null mutants exhibit significant expansion of the hem [14,15]. Recently, Godbole and colleagues showed that *Foxg1* functions upstream of *Lhx2* to regulate the specification and positioning of the hem, providing new insight into the mechanisms regulating the hem formation [16]. On the other hand, there are obvious phenotype differences between *Foxg1* and *Lhx2* knockouts. The hem expands to the whole cortex accompanied with a complete loss of the dorsal pallium after constitutive disruption of *Lhx2* [8], while part of the hippocampal primordium remains in the *Foxg1*-deficient telencephalon [13,17], suggesting a complicated mechanism of *Foxg1* and *Lhx2* to regulate the hem. Meanwhile, the hem itself displays a heterogeneous feature with its dorsal and ventral parts displaying distinct expression patterns of Wnts and BMPs [4], and the more detailed mechanism underlying the maintaining of the hem still needs to be further elucidated.

The cortical hem forms before cortical neurogenesis [3,4]. Although constitutive loss of Foxg1 from the very early developmental stage results in a severe expansion of the hem, removal of Foxg1 at E13 does not cause the hem expansion [18], suggesting Foxg1 may regulate the hem in a spatiotemporal manner. In this study, we focused on the role of Foxg1 during the time window of E10.5 onwards when the formation of the hem is almost completed. We have generated a transgenic mouse line *CAG-loxp-stop-loxp-Foxg1-IRES-EGFP* in which *Foxg1* cDNA is driven by *CAG* promoter, and the overexpression commenced upon the deletion of stop codon by Cre medicated recombination. By crossing with *Fzd10CreER*^{TM}, Foxg1 was ectopically overexpressed in the hem. We found a large population of the hem-derived CR cells switched their fate into dentate granule neurons, accompanied with upregulation of Lhx2, suggesting Foxg1 is sufficient to induce Lhx2 and specify granule cell fates. However, it seems the morphology of the hem was normal with only slightly decreased expression levels of *Wnt3a* and *Wnt2b*. Our study will help to understand the mechanism underlying the maintenance of the hem.

2. Materials and Methods

2.1. *Generation of* CAG-loxp-stop-loxp-Foxg1-IRES-EGFP *Mouse Line*

Foxg1 cDNA was subcloned into an *IRES2-EGFP* vector (Addgene, cat6029-1, Cambridge, MA, USA). The *loxp-stop-loxp* fragment was inserted ahead of the *Foxg1* cDNA fragment. *CAG* promoter was cut out from a *pCAGIG* vector and inserted into the *loxp-stop-loxp-Foxg1-IRES-EGFP* vector. The transgenic construct was then linearized, purified with a gel extraction kit (Qiagen, 20021, Duesseldorf, Germany) and microinjected into the B6/FVB oocytes according to the standard procedures. Genotypes of the offspring were determined by PCR analysis using primers (5′-AAGGACGACGGCAACTACAAG-3′, and 5′-AAGGACGACGGC AACTACAAG-3′) to amplify a 378 bp GFP fragment.

2.2. *Mouse Breeding and Tamoxifen Administering*

All mice were maintained on an outbred CD1 genetic background and were housed in the animal facility at the Southeast University. All experimental procedures followed the guidelines approved by Southeast University. The day the vaginal plug was found at noon was considered E0.5. To overexpress Foxg1 in the hem, the *CAG-loxp-stop-loxp-Foxg1-IRES-EGFP* males were crossed with *Fzd10CreER*^{TM} females and Tamoxifen (TM, Sigma, T5648, St. Louis, MO, USA) was administered by oral gavage to pregnant females at E10.5. TM was dissolved in corn oil (Sigma-Aldrich, C8267, St. Louis, MO, USA) at a concentration of 15 mg/mL and the dose administered was 75 μg/gm body weight.

2.3. Tissue Processing

E12.5 brains were rinsed with cold phosphate buffered saline (PBS) then fixed in 4% paraformaldehyde (PFA, Sigma-Aldrich, 441244, St. Louis, MO, USA) at 4 °C overnight, while E18.5 brains were perfused by 4% PFA and post-fixed in 4% PFA at 4 °C for 12–16 h. Brains were cryoprotected in 30% sucrose, embedded in OCT. The tissues applied in situ hybridization were rinsed with DEPC-PBS and the sucrose were dissolved in DEPC-PBS. The coronal sections (12 μm thick) were obtained using a Leica cryostat (CM 3050S) and stored at −70 °C until use.

2.4. In Situ Hybridization

Digoxigenin (DIG) UTP-labelled riboprobes were used. Probes were obtained by PCR amplification. The in situ hybridization was performed as previously described [19,20].

2.5. Immunofluorescence

Immunofluorescence was performed as previously described [21,22]. The primary antibodies and dilutions were as follows: anti-Calretinin (Millipore, AB5054, 1:500, Billerica, MA, USA); anti-Ctip2 (Abcam, ab18465, 1:2000, Cambridge, MA, USA); anti-Foxg1 (Abcam, ab18259, 1:1000, Cambridge, MA, USA); anti-GFP (Abcam, ab13970, 1:1000, Cambridge, MA, USA); anti-Lhx2 (Abcam, ab184337, 1:500, Cambridge, MA, USA); anti-P73 (Abcam, ab40658, 1:500, Cambridge, MA, USA); anti-Prox1 (Millipore, AB5475, 1:1000, Billerica, MA, USA); and anti-Reelin (Millipore, MAB5364, 1:1000, Billerica, MA, USA). The secondary antibodies used were Alexa Fluro 488 donkey anti-chicken (Jackson Lab, 703-545-155, 1:500, West Grove, PA, USA), Alexa Fluor 546 donkey anti-rabbit (Life, A10040, 1:500, Gaithersburg, MD, USA), Alexa Fluro 647 donkey anti-rabbit (Life, A31573, 1:500, Gaithersburg, MD, USA), Alexa Fluor 546 donkey anti-rat (Life, A10040, 1:500, Gaithersburg, MD, USA), CF 568 donkey anti-rat (Sigma-Aldrich, SAB4600077, 1:500, St. Louis, MO, USA), CF 633 donkey anti-rat (Sigma-Aldrich, SAB4600133, 1:500, St. Louis, MO, USA), and Alexa Fluro 647 donkey anti-mouse (Invitrogen, A21236, 1:500, Carlsbad, CA, USA).

3. Results

3.1. Forced Expression of Foxg1 in the Cortical Hem and Impaired Development of the DG

To elucidate the role of Foxg1 in the maintaining of the hem from E10.5 onwards, we first generated a transgenic mouse line in which full-length *Foxg1* cDNA was driven by the *CAG* promoter. A Floxed stop codon and a reporter *IRES-EGFP* fragment were also introduced into the transgenic vector (Figure 1A). Previously, we have reported that one of the Wnts receptors, *Frizzled10* (*Fzd10*), is specifically expressed in the hem [23] and generated transgenic mouse lines of *Fzd10-TauLacZ* and *Fzd10-CreER^{TM}* using 5′ untranslated regions of the *Fzd10* gene demonstrated that a large population of CR cells are derivatives of Fzd10-positive cells in the hem [23–25]. In this study, the overexpression of Foxg1 in the hem was achieved by crossing *Fzd10CreER^{TM}* with the *CAG-loxp-stop-loxp-Foxg1-IRES-EGFP* line combined with tamoxifen induction. We first performed tamoxifen administration in *Fzd10CreER^{TM};CAG-loxp-stop-loxp-Foxg1-IRES-EGFP* at E10.5, the time window when the formation of the hem is almost completed. As shown in Figure 1B–C′, at E12.5, the reporter EGFP specifically labelled the hem and the dorsal thalamus where endogenous *Fzd10* is expressed. Strong ectopic expression of Foxg1 in the hem was observed to be completely co-localized with EGFP (Figure 1C′–E′), indicating a successful forced expression of Foxg1.

Since the hem is critical for the development of the hippocampus, we then analysed the hippocampal morphology at E18.5. In the *CAG-loxp-stop-loxp-Foxg1-IRES-EGFP* control brain, no GFP was detected (Figure 1F), and Foxg1 was expressed in both cornu ammonis (CA) and DG regions, but specifically excluded from the fimbria, which contains the residue of the hem (Figure 1G). However, in the transgenic *Fzd10CreER^{TM};CAG-loxp-Foxg1-IRES-EGFP* mice, in the future dentate blade, besides the granule cells, many GFP^{+} cells were detected to be Foxg1^{+}. Several GFP^{+} clusters were found to be distributed outside the DG region (Figure 1F′–H′, I–J). Cells in clusters displayed strong ectopic expression of Foxg1 (Figure 1I–K). A mass of hem-derived GFP$^+$ tissue was ectopically located at the area between the fimbria and the developing DG, and Foxg1 was detected to be located in cell bodies within the mass with the processes of these cells were Foxg1$^-$ (Figure 1F′–H′). The fimbria, the structure mainly consisting of neuronal projections, was also found to contain Foxg1-expressing cells migrating towards the future DG (Figure 1L–N). Thus, Foxg1 was successfully overexpressed in the hem and its derivatives.

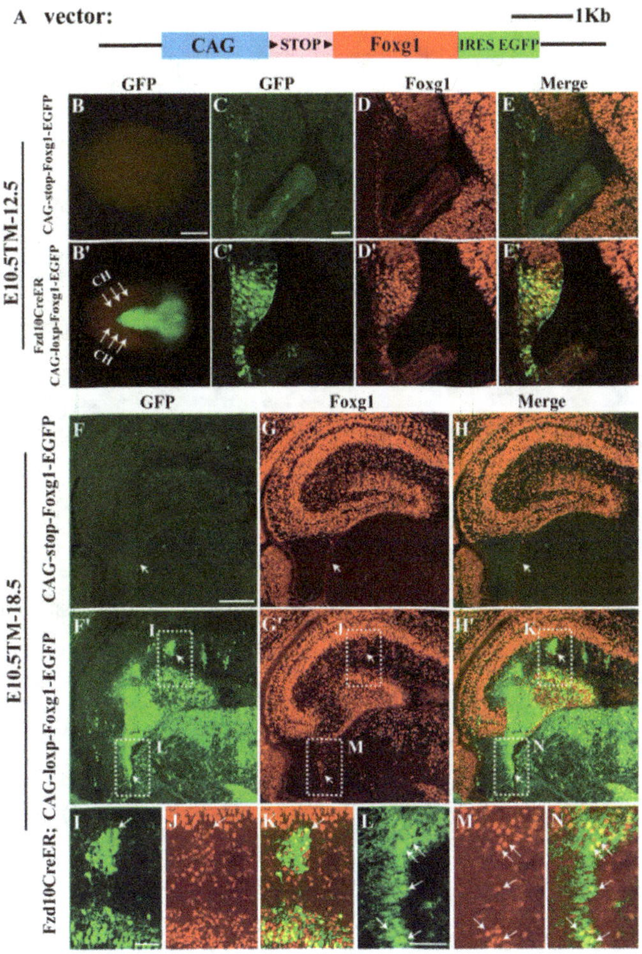

Figure 1. Forced expression of Foxg1 in the cortical hem and abnormal development of the dentate gyrus. (**A**) Schematic representation of the construction of the transgenic vector. (**B**,**B′**) Dorsal view of an E12.5 *Fzd10CreER*^{TM};*CAG-loxp-Foxg1-IRES-EGFP* hemisphere after E10.5 TM-induction. Reporter EGFP specifically labelled the hem and the dorsal thalamus where endogenous *Fzd10* is expressed. Arrows in B′ indicate the GFP-expressing cortical hem (CH). (**C**–**E′**) Double immunostaining for GFP with Foxg1 at E12.5 coronal sections. In controls (**C**–**E**), GFP was not expressed, and Foxg1 expression is excluded from the hem, no co-localization of Foxg1 and GFP to be detected. In (**C′**–**E′**), Foxg1 was ectopically expressed in the hem and co-expressing GFP. (**F**–**H**) At E18.5, in the *CAG-loxp-stop-loxp-Foxg1-IRES-EGFP* control, no GFP was detected, and Foxg1 was expressed in both CA and DG regions but specifically excluded from the fimbria. (**F′**–**H′**) In addition to the normal DG granule cells, the expression of Foxg1 was also detected in GFP$^+$ cells. There were several clusters that were distributed outside the DG region. A mass of GFP$^+$ tissue was ectopically located at the area between the fimbria and the developing DG. (**I**–**N**) Magnified views of boxed areas in (**F′**–**H′**) showing the ectopic expression of Foxg1 in clusters and the fimbria. Scale bars: (**B**,**B′**) 2 mm; (**C**–**H′**) 200 μm; (**I**–**N**) 50 μm.

3.2. Most Hem-Derived Cells Lost CR Cell Fate and Mis-Distributed in the DG Area

By cell linage mapping, we previously demonstrated that a distinct population of CR cells originated from Fzd10-expressing progenitors in the hem and preferentially distributed to the hippocampal marginal zone (MZ) [23,24,26]. To examine whether the mis-located GFP$^+$ cells in *Fzd10CreER*^{TM}*;CAG-loxp-Foxg1-IRES-EGFP* still owned the CR cell fate, immunostaining of anti-Reelin and P73, two specific markers for the hem-derived CR cells, was then carried out. Since Fzd10 itself is only expressed in the hem but not expressed in its derivatives of CR cells, and previously we have shown the reporter gene *Tau-LacZ* driven by *Fzd10* promoter can act as a linage tracer during the embryonic developmental stages [25], the mouse lines of *Fzd10-EGFP* and *CAG-loxp-stop-loxp-Foxg1-IRES-EGFP* were used as controls in this study. In the *Fzd10-EGFP*

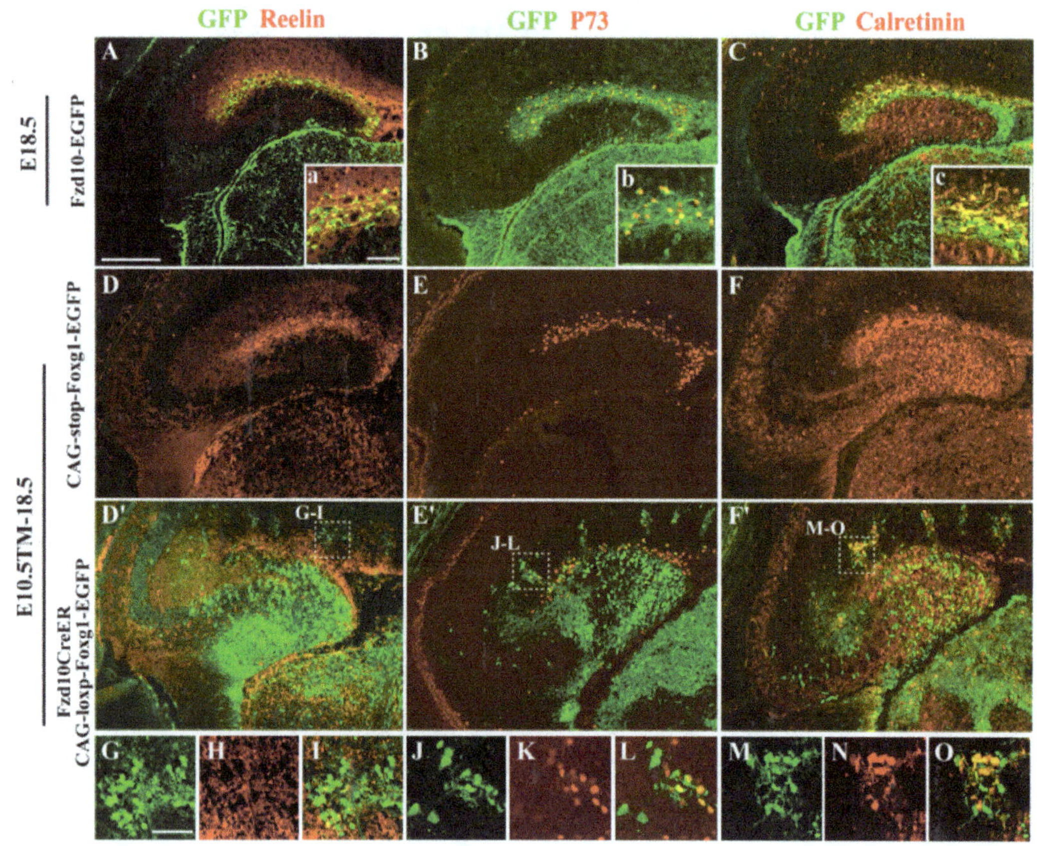

Figure 2. Most hem-derived cells lost CR cell fate and mis-distributed in the DG area. (**A**–**C**) Double immunofluorescence analysis of GFP and Reelin (**A**), P73 (**B**), and Calretinin (**C**) on coronal brain sections of *Fzd10-EGFP* brains at E18.5. The hem-derived GFP cells are both Reelin$^+$, P73$^+$ and CR$^+$, indicating these are Cajal-Retzius cells. (**a**–**c**) Magnified views of the double immunostaining in the hippocampal MZ. (**D**–**F**) Confocal images of immunofluorescence of GFP and Reelin (**D**), P73 (**E**), and Calretinin (**F**) on *CAG-loxp-stop-loxp-Foxg1-IRES-EGFP* control brain when TM was administered at E10.5. (**D′**,**E′**) In the Foxg1 overexpressed transgenic brain, only a small portion of GFP$^+$ cells distributed to the MZ and co-expressed Reelin and P73. The majority of GFP$^+$ cells were Reelin$^-$ and P73$^-$ and dispersed within the DG area. (**F′**) In addition to immature granule neurons and CR cells, Calretinin was also found to be expressed in most GFP$^+$ cells within clusters, as well as these dispersed cells in the DG area. (**G**–**O**) Magnified views of boxed areas in (**D′**–**F′**). Scale bars: (**A**–**F′**) 200 μm; (**G**–**O**) 50 μm.

control, GFP$^+$ cells were observed to originate from the hem and populated only to the MZ, and these cells are both Reelin$^+$ and P73$^+$ (Figure 2A,B). In the *CAG-loxp-stop-loxp-Foxg1-IRES-EGFP* control, Reelin$^+$ and P73$^+$ CR cells are also observed to be distributed in the MZ and the migration route from the fimbria to the developing DG (Figure 2D,E). However, in the Foxg1 overexpressed transgenic brain, only a small portion of GFP$^+$ cells distributed to the MZ and co-expressed Reelin and P73 (Figure 2D′,E′). The majority of GFP$^+$ cells were Reelin$^-$ and P73$^-$ and dispersed within the DG area (Figure 2D′,E′), suggesting most of them lost CR cell fate. Interestingly, GFP$^+$ clusters outside the DG showed heterogeneous property. Some cells within clusters expressing Reelin and P73, displaying a CR cell fate, while many cells were Reelin$^-$ and P73$^-$, indicating the clusters contained different cell types (Figure 2G–L).

To further demonstrate the cell fate alteration caused by Foxg1 forced expression, immunostaining of anti-Calretinin, a calcium binding protein commonly used as a marker for CR cells and immature granule neurons [27–29], was employed. As shown in Figure 2C,F,F′,M–O, in both controls, Calretinin was expressed in immature granule neurons in the developing DG and CR cells located in the MZ (Figure 2C,F), while, in the *Fzd10CreER*^{TM}*;CAG-loxp-Foxg1-IRES-EGFP* brain, besides immature granule neurons and CR cells, Calretinin was also found to be expressed in most GFP$^+$ cells within clusters as well as these dispersed cells in the DG area (Figure 2F′,M–O). These Reelin$^-$P73$^-$GFP$^+$Calretinin$^+$ cells might represent a cell population of immature granule neurons.

3.3. Hem-Derived CR Cells Switched Their Fates into Dentate Granule Neurons

To further analyse the feature of the population of Reelin$^-$P73$^-$GFP$^+$Calretinin$^+$ cells, immunostaining of anti-Prox1 and Ctip2, two transcription factors that are critical for the development of granule neurons [30,31], was carried out. At E18.5, in the *Fzd10-EGFP* control, there was no co-localization of GFP with Prox1 or Ctip2, and Prox1$^+$ and Ctip2$^+$ granule neurons were located inside the DG MZ, forming the upper blade of the DG (Figure 3A,B). On the contrary, in the *Fzd10CreER*^{TM}*;CAG-loxp-Foxg1-IRES-EGFP* brain, both Prox1 and Ctip2 were found to be expressed in clusters outside the DG (Figure 3C,D,H–J,N–P). Within the DG, many GFP$^+$ cells co-expressed Prox1 and Ctip2 (Figure 3C–G,K–M), demonstrating these ectopically-distributed hem-derived cells transformed into granule neurons.

Steel and Ephb1 have been used to label developing granule cells in the DG [32–34]. To confirm the cell fate transformation, in situ hybridization of *Steel* and *Ephb1* was performed on E18.5 brains after tamoxifen administration at E10.5. As shown in Figure 4A,B, in the control, strong staining for *Steel* was detected in the DG, and *Ephb1* was strongly detected in migrating granule cells. In Foxg1 overexpressed transgenic brains, in the migration route increased staining of *Steel* was observed (Figure 4A,A′), and the staining of *Ephb1* seemed a little thicker than that of the controls. Ectopic staining of *Steel* and *Ephb1* was also detected in clusters outside the DG (Figure 4A′,B′,D,E). Since *Lef1*, a mediator in the Wnt signalling pathway, plays an important role during the development of the DG [7], we then examined its expression. Compared to the control, in *Fzd10CreER*^{TM}*;CAG-loxp-Foxg1-IRES-EGFP* brain, similar to that of *Steel* and *Ephb1*, clusters also displayed ectopic strong staining for *Lef1* (Figure 4C–C′,F). A little thicker staining of *Lef1* in the migration route was also detected (Figure 4C,C′). Taken together, our data demonstrated hem-derived CR cells switched their fates into dentate granule neurons.

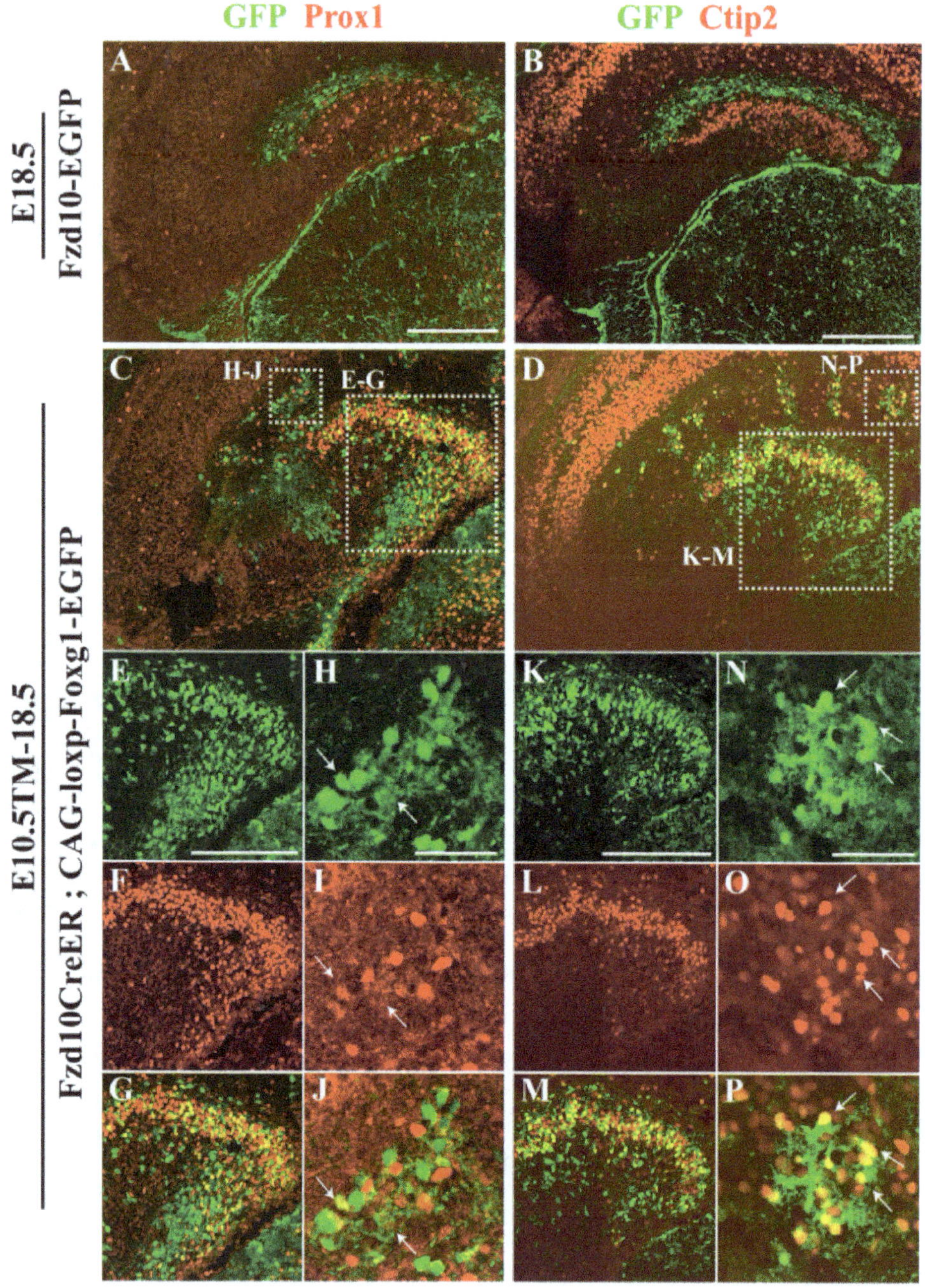

Figure 3. Hem-derived CR cells switched their fates into dentate granule neurons. (**A**,**B**) Double immunofluorescence analysis of GFP and Prox1 (**A**), Ctip2 (**B**) on coronal brain sections of *Fzd10-EGFP* embryos at E18.5. Hem-derived GFP cells are negative for Prox1 and Ctip2. (**C**,**D**) Confocal images of immunofluorescence of GFP and Prox1 (**C**), Ctip2 (**D**) on coronal E18.5 brain sections when TM was administered at E10.5. The ectopically-located hem-derived cells within the DG transformed to the granule neurons. (**E**–**P**) Magnified views of boxed areas in (**C**,**D**). Arrows in (**I**,**J**) indicate the GFP^+; $Prox1^+$ cells, arrows in (**O**,**P**) indicate the GFP^+;$Ctip2^+$ cells. Scale bars: (**A**–**D**) 200 μm; (**E**–**P**) 50 μm.

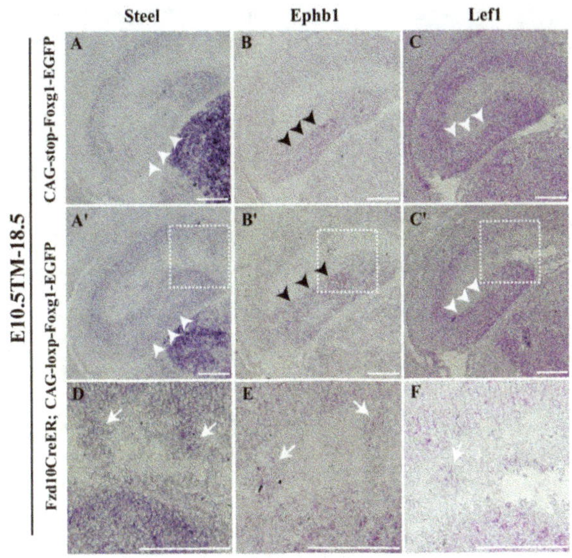

Figure 4. Misdistributed hem-derived cells outside the DG has the property of DG granule neurons. (**A–C′**) In situ hybridization assay of *Steel* (**A**,**A′**), *Ephb1* (**B**,**B′**), and *Lef1* (**C**,**C′**) on coronal sections of E18.5 *CAG-loxp-stop-loxp-Foxg1-IRES-EGFP* (**A–C**) and *Fzd10CreER*^{TM}*;CAG-loxp-Foxg1-IRES-EGFP* embryos (**A′–C′**) showing increased staining for *Steel* ((**A′**), white arrowheads) and a little thicker staining for *Ephb1* ((**B′**), black arrowheads) on the migration route. Thicker staining for *Lef1* was also observed in the migration route ((**C′**), white arrowheads). (**D–F**) Magnified views of boxed areas in (**A′–C′**). Short white arrows indicate the ectopically located clusters. Scale bars: 200 μm.

3.4. Normal Morphology of the Hem and Ectopic Expression of Lhx2 after Foxg1 Overexpression

Next, we analysed whether the hem itself was affected by Foxg1 overexpression. First, in situ hybridization for *Wnt3a* and *Wnt2b*, two members of the Wnt family, which are specifically expressed in the hem, was performed. Interestingly, the morphology of the hem delineated as the staining of *Wnt3a* and *Wnt2b* seemed normal, while the expression level of *Wnt3a* and *Wnt2b* were slightly reduced (Figure 5A–B′). We then examined *Wnt8b*, another Wnt family member normally expressed in the medial pallium, and found *Wnt8b* was also downregulated (Figure 5C,C′), consistent with previously-reported results that *Foxg1* suppresses *Wnt8b* [35].

Previous studies have shown that Lhx2-expressing tissues near the hem can be specified into the hippocampi [8]. Considering the transformation of CR cells into dentate granule neurons in our transgenic mice, we suspected that the expression of Lhx2 was induced upon forced expression of Foxg1 in the hem. To confirm the hypothesis, double immunostaining of anti-Lhx2 with GFP was carried out. At E12.5, in the *Fzd10-EGFP* control, Lhx2 was strongly expressed in the medial pallium, but specifically excluded from the GFP$^+$ hem area, there was no co-localization of Lhx2 with GFP (Figure 5D). The same result was obtained in the *CAG-loxp-stop-loxp-Foxg1-IRES-EGFP* control. There was no Lhx2 expression detected in the hem either (Figure 5E). However, in the *Fzd10CreER*^{TM}*;CAG-loxp-Foxg1-IRES-EGFP* transgenic line, Lhx2 was found to be ectopically expressed in the GFP$^+$ hem (Figure 5F–H), demonstrating that Foxg1 is sufficient to induce the expression of Lhx2. Interestingly, it seemed that the expression of Lhx2 was only induced in the dorsal Wnt-rich part of the hem, while, in the ventral BMP-rich subarea, Lhx2 expression was not detected (Figure 5F–H), although Foxg1 was ectopically expressed there as well (Figure 1D′,E′). At E18.5, compared with Lhx2 expression observed in the DG area in both *Fzd10-EGFP* and *CAG-loxp-stop-loxp-Foxg1-IRES-EGFP* controls (Figure 5I,J), in the Foxg1 overexpressed transgenic brain, ectopic Lhx2 was additionally observed to be expressed in the clusters outside the DG, and co-localized with GFP and Ctip2 (Figure 5K–O). Taken together, by inducing the expression of Lhx2, overexpressed Foxg1 transformed the hem-derived CR cells into granule neurons.

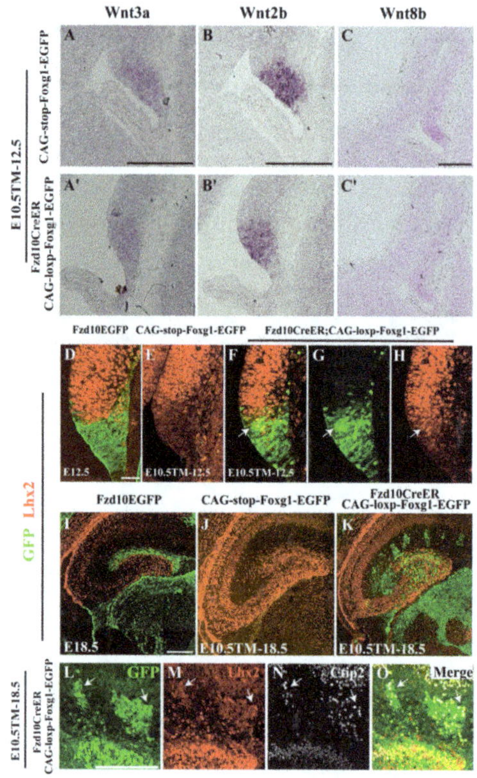

Figure 5. Normal morphology of the hem and ectopic expression of Lhx2. (**A–C′**) In situ hybridization assay of *Wnt3a* (**A**,**A′**), *Wnt2b* (**B**,**B′**), and *Wnt8b* (**C**,**C′**) on coronal sections of E12.5 *CAG-loxp-stop-loxp-Foxg1-IRES-EGFP* (**A–C**) and *Fzd10CreER*^{TM}*; CAG-loxp-Foxg1-IRES-EGFP* embryos (**A′–C′**). (**D–K**) Double-immunofluorescence analysis of GFP with Lhx2 at E12.5 (**D–H**) and E18.5 (**I–K**). (**L–O**) Triple immunostaining for GFP/Lhx2/Ctip2 in E12.5 *Fzd10CreER*^{TM}*; CAG-loxp-Foxg1-IRES-EGFP* brains. The arrows show the triple-labelled cells. Scale bars: 200 μm ((**D–G**) 50 μm).

4. Discussion

The cortical hem is one of the cortical organizers, which plays critical roles in the pattern formation of the telencephalon, and it is particularly required for the induction of the hippocampus [5,6,8]. Loss of the hem severely impairs the dorsomedial neocortex [5]. The hem is also one of the sources of CR cells [9,36]. A subpopulation of CR cells originates from the hem and settle in the media-caudal surface of the developing telencephalon [24,26]. However, how the hem is regulated remains unclear. To gain insight of the mechanism underlying the maintenance of the hem, in this study, Foxg1 was ectopically expressed in the hem from E10.5 onwards. Hem-derived CR cells switched their fate into dentate granule neurons after forced expression of Foxg1. However, the morphology of the hem seemed normal and only showed slightly down-regulated Wnts. We also found Lhx2 expression was induced, which may be responsible for the cell fate transformation. Our result will help to understand the mechanism of the maintenance of the hem.

Foxg1 has been shown to play important roles in pattern formation, cell proliferation and cell specification [15,22,37–39]. Here, by overexpression of Foxg1 in the hem from E10.5 onwards, we detected a cell fate switch of the hem-derived CR cells to dentate granule neurons, accompanied with ectopic expression of Lhx2. Previous study has reported that loss of Foxg1 leads to a reduction of Lhx2 expression, and *Foxg1* can directly bind to *Lhx2* locus [16]. In this study, we demonstrated that forced expression of Foxg1 is sufficient to induce Lhx2 expression in the hem. It has been reported that Lhx2-expressing cells can be specified into hippocampal cells and this only happens in tissues adjacent to the hem [16]. Here, induced expression of Lhx2 in our transgenic mice may be responsible for the transformation of CR cells to their granular fate. Interestingly, we have observed that the induction of

Lhx2 occurs only in the dorsal part of the hem. Although Foxg1 is also overexpressed in the ventral part, the expression of Lhx2 is not detected. Since the dorsal part of the hem is Wnt rich, while the ventral is BMP rich, one possible explanation is that other signalling pathways are also involved in the maintenance of the hem. Further studies are required to elucidate the mechanism underlying the hem heterogeneity.

Early constitutive loss of Foxg1 (around E8.5–9.0) results in severe expansion of the hem, while removal of Foxg1 at E13.5 has no obvious influences [13,16], suggesting that Foxg1 may regulate the hem in a spatiotemporal manner. In this study, Foxg1 was forced to be expressed in the hem from E10.5 onwards. To our surprise, the morphology of the hem seemed normal, and *Wnt3a* and *Wnt2b* were only slightly down-regulated, indicating the suppression of *Foxg1* on the hem and Wnt signalling is gradually weakened. *Wnt8b*, expressed in the medial pallium, was down-regulated, consistent with previously-reported results [35]. We also found *Lef1*, the downstream transcription factor in the Wnt signalling pathway, was ectopically expressed in clusters outside the DG, suggesting that Wnt signalling may contribute to the cell fate switch as well.

Author Contributions: Designed the study: B.L. and C.Z. Conducted the experiments: B.L. and H.X. Analysed and interpreted the data: B.L., H.X. and C.Z. Prepared the manuscript: C.Z. and B.L. All authors read and approved the final manuscript.

Funding: This study was supported by grant 2016YFA0501001 from the Ministry of Science and Technology of China and grants 91232301 and 31471041 from the National Natural Science Foundation of China to C.Z.

Acknowledgments: We would like to thank Yiquan Wei and Li Liu for their assistance with the laboratory and animal care.

Conflicts of Interest: The authors declare no conflict of interest.

References

1. Borello, U.; Pierani, A. Patterning the cerebral cortex: Traveling with morphogens. *Curr. Opin. Genet. Dev.* **2010**, *20*, 408–415. [CrossRef] [PubMed]
2. Shimogori, T.; Banuchi, V.; Ng, H.Y.; Strauss, J.B.; Grove, E.A. Embryonic signaling centers expressing BMP, WNT and FGF proteins interact to pattern the cerebral cortex. *Development* **2004**, *131*, 5639–5647. [CrossRef] [PubMed]
3. Furuta, Y.; Piston, D.W.; Hogan, B.L. Bone morphogenetic proteins (BMPs) as regulators of dorsal forebrain development. *Development* **1997**, *124*, 2203–2212. [PubMed]
4. Grove, E.A.; Tole, S.; Limon, J.; Yip, L.; Ragsdale, C.W. The hem of the embryonic cerebral cortex is defined by the expression of multiple Wnt genes and is compromised in Gli3-deficient mice. *Development* **1998**, *125*, 2315–2325. [PubMed]
5. Yoshida, M.; Assimacopoulos, S.; Jones, K.R.; Grove, E.A. Massive loss of Cajal-Retzius cells does not disrupt neocortical layer order. *Development* **2006**, *133*, 537–545. [CrossRef] [PubMed]
6. Lee, S.M.; Tole, S.; Grove, E.; McMahon, A.P. A local Wnt-3a signal is required for development of the mammalian hippocampus. *Development* **2000**, *127*, 457–467. [PubMed]
7. Galceran, J.; Miyashita-Lin, E.M.; Devaney, E.; Rubenstein, J.L.; Grosschedl, R. Hippocampus development and generation of dentate gyrus granule cells is regulated by LEF1. *Development* **2000**, *127*, 469–482. [PubMed]
8. Mangale, V.S.; Hirokawa, K.E.; Satyaki, P.R.; Gokulchandran, N.; Chikbire, S.; Subramanian, L.; Shetty, A.S.; Martynoga, B.; Paul, J.; Mai, M.V.; et al. Lhx2 selector activity specifies cortical identity and suppresses hippocampal organizer fate. *Science* **2008**, *319*, 304–309. [CrossRef] [PubMed]
9. Meyer, G.; Perez-Garcia, C.G.; Abraham, H.; Caput, D. Expression of p73 and Reelin in the developing human cortex. *J. Neurosci. Off. J. Soc. Neurosci.* **2002**, *22*, 4973–4986. [CrossRef]
10. Takiguchi-Hayashi, K.; Sekiguchi, M.; Ashigaki, S.; Takamatsu, M.; Hasegawa, H.; Suzuki-Migishima, R.; Yokoyama, M.; Nakanishi, S.; Tanabe, Y. Generation of reelin-positive marginal zone cells from the caudomedial wall of telencephalic vesicles. *J. Neurosci. Off. J. Soc. Neurosci.* **2004**, *24*, 2286–2295. [CrossRef] [PubMed]

11. Bielas, S.; Higginbotham, H.; Koizumi, H.; Tanaka, T.; Gleeson, J.G. Cortical neuronal migration mutants suggest separate but intersecting pathways. *Annu. Rev. Cell Dev. Biol.* **2004**, *20*, 593–618. [CrossRef] [PubMed]
12. Rice, D.S.; Curran, T. Role of the reelin signaling pathway in central nervous system development. *Annu. Rev. Neurosci.* **2001**, *24*, 1005–1039. [CrossRef] [PubMed]
13. Xuan, S.; Baptista, C.A.; Balas, G.; Tao, W.; Soares, V.C.; Lai, E. Winged helix transcription factor bf-1 is essential for the development of the cerebral hemispheres. *Neuron* **1995**, *14*, 1141–1152. [CrossRef]
14. Bulchand, S.; Grove, E.A.; Porter, F.D.; Tole, S. Lim-homeodomain gene lhx2 regulates the formation of the cortical hem. *Mech. Dev.* **2001**, *100*, 165–175. [CrossRef]
15. Muzio, L.; Mallamaci, A. Foxg1 confines Cajal-Retzius neuronogenesis and hippocampal morphogenesis to the dorsomedial pallium. *J. Neurosci. Off. J. Soc. Neurosci.* **2005**, *25*, 4435–4441. [CrossRef] [PubMed]
16. Godbole, G.; Shetty, A.S.; Roy, A.; D'Souza, L.; Chen, B.; Miyoshi, G.; Fishell, G.; Tole, S. Hierarchical genetic interactions between FOXG1 and LHX2 regulate the formation of the cortical hem in the developing telencephalon. *Development* **2018**, *145*. [CrossRef] [PubMed]
17. Martynoga, B.; Morrison, H.; Price, D.J.; Mason, J.O. Foxg1 is required for specification of ventral telencephalon and region-specific regulation of dorsal telencephalic precursor proliferation and apoptosis. *Dev. Biol.* **2005**, *283*, 113–127. [CrossRef] [PubMed]
18. Hanashima, C.; Fernandes, M.; Hebert, J.M.; Fishell, G. The role of foxg1 and dorsal midline signaling in the generation of cajal-retzius subtypes. *J. Neurosci. Off. J. Soc. Neurosci.* **2007**, *27*, 11103–11111. [CrossRef] [PubMed]
19. Correia, K.M.; Conlon, R.A. Whole-mount in situ hybridization to mouse embryos. *Methods* **2001**, *23*, 335–338. [CrossRef] [PubMed]
20. Liu, B.; Zhou, K.; Wu, X.; Zhao, C. Foxg1 deletion impairs the development of the epithalamus. *Mol. Brain* **2018**, *11*, 5. [CrossRef] [PubMed]
21. Yang, Y.; Shen, W.; Ni, Y.; Su, Y.; Yang, Z.; Zhao, C. Impaired interneuron development after Foxg1 disruption. *Cereb. Cortex* **2017**, *27*, 793–808. [PubMed]
22. Tian, C.; Gong, Y.; Yang, Y.; Shen, W.; Wang, K.; Liu, J.; Xu, B.; Zhao, J.; Zhao, C. Foxg1 has an essential role in postnatal development of the dentate gyrus. *J. Neurosci. Off. J. Soc. Neurosci.* **2012**, *32*, 2931–2949. [CrossRef] [PubMed]
23. Yan, Y.; Li, Y.; Hu, C.; Gu, X.; Liu, J.; Hu, Y.A.; Yang, Y.; Wei, Y.; Zhao, C. Expression of Frizzled10 in mouse central nervous system. *Gene Expr. Patterns* **2009**, *9*, 173–177. [CrossRef] [PubMed]
24. Gu, X.; Liu, B.; Wu, X.; Yan, Y.; Zhang, Y.; Wei, Y.; Pleasure, S.J.; Zhao, C. Inducible genetic lineage tracing of cortical hem derived Cajal-Retzius cells reveals novel properties. *PLoS ONE* **2011**, *6*, e28653. [CrossRef] [PubMed]
25. Zhao, C.; Guan, W.; Pleasure, S.J. A transgenic marker mouse line labels Cajal-Retzius cells from the cortical hem and thalamocortical axons. *Brain Res.* **2006**, *1077*, 48–53. [CrossRef] [PubMed]
26. Gu, X.; Yan, Y.; Li, H.; He, D.; Pleasure, S.J.; Zhao, C. Characterization of the Frizzled10-CreER transgenic mouse: An inducible Cre line for the study of Cajal-Retzius cell development. *Genesis* **2009**, *47*, 210–216. [CrossRef] [PubMed]
27. Verney, C.; Derer, P. Cajal-Retzius neurons in human cerebral cortex at midgestation show immunoreactivity for neurofilament and calcium-binding proteins. *J. Comp. Neurol.* **1995**, *359*, 144–153. [CrossRef] [PubMed]
28. Martinez-Galan, J.R.; Moncho-Bogani, J.; Caminos, E. Expression of calcium-binding proteins in layer 1 reelin-immunoreactive cells during rat and mouse neocortical development. *J. Histochem. Cytochem.* **2014**, *62*, 60–69. [CrossRef] [PubMed]
29. Brandt, M.D.; Jessberger, S.; Steiner, B.; Kronenberg, G.; Reuter, K.; Bick-Sander, A.; von der Behrens, W.; Kempermann, G. Transient calretinin expression defines early postmitotic step of neuronal differentiation in adult hippocampal neurogenesis of mice. *Mol. Cell. Neurosci.* **2003**, *24*, 603–613. [CrossRef]
30. Lavado, A.; Lagutin, O.V.; Chow, L.M.; Baker, S.J.; Oliver, G. Prox1 is required for granule cell maturation and intermediate progenitor maintenance during brain neurogenesis. *PLoS Biol.* **2010**, *8*, e1000460. [CrossRef] [PubMed]
31. Simon, R.; Brylka, H.; Schwegler, H.; Venkataramanappa, S.; Andratschke, J.; Wiegreffe, C.; Liu, P.; Fuchs, E.; Jenkins, N.A.; Copeland, N.G.; et al. A dual function of Bcl11b/Ctip2 in hippocampal neurogenesis. *EMBO J.* **2012**, *31*, 2922–2936. [CrossRef] [PubMed]

32. Motro, B.; van der Kooy, D.; Rossant, J.; Reith, A.; Bernstein, A. Contiguous patterns of c-kit and steel expression: Analysis of mutations at the W and Sl loci. *Development* **1991**, *113*, 1207–1221. [PubMed]
33. Wisden, W.; Seeburg, P.H. A complex mosaic of high-affinity kainate receptors in rat brain. *J. Neurosci. Off. J. Soc. Neurosci.* **1993**, *13*, 3582–3598. [CrossRef]
34. Tole, S.; Goudreau, G.; Assimacopoulos, S.; Grove, E.A. Emx2 is required for growth of the hippocampus but not for hippocampal field specification. *J. Neurosci. Off. J. Soc. Neurosci.* **2000**, *20*, 2618–2625. [CrossRef]
35. Danesin, C.; Peres, J.N.; Johansson, M.; Snowden, V.; Cording, A.; Papalopulu, N.; Houart, C. Integration of telencephalic Wnt and hedgehog signaling center activities by Foxg1. *Dev. Cell* **2009**, *16*, 576–587. [CrossRef] [PubMed]
36. Bielle, F.; Griveau, A.; Narboux-Neme, N.; Vigneau, S.; Sigrist, M.; Arber, S.; Wassef, M.; Pierani, A. Multiple origins of Cajal-Retzius cells at the borders of the developing pallium. *Nat. Neurosci.* **2005**, *8*, 1002–1012. [CrossRef] [PubMed]
37. Hanashima, C.; Shen, L.; Li, S.C.; Lai, E. Brain factor-1 controls the proliferation and differentiation of neocortical progenitor cells through independent mechanisms. *J. Neurosci. Off. J. Soc. Neurosci.* **2002**, *22*, 6526–6536. [CrossRef]
38. Hanashima, C.; Li, S.C.; Shen, L.; Lai, E.; Fishell, G. Foxg1 suppresses early cortical cell fate. *Science* **2004**, *303*, 56–59. [CrossRef] [PubMed]
39. Miyoshi, G.; Fishell, G. Dynamic FoxG1 expression coordinates the integration of multipolar pyramidal neuron precursors into the cortical plate. *Neuron* **2012**, *74*, 1045–1058. [CrossRef] [PubMed]

8

Loss of Suppressor of Fused in Mid-Corticogenesis Leads to the Expansion of Intermediate Progenitors

Odessa R. Yabut [1], Hui Xuan Ng [1], Gloria Fernandez [1], Keejung Yoon [1,2], Jeremy Kuhn [1] and Samuel J. Pleasure [1,3,*]

1 Department of Neurology, University of California San Francisco, San Francisco, CA 94158, USA; odessa.yabut@ucsf.edu (O.R.Y.); huixuan.ng@ucsf.edu (H.X.N.); gloria__fg@hotmail.com (G.F.); keejung@skku.edu (K.Y.); jeremydavidkuhn@gmail.com (J.K.)
2 College of Biotechnology and Bioengineering, Sungkyunkwan University, Suwon, Gyeonggi-do 16419, Korea
3 Programs in Neuroscience and Developmental Biology, Eli and Edythe Broad Center of Regeneration Medicine and Stem Cell Research, University of California San Francisco, San Francisco, CA 94143, USA
* Correspondence: sam.pleasure@ucsf.edu; Tel.: +1-415-502-5683

Academic Editor: Henk Roelink

Abstract: Neural progenitors in the embryonic neocortex must be tightly regulated in order to generate the correct number and projection neuron subtypes necessary for the formation of functional neocortical circuits. In this study, we show that the intracellular protein Suppressor of Fused (Sufu) regulates the proliferation of intermediate progenitor (IP) cells at later stages of corticogenesis to affect the number of Cux1+ upper layer neurons in the postnatal neocortex. This correlates with abnormal levels of the repressor form of Gli3 (Gli3R) and the ectopic expression of Patched 1 (Ptch1), a Sonic Hedgehog (Shh) target gene. These studies reveal that the canonical role of Sufu as an inhibitor of Shh signaling is conserved at later stages of corticogenesis and that Sufu plays a crucial role in regulating neuronal number by controlling the cell cycle dynamics of IP cells in the embryonic neocortex.

Keywords: Suppressor of Fused; Sonic Hedgehog; Gli3; corticogenesis; neurogenesis

---

1. Introduction

The generation of molecularly diverse glutamatergic projection neurons in the adult mammalian neocortex is dynamically regulated by spatial and temporal events during embryonic cortical development. Virtually all glutamatergic projection neurons originate from neural progenitors in the ventricular (VZ) and subventricular zone (SVZ), which are derived from multipotent radial glial (RG) cells [1]. At early stages of corticogenesis, the majority of RGs initially expand through self-renewing symmetric divisions at the apical surface of the VZ. As corticogenesis progresses, RG cells undergo asymmetric neurogenic divisions to self-renew and generate intermediate progenitors (IP) or short neural precursors (SNP) [2,3]. IP cells that populate the SVZ at later stages of corticogenesis will undergo one or more symmetric divisions before terminally differentiating to produce a pair of (or four) neurons that will largely populate the upper layers of the neocortex [4]. This temporal sequence is a fundamental feature of neocortical development, which ensures the generation of correct number of projection neurons, necessary for precise formation of local and long-range circuits in the adult neocortex.

Specialized molecular cascades incorporate intrinsic and extrinsic cues to direct the expansion and differentiation of neural progenitors in the telencephalon. Sonic Hedgehog (Shh) signaling is centrally involved in this process and is required during distinct developmental windows to direct patterning, specification, and proliferation of progenitors [5]. In the dorsal telencephalon, Shh signaling activity is relatively low during corticogenesis, yet deregulation of Shh signaling in the embryonic neocortex profoundly affects the behavior of neural stem/progenitor cells. In particular, we recently found that when modulation of Shh signaling is lost at early stages of corticogenesis, neocortical progenitors are improperly specified, which results in the misspecification of projection neurons [6]. Tightly regulating Shh signaling is also likely crucial in late corticogenesis, since loss of transcription repressor activity of Gli3, a downstream effector of Shh signaling, leads to abnormal proliferation and specification of cortical progenitors [7]. Similarly, forced overexpression of Shh at this stage results in the abnormal specification of cortical progenitors [8]. However, the implications of deregulating endogenous Shh signaling in late corticogenesis have not been yet fully examined.

Suppressor of Fused (Sufu) is a critical antagonist of Shh signaling that influences patterning, specification, differentiation, and migration of neural progenitors and their progenies in the developing central nervous system [6,9–11]. Sufu exerts this role by directly acting on Gli transcription factors downstream of the Shh signaling cascade via control of their stability, degradation, and the formation of Gli repressors [11–14]. At early stages of corticogenesis, we have shown that Sufu modulates Shh signaling by controlling the stability of full-length Gli2 and Gli3 [6]. These defects were absent at later stages of corticogenesis, where Gli3, the predominant Gli at this stage, was not similarly affected and all projection neurons of the neocortex were properly specified. However, in-depth studies on the state of cortical progenitors at later stages of corticogenesis and the effect of endogenous Shh signaling are needed, particularly since additional Shh-expressing cells are present in a subset of progenitors lining the lateral ventricles, Cajal-Retzius neurons, and in ventrally derived interneurons that have migrated into the neocortex [15]. Furthermore, loss of function studies indicate that Shh signaling continues to play a role in regulating the cell cycle dynamics of cortical progenitors [15]. In this study, we show that Sufu continues to function as a regulator of Shh signaling activity at later stages of corticogenesis. Specifically, loss of Sufu deregulates the proliferation program of basally dividing cortical progenitors resulting in an increase in the number of upper layer neurons in the postnatal neocortex. These observations correlate with an increase in Shh activity, in part due to an imbalance between the full-length Gli3 activator (Gli3FL) and the cleaved Gli3R repressor (Gli3R) levels. These studies elucidate an important role for Sufu in regulating Shh signaling activity in late corticogenesis.

2. Materials and Methods

2.1. Animals

Mice carrying the floxed Sufu allele (*Sufu*^{fl}) were kindly provided by Dr. Chi-Chung Hui (University of Toronto, Toronto, ON, Canada) and were genotyped as described [16]. The *hGFAP-Cre* (Stock #004600) was obtained from Jackson Laboratories (Bar Harbor, ME, USA). All animal protocols were in accordance with the regulations of the National Institute of Health and approved by the University of California San Francisco Institutional Animal Care and Use Committee (IACUC).

2.2. Immunohistochemistry and DiI Labeling

Perfusion, dissection, and immunofluorescence staining were conducted according to standard protocols as previously described [17]. The following are the antibodies used: mouse anti-BrdU (1:50 dilution; BD Pharmingen, Franklin Lakes, NJ, USA), rabbit anti-Cux1 (1:100 dilution; Santa Cruz Biotechnology, Dallas, TX, USA), rabbit anti-Phospho-Histone H3 (1:250 dilution; Millipore, Billerica, MA, USA), rabbit anti-Pax 6 (1:500 dilution; Covance, Princeton, NJ, USA), rabbit anti-Tbr2 (1:500 dilution; Abcam, Cambridge, UK), mouse anti-Tuj1 (1:500 dilution; Covance, Princeton, NJ, USA), rabbit anti-Gli3 (1:100 dilution; Santa Cruz Biotechnology, Dallas, TX, USA), and rabbit anti-Cleaved Caspase 3 (1:300 dilution; Cell Signaling, Madison, WI, USA).

For 5-bromo-2-deoxyuridine (BrdU, Sigma, St. Louis, MO, USA) labeling, pregnant dams were treated with 50 μg/g BrdU by intraperitoneal injection for 4 h prior to dissection at E16.5. DiI labeling was conducted by placing small crystals of the lipophilic tracer (1,1′-dioctadecyl-3,3,3′,3′-tetramethylindocarbocyanine; Invitrogen, Waltham, MA, USA) in the neocortex to target the upper layer (2/3) and then remained in 4% paraformaldehyde (PFA). After 6 weeks, brains were sectioned at 100 μm, counterstained with bisbenzimide, mounted, and imaged.

2.3. *In Situ Hybridization*

Patched 1 (Ptch1) in situ hybridization (ISH) was conducted according to standard protocols [17] using probes generated from mouse ptc probe M2-3, a gift from Matthew Scott (Addgene plasmid #58701).

2.4. *Image Acquisition and Analysis*

Images were acquired using a Nikon E600 microscope equipped with a QCapture Pro camera (QImaging) or Zeiss Axioscan Z.1 (Zeiss, Thornwood, NY, USA) using the Zen 2 blue edition software (Zeiss, Thornwood, NY, USA. Z-stack images were acquired using a Nikon Spectral C1si Laser Scanning Confocal (Nikon Imaging Center, UCSF) and scanned a 10 μm section at a resolution of 1024 × 1024 pixels using the Nikon EZ-C1 software v3.8 (Nikon Instruments, Melville, NY, USA). Adobe Photoshop CS6 was used for image editing with brightness/contrast levels adjusted equally in control and mutant tissues. NIH Image J was used to quantify raw, unedited images. For measurement of VZ/SVZ thickness, the length of densely populated 4′,6-diamidino-2-phenylindole (DAPI)+ regions adjacent to the lateral ventricles to the SVZ/intermediate zone (IZ) boundary was measured. For measurement of BrdU-labeled cells in the VZ or SVZ, the VZ was defined as the area within 70 μm from the lateral ventricles, while the SVZ was defined as the rest of the DAPI-dense regions along the progenitor zones. Pax6+, Tbr2+, and BrdU+, and phospho-Histone H3+ cells were quantified within the VZ/SVZ, which was characterized by the densely populated DAPI+ regions adjacent to the lateral ventricles or by where Tuj1 staining was sparse. Fluorescently labeled cells within this region were quantified to measure the number of cells per 100 μm^2. Apical progenitors were quantified by measuring the number of labeled cells along the length of the ventricular surface. For quantification of labeled Cux1+ neurons at postnatal stages, layers 2/3/4 were measured to determine the area. Fluorescently labeled cells within this region were quantified to measure the number of cells per 100 μm^2. Two to three optical sections at 10 μm-thick each, which were histologically matched for rostral-caudal level between genotypes, were analyzed.

2.5. *Quantitative PCR Analysis*

Total RNA was isolated from dissected E16.5 cortical tissues using the RNEasy Mini Kit (Qiagen, Venlo, The Netherlands). cDNA was generated using the High Capacity cDNA Reverse Transcription Kit (Applied Biosystems, Waltham, MA, USA). Transcript expression was measured via the incorporation of SYBR Green (Life Technologies, Carlsbad, CA, USA) using the Applied Biosystem 7500 Real-Time PCR System (Life Technologies, Carlsbad, CA, USA). Primers for Ptch1 and Gli1 have been previously described [18]. Quantitative PCR (qPCR) data were analyzed using the comparative C_T or the relative standard curve method, with β-actin [19] used as control.

2.6. Western Blot Analysis

Cortical tissues were lysed in radioimmunoprecipitation assay (RIPA) buffer (Sigma) supplemented with protease (Complete Mini, Roche, Basel, Switzerland) and phosphatase (PhosStop, Roche) inhibitors according to standard protocols. Western blot analyses were conducted according to standard protocols. Briefly, soluble extracts were loaded onto Criterion, 4%–15% Tris-HCI 4 sodium dodecyl sulfate polyacrylamide gel electrophoresis (SDS-PAGE) gels (Bio-Rad, Berkeley, CA, USA), separated at 120 V, and transferred to polyvinylidene difluoride (PVDF) membrane (Bio-Rad) at 30 V for 2 h or overnight at 4 °C. Membranes were blocked with 5% BSA/1X TBS-T (Tris-buffered saline with 0.1% Tween 20) for 1 h at room temperature, and incubated with primary antibodies diluted in blocking buffer overnight at 4 °C, and secondary antibodies (1:5000 dilution; IR-Dye antibodies, LI-COR, Lincoln, NE, USA) for 1 h at RT. Membranes were washed in 1X TBS-T and scanned using the Odyssey Infrared Imaging System (LI-COR). Primary antibodies were used as follows: goat anti-Gli3 (1:500; R&D Systems) and α-Tubulin (1:5000, Abcam). Quantification and analysis were conducted using the Odyssey System and Image Studio Software version 4.0.21 (LI-COR, Lincoln, NE, USA).

2.7. Statistics

All experiments were conducted in triplicate with a sample size of $n = 3-6$ embryos/animals per genotype. For pairwise analysis of control and mutant genotypes, the Student t-test was used. Graphs display the mean ± standard error of the mean (SEM).

3. Results

3.1. Expansion of the VZ and SVZ in the E16.5 Neocortex of $hGFAP^{cre/+};Sufu^{fl/fl}$ Mice

We previously determined that Sufu has critical functions at early stages of corticogenesis particularly in the specification of cortical progenitors into distinct neuronal lineages [6]. Since Sufu remains specifically expressed in the VZ/SVZ of the embryonic neocortex mid-corticogenesis [6], we postulated that it continues to play important roles in cortical progenitors at this stage. To identify the function of Sufu, we examined how loss of Sufu affected the behavior of cortical progenitors within the dorsolateral regions of the rostral neocortex (Figure 1A) in $hGFAP^{cre/+};Sufu^{fl/fl}$ mice, in which Sufu deletion in all cortical progenitors occurred mid-corticogenesis (~E13.5) [20]. In $hGFAP^{cre/+};Sufu^{fl/fl}$ mice, the specification of cortical progenitors was relatively normal [6]. However, we observed an expansion of the VZ/SVZ at E16.5 in the dorsolateral neocortex of $hGFAP^{cre/+};Sufu^{fl/fl}$ mice (Figure 1B). The thickness of the VZ/SVZ, defined as the DAPI-dense regions with sparse Tuj1 staining along the lateral ventricles, was significantly increased in the $hGFAP^{cre/+};Sufu^{fl/fl}$ neocortex compared to controls (Figure 1C). The expansion of VZ/SVZ did not correlate with an increase in the thickness of Pax6+ VZ regions (Figure 1D,E). Rather, we observed an increase in the thickness of Tbr2+ SVZ regions (Figure 1F,G). These findings prompted examination of the proliferative state of the neocortical progenitors in the VZ/SVZ.

3.2. Abnormal Proliferation of Cortical Progenitors in the E16.5 Neocortex of $hGFAP^{cre/+};Sufu^{fl/fl}$ Mice

At E16.5, the VZ and SVZ are composed of RG and IP cells, respectively [21]. The embryonic neocortex has accumulated a larger population of IPs at this stage as a result of RG cell expansion at earlier stages of corticogenesis. IP cells will undergo one or more cell divisions prior to differentiating into projection neurons that will eventually populate the upper cortical layers [2,4]. To determine whether the expansion of VZ/SVZ in the E16.5 $hGFAP^{cre/+};Sufu^{fl/fl}$ neocortex was due to deregulated proliferation of cortical progenitors, we conducted BrdU treatments to label proliferating cells. Examination of BrdU-treated embryos 4 h after treatment showed a visible increase in BrdU-labeled cells in the SVZ of the E16.5 $hGFAP^{cre/+};Sufu^{fl/fl}$ neocortex compared to control (Figure 2A,B). Quantification of BrdU-labeled cells verified these observations in which a significantly higher density of BrdU+ cells was evident in the SVZ of the E16.5 neocortex of *hGFAP*$^{cre/+}$;*Sufu*$^{fl/fl}$ mice, while a slight reduction in BrdU+ cells was observed in the VZ when compared to controls (Figure 2C). These findings indicated that a larger proportion of progenitors were actively dividing in this period in the *hGFAP*$^{cre/+}$;*Sufu*$^{fl/fl}$ neocortical SVZ.

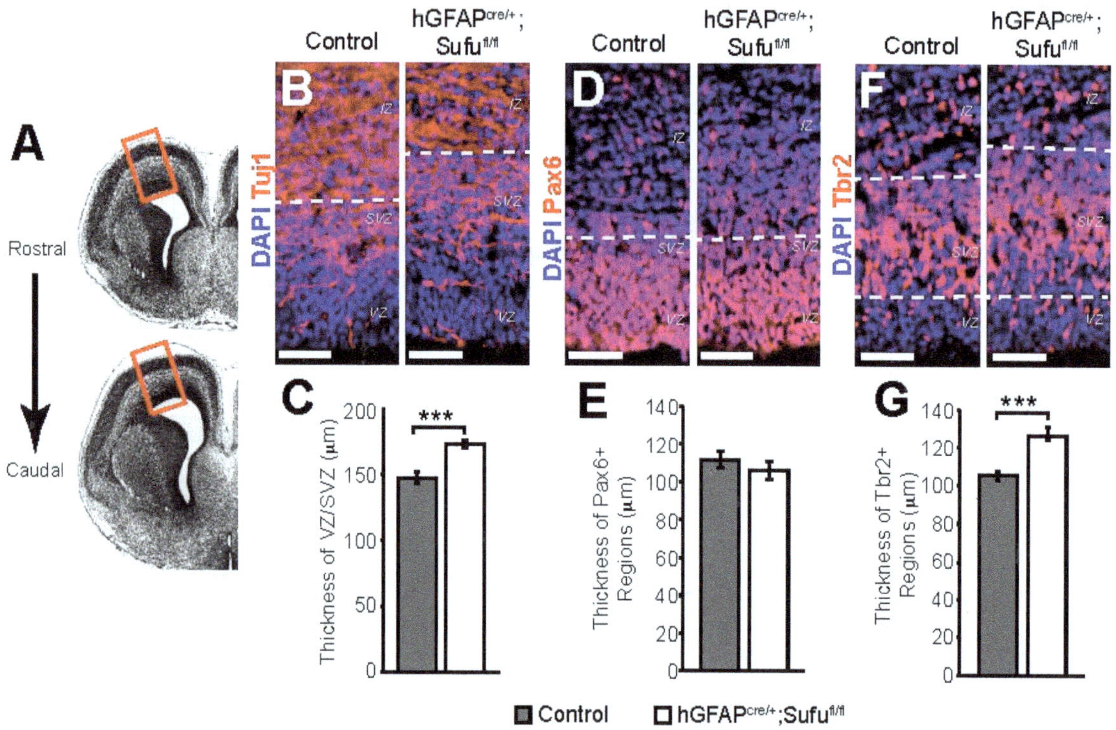

Figure 1. Expansion of the progenitor zones in E16.5 *hGFAP*$^{cre/+}$;*Sufu*$^{fl/fl}$ mice. (**A**) The experiments presented here analyzed the dorsolateral region of the neocortex (boxed region), obtained from the rostral forebrain spanning between these representative images; (**B**,**C**) Targeted deletion of Suppressor of Fused (Sufu) in cortical progenitors at E13.5 (*hGFAP*$^{cre/+}$;*Sufu*$^{fl/fl}$) led to the expansion of the VZ/SVZ by E16.5 as defined by immunofluorescence staining. Regions marked by DAPI-dense (blue) cells and reduced Tuj1 (red) immunostaining along the lateral ventricles showed expansion of the *hGFAP*$^{cre/+}$;*Sufu*$^{fl/fl}$ VZ/SVZ compared to controls (**B**); The thickness of the VZ/SVZ was determined as the distance between the ventricular lining along the lateral ventricles and the IZ, which was identified by densely packed DAPI+ cells and minimal Tuj1 staining within the dorsolateral neocortex. These measurements confirmed that the *hGFAP*$^{cre/+}$;*Sufu*$^{fl/fl}$ VZ/SVZ was significantly thicker than those of control littermates (**C**); (**D**,**E**) Immunofluorescence staining against the RG cell marker, Pax6 (red), showed no visible differences between the E16.5 control and *hGFAP*$^{cre/+}$;*Sufu*$^{fl/fl}$ Pax6+ regions as defined by the dashed lines (**D**); This observation was verified when the thickness of Pax6+ regions was measured (**E**); (**F**,**G**) Immunofluorescence staining against the intermediate progenitor (IP) cell marker, Tbr2 (red), showed a visible expansion of Tbr2+ regions (between dashed lines) in the E16.5 *hGFAP*$^{cre/+}$;*Sufu*$^{fl/fl}$ VZ/SVZ when compared to controls (**F**); Quantification of the thickness of Tbr2+ regions verified these observations (**G**); *** *p*-value < 0.01. Scale bars = 50 μm; VZ, ventricular zone; SVZ, subventricular zone; IZ, intermediate zone; DAPI, 4′,6-diamidino-2-phenylindole.

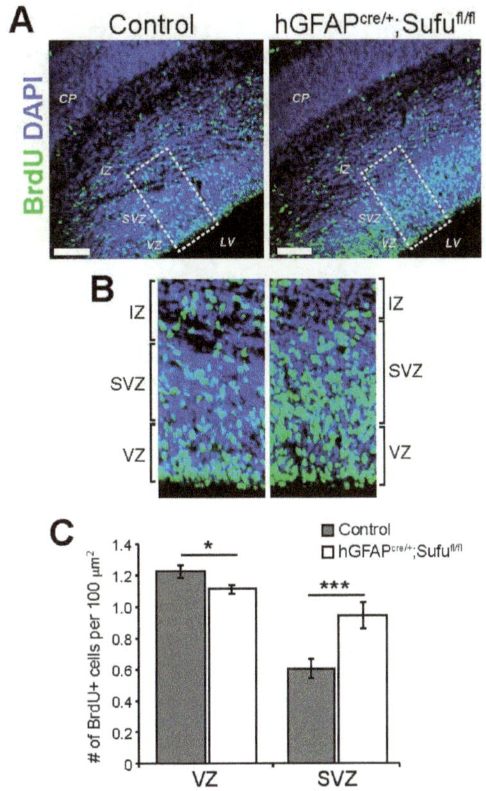

Figure 2. Abnormal proliferation of cortical progenitors in the E16.5 *hGFAP*$^{cre/+}$;*Sufu*$^{fl/fl}$ neocortex; (**A**,**B**) Immunofluorescence staining with anti-BrdU showed that after 4 h of BrdU-treatment, BrdU+ cells were greater in the E16.5 *hGFAP*$^{cre/+}$;*Sufu*$^{fl/fl}$ neocortex compared to control littermates (**A**); Boxed areas in (**A**) showed that this increase was apparent in the SVZ of the E16.5 *hGFAP*$^{cre/+}$;*Sufu*$^{fl/fl}$ neocortex (**B**); Scale bars = 100 μm; CP, cortical plate; BrdU, 5-bromo-2-deoxyuridine (**C**) Quantification of BrdU+ cells within the VZ were mildly, yet significantly reduced, in the E16.5 *hGFAP*$^{cre/+}$;*Sufu*$^{fl/fl}$ neocortex. In contrast, a significant increase in the density of BrdU+ cells were quantified in the SVZ of the E16.5 *hGFAP*$^{cre/+}$;*Sufu*$^{fl/fl}$ neocortex compared to controls; * p-value < 0.05, *** p-value < 0.01.

3.3. Increase in Basally Dividing Progenitors in the E16.5 hGFAP$^{cre/+}$;Sufu$^{fl/fl}$ Neocortex

RGs are typically undergoing asymmetric cell division in the apical lining of the VZ, while IPs are undergoing symmetric division in the basal SVZ regions, prior to differentiating into upper layer projection neurons mid-corticogenesis [2]. The increase in density of BrdU+ cells in the SVZ of *hGFAP*$^{cre/+}$;*Sufu*$^{fl/fl}$ mice (Figure 2B) indicates a potential increase in basally dividing progenitors. However, because BrdU is incorporated during the S-phase of the cell cycle along the VZ/SVZ, including RGs that may be undergoing interkinetic nuclear migration [1], BrdU-labeled cells in the SVZ may not be exclusively IPs. We therefore conducted immunofluorescence staining against the mitosis marker, Phospho-Histone H3, to distinguish between RGs undergoing mitosis in the apical lining of the VZ, and IPs that undergo mitosis in the basal or SVZ regions of the neocortical progenitor zone. These experiments showed a dramatic increase in basally dividing cells in the E16.5 *hGFAP*$^{cre/+}$;*Sufu*$^{fl/fl}$ neocortex (boxed inset, Figure 3A). Indeed, we quantified a significant increase in basally dividing progenitors in the E16.5 *hGFAP*$^{cre/+}$;*Sufu*$^{fl/fl}$ neocortex but not in apically dividing cells (Figure 3B). These findings indicated that loss of Sufu promoted the mitotic division of progenitors within the SVZ.

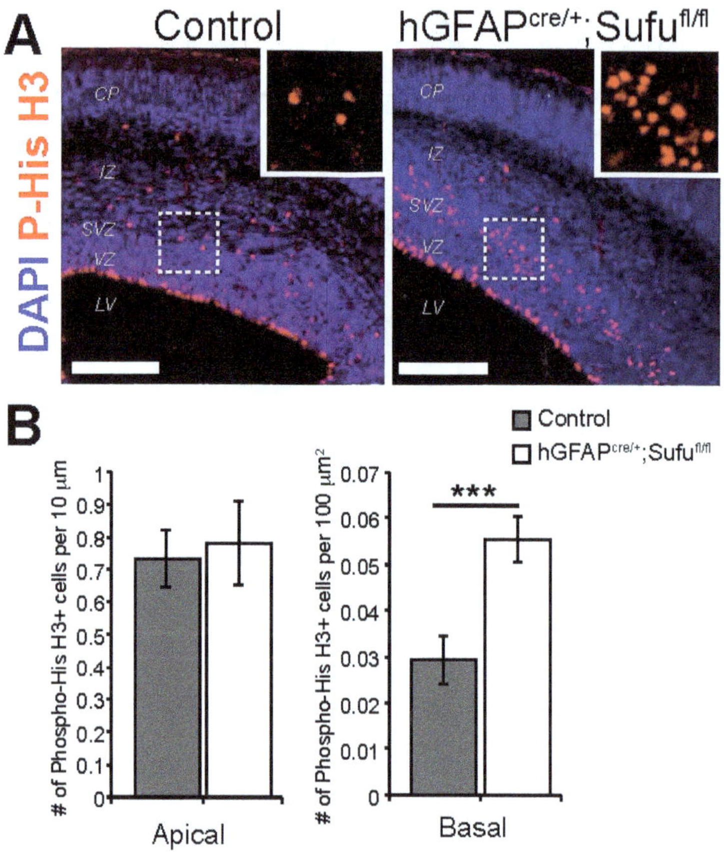

Figure 3. Basally dividing progenitors dramatically increased in the neocortex of E16.5 *hGFAP$^{cre/+}$;Sufu$^{fl/fl}$* mice. (**A**) Labeling of mitotically active cells by immunofluorescence staining against Phospho-Histone H3 (P-His H3) showed an obvious expansion of cells undergoing mitosis in the SVZ of the *hGFAP$^{cre/+}$;Sufu$^{fl/fl}$* neocortex compared to control littermates (boxed inset). Scale bars = 200 μm; (**B**) Quantification of mitotically active cells along the apical lining of the VZ, as labeled by P-His H3, showed no difference between control and *hGFAP$^{cre/+}$;Sufu$^{fl/fl}$* mice. However, a significant almost two-fold increase in P-His H3+ cells were quantified in the SVZ of the *hGFAP$^{cre/+}$;Sufu$^{fl/fl}$* neocortex compared to controls; *** p-value < 0.01.

To determine the identity of proliferating cells in the VZ/SVZ, we conducted double immunofluorescence staining with antibodies against BrdU and Pax6, a marker for both apical RG cells and basal RG cells (those that localize in the SVZ), or Tbr2, a specific marker for IPs [4]. As expected, we found an increase in BrdU-labeled Tbr2+ cells in the VZ/SVZ of the E16.5 *hGFAP$^{cre/+}$;Sufu$^{fl/fl}$* neocortex (Figure 4D,F). Surprisingly, we also found an increase in BrdU-labeled Pax6+ cells (Figure 4C), particularly those that localized in the SVZ (boxed areas in Figure 4A). Despite this, we did not observe any changes in the density of Pax6+ RG cells or Tbr2+ IP cells in the VZ/SVZ regions (Figure 4B,E). Since we observed an expansion of Tbr2+ regions (Figure 1F,G), an increase in basally dividing cells (Figure 3), and that Pax6 expression is typically found in newly generated IP cells [21], our findings suggested that loss of Sufu not only promoted the proliferation of basally dividing progenitors, but also likely triggered the transition of Pax6+ RG cells into Tbr2+ IP cells, in the E16.5 neocortex.

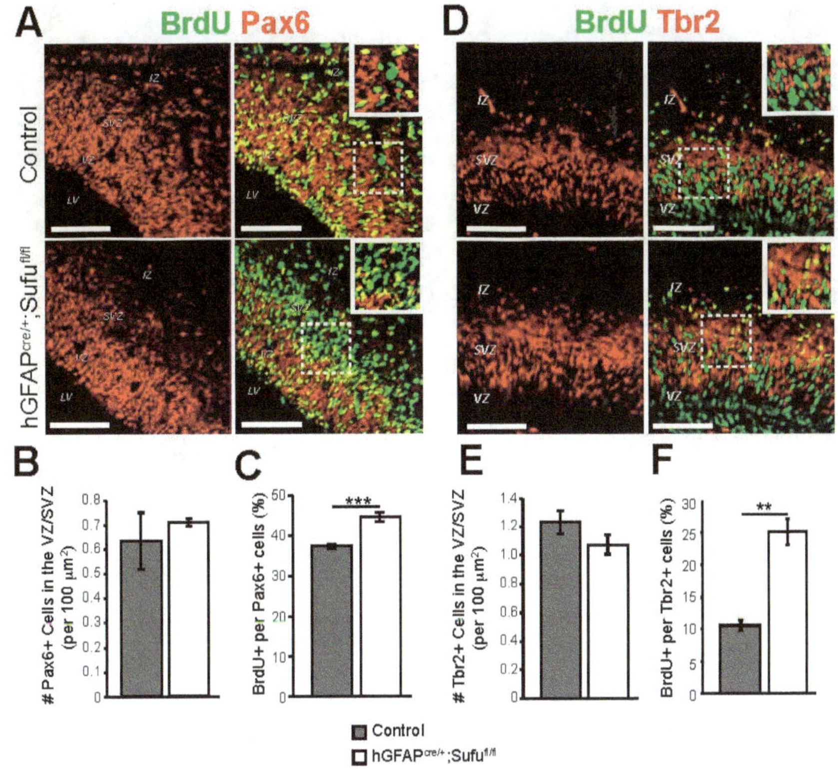

Figure 4. Loss of Sufu led to expansion of Pax6+ and Tbr2+ basally dividing cells. (**A**–**C**) Pax6 and BrdU double immunostaining (**A**) did not show obvious differences in the density of Pax6+ RG cells in the VZ/SVZ between control and *hGFAP*$^{cre/+}$;*Sufu*$^{fl/fl}$ mice at E16.5 and verified by quantification (**B**); On the other hand, BrdU-labeled Pax6+ cells were visibly obvious (boxed insets, **A**) and significantly increased in the E16.5 *hGFAP*$^{cre/+}$;*Sufu*$^{fl/fl}$ neocortex compared to controls (**C**); (**D**–**F**) Tbr2 and BrdU double immunostaining (**D**) did not show obvious differences in the density of Tbr2+ IP cells in the VZ/SVZ between control and *hGFAP*$^{cre/+}$;*Sufu*$^{fl/fl}$ mice at E16.5. This was verified by quantification of Tbr2+ cell density (**E**); However, a greater number of BrdU-labeled Tbr2+ cells was observed within the VZ/SVZ of *hGFAP*$^{cre/+}$;*Sufu*$^{fl/fl}$ mice at E16.5 (boxed inset, **D**); Quantification of double-labeled IP cells was measured and yielded a significantly higher percentage of BrdU-labeled Tbr2+ cells in the E16.5 *hGFAP*$^{cre/+}$;*Sufu*$^{fl/fl}$ VZ/SVZ compared to control littermates (**F**); ** p-value < 0.03; *** p-value < 0.01. Scale bars = 200 μm; LV, lateral ventricle.

3.4. Reduced Levels of Gli3R in the E16.5 hGFAP$^{cre/+}$;Sufu$^{fl/fl}$ Neocortex

Deletion of Sufu in the mid-corticogenesis stage does not affect the stability of total Gli3 protein because both Gli3FL and its cleaved Gli3R isoform were detectable in the E16.5 *hGFAP*$^{cre/+}$;*Sufu*$^{fl/fl}$ neocortex [6]. We conducted immunostaining against Gli3 in the E16.5 control and *hGFAP*$^{cre/+}$;*Sufu*$^{fl/fl}$ neocortex and found that Gli3 was indeed expressed in the progenitor zones of mice from both genotypes (Figure 5A). Examination of the subcellular localization of Gli3 showed that in both control and mutant neocortex, Gli3 was largely cytoplasmic along the VZ lining (white arrows, Figure 5B) but usually localized in the nucleus as cells transition into the SVZ (yellow arrows, Figure 5B). Western blot analysis of Gli3 proteins showed that both Gli3FL and Gli3R were present in the E16.5 *hGFAP*$^{cre/+}$;*Sufu*$^{fl/fl}$ neocortex, as we previously observed (Figure 5B) [6]. However, quantification of these isoforms showed that Gli3R protein levels were significantly reduced in the E16.5 *hGFAP*$^{cre/+}$;*Sufu*$^{fl/fl}$ neocortex. These findings indicated that in the mid-corticogenesis stage, Sufu does not affect the stability of Gli3FL, nor did it affect the subcellular localization of Gli3R, the predominant form of Gli3 in the embryonic neocortex [7]. Rather, loss of Sufu mid-corticogenesis resulted in the failure to maintain levels of Gli3R in the progenitor zones of the E16.5 *hGFAP*$^{cre/+}$;*Sufu*$^{fl/fl}$ neocortex.

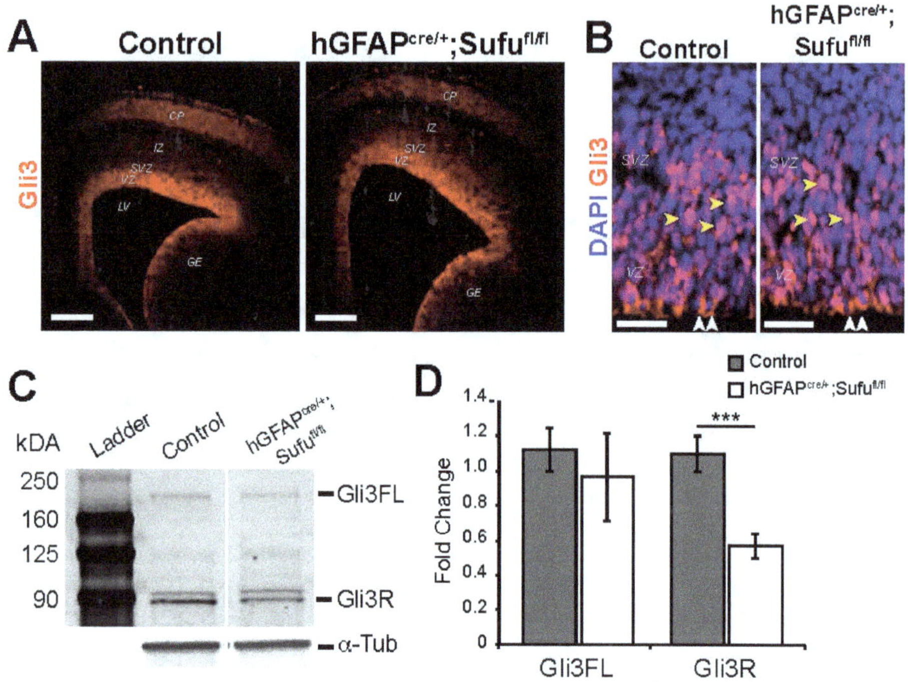

Figure 5. Reduced Gli3R levels in the VZ/SVZ of the E16.5 *hGFAP*$^{cre/+}$*;Sufu*$^{fl/fl}$ neocortex. (**A**,**B**) Gli3 immunostaining showed highly specific expression in the VZ/SVZ of the E16.5 neocortex of control and *hGFAP*$^{cre/+}$*;Sufu*$^{fl/fl}$ mice (**A**); Confocal z-stack imaging showed that the subcellular localization of Gli3 did not significantly differ between control and *hGFAP*$^{cre/+}$*;Sufu*$^{fl/fl}$ VZ/SVZ, where Gli3 appeared to be cytoplasmic along the ventricular lining (white arrows) and nuclear localized in the SVZ (yellow arrows) (**B**); Scale bars = 200 μm in (**A**); 50 μm in (**B**); GE, ganglionic eminence; (**C**) Western blot analysis of protein extracts from the E16.5 neocortex showed the presence of both Gli3FL and cleaved Gli3R proteins, with Gli3R being predominant in both control and *hGFAP*$^{cre/+}$*;Sufu*$^{fl/fl}$ mice. However, quantification of Gli3FL and Gli3R (**D**) showed significantly reduced Gli3R protein levels in the *hGFAP*$^{cre/+}$*;Sufu*$^{fl/fl}$ neocortex compared to controls. α-Tubulin (α-Tub) was used as a loading control; *** *p*-value < 0.01; Gli3FL, full-length Gli3; Gli3R, cleaved Gli3R repressor.

3.5. Activation of Shh Signaling in the E16.5 hGFAP$^{cre/+}$*;Sufu*$^{fl/fl}$ *Neocortex*

Since Gli3R is known to inhibit transcription of Shh targets, we examined the activity of Shh signaling in the E16.5 *hGFAP*$^{cre/+}$*;Sufu*$^{fl/fl}$ neocortex. We previously did not observe ectopic Gli1 expression in the E16.5 *hGFAP*$^{cre/+}$*;Sufu*$^{fl/fl}$ neocortex in mice that also carried the Shh-responsive Gli1-LacZ transgene [6,22]. Because the responsiveness of Gli1-LacZ might be tissue- and age- specific, we examined the expression of another Shh target, Ptch1, in the E16.5 neocortex of control and mutant mice by conducting qPCR. As a control, we conducted qPCR using Sufu-specific probes, which verified reduced mRNA expression levels of Sufu in the E16.5 *hGFAP*$^{cre/+}$*;Sufu*$^{fl/fl}$ neocortex (Figure 6A). We also confirmed our previous observation that Gli1 was not ectopically expressed in the E16.5 *hGFAP*$^{cre/+}$*;Sufu*$^{fl/fl}$ neocortex (Figure 6B). However, a significant increase in Ptch1 expression was observed in the E16.5 *hGFAP*$^{cre/+}$*;Sufu*$^{fl/fl}$ neocortex compared to controls (Figure 6C). This was verified by in situ hybridization using Ptch1-specific riboprobes showing ectopic expression of Ptch1 in the VZ/SVZ of the E16.5 *hGFAP*$^{cre/+}$*;Sufu*$^{fl/fl}$ neocortex (Figure 6D). These findings indicated that loss of Sufu resulted in the upregulation of the Shh target Ptch1, a potential consequence of reduced Gli3R levels. Thus, in the E16.5 neocortex, Sufu continues to function as a modulator of Shh signaling in cortical progenitors within the neurogenic zones.

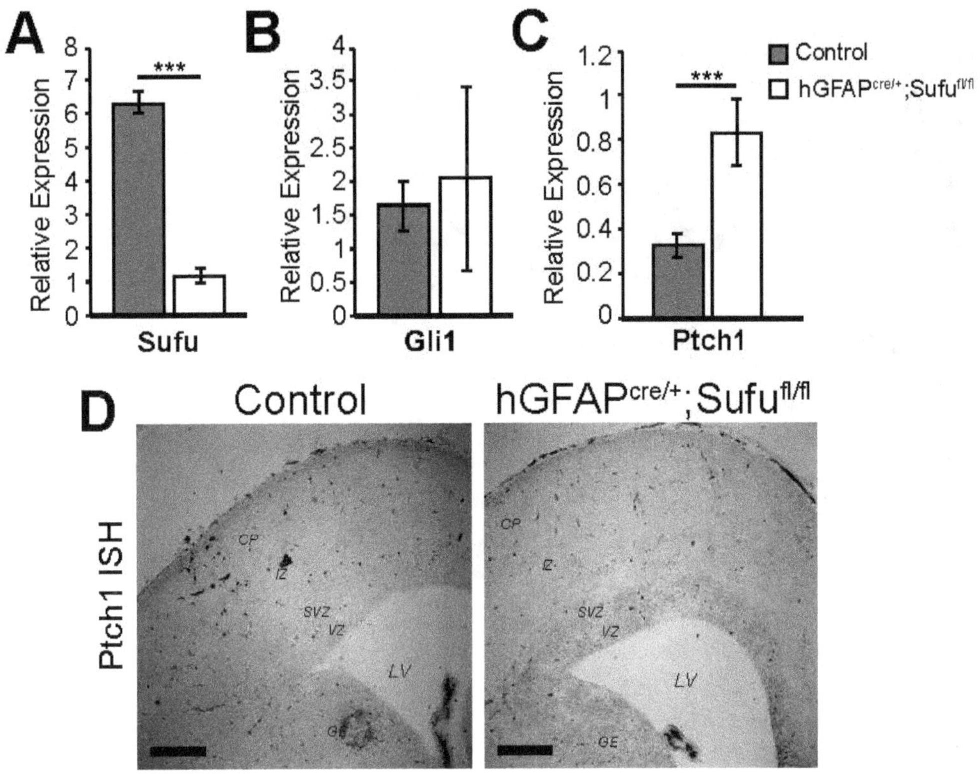

Figure 6. Ectopic expression of Ptch1 in the E16.5 *hGFAP*$^{cre/+}$*;Sufu*$^{fl/fl}$ VZ/SVZ. (**A–C**) quantitative PCR (qPCR) analysis of mRNA extracted from the neocortex showed downregulation of Sufu as expected in the E16.5 *hGFAP*$^{cre/+}$*;Sufu*$^{fl/fl}$ neocortex (**A**); No changes in Gli1 expression were detected between control and mutant neocortex as previously reported (**B**); However, a significant increase in the expression of Ptch1 was detected in the E16.5 *hGFAP*$^{cre/+}$*;Sufu*$^{fl/fl}$ neocortex compared to controls (**C**) indicating activation of Sonic Hedgehog (Shh) signaling; *** *p*-value < 0.01; (**D**) In situ hybridization using Patched-1 (Ptch1)-specific riboprobes showed ectopic expression of Ptch1 in the VZ/SVZ of the E16.5 *hGFAP*$^{cre/+}$*;Sufu*$^{fl/fl}$ neocortex. Scale bars = 200 μm.

3.6. Increase in Cell Death in the E16.5 hGFAP$^{cre/+}$*;Sufu*$^{fl/fl}$ *Neocortex*

We previously observed that loss of Sufu at early stages of corticogenesis resulted in cortical progenitors or neurons that were unable to survive [6]. Given that there is a significant increase in IP cells in the E16.5 *hGFAP*$^{cre/+}$*;Sufu*$^{fl/fl}$ neocortex, we examined whether these progenitors and their progenies were able to survive in the developing neocortex. By immunolabeling with the cell death marker, cleaved-Caspase 3, we found that there was a dramatic increase in apoptotic cells in the dorsolateral neocortex of the E16.5 *hGFAP*$^{cre/+}$*;Sufu*$^{fl/fl}$ mice (within dashed lines, Figure 7A) compared to controls, particularly in the VZ/SVZ regions (Figure 7B). Quantification of cleaved-Caspase 3+ cells confirmed these observations, in which a significantly higher number of apoptotic cells were found in the dorsolateral neocortex of the E16.5 *hGFAP*$^{cre/+}$*;Sufu*$^{fl/fl}$ mice compared to controls (Figure 7C). These findings indicated that the deletion of Sufu in cortical progenitors affected its ability to survive during corticogenesis.

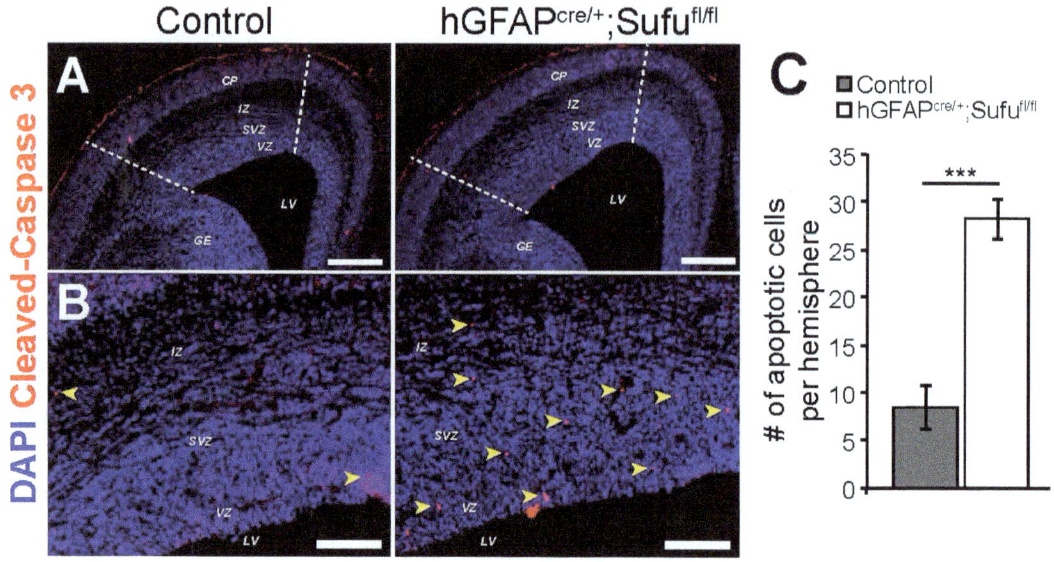

Figure 7. Increased cell death in the E16.5 *hGFAP*$^{cre/+}$*;Sufu*$^{fl/fl}$ dorsolateral neocortex. (**A**,**B**) Immunostaining against the cell death marker, cleaved-Caspase 3 showed an increase in apoptotic cells within the dorsolateral neocortex (defined by dashed lines in **A**); and as marked by yellow arrows in high magnification images (**B**); in the E16.5 *hGFAP*$^{cre/+}$*;Sufu*$^{fl/fl}$ neocortex. Scale bars = 500 μm (**A**); 100 μm (**B**); (**C**) Quantification of cleaved-Caspase 3–positive cells in the dorsolateral region of the neocortical hemispheres confirmed the significant and dramatic increase in apoptotic cells in the E16.5 *hGFAP*$^{cre/+}$*;Sufu*$^{fl/fl}$ neocortex compared to controls; *** *p*-value < 0.01.

3.7. Increase in the Number of Cux1+ Upper Layer Neurons in the P7 hGFAP$^{cre/+}$*;Sufu*$^{fl/fl}$ *Neocortex*

At E16.5, RGs and IPs largely generate projection neurons that eventually populate upper layers 2–4 of the postnatal neocortex. We wondered whether the expansion of IPs at E16.5 resulted in an increase in the number of upper layer neurons in the postnatal *hGFAP*$^{cre/+}$*;Sufu*$^{fl/fl}$ neocortex. To test this, we conducted immunofluorescence staining against Cux1, a marker for upper neocortical layers as observed in the control brains. Cux1+ neurons in the P7 *hGFAP*$^{cre/+}$*;Sufu*$^{fl/fl}$ neocortex were properly localized as with controls (Figure 8A). However, we detected a modest increase in the density of Cux1+ neurons in layers 2–4, particularly in layer 4, of the P7 *hGFAP*$^{cre/+}$*;Sufu*$^{fl/fl}$ neocortex compared to controls (Figure 8B). Thus, despite the increase in cell death at E16.5 (Figure 7), the expansion of basally-dividing progenitors in the E16.5 *hGFAP*$^{cre/+}$*;Sufu*$^{fl/fl}$ neocortex still resulted in an increase in Cux1+ projection neurons. These observations also verified that the differentiation capacity of surviving cortical progenitors in the E16.5 *hGFAP*$^{cre/+}$*;Sufu*$^{fl/fl}$ neocortex were not compromised. These Cux1+ projection neurons were able to migrate out of the progenitor zones and setlle into the upper cortical layers, particularly in layer 4, of the postnatal *hGFAP*$^{cre/+}$*;Sufu*$^{fl/fl}$ neocortex.

3.8. Normal Formation of Callosal Projections in the P15 hGFAP$^{cre/+}$*;Sufu*$^{fl/fl}$ *Neocortex*

Cux1+ projection neurons of the upper layers of the neocortex form callosal projections, which interconnect the two cortical hemispheres via the corpus callosum [23]. To examine whether callosal projections were affected in the *hGFAP*$^{cre/+}$*;Sufu*$^{fl/fl}$ neocortex as a consequence of the increase in Cux1+ cells, we embedded DiI crystals in the upper cortical layers of the P15 control and mutant mice to anterogradely label axons crossing the midline along the corpus callosum. As with the controls, we found that DiI-labeled projections in the *hGFAP*$^{cre/+}$*;Sufu*$^{fl/fl}$ neocortex were present in the corpus callosum (Figure 8C) and were able to cross the midline (Figure 8D). These findings indicated that callosal projections typically formed by Cux1+ upper layer neurons were properly generated and projected into the contralateral hemisphere of the $hGFAP^{cre/+};Sufu^{fl/fl}$ neocortex similar to control littermates.

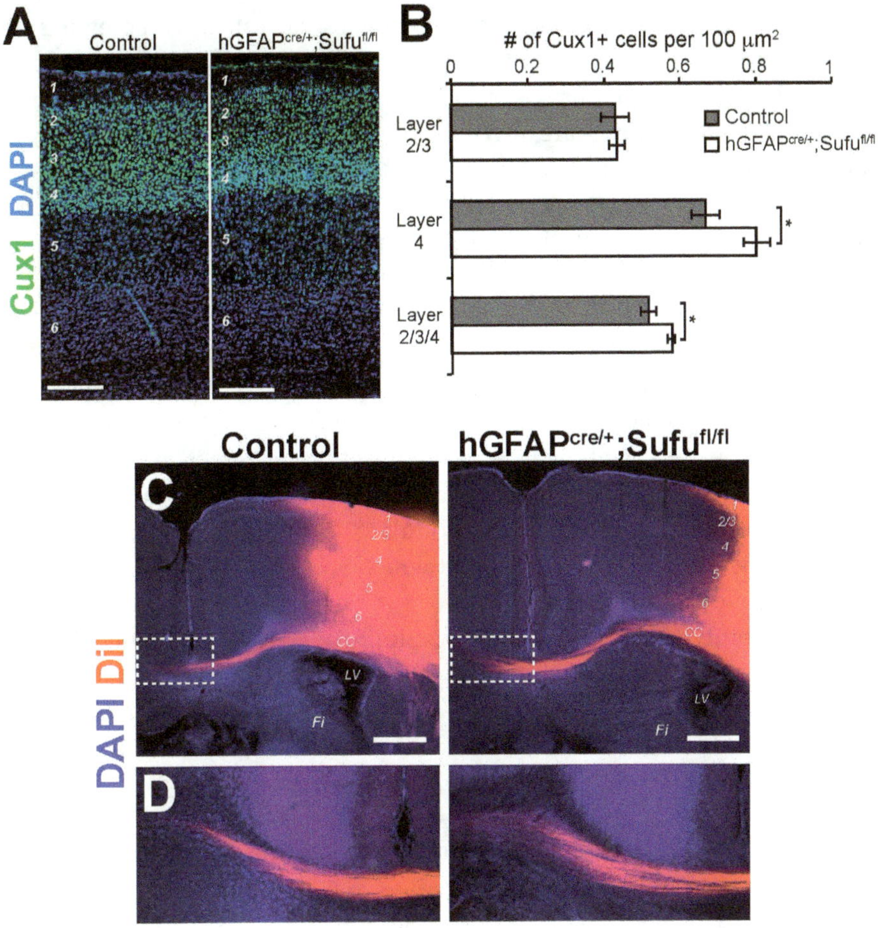

Figure 8. Mild increase in Cux1+ projection neurons and normal callosal projections in the postnatal $hGFAP^{cre/+};Sufu^{fl/fl}$ neocortex. (**A**,**B**) Cux1 immunolabeling of upper layer projection neurons showed a comparable distribution of Cux1+ neurons, with a densely packed layer 4, in the P7 $hGFAP^{cre/+};Sufu^{fl/fl}$ neocortex (**A**); Quantification of Cux1+ projection neurons (**B**) yielded comparable densities of Cux1+ neurons in Layer 2/3 between control and $hGFAP^{cre/+};Sufu^{fl/fl}$ neocortex. A borderline increase in the density of Cux1+ neurons were quantified in layers 2–4 of the P7 $hGFAP^{cre/+};Sufu^{fl/fl}$ neocortex (p-value = 0.054) and appears to be due to a mild increase in the density of Cux1+ neurons in layer 4 (p-value = 0.058). Scale bars = 200 µm. CC, corpus callosum; LV, lateral ventricles; FI, fimbria; (**C**,**D**) DiI tracing showed that callosal projections that originate from upper cortical layers were grossly unaffected in the P15 $hGFAP^{cre/+};Sufu^{fl/fl}$ neocortex and is comparable with control littermates (**C**); Callosal projections were able to project across the midline into the contralateral hemisphere (boxed, **D**); Scale bars = 500 µm. DiI, 1,1′-dioctadecyl-3,3,3′,3′-tetramethylindocarbocyanine.

4. Discussion

During neocortical development, the proliferation and differentiation of neural progenitors are tightly controlled to generate predetermined numbers of specific projection neuron subtypes [24]. Although key developmental signaling pathways are known to regulate these processes, modulators of these signaling events have yet to be identified and characterized, particularly those that exert fine control over the behavior of specific progenitor populations. In this study, we provide evidence that the cytoplasmic protein, Sufu, plays a significant role in controlling IP cell number in the progenitor zones of the embryonic neocortex, possibly through its antagonistic role on the Shh signaling pathway.

Our studies show that loss of Sufu specifically affects the generation and proliferation of IPs. At E16.5, most IP cells generate Cux1+ projection neurons that migrate and settle into cortical layers 2–4 of the postnatal neocortex. In the $hGFAP^{cre/+};Sufu^{fl/fl}$ embryonic neocortex, Sufu deletion led to the expansion of Tbr2+ IP cells in the SVZ. These observations are consistent with previous studies, where elevated or deregulated Shh signaling resulted in the expansion of basal radial glial cells and Tbr2+ IP cells, leading to changes in the mouse neocortex that resemble processes underlying gyrification in the human brain [25]. Our findings are also in line with previous findings by Shikata et al. where the upregulation of Shh signaling enhanced the transition of RG progenitors into IPs [8]. The increase in IP cells contributed to the expansion of VZ/SVZ regions of the E16.5 $hGFAP^{cre/+};Sufu^{fl/fl}$ neocortex. Therefore, our studies pinpoint a specific role for Sufu in regulating the cell cycle dynamics of RG and IP cells during mid-corticogenesis.

Shh signaling is essential in the developing forebrain, with established roles in axis formation and maintenance of cortical interneuron progenitor identity [5]. In the dorsal forebrain, Shh activity is maintained at low levels since an increase in Shh activity, particularly at early stages of corticogenesis, leads to severe malformation of the neocortex [6,26]. However, in the mid-corticogenesis stage (E14.5 to E17.5), there are more sources of Shh ligands in the neocortex where it functions as a mitogen for cortical progenitors and is subject to modulation by known Shh antagonists [27]. Our findings show that Sufu continues to be an important modulator of Shh activity in the neocortex during mid-corticogenesis. This provides another level at which Shh signaling is tightly regulated by Sufu during cortical development.

Regulating the stability, localization, and transcriptional activity of Gli transcription factors are the primary modes by which Sufu exerts its antagonistic function on Shh signaling [10,13,28–31]. During corticogenesis, the repressor activity of Gli3 is predominant [7]. Our studies showed that although both Gli3FL and Gli3R were generated in the $hGFAP^{cre/+};Sufu^{fl/fl}$ neocortex at later stages of corticogenesis, a reduction in Gli3R levels were apparent in the VZ/SVZ. We did not observe any changes in Gli3FL levels suggesting that the stability of full-length Gli3 was not compromised in the absence of Sufu, in contrast to its role in early corticogenesis [6]. Additionally, we also did not observe differences in the nuclear localization of Gli3 in control and mutant mice. Thus, Sufu primarily functions to maintain Gli3R levels, perhaps by preventing its degradation [13,29,32] during mid-corticogenesis.

The reduction in Gli3R levels in the E16.5 $hGFAP^{cre/+};Sufu^{fl/fl}$ neocortex correlated with an increase in the proliferation of IP cells, which subsequently resulted in the expansion of Cux1+ upper layer neurons. These findings are contrary to the effect of completely abolishing Gli3 early in corticogenesis when there were fewer Cux1+ upper layer neurons because IPs exited the cell cycle prematurely [7]. This discrepancy is likely due to the timing of Gli3 ablation. Furthermore, our findings are consistent with the growing evidence that neural progenitors elicit differential responses to Gli3R or Shh signaling in a dose-dependent manner [33,34]. It would be interesting to determine whether reduced Gli3R levels (i.e., in Gli3 heterozygous knockout mice) during mid-corticogenesis similarly result in the expansion of IP cells. Based on our results, we postulate that the effect of Shh signaling in neural progenitor cells in the embryonic neocortex is crucially dependent on the appropriate Gli3R protein levels that must be maintained by Sufu throughout corticogenesis. However, we also cannot exclude the possibility that Sufu regulates additional signaling pathways, besides its regulatory role in Gli3, to control the behavior of cortical progenitors. Indeed, Sufu has been shown to regulate components of Wnt signaling [35,36], another important regulator of progenitor proliferation and differentiation in the neocortex [37].

In the postnatal $hGFAP^{cre/+};Sufu^{fl/fl}$ neocortex, we found only a modest increase of Cux1+ projection neurons. This modest increase is likely the result of the dramatic increase in apoptotic cells in the E16.5 $hGFAP^{cre/+};Sufu^{fl/fl}$ neocortex. Although it is possible that Sufu directly regulates cell death signaling pathways, it is also likely that the limited amounts of growth factors within the VZ/SVZ are unable to sustain the dramatic increase in IPs in the VZ/SVZ. Nevertheless, a significant proportion of progenitors were able to differentiate particularly into Cux1+ layer 4 projection neurons. This hints at the possibility that Sufu specifically affects a subset of Tbr2+ cortical progenitors that are destined to become layer 4 projection neurons. Indeed, recent studies have shown that upper layer neurons originate from distinct IP subpopulations [38,39]. Future studies should take into consideration that distinct subsets of IPs and their neuronal progenies may be differentially regulated by molecular factors, such as Sufu.

Investigating the functional consequences of an increase in projection neurons in the neocortex could prove useful in a number of neurodevelopmental disorders, such as autism, in which alterations in neuronal numbers have been observed [40,41]. Our gross analysis of callosal axons showed no obvious defects in the ability of Cux1+ neurons to form these projections. However, we cannot rule out the possibility that callosal projections in the $hGFAP^{cre/+};Sufu^{fl/fl}$ neocortex inappropriately target neurons in the contralateral hemisphere, disrupt local connections, or exhibit electrophysiological deficits. Furthermore, Cux1+ upper layer neurons represent several subpopulations of projection neurons in the neocortex [23]. Overall, in-depth analysis is necessary to determine how altered production of Cux1+ neurons during corticogenesis, due to Sufu dysfunction and deregulated Shh signaling, affect neuronal connectivity, function, and behavior. These studies could provide a better understanding and diagnosis of neurological abnormalities that relate to abnormal callosal projection neuron development and function.

Acknowledgments: We thank members of the Pleasure Lab for helpful discussions, Trung Huynh for genotyping, and Kurt Thorn and DeLaine Larsen at the UCSF Nikon Imaging Center for assistance with imaging. This work was supported by the NIH R01 NS075188 (S.J.P.) and a Ruth L. Kirschstein National Research Service Award F32NS087719 (O.R.Y.).

Author Contributions: O.R.Y. and S.J.P. conceived and designed or performed experiments, analyzed the data, and wrote the manuscript. H.X.N., G.F., K.Y. and J.K. performed experiments and analyzed the data.

Conflicts of Interest: The authors declare no conflict of interest.

References

1. Noctor, S.C.; Flint, A.C.; Weissman, T.A.; Dammerman, R.S.; Kriegstein, A.R. Neurons derived from radial glial cells establish radial units in neocortex. *Nature* **2001**, *409*, 714–720. [CrossRef] [PubMed]
2. Noctor, S.C.; Martínez-Cerdeño, V.; Ivic, L.; Kriegstein, A.R. Cortical neurons arise in symmetric and asymmetric division zones and migrate through specific phases. *Nat. Neurosci.* **2004**, *7*, 136–144. [CrossRef] [PubMed]
3. Stancik, E.K.; Navarro-Quiroga, I.; Sellke, R.; Haydar, T.F. Heterogeneity in ventricular zone neural precursors contributes to neuronal fate diversity in the postnatal neocortex. *J. Neurosci.* **2010**, *30*, 7028–7036. [CrossRef] [PubMed]
4. Kowalczyk, T.; Pontious, A.; Englund, C.; Daza, R.A.M.; Bedogni, F.; Hodge, R.; Attardo, A.; Bell, C.; Huttner, W.B.; Hevner, R.F. Intermediate neuronal progenitors (basal progenitors) produce pyramidal-projection neurons for all layers of cerebral cortex. *Cereb. Cortex* **2009**, *19*, 2439–2450. [CrossRef] [PubMed]
5. Sousa, V.H.; Fishell, G. Sonic hedgehog functions through dynamic changes in temporal competence in the developing forebrain. *Curr. Opin. Genet. Dev.* **2010**, *20*, 391–399. [CrossRef] [PubMed]
6. Yabut, O.R.; Fernandez, G.; Huynh, T.; Yoon, K.; Pleasure, S.J. Suppressor of Ffused is critical for maintenance of neuronal progenitor identity during corticogenesis. *Cell Rep.* **2015**, *12*, 2021–2034. [CrossRef] [PubMed]
7. Wang, H.; Ge, G.; Uchida, Y.; Luu, B.; Ahn, S. Gli3 is required for maintenance and fate specification of cortical progenitors. *J. Neurosci.* **2011**, *31*, 6440–6448. [CrossRef] [PubMed]
8. Shikata, Y.; Okada, T.; Hashimoto, M.; Ellis, T.; Matsumaru, D.; Shiroishi, T.; Ogawa, M.; Wainwright, B.; Motoyama, J. Ptch1-mediated dosage-dependent action of Shh signaling regulates neural progenitor development at late gestational stages. *Dev. Biol.* **2011**, *349*, 147–159. [CrossRef] [PubMed]
9. Pozniak, C.D.; Langseth, A.J.; Dijkgraaf, G.J.P.; Choe, Y.; Werb, Z.; Pleasure, S.J. Sox10 directs neural stem cells toward the oligodendrocyte lineage by decreasing Suppressor of Fused expression. *Proc. Natl. Acad. Sci. USA* **2010**, *107*, 21795–21800. [CrossRef] [PubMed]

10. Liu, J.; Heydeck, W.; Zeng, H.; Liu, A. Dual function of suppressor of fused in Hh pathway activation and mouse spinal cord patterning. *Dev. Biol.* **2012**, *362*, 141–153. [CrossRef] [PubMed]
11. Kim, J.J.; Gill, P.S.; Rotin, L.; van Eede, M.; Henkelman, R.M.; Hui, C.-C.; Rosenblum, N.D. Suppressor of fused controls mid-hindbrain patterning and cerebellar morphogenesis via GLI3 repressor. *J. Neurosci.* **2011**, *31*, 1825–1836. [CrossRef] [PubMed]
12. Lin, C.; Chen, M.-H.; Yao, E.; Song, H.; Gacayan, R.; Hui, C.; Chuang, P.-T. Differential regulation of Gli proteins by Sufu in the lung affects PDGF signaling and myofibroblast development. *Dev. Biol.* **2014**, *392*, 324–333. [CrossRef] [PubMed]
13. Wang, C.; Pan, Y.; Wang, B. Suppressor of fused and Spop regulate the stability, processing and function of Gli2 and Gli3 full-length activators but not their repressors. *Development* **2010**, *137*, 2001–2009. [CrossRef] [PubMed]
14. Shahi, M.H.; Rey, J.A.; Castresana, J.S. The sonic hedgehog-GLI1 signaling pathway in brain tumor development. *Expert Opin. Ther. Targets* **2012**, *16*, 1227–1238. [CrossRef] [PubMed]
15. Komada, M.; Saitsu, H.; Kinboshi, M.; Miura, T.; Shiota, K.; Ishibashi, M. Hedgehog signaling is involved in development of the neocortex. *Development* **2008**, *135*, 2717–2727. [CrossRef] [PubMed]
16. Pospisilik, J.A.; Schramek, D.; Schnidar, H.; Cronin, S.J.F.; Nehme, N.T.; Zhang, X.; Knauf, C.; Cani, P.D.; Aumayr, K.; Todoric, J.; et al. Drosophila genome-wide obesity screen reveals hedgehog as a determinant of brown versus white adipose cell fate. *Cell* **2010**, *140*, 148–160. [CrossRef] [PubMed]
17. Siegenthaler, J.A.; Ashique, A.M.; Zarbalis, K.; Patterson, K.P.; Hecht, J.H.; Kane, M.A.; Folias, A.E.; Choe, Y.; May, S.R.; Kume, T.; et al. Retinoic acid from the meninges regulates cortical neuron generation. *Cell* **2009**, *139*, 597–609. [CrossRef] [PubMed]
18. Regard, J.B.; Malhotra, D.; Gvozdenovic-Jeremic, J.; Josey, M.; Chen, M.; Weinstein, L.S.; Lu, J.; Shore, E.M.; Kaplan, F.S.; Yang, Y. Activation of Hedgehog signaling by loss of GNAS causes heterotopic ossification. *Nat. Med.* **2013**, *19*, 1505–1512. [CrossRef] [PubMed]
19. Lobo, S.; Wiczer, B.M.; Bernlohr, D.A. Functional analysis of long-chain acyl-CoA synthetase 1 in 3T3-L1 adipocytes. *J. Biol. Chem.* **2009**, *284*, 18347–18356. [CrossRef] [PubMed]
20. Zhuo, L.; Theis, M.; Alvarez-Maya, I.; Brenner, M.; Willecke, K.; Messing, A. hGFAP-cre transgenic mice for manipulation of glial and neuronal function in vivo. *Genesis* **2001**, *31*, 85–94. [CrossRef] [PubMed]
21. Englund, C.; Fink, A.; Lau, C.; Pham, D.; Daza, R.A.M.; Bulfone, A.; Kowalczyk, T.; Hevner, R.F. Pax6, Tbr2, and Tbr1 are expressed sequentially by radial glia, intermediate progenitor cells, and postmitotic neurons in developing neocortex. *J. Neurosci.* **2005**, *25*, 247–251. [CrossRef] [PubMed]
22. Ahn, S.; Joyner, A.L. In vivo analysis of quiescent adult neural stem cells responding to Sonic hedgehog. *Nature* **2005**, *437*, 894–897. [CrossRef] [PubMed]
23. Fame, R.M.; MacDonald, J.L.; Macklis, J.D. Development, specification, and diversity of callosal projection neurons. *Trends Neurosci.* **2011**, *34*, 41–50. [CrossRef] [PubMed]
24. Greig, L.C.; Woodworth, M.B.; Galazo, M.J.; Padmanabhan, H.; Macklis, J.D. Molecular logic of neocortical projection neuron specification, development and diversity. *Nat. Rev. Neurosci.* **2013**, *14*, 755–769. [CrossRef] [PubMed]
25. Wang, L.; Hou, S.; Han, Y.-G. Hedgehog signaling promotes basal progenitor expansion and the growth and folding of the neocortex. *Nat. Neurosci.* **2016**, *19*, 888–896. [CrossRef] [PubMed]
26. Lien, W.-H.; Klezovitch, O.; Fernandez, T.E.; Delrow, J.; Vasioukhin, V. alphaE-catenin controls cerebral cortical size by regulating the hedgehog signaling pathway. *Science* **2006**, *311*, 1609–1612. [CrossRef] [PubMed]
27. Dahmane, N.; Sanchez, P.; Gitton, Y.; Palma, V.; Sun, T.; Beyna, M.; Weiner, H.; Ruiz i Altaba, A. The Sonic Hedgehog-Gli pathway regulates dorsal brain growth and tumorigenesis. *Development* **2001**, *128*, 5201–5212. [PubMed]
28. Humke, E.W.; Dorn, K.V.; Milenkovic, L.; Scott, M.P.; Rohatgi, R. The output of Hedgehog signaling is controlled by the dynamic association between Suppressor of Fused and the Gli proteins. *Genes Dev.* **2010**, *24*, 670–682. [CrossRef] [PubMed]
29. Chen, M.-H.; Wilson, C.W.; Li, Y.-J.; Law, K.K.L.; Lu, C.-S.; Gacayan, R.; Zhang, X.; Hui, C.; Chuang, P.-T. Cilium-independent regulation of Gli protein function by Sufu in Hedgehog signaling is evolutionarily conserved. *Genes Dev.* **2009**, *23*, 1910–1928. [CrossRef] [PubMed]

30. Liu, X.; Wang, X.; Du, W.; Chen, L.; Wang, G.; Cui, Y.; Liu, Y.; Dou, Z.; Wang, H.; Zhang, P.; et al. Suppressor of fused (Sufu) represses Gli1 transcription and nuclear accumulation, inhibits glioma cell proliferation, invasion and vasculogenic mimicry, improving glioma chemo-sensitivity and prognosis. *Oncotarget* **2014**, *5*, 11681–11694. [CrossRef] [PubMed]
31. Kise, Y.; Morinaka, A.; Teglund, S.; Miki, H. Sufu recruits GSK3β for efficient processing of Gli3. *Biochem. Biophys. Res. Commun.* **2009**, *387*, 569–574. [CrossRef] [PubMed]
32. Makino, S.; Zhulyn, O.; Mo, R.; Puviindran, V.; Zhang, X.; Murata, T.; Fukumura, R.; Ishitsuka, Y.; Kotaki, H.; Matsumaru, D.; et al. T396I mutation of mouse Sufu reduces the stability and activity of Gli3 repressor. *PLoS ONE* **2015**, *10*, e0119455. [CrossRef] [PubMed]
33. Petrova, R.; Garcia, A.D.R.; Joyner, A.L. Titration of GLI3 repressor activity by Sonic hedgehog signaling is critical for maintaining multiple adult neural stem cell and astrocyte functions. *J. Neurosci.* **2013**, *33*, 17490–17505. [CrossRef] [PubMed]
34. Tong, C.K.; Fuentealba, L.C.; Shah, J.K.; Lindquist, R.A.; Ihrie, R.A.; Guinto, C.D.; Rodas-Rodriguez, J.L.; Alvarez-Buylla, A. A Dorsal SHH-Dependent Domain in the V-SVZ Produces Large Numbers of Oligodendroglial Lineage Cells in the Postnatal Brain. *Stem Cell Rep.* **2015**, *5*, 461–470. [CrossRef] [PubMed]
35. Min, T.H.; Kriebel, M.; Hou, S.; Pera, E.M. The dual regulator Sufu integrates Hedgehog and Wnt signals in the early Xenopus embryo. *Dev. Biol.* **2011**, *358*, 262–276. [CrossRef] [PubMed]
36. Meng, X.; Poon, R.; Zhang, X.; Cheah, A.; Ding, Q.; Hui, C.C.; Alman, B. Suppressor of fused negatively regulates beta-catenin signaling. *J. Biol. Chem.* **2001**, *276*, 40113–40119. [CrossRef] [PubMed]
37. Harrison-Uy, S.J.; Pleasure, S.J. Wnt signaling and forebrain development. *Cold Spring Harb. Perspect. Biol.* **2012**, *4*, a008094. [CrossRef] [PubMed]
38. Tyler, W.A.; Medalla, M.; Guillamon-Vivancos, T.; Luebke, J.I.; Haydar, T.F. Neural precursor lineages specify distinct neocortical pyramidal neuron types. *J. Neurosci.* **2015**, *35*, 6142–6152. [CrossRef] [PubMed]
39. Mihalas, A.B.; Elsen, G.E.; Bedogni, F.; Daza, R.A.M.; Ramos-Laguna, K.A.; Arnold, S.J.; Hevner, R.F. Intermediate Progenitor Cohorts Differentially Generate Cortical Layers and Require Tbr2 for Timely Acquisition of Neuronal Subtype Identity. *Cell Rep.* **2016**, *16*, 92–105. [CrossRef] [PubMed]
40. Courchesne, E.; Mouton, P.R.; Calhoun, M.E.; Semendeferi, K.; Ahrens-Barbeau, C.; Hallet, M.J.; Barnes, C.C.; Pierce, K. Neuron number and size in prefrontal cortex of children with autism. *JAMA* **2011**, *306*, 2001–2010. [CrossRef] [PubMed]
41. Casanova, M.F.; van Kooten, I.A.J.; Switala, A.E.; van Engeland, H.; Heinsen, H.; Steinbusch, H.W.M.; Hof, P.R.; Trippe, J.; Stone, J.; Schmitz, C. Minicolumnar abnormalities in autism. *Acta Neuropathol.* **2006**, *112*, 287–303. [CrossRef] [PubMed]

9

Mechanisms of Specificity for Hox Factor Activity

Arya Zandvakili [1,2] and Brian Gebelein [3,*]

1 Molecular and Developmental Biology Graduate Program, Cincinnati Children's Hospital Medical Center, Cincinnati, OH 45229, USA
2 Medical-Scientist Training Program, University of Cincinnati College of Medicine, Cincinnati, OH 45267, USA; zandvaaa@mail.uc.edu
3 Division of Developmental Biology, Cincinnati Children's Hospital Medical Center, Cincinnati, OH 45229, USA
* Correspondence: brian.gebelein@cchmc.org

Academic Editors: Vincenzo Zappavigna and Simon J. Conway

Abstract: Metazoans encode clusters of paralogous Hox genes that are critical for proper development of the body plan. However, there are a number of unresolved issues regarding how paralogous Hox factors achieve specificity to control distinct cell fates. First, how do Hox paralogs, which have very similar DNA binding preferences *in vitro*, drive different transcriptional programs *in vivo*? Second, the number of potential Hox binding sites within the genome is vast compared to the number of sites bound. Hence, what determines where in the genome Hox factors bind? Third, what determines whether a Hox factor will activate or repress a specific target gene? Here, we review the current evidence that is beginning to shed light onto these questions. In particular, we highlight how cooperative interactions with other transcription factors (especially PBC and HMP proteins) and the sequences of *cis*-regulatory modules provide a basis for the mechanisms of Hox specificity. We conclude by integrating a number of the concepts described throughout the review in a case study of a highly interrogated Drosophila *cis*-regulatory module named "The Distal-less Conserved Regulatory Element" (DCRE).

Keywords: Hox; transcription factor; *cis*-regulatory modules; DNA binding specificity

1. Introduction

Hox proteins are a family of homeodomain-containing Transcription Factors (TFs) that are critical for regulating developmental processes in metazoans [1,2]. The stark homeotic changes occurring in Hox mutant animals, such as the classic antenna-to-leg transformations in *Drosophila*, testify to the importance of Hox factors for proper development [3]. In addition, more recent studies have demonstrated that Hox factors not only drive stereotypic developmental programs, but also have a role in maintaining differentiated cell populations [4,5]. Given that Hox factors are conserved from *C. elegans* to *H. sapiens*, a fundamental understanding of how Hox factors function will yield significant insights into both the development and evolution of body plans.

The majority of metazoan genomes encode clusters of paralogous Hox genes (Figure 1A). While invertebrates generally have one Hox cluster, vertebrates have multiple Hox clusters owing to duplications of the entire cluster during evolution [6]. The order of Hox genes within the clusters typically correlates with their expression pattern along the Anterior-Posterior (A-P) axis of the embryo [7–9]. Genes at the 3′ end of a cluster are expressed in anterior regions of an embryo, whereas genes toward the 5′ end are expressed in progressively more posterior regions of the embryo. Thus, the differential expression of Hox genes is a key step in the specification of distinct cell fates along the A-P axis.

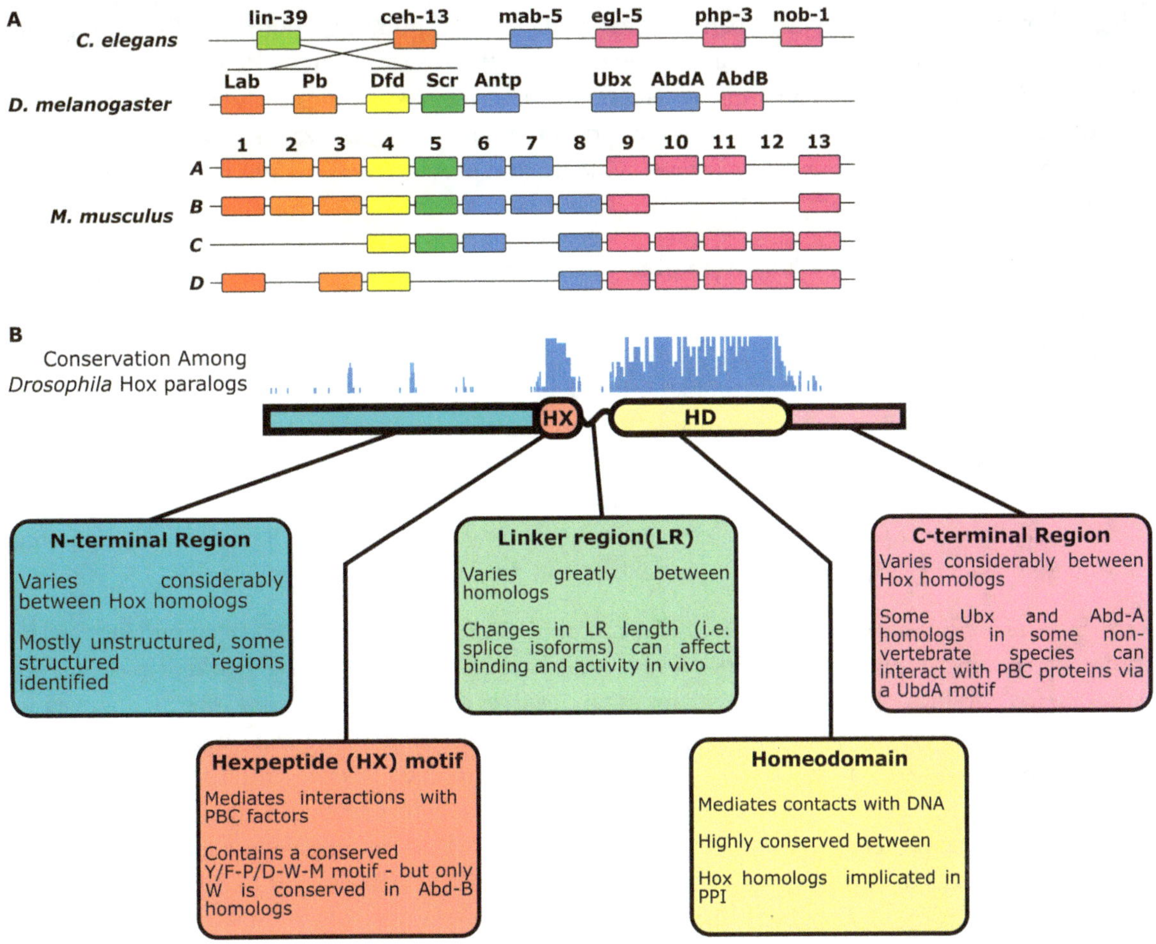

Figure 1. Schematic of Hox gene locus and Hox proteins. (**A**) Schematic of *Hox* gene clusters in *C. elegans*, *D. melanogaster* and *M. musculus*. Genes with similar colors are thought to derive from a common ancestor [10]; (**B**) Schematic of a Hox protein with regions labeled and described in boxes (bottom) [11–18]. Per-amino-acid conservation score across *Drosophila* Hox protein sequences demonstrates that the HX and homeodomain regions are the most conserved regions across paralogous Hox proteins. Multiple sequence alignment was produced via ClustalΩ (top) [19,20]. Note: The size of each Hox protein region is not to scale.

At the sequence level, Hox proteins share two stereotypic domains: a conserved homeodomain and a conserved Hexapeptide (HX) motif that is located N-terminal to the homeodomain (Figure 1B). Hox factors utilize their homeodomains to directly bind DNA, and they share very similar binding preferences *in vitro* as monomers [21–23]. The HX motif mediates direct interactions with another family of TFs (PBC proteins), and it is separated from the homeodomain by a flexible and highly variable linker region [12,13]. Outside of the homeodomain and the HX motif, Hox protein sequences diverge substantially; and while we largely do not understand their functions, these non-conserved regions contain residues that can be post-translationally modified and/or have been implicated in protein-protein interactions and the regulation of transcriptional outputs [24–27].

Hox factors specify cell-fates based on their ability to interact with *Cis*-Regulatory Modules (CRMs) and to regulate transcription [28,29]. CRMs provide a DNA platform to organize interactions between Hox factors and other proteins. Although Hox interactions with DNA and other proteins have been studied extensively, we still lack a comprehensive understanding of how specificity is achieved to accurately recognize and regulate target genes required to direct specific cell fates. In this review, we explore three questions related to Hox specificity: First, how do paralogous Hox proteins drive different cell fates even though they utilize conserved homeodomains with highly similar DNA binding preferences? Second, metazoan genomes contain a large number of potential Hox binding sites, yet only a subset of these sites are bound at any one time. What factors determine which sequences are bound and regulated by Hox proteins, whereas other sequences containing Hox binding sites remain unbound? Third, once bound, how does an individual Hox factor activate some target genes and repress others? Below, we present emerging evidence that provides new insight into the mechanisms that contribute to the specificity of Hox action.

2. Differentiation of Hox Paralog Activities

A fundamental problem in the study of TF specificity is that most TFs are members of large protein families that have highly similar DNA binding properties yet distinct *in vivo* functions [30]. The Hox family of TFs is an exemplar of this problem. While paralogous Hox factors bind highly similar DNA sequences, genetic loss- and gain-of-function studies demonstrate that Hox factors control diverse cell fates along the A-P axis of metazoans [1,31]. Furthermore, in *C. elegans*, it has been demonstrated that paralogous Hox proteins have highly divergent genomic binding patterns *in vivo* [32]. However, it is not immediately clear whether this difference in *in vivo* function is due to differences in Hox paralogs or due to the fact that Hox paralogs are functioning in different cellular contexts.

Several studies have controlled for cellular context and have demonstrated that paralogous Hox proteins have different *in vivo* activities within the same cell types. First, over- or under-expressing specific Hox paralogs within the same cell types can result in different phenotypes [33–35]. For instance, in *C. elegans*, when the Hox genes *egl-5* and *lin-39* are expressed under the control of regulatory elements from the *mab-5* locus (a different Hox gene), they do not rescue the *mab-5* mutant phenotype [35]. Second, the misexpression of different Hox genes results in distinct changes in global gene expression patterns. For example, six *Drosophila* Hox genes were individually expressed in a ubiquitous pattern with the same Gal4 driver line in *Drosophila* embryos, and RNA was isolated to compare changes in gene expression [36]. Of the genes that changed in expression, the majority (nearly 70%) changed in response to a single Hox factor, while only 1.3% of the genes changed in response to all six Hox factors [36]. Third, a growing number of Hox-regulated CRMs has been identified, and many are regulated by only one or a small subset of paralogous Hox factors when tested in the same cellular contexts [37–41]. While these studies all support the notion that Hox factors largely control distinct cell fates, it has been found that, at least in some contexts, Hox factors can produce very similar phenotypes when expressed in the same cell types. For example, in *Drosophila*, all Hox factors (except abdominal-B) can substitute for Labial (Lab) in specifying tritocerebral neuromere [42]. Furthermore, in vertebrates, Hox mutant phenotypes become more severe when multiple Hox genes with overlapping expression are mutated, suggesting that Hox genes have some functional redundancy [9,43]. Thus, each Hox factor is likely to regulate both common and paralog-specific target genes and cell fates *in vivo*.

Here, we focus on how the cooperation with other sequence-specific TFs aids in the differentiation of the binding preferences and activities of Hox paralogs. We describe how interactions between Hox factors and the PBC family of homeodomain proteins uncover the importance of latent DNA binding specificity and low affinity binding sites in generating Hox-specific outputs. We subsequently explore how the HMP family of homeodomain proteins selectively interacts with posterior Hox factors. In addition, HMP TFs also directly interact with the PBC proteins and thereby form higher-order TF complexes with Hox factors that can enhance both DNA binding affinity and specificity. Next, we provide a few examples of the diversity of other TFs that interact with subsets of Hox factors, either directly or via nearby binding sites encoded within CRMs. Finally, we describe how post-translational modifications differentially affect the activities of paralogous Hox proteins. A summary of these mechanisms is provided in Figure 2.

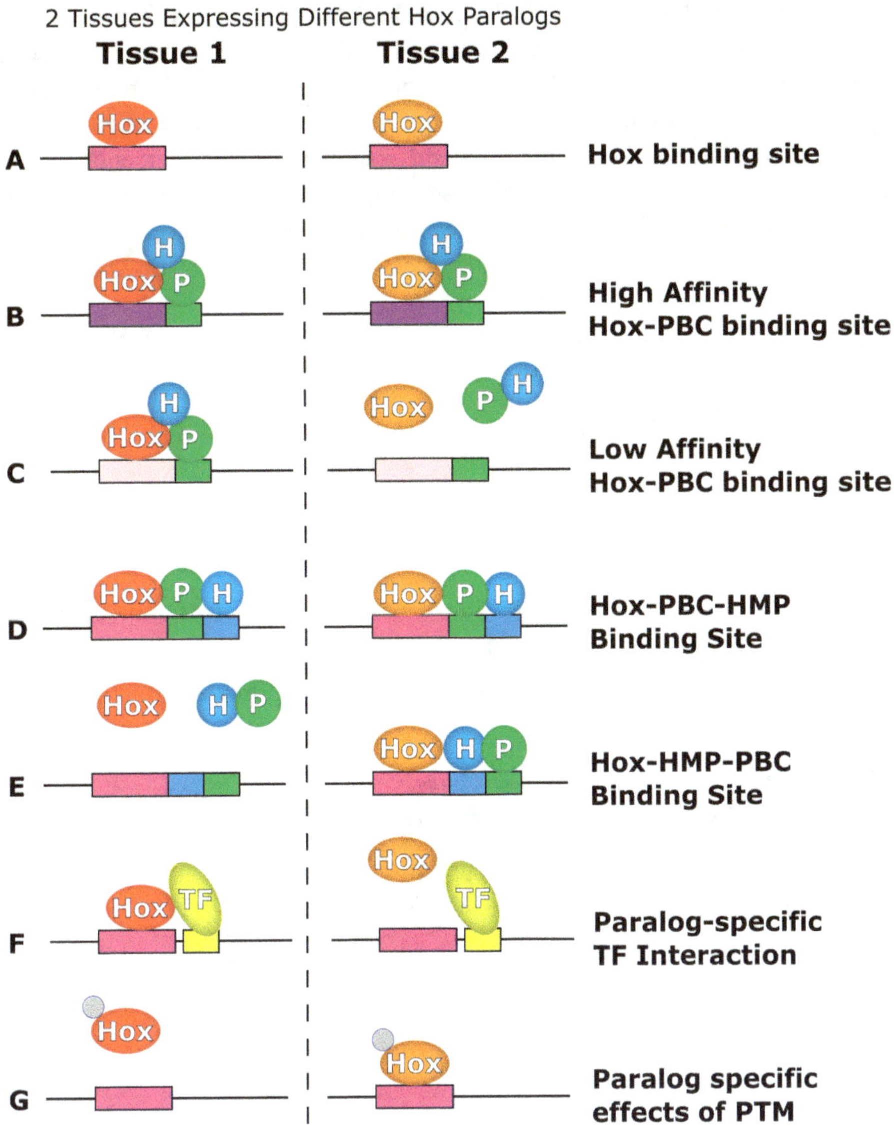

Figure 2. Summary of the mechanisms of Hox paralog specificity. (**A**) A single Hox site is unlikely to differentiate between two different Hox paralogs; (**B**,**C**) Low-affinity Hox-PBC binding sites are more likely to differentiate between Hox paralogs than high affinity Hox-PBC binding sites [23,41]; (**D**,**E**) All Hox paralogs can typically bind adjacent Hox-PBC sites, but adjacent Hox-HMP sites have a preference for posterior Hox paralogs [26,44,45]; (**F**) Interactions with a nearby binding site for another TF can cause a CRM to be paralog specific [46,47]; (**G**) The same post-translational modifications (gray circle) can affect paralogous Hox proteins differently [48–51]. "P" (PBC protein) and "H" (HMP protein).

2.1. Interactions with PBC Factors Reveal Hox Latent Specificity

First identified in 1990, PBC proteins were nearly simultaneously discovered as modulators of Hox function in *Drosophila* and proto-oncogenes in mammals [52–54]. The PBC family includes Extradenticle (Exd) in *Drosophila*, CEH-20 and CEH-40 in *C. elegans* and Pbx factors in mammals; and all PBC proteins contain a highly-conserved homeodomain that differs from a canonical homeodomain by the addition of a Three Amino-acid Loop Extension (TALE) motif between helix 1 and helix 2 of the homeodomain [55] (Figure 3A). It was recognized early on that Hox and PBC proteins bind DNA cooperatively [56–58] and that PBC proteins were essential for Hox function [59–61]. Subsequent structural studies determined that Hox-PBC interactions on DNA were mediated via insertion of the Hox HX motif into a hydrophobic pocket of the PBC homeodomain that is composed of residues from the TALE motif, helix 1 and helix 3 [12,13] (Figure 3B). The Hox HX motif is located N-terminal to the homeodomain and contains a highly-conserved Y/F-P/D-W-M sequence (Figure 1B), where the W residue is critical for making hydrophobic interactions with the PBC TALE motif [12]. While the majority of Hox factors have a defined HX motif, the posterior Abdominal-B (Abd-B) or Hox paralog group 9–13 factors only rely upon a conserved W residue to mediate this interaction [62]. Importantly, the interaction between Hox and PBC factors occurs through nearby DNA binding sites for each factor and, thereby, results in both enhanced DNA binding specificity and affinity [23,56,63].

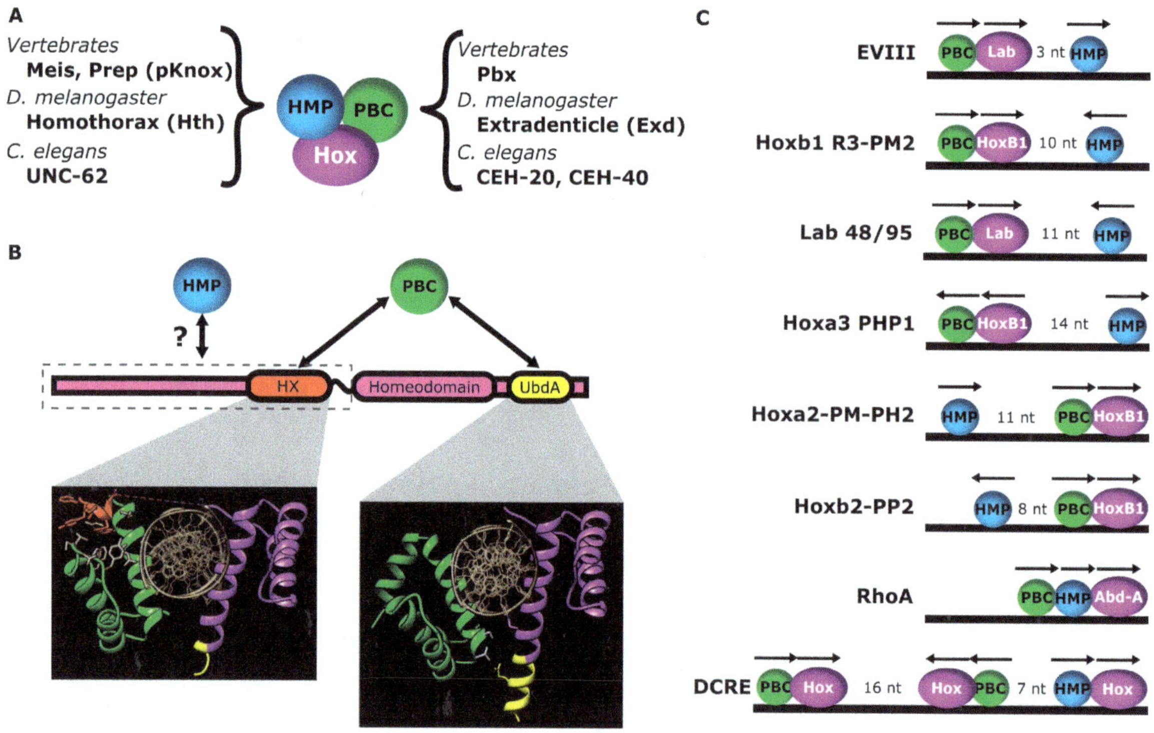

Figure 3. Interaction between Hox factors and PBC/HMP proteins. (**A**) Names of PBC and HMP homeodomain proteins in *C. elegans*, *Drosophila* and vertebrates; (**B**) Motifs in Hox factors used to mediate interactions with PBC and HMP proteins (top). Yeast-2-hybrid data suggest sequences N-terminal to the homeodomain mediate interactions with homothorax [26]. Hox proteins can mediate interactions with PBC proteins via the HX motif or, in the case of non-vertebrate Abd-A and Ubx homologs, the UbdA motif. *Structural panels*: Structures produced by Foos *et al.*, 2015, demonstrating both HX and UbdA interaction modes between *Drosophila* Ubx and Exd. PBC protein in green; Ubx protein in purple; HX motif in red; UbdA motif in yellow [18]. When bound to a canonical Hox-PBC binding site, Hox and PBC proteins bind on opposite sides of the DNA. *Left structural panel*: The HX motif (red) mediates interactions with a hydrophobic pocket formed by helix 1, helix 3 and the TALE motif of the Exd homeodomain. *Right structural panel*: The UbdA motif mediates interactions with helix 3 of the PBC homeodomain. UbdA can either be unstructured or form an α-helical extension of the third helix of the Hox homeodomain. Note: The Ubx protein fragment in the left panel contains both the HX and UbdA motifs, while the Ubx protein fragment in the right panel only contains the UbdA motif. (**C**) Examples of CRMs with Hox-PBC-HMP binding sites, demonstrating variations in the order, orientations and spacing of the binding sites. The space between HMP and Hox-PBC sites is indicated. EVIII [64]; Lab 48/95 [65]; R3-PM2 [66]; PP2 [67]; PM-PH2 [68]; PHP1 [69]; RhoA [40]; DCRE [70].

In addition to enhancing DNA binding affinity, interactions between TFs can result in changes in binding preferences via allosteric changes in conformation of the TFs and/or the DNA [71–73]. The concept of revealing differences in binding preferences between similar TFs is known as latent specificity. The latent specificity mechanism is well-established for interactions between Hox and

PBC proteins. Soon after the discovery of PBC proteins, biochemical evidence demonstrated that PBC proteins modify the DNA binding specificity of *Drosophila* and mammalian Hox paralogs [56,57,74,75]. This concept was most comprehensively tested *in vitro* by a 2011 SELEX-seq (Systematic Evolution of Ligands by EXponential enrichment followed by sequencing) study, which demonstrated that *Drosophila* Exd-Hox heterodimers have a greater diversity of binding site preferences than Hox monomers [23]. Specifically, it was found that interactions between Hox factors and Exd result in the selection of distinct sequences within six core nucleotides (A_5YNNAY_{10}) of the Hox binding site (note, the combined PBC-Hox site typically consists of 12 nucleotides, nnTGAYnnAYnn). Interestingly, the binding preference for Exd-Hox dimers segregated with the expression of Hox factors along the A-P axis; thus, the preference for a core motif can be grouped into distinct clusters of anterior (Labial and Proboscipedia), middle (Deformed and Sex combs reduced) and posterior Hox factors (Antennapedia, Ultrabithorax, Abdominal-A and Abdominal-B). These preferences also segregated with the predicted width of the DNA minor groove within the Exd-Hox core-motif. Specifically, it was found that anterior Hox factors select sequences with a narrow minor groove in position A_9Y_{10}, while posterior Hox factors select sequences with wider minor grooves [23,76]. In addition to differentiating preferences within the Exd-Hox core-motif, there were distinct preferences between Hox factors for sequences flanking the Hox half of the Exd-Hox motif, especially for anterior Hox factors [23]. However, there is an exception to the rule of Exd differentiating preferences; interaction with Exd actually made the preferences of Ultrabithorax (Ubx) and Abd-B more similar [23].

Structural studies have revealed that interactions with Exd on DNA allow protein sequences that are more variable between Hox paralogs (namely the linker region and N-terminal arm of the homeodomain) to make a greater contribution to DNA binding. Comparing crystal structures of Exd and the Sex Combs Reduced (Scr) Hox factor bound to either a selective sequence specifically bound by Scr (*fkh250*) or a non-selective consensus Exd-Hox site bound by many Hox factors (*fkh250con*) revealed significant differences in the mechanism of DNA binding [71]. On the selective sequence, the Scr N-terminal arm is structured and inserted into the minor groove, whereas these same amino acids are disordered and make a minimal contribution to Scr binding on the *fkh250con* sequence [71]. Selectivity was found to depend upon two residues—Histidine-12 (His-12, in the Linker Region) and Arginine 3 (Arg3, in the N-terminal arm)—that insert into the minor groove of *fkh250*. His-12 and Arg3 are highly conserved among Scr homologs, suggesting that these residues are important to Scr function. Since His-12 and Arg3 do not make direct hydrogen bonds with bases, it was proposed that these two residues recognize the shape of the DNA. Indeed, a recent SELEX-seq study using Hox-Exd dimers demonstrated that mutating His-12 and Arg3 in Scr results in less selectivity for sequences that are predicted to have narrow minor grooves at position A_8Y_9 [76]. Furthermore, mutating the linker region and the N-terminal arm of the homeodomain in Antennapedia (Antp) to match those of Scr not only changed Exd-Antp binding preferences to be very similar to those of Exd-Scr, but it also allowed the mutant Antp protein to activate the Scr-specific enhancer (*fkh250*) *in vivo* [76]. This SELEX-seq study is also notable for demonstrating that using both DNA shape-readout, as well as DNA base-readout greatly improves the predictions of sequence-specific affinity of Hox-Exd complexes.

Importantly, the concept of latent specificity may be broadly applicable to the study of TFs in general. A recent large-scale SELEX-based screen of cooperative DNA binding between TF pairs identified additional examples of latent specificity. For instance, HoxB2 was found to diversify the binding preferences of a subset of ETS factors (E-26 transcription factors) [45]. Interestingly, this screen demonstrated that of the 315 TF-TF heterodimers that were analyzed, 2/3 of the heterodimeric binding preferences were substantially different than the monomeric binding preferences of the individual TFs. However, in many cases, when a set of paralogous TFs interacted with the same partner, the resultant binding preference was very similar, suggesting that latent specificity is unlikely to be the sole source of distinguishing the activities of paralogous TFs.

2.2. Low-Affinity Binding Sites Distinguish Hox-PBC Dimers

The latent specificity mechanism demonstrates that adjacent Hox-PBC binding sites in CRMs can contribute to paralog-specific binding of CRMs. However, many of the highest affinity sites identified from these assays can be bound by several paralogous Hox factors, suggesting that high-affinity sites may be regulated by many Hox factors. More recently, a study proposed that CRMs distinguish between Hox paralogs via low-affinity binding sites [41]. The idea behind this proposal is that TFs within a family have very similar DNA binding domains that vary only in sites that make weak interactions with DNA. Therefore, the variations in DNA-binding domains only make minor contributions to the overall binding energy when binding high affinity sites. By contrast, subtle differences in DNA binding domains can make a larger contribution to the overall binding energy when interacting with low-affinity sites. The binding of the *fkh250* sequence by Scr described above is an example of how a lower affinity site depends on Scr-specific amino acid contacts for DNA binding, whereas a higher affinity site (*fkh250con*) only requires homeodomain amino acids that are conserved in many Hox paralogs [71]. In addition, Crocker *et al.* identified a cluster of low-affinity Ubx-Exd binding sites within enhancers in the *Drosophila shavenbaby* locus, and they found that replacing these sites with higher affinity Hox-Exd sites allowed the activation of the enhancer by other Hox factors in reporter assays [41]. Interestingly, it is not uncommon to find functional, low-affinity binding sites within developmentally-important enhancers [41,77–80], suggesting that low affinity might provide a selective advantage over high-affinity binding sites. However, the prevalence of low-affinity binding sites in developmentally-important enhancers may also be a function of the fact that for any given TF, a greater number of sequences can yield low-affinity binding sites than high affinity binding sites. Crocker *et al.* provide a recent review of low-affinity binding sites and their potential functions [81].

2.3. Additional PBC Interaction Surfaces Differentiate Hox Paralogs

While it has been widely demonstrated that the Hox HX motif mediates Hox-PBC interactions, results from several biochemical experiments suggest that it is not the sole mechanism for Hox-PBC interactions. In fact, mutating or removing the HX motif in Lab, Ubx or Abdominal-A (Abd-A) does not abolish the ability of these Hox factors to interact with Exd on DNA [15,56,82–87]. Furthermore, *in vivo* Exd-dependent activities for Lab, Ubx and Abd-A were largely not abolished by HX mutations [15,57,83–87]. In addition, recent *in vivo* fluorescence complementation assays (Bimolecular Fluorescence (BiFC)) have demonstrated that the majority of Hox factors are not completely dependent on the HX motif to form complexes with Exd [88]. These findings indicate that Hox factors can use alternative mechanisms to interact with PBC factors to regulate gene expression.

An alternative PBC interaction motif has been identified in the *Drosophila* Ubx and Abd-A Hox factors. This motif is conserved among arthropod and cnidarian Ubx and Abd-A homologs and, hence, has been named the UbdA motif [18]. The UbdA motif is located immediately C-terminal to the homeodomain. Biochemical and *in vivo* BiFC assays have demonstrated that the UbdA motif contributes to interactions with Exd [15,84,86,88]. Moreover, mutation of the UbdA motif in Abd-A and Ubx results in decreased Exd-dependent activities *in vivo* [15,18,84,86]. More recently, a crystal structure of Ubx interacting with Exd via its UbdA domain has been described. Foos *et al.* crystalized two types of Ubx/Exd/DNA complexes, one using Ubx proteins with both the HX and UbdA motif and one with only the UbdA motif [18]. In both complexes, the homeodomains of Ubx and Exd were situated on opposite sides of the DNA. When both the HX and UbdA motifs were present in Ubx, only the HX motif made contacts Exd, whereas the UbdA motif was mostly disordered. By contrast, if only the UbdA only is present in Ubx, the UbdA forms a coiled extension off of the homeodomain's third helix and makes contact with the loop region between the second and third helices of the Exd homeodomain (Figure 3B). The UbdA-Exd interaction occurs on the opposite side of the DNA relative to the HX-Exd interaction. Based on these structures, it is not clear what regulates whether the HX and/or UbdA mediates interactions between Ubx/Abd-A and PBC proteins. However, Saadaoui *et al.* (2011) demonstrated that Ubx isoforms with longer linker region lengths are more resistant to HX mutations in regulating a CRM containing adjacent Exd-Ubx binding sites [15]. Given the topology, this may be due to the fact that the longer linker region length gives flexibility to Ubx and Abd-A proteins to mediate contacts on both sides of the DNA via the HX and UbdA motifs.

2.4. Interactions with HMP Proteins Differentiate Hox Paralogs

In addition to PBC factors, another group of TALE homeodomain proteins directly associates with Hox factors on DNA and contributes to Hox paralog specificity. The HMP family of proteins includes Homothorax (Hth) in *Drosophila*, UNC-62 in *C. elegans* and Meis and Prep (pKnox) in vertebrates. Hth was originally identified as a gene important to larval cuticle development in *Drosophila*, while Meis was originally identified as a proto-oncogene in mammals [89,90]. These genes were later shown to interact with PBC and Hox proteins and thereby contribute to Hox-mediated outcomes [26,44,91].

The vertebrate HMP proteins, Meis1A and 1B, have been shown to interact with Hox factors from multiple paralog groups (murine Hox paralog groups 2, 4, 5, 8, 9, 10, 11, 12 and 13) using yeast two-hybrid assays [26]. However, DNA binding assays indicate that only posterior Hox factors (Abd-B family members in particular) bind cooperatively with HMP proteins to DNA containing adjacent Hox-HMP binding sites [26,44,45]. In *Drosophila*, the Ubx and Abd-A Hox factors have also been shown to directly cooperate with Hth by binding and regulating CRMs containing adjacent Hox-Hth binding sites [40,46]. These findings suggest that the arrangement of adjacent Hox-HMP binding sites may contribute to paralog specificity by favoring posterior Hox factors over anterior ones. However, the crystal structures of Hox-HMP-DNA complexes have not been solved, and therefore, the mechanism and specificity of interaction between Hox and HMP factors remains unclear.

In addition to interacting with Hox factors, the TALE PBC and HMP proteins heterodimerize through highly-conserved domains found N-terminal to their homeodomains. Importantly, PBC-HMP heterodimer formation increases the nuclear import of PBC, which is largely cytoplasmic in the absence of HMP proteins, and enhances the stability of HMP proteins [91–95]. More recently, it has been shown that the Hox-PBC-HMP interactions are ancient and evolutionarily conserved among cnidarians and bilaterians and that Hox, PBC and HMP from different species can interact with one another and substitute for one another in functional assays [96]. Molecularly, Hox-PBC-HMP interactions result in the formation of higher order complexes on DNA [97–100]. In fact, several CRMs containing binding sites for all three factors have been identified, especially in the vertebrate hindbrain and in *Drosophila* [40,46,58,64–67,69] (Figure 3C). In general, these CRMs contain either an adjacent PBC/Hox site with a nearby HMP site or an adjacent HMP/Hox site with a nearby PBC site. Analysis of binding site patterns in these CRMs reveals a variety of spacing and orientations of Hox, PBC and HMP binding sites that are sufficient to mediate the formation of Hox/PBC/HMP complexes on DNA [101] (Figure 3C). Moreover, one recent study found that hindbrain enhancers in zebrafish are enriched for HMP binding sites within 50 bp of combined Hox/PBC sites [102]. While this study did not conclusively demonstrate that these CRMs were directly regulated by Hox/PBC/HMP complexes, it does illustrate the potential flexibility of spacing for interacting HMP, Hox and PBC binding sites. In addition, the fact that certain interactions between Hox and PBC/HMP are paralog specific, *i.e.*, the UbdA interaction motif in Ubx/Abd-A, the non-conserved HX motif in Abd-B homologs, the cooperative DNA binding between HMP proteins and posterior Hox factors, suggests the possibility that particular Hox-PBC-HMP binding site arrangements are paralog specific. Future studies focused on analyzing a large number of Hox-regulated CRMs will be needed to determine if specific arrangements of binding sites are an important contributor to Hox paralog specificity.

2.5. Hox Paralogs Have Different Partner Preferences

As described for the PBC, HMP and Hox factors, interactions between TFs can yield cooperative complex formation on DNA, and the binding of each TF contributes to the overall binding affinity and specificity. In addition to PBC and HMP proteins, other sequence-specific TFs have been shown to interact with Hox factors, and several of these interactions are paralog specific. Thus, a CRM may encode Hox-paralog-specific activities by including a nearby binding site for a paralog-specific interacting TF.

Large-scale screens have identified additional TFs that interact with specific Hox factors [45,103]. For instance, Baëza *et al.* tested the ability of five *Drosophila* Hox factors (Scr, Antp, Ubx, Abd-A and Abd-B) to interact with 35 other TFs using an *in vivo* BiFC assay [103]. This study demonstrated that each tested Hox factor interacts with a distinct combination of the 35 TFs. On average, there was a 59% pairwise similarity between the sets of TFs that interacted with each Hox factor. Presumably these differences in affinity for other TFs will contribute to the differential binding of Hox factors to genomic sequences. However, there is currently limited data showing that the interactions between the Hox factors and the 35 other TFs occur on DNA, much less affect DNA binding specificity.

The data above suggest that Hox-regulated CRMs will require additional TF binding sites to yield paralog-specific outputs. This model is supported by examples of specific CRMs [40,46]. For instance, a *rhomboid* (*rho*) CRM (*RhoA*) in *Drosophila* contains binding sites for both an abdominal Hox complex composed of Exd/Hth/Abd-A, as well as the Pax2 transcription factor [47]. Importantly, mutations in either the Hox site or the Pax2 site disrupt activation, demonstrating that both the abdominal Hox complex and Pax2 protein need to bind the CRM for proper output. Moreover, the *RhoA* CRM is only activated by one specific abdominal Hox factor (Abd-A) that interacts with Pax2, whereas a thoracic Hox factor, Antp, does not interact with Pax2 and is unable to stimulate this CRM. This difference in ability to form complexes with Pax2 is thought to contribute to the paralog-specific activity of *RhoA*. Moreover, the interaction between specific posterior Hox factors and Pax2 may extend to vertebrates, as well. HoxA11 has been shown to interact with Pax2, and binding sites for both factors are required for the expression of a Six2 CRM in the developing mouse kidney [104].

2.6. Post-Translational Modifications Differentially Affect Hox Paralogs

Several different enzymes post-translationally modify Hox factors [48–51,105–111]. Here, we review those modifications that are known to differentially affect Hox paralogs.

Poly(ADP)-Ribose Polymerase-1 (PARP-1) was shown to poly(ADP)-ribosylate several mammalian Hox factors (HoxA5, HoxA7, HoxB6, HoxB7, HoxC6, HoxC8), but this modification only reduces the *in vitro* DNA binding and transcriptional activity of HoxA7 and HoxB7 [48]. This study demonstrated that HoxA7 and HoxB7 have a long glutamate-rich repeat in their C-terminal motif that is necessary for poly(ADP)-ribosylation and that the addition of a long glutamate-rich repeat in HoxB6 was sufficient to allow poly(ADP)-ribosylation to disrupt HoxB6 activity *in vitro* [48]. Similarly, it has been shown that phosphorylation by Casein Kinase II (CKII) differentially affects Hox paralogs. Phosphorylation of the HoxA9 homeodomain by CKII decreases DNA binding and thereby diminishes the ability of HoxA9 to keep cultured hematopoietic progenitors from differentiating [49,50]. By contrast, HoxB7 appears to gain activity after phosphorylation. Two CKII phosphorylation sites (S132 and T203) were shown to flank the HoxB7 homeodomain, and mutating these sites to alanine inhibited the ability of HoxB7 to maintain hematopoietic cell culture in an undifferentiated state [51]. However, in both of these cases, it has not yet been delineated what mechanisms signal for the Hox paralogs to be post-translationally modified by these enzymes. An intriguing possibility is that post-translational modifications may occur at CRMs. Previous studies have shown that post-translational modifiers, such as PARP and CKII, can be recruited to CRMs and regulate gene expression [112–114], but it is unclear whether post-translational modifications of Hox factors is a CRM-dependent process or a global mechanism of regulating Hox factor activity. If the former is true, recruitment of post-translational modifiers to DNA could potentially contribute to paralog-specific activity of individual CRMs as well.

3. Target Accessibility

In order for Hox factors to regulate gene expression, they must gain access to target CRMs. There is substantial evidence that the binding patterns of most TFs are primarily determined by nucleosome positioning [115]. Are Hox factors like most TFs in that they only bind genomic sequences found in open chromatin or do they have pioneer activity capable of interacting with DNA wrapped in nucleosomes? Here, we present recent evidence that the Hox factor binding may be largely constrained to open chromatin and provide examples demonstrating that Hox factors rely on other TFs to open chromatin prior to Hox binding.

If Hox factors significantly alter the genomic chromatin landscape, one would predict that serially homologous tissues that express different Hox factors along the body plan would differ in their chromatin accessibility profiles. In *Drosophila*, the wing and haltere are serially homologous appendages with the wing arising from the second thoracic segment (T2) and the haltere developing from the third thoracic segment (T3) [116]. Previous studies demonstrated that the Hox factor, Ubx, is necessary to specify the haltere. In fact, genetic removal of Ubx can transform the haltere into wing tissue, whereas the misexpression of Ubx in the second thoracic segment transforms the wing into a haltere-like structure [7,117]. Comparisons of genome accessibility between the haltere and wing discs, however, revealed a largely similar chromatin landscape [118]. This finding suggests that the expression of Ubx does not alter cell fates by global changes in chromatin structure, but by acting upon accessible regulatory elements that are present in both the wing and haltere imaginal discs.

As a more direct test of the ability of Hox factors to alter genome accessibility, recent studies in cell culture revealed that Hox factors may differ in their ability to bind closed *versus* open chromatin. Chromatin-immunoprecipitation of Hox factors followed by genomic sequencing (ChIP-seq) in *Drosophila* cell culture found that two Hox factors are largely constrained by nucleosome positioning [119]. In this study, ChIP was performed for transiently transfected Ubx or Abd-A in *Drosophila* Kc167 embryonic cell lines, which revealed that more than 94% of Ubx and Abd-A ChIP peaks occurred within pre-existing DNaseI hypersensitive regions of the genome. The findings for Ubx are consistent with the global accessibility profiles found in the haltere *vs.* wing described above [118]. In contrast to Ubx and Abd-A, however, a substantial proportion (25%) of Abd-B-associated ChIP peaks in *Drosophila* cell culture were located in previously closed chromatin [119]. These studies suggest that Hox factors may fundamentally differ in their ability to associate with nucleosome-bound genomic regions.

The finding that at least a subset of Hox factors bind predominantly open chromatin leads to an interesting question: how do Hox factors gain access to CRMs that exist in closed genomic states? At least part of the answer appears to be through interactions with PBC and HMP proteins. As described earlier, heterodimerization of PBC and HMP allows translocation of the PBC proteins from the cytoplasm to the nucleus [91]. Since Kc167 cells lack Hth expression, Exd remains cytoplasmic, and thus, the above ChIP-seq studies performed on Ubx and Abd-A were done in the absence of PBC and HMP proteins. To determine what impact Exd and Hth expression has on Hox binding profiles, the authors performed complimentary ChIP-seq studies in cells co-transfected with Hth [119]. These experiments revealed that the proportion of Ubx ChIP peaks in DNaseI insensitive regions expanded from 5% to 17%, and the total number of Ubx ChIP peaks doubled. Moreover, the additional sites that were bound by Ubx were enriched for PBC and HMP binding motifs. Altogether these studies suggest that by itself, Ubx does not substantially change nucleosome positioning; and that PBC and HMP proteins enhance Ubx binding to a greater number of sites, including those in closed chromatin.

Mechanistic support for the idea that PBC and HMP proteins can change the chromatin status of Hox targets prior to Hox binding comes from recent studies in zebrafish [120]. A complex of HoxB1b-Pbx-Prep/Meis activates *hoxb1a* transcription in the early zebrafish embryo [66]. Maternally-loaded Pbx and Prep bind the *hoxb1a* locus and cause histone acetylation of the locus prior to the expression of HoxB1b or transcription of *hoxb1a* [121]. Additionally, Pbx and Prep were shown to recruit RNA polymerase II to the *hoxb1a* promoter and maintain RNA polymerase II in a paused state. Slightly later in embryogenesis, HoxB1b is expressed and binds the *hoxb1a* locus, where it recruits factors that phosphorylate RNA polymerase II to promote transcriptional elongation [120].

Given the role of histone acetylation in changing nucleosome positioning, these studies suggest that PBC and HMP open chromatin and allow Hox factors to gain access to targets [122].

In total, these studies demonstrate that PBC and HMP proteins can contribute to Hox genomic binding patterns (Figure 4). Hence, PBC and HMP binding sites in Hox-regulated CRMs may provide Hox factors greater access to CRMs. Consistent with this idea, several genome-wide binding studies demonstrate substantial overlap between the binding of Hox factors and PBC/HMP proteins [123–125]. However, these studies did not determine if the PBC/HMP factors were bound to genomic regions prior to Hox factor recruitment *versus* direct cooperative binding of PBC/HMP/Hox complexes to DNA. Thus, future studies on the potential role of PBC/HMP proteins as pioneer factors are needed to determine if PBC/HMP proteins broadly define Hox target selection via the opening of chromatin prior to Hox factor recruitment.

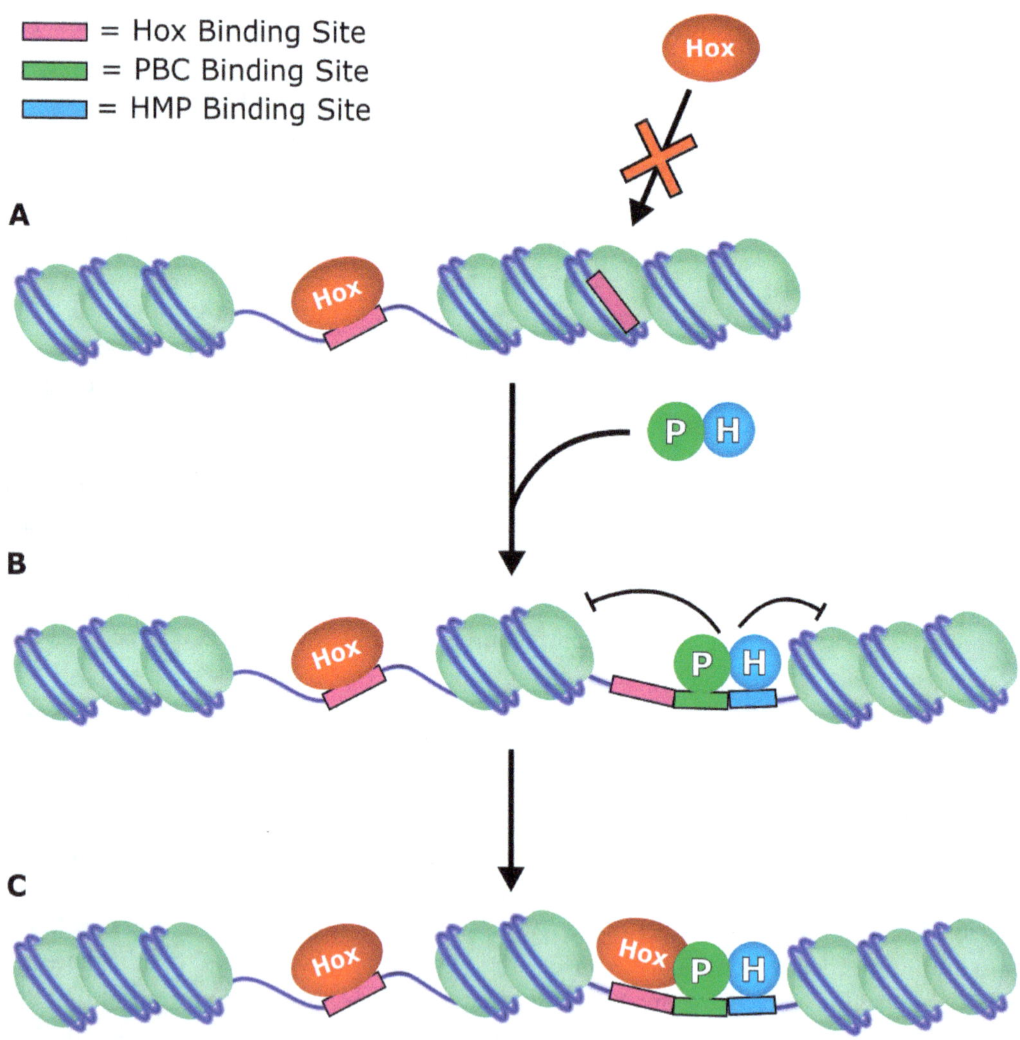

Figure 4. PBC and HMP can provide Hox factors access to targets. (**A**) Evidence suggests that Hox factors are typically limited to binding DNA in open chromatin regions [119]. (**B**,**C**) Expression of PBC and HMP protein has been shown to allow binding of Hox proteins to previously closed chromatin [119]. Prior to Hox binding, HMP and PBC proteins can bind DNA to promote chromatin modifications associated with chromatin opening, such as acetylation [120].

4. Effect on Transcription

Like many TFs, individual Hox factors can both activate and repress target expression [38,126]. Although, Hox factors have been shown to recruit and/or interact with factors known to affect transcriptional outcomes, *i.e.*, chromatin remodelers, the mediator complex and the general transcriptional machinery [16,17,124,127–131], it is largely unclear what determines whether a Hox factor will act as an activator or a repressor on a given target gene. One example that explores the mechanism of Hox-mediated repression *versus* activation is the mammalian *osteocalcin* promoter [128,132,133]. The *osteocalcin* promoter contains adjacent Hox-Pbx binding sites, and in pre-osteoblasts, Pbx1 complexes with HoxA10 and recruits HDACs to the *osteocalcin* promoter, which represses gene expression. However, as pre-osteoblasts differentiate into osteoblasts, Pbx1 expression decreases, and HoxA10 then recruits CBP/p300 to acetylate *osteocalcin* and activate gene expression. These data suggest that the same PBC-Hox binding site can function in either gene activation or repression dependent on the presence of a PBC protein.

The binding of other TFs besides PBC and HMP can also influence whether a Hox factor activates or represses a target gene. For example, the Abd-A Hox factor can both activate the expression of *rhomboid* in abdominal sensory cells and repress the expression of Distal-less (Dll) in the abdominal ectoderm [40,46]. As described above, Abd-A activates the *rhoA* CRM via interactions with Pax2, and the RhoA CRM contains nearby binding sites for both of these factors [47]. However, Abd-A represses Dll expression via multiple Hox/PBC and Hox/HMP binding sites, and this repression activity requires the nearby binding sites for two additional transcription factors, Sloppy-paired (Slp, a FoxG homolog) and Engrailed (En), which are known to recruit the Groucho co-repressor protein [46,134–136]. Hence, the decision for this Hox factor to activate *versus* repress is dependent on which additional TF sites are located nearby. Since we know a great deal about how the Hox factors regulate Dll expression, we use this example as a case study below to highlight different mechanisms of Hox specificity.

5. Case Study: The Distal-less Conserved Regulatory Element

A number of the concepts underlying Hox specificity are represented in a single case study of how different Hox factors bind to and regulate the *DCRE*, a Hox-regulated CRM that controls *Distal-less* (*Dll*) expression in *Drosophila* embryos. *Dll* expression is essential for the specification of appendages, such as legs from the thorax [137]. Therefore, the viability of *Drosophila* depends on activating *Dll* expression in thoracic segments and repressing its expression in the abdomen. Several *Dll* CRMs have been identified, and the DMX is the best-characterized regulatory element that is active in the early leg precursor cells [46,70,138]. Initial studies suggested the DMX is composed of separable activation (DMEact) and repression elements (DCRE) [39,46]. More recent studies paint a more complex picture with the DMEact and the DCRE contributing to both thoracic activation and abdominal repression [70]. For the purpose of this review, we will focus on how the DCRE utilizes several different Hox/PBC and Hox/HMP sites, as well as additional TF binding sites to contribute to paralog-specific activation *versus* paralog-specific repression.

Functional studies have revealed that the *DCRE* contains two Hox/PBC and one Hox/HMP site that are bound and regulated differentially by the Antp, Ubx, Abd-A and Abd-B Hox factors (Figure 5A,B) [46,70,138,139]. In the thorax, Antp cooperates with Exd/Hth to stimulate enhancer activity via binding the two Hox/PBC sites of the DCRE [70]. However, the Hox/HMP sites are not required for this activity, indicating that Antp may only specifically work on Hox/PBC sites and not Hox/HMP sites. While genetic and reporter studies have revealed that Antp is necessary for enhancing thoracic gene expression, it is currently unclear how Antp activates gene expression and whether additional binding sites in the DCRE are required for this activity.

In the abdomen, the abdominal Hox factors (Ubx, Abd-A and Abd-B) repress gene expression through the DCRE using two different cell-specific mechanisms. The *Drosophila* embryo is segmented, and each segment is composed of distinct anterior and posterior compartment cell types [140]. Mutagenesis studies on the DCRE revealed that mutations in the Hox/PBC and Hox/HMP sites resulted in a loss in repression activity in both cell types [46,70]. In contrast, a subset of mutations outside of the Hox/PBC and Hox/HMP sites revealed that distinct TF binding sites were required for anterior- *versus* posterior-compartment-mediated repression [46]. In anterior compartment cells, Slp binding sites are necessary to repress gene expression, whereas in posterior compartment cells, an En binding site is needed for repression [46]. Hence, both Slp and En are necessary for complete repression in the abdomen, because Slp expression is limited to the anterior compartment of embryonic segments, while En expression in limited to the posterior compartment of embryonic segments [141,142]. Slp and En both directly interact with Groucho, a histone deacetylase, and recruit it to CRMs [134–136]. Thus, it is thought that abdominal Hox factors promote a repressive chromatin state on the DCRE. Moreover, chromatin studies of the *Dll* locus demonstrate that in the thorax, the *DMEact* and *Dll* promoter physically communicate (*i.e.*, DNA looping), but *DCRE*-mediated repression in the abdomen disrupts this communication [143] (Figure 4B). These data help to demonstrate how the same Hox-regulated CRM can mediate multiple outputs depending on the cellular context.

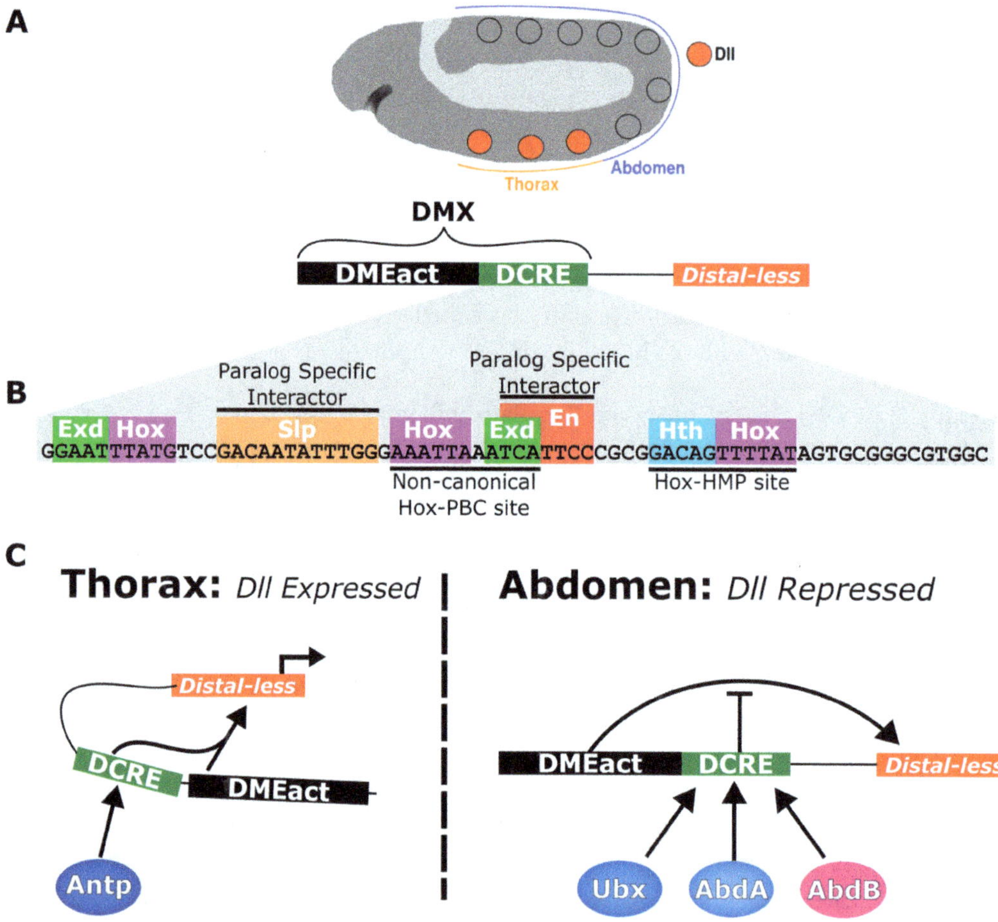

Figure 5. Regulation of *Distal-less* (*Dll*) gene expression by the *DCRE* in *Drosophila*. *Dll* is expressed in the *Drosophila* embryonic thorax, but not the abdomen. (**A**) *Dll* expression is controlled by two elements: the *DMEact* (an activator) and the *DCRE* (a mixed repressor and activator). *DMEact* is sufficient to drive expression in both the abdomen and the thorax, whereas the *DCRE* inhibits abdominal expression and enhances *DMEact* driven thoracic expression [70]; (**B**) Sequence of the *DCRE*, with known binding sites labeled. Slp and En binding sites are necessary for repressive *DCRE* activity in the abdomen [46,70]; (**C**) In the thorax, the chromatin structure of the *Dll* locus places the *DMEact* in proximity to the *Dll* promoter, whereas DNA looping between the DMEact regulatory region and the *Dll* promoter is not observed in abdominal segments [143]. These findings are consistent with results demonstrating that the DCRE "boosts" thoracic *DMEact* activation of *Dll* transcription via a mechanism dependent on Antp (the thoracic Hox factor) [70]. In contrast, the DCRE represses *Dll* transcription via mechanisms dependent on the abdominal Hox factors (Abd-A, Ubx and Abd-B) in the abdomen [46,70,138,139]. However, it is currently unclear if the thoracic and/or abdominal Hox factors directly regulate DNA looping between the distal leg enhancer and the proximal Dll promoter region.

Our advanced understanding of how the DCRE mediates three distinct functions (thoracic activation *vs.* anterior and posterior compartment repression) raises multiple questions in regards to Hox specificity. First, since Slp and En are expressed in both the abdomen and the thorax, so why does Antp not repress gene expression via the *DCRE*? Different preferences for TF interactors provide insight into this question: abdominal Hox factors have been shown to form complexes with En on DNA and to interact with Slp via fluorescent complementation assays, but Antp fails to do so in the same assays [46,103]. Hence, without being able to interact with Slp an En, Antp is unable to repress gene expression via the *DCRE*.

Second, how do the specific sequences found within the two Hox/PBC and one Hox/HMP binding sites contribute to paralog specificity? As described above, the Hox/HMP site only contributes to abdominal repression and not to thoracic activation [70]. This finding is consistent with previous studies that suggest that only posterior Hox factors mediate strong interactions with HMP proteins on Hox/HMP sites [26,44,45]. In addition, the analysis of the two Hox/PBC sites suggests that both are non-optimal sequences with either an abnormal spacing (inclusion of an additional nucleotide between the PBC and Hox site) or with several mismatches from consensus PBC/Hox sites [23]. Moreover, Electromobility Shift Assays (EMSAs) demonstrate that Abd-A forms stronger complexes with Exd and Hth on the *DCRE* sequence than does Antp [70]. This may be due to the fact that the *DCRE* contains adjacent Hox-Hth binding sites that favor more posterior Hox factors [26,44,45]. Moreover, the molecular basis of this difference in affinity may be related to the presence of the UbdA motif in Abd-A (and Ubx), whereas Antp lacks this motif. For example, mutations in the UbdA motif in Ubx disrupt *DCRE* binding and repression more than mutations in the HX motif [86]. These findings suggest that the DCRE sequence achieves paralog specificity by favoring Hox paralogs that contain the UbdA motif (Ubx and Abd-A, not Antp).

Third, does the arrangement of Hox-PBC-HMP binding sites contribute to Hox paralog specificity? It has been previously suggested that the arrangement of Hox-PBC-HMP binding sites may contribute to the activity of Hox-regulated CRM independent of paralog specificity [144]. To test this idea, we recently altered the configuration of Hox sites within the DCRE by swapping them with the Exd/Hth/Hox site found within the RhoA CRM [70]. As described, Abd-A/Exd/Hth complexes repress gene expression via the *DCRE* and activate gene expression via the *RhoA*. We found that this hybrid CRM is able to mediate gene repression equivalently to wild-type *DCRE* in the anterior compartments of the abdomen. These results demonstrate that, at least in this case, the Hox-PBC-HMP binding site arrangement does not determine the activity of the CRM.

Fourth, does post-translational modifications of Hox factors alter their ability to regulate the DCRE? In *Drosophila*, Ubx inhibits limb development in the abdomen by repressing expression of *Dll* via the DCRE [46,138]. However, Ubx orthologs in other arthropods have been shown to lack this repression activity [145,146]. In fact, Ubx is expressed in cells that give rise to legs along the trunk in *Artemia* (brine shrimp) [147,148]. In contrast to *Drosophila* Ubx, *Artemia* Ubx can be phosphorylated by CKII at serine and threonine residues in its C-terminal region [106]. Mutating phosphorylation sites in *Artermia* Ubx results in an increased ability to repress *Dll* when misexpressed in *Drosophila* embryos. Moreover, mutating *Drosophila* Ubx to add phosphorylation sites to its C-terminal region reduces the ability of the mutant Ubx to repress *Dll* expression [108]. These studies suggest that an evolutionary divergence of legs along the trunk of *Artemia* and *Drosophila* may be in part due to divergence in the ability of Ubx to be phosphorylated in these two species.

Fifth, do all of the Hox factors that regulate Dll expression require the function of Exd and Hth? Exd and Hth cooperate with Antp, as well as Ubx/Abd-A to mediate gene activation and repression, respectively [46]. However, Exd and Hth have been found to antagonize Abd-B-mediated repression at the DCRE [139]. In fact, Abd-B represses Hth expression in the most posterior abdominal segments during embryogenesis. Since Hth is necessary for nuclear localization of Exd, the repression of Hth expression by Abd-B results in a PBC/HMP-free region. Hence, Abd-B creates an Exd/Hth-free area and represses the *DCRE* in an Exd/Hth independent manner [91,139].

6. Conclusions

Throughout this review, we have described multiple mechanisms by which Hox paralogs may differentiate their binding specificities and activities. These include: (1) latent specificity revealed by PBC proteins; (2) utilization of low affinity sites; (3) multiple interaction surfaces for PBC proteins; (4) preference of HMP proteins to cooperatively bind DNA with posterior Hox paralogs; (5) differential affinity for other TFs; and (6) differential regulation by post-translational modifiers (Figure 2). Additionally, we reviewed how interactions with other TFs regulate genomic binding patterns of Hox factors, as well as their effect on transcription. While we focused on Hox factors in this review, many of the same problems of specificity presented here apply to other families of transcription factors. However, it is unclear whether the different TF families primarily rely on similar strategies or distinct strategies to overcome these problems of specificity. Elucidating the molecular mechanisms of how TFs bind specific targets and produce specific regulatory outcomes will be an important step in our understanding of how eukaryotes produce robust and specific gene expression patterns throughout development.

Acknowledgments: The authors wish to acknowledge the National Institutes of Health (NIH) grant (GM079428) for funding this work.

Conflicts of Interest: The authors declare no conflict of interest. The funding sponsors had no role in the design of the study; in the collection, analyses or interpretation of data; in the writing of the manuscript; nor in the decision to publish the results.

Abbreviations

The following abbreviations are used in this manuscript:

CRM	*cis*-regulatory module
Lab	Labial
Scr	Sex-combs Reduced
Antp	Antennapedia
Ubx	Ultrabithorax
Abd-A	Abdominal-A
Abd-B	Abdominal-B
TALE	Three Amino-acid Loop Extension
Exd	Extradenticle
Hth	Homothorax

References

1. McGinnis, W.; Krumlauf, R. Homeobox genes and axial patterning. *Cell* **1992**, *68*, 283–302. [CrossRef]
2. Cerdá-Esteban, N.; Spagnoli, F.M. Glimpse into Hox and tale regulation of cell differentiation and reprogramming. *Dev. Dyn.* **2014**, *243*, 76–87. [CrossRef] [PubMed]
3. Schneuwly, S.; Klemenz, R.; Gehring, W.J. Redesigning the body plan of *Drosophila* by ectopic expression of the homoeotic gene Antennapedia. *Nature* **1987**, *325*, 816–818. [CrossRef] [PubMed]
4. Wang, K.C.; Helms, J.A.; Chang, H.Y. Regeneration, repair and remembering identity: The three Rs of Hox gene expression. *Trends Cell Biol.* **2009**, *19*, 268–275. [CrossRef] [PubMed]
5. Friedrich, J.; Sorge, S.; Bujupi, F.; Eichenlaub, M.P.; Schulz, N.G.; Wittbrodt, J.; Lohmann, I. Hox Function is Required for the Development and Maintenance of the Drosophila Feeding Motor Unit. *Cell Rep.* **2016**, *14*, 850–860. [CrossRef] [PubMed]
6. Pascual-Anaya, J.; D'Aniello, S.; Kuratani, S.; Garcia-Fernàndez, J. Evolution of Hoxgene clusters in deuterostomes. *BMC Dev. Biol.* **2013**, *13*, 26. [CrossRef] [PubMed]
7. Lewis, E.B. A gene complex controlling segmentation in Drosophila. *Nature* **1978**, *276*, 565–570. [CrossRef] [PubMed]

8. Monteiro, A.S.; Ferrier, D.E.K. Hox genes are not always Colinear. *Int. J. Biol. Sci.* **2006**, *2*, 95–103. [CrossRef] [PubMed]
9. Wellik, D.M. Hox Genes and Vertebrate Axial Pattern. *Curr. Top. Dev. Biol.* **2009**, *88*, 257–278. [PubMed]
10. Swalla, B.J. Building divergent body plans with similar genetic pathways. *Heredity* **2006**, *97*, 235–243. [CrossRef] [PubMed]
11. Bondos, S.E.; Swint-Kruse, L.; Matthews, K.S. Flexibility and Disorder in Gene Regulation: LacI/GalR and Hox Proteins. *J. Biol. Chem.* **2015**, *290*, 24669–24677. [CrossRef] [PubMed]
12. Passner, J.M.; Ryoo, H.D.; Shen, L.; Mann, R.S.; Aggarwal, A.K. Structure of a DNA-bound Ultrabithorax–Extradenticle homeodomain complex. *Nature* **1999**, *397*, 714–719. [CrossRef] [PubMed]
13. Piper, D.E.; Batchelor, A.H.; Chang, C.-P.; Cleary, M.L.; Wolberger, C. Structure of a HoxB1–Pbx1 Heterodimer Bound to DNA: Role of the Hexapeptide and a Fourth Homeodomain Helix in Complex Formation. *Cell* **1999**, *96*, 587–597. [CrossRef]
14. Papadopoulos, D.K.; Reséndez-Pérez, D.; Cárdenas-Chávez, D.L.; Villanueva-Segura, K.; Canales-del-Castillo, R.; Felix, D.A.; Fünfschilling, R.; Gehring, W.J. Functional synthetic Antennapedia genes and the dual roles of YPWM motif and linker size in transcriptional activation and repression. *Proc. Natl. Acad. Sci. USA* **2011**, *108*, 11959–11964. [CrossRef] [PubMed]
15. Saadaoui, M.; Merabet, S.; Litim-Mecheri, I.; Arbeille, E.; Sambrani, N.; Damen, W.; Brena, C.; Pradel, J.; Graba, Y. Selection of distinct Hox-Extradenticle interaction modes fine-tunes Hox protein activity. *Proc. Natl. Acad. Sci. USA* **2011**, *108*, 2276–2281. [CrossRef] [PubMed]
16. Prince, F.; Katsuyama, T.; Oshima, Y.; Plaza, S.; Resendez-Perez, D.; Berry, M.; Kurata, S.; Gehring, W.J. The YPWM motif links Antennapedia to the basal transcriptional machinery. *Development* **2008**, *135*, 1669–1679. [CrossRef] [PubMed]
17. Boube, M.; Hudry, B.; Immarigeon, C.; Carrier, Y.; Bernat-Fabre, S.; Merabet, S.; Graba, Y.; Bourbon, H.-M.; Cribbs, D.L. Drosophila melanogaster Hox transcription factors access the RNA polymerase II machinery through direct homeodomain binding to a conserved motif of mediator subunit Med19. *PLoS Genet.* **2014**, *10*, e1004303. [CrossRef] [PubMed]
18. Foos, N.; Maurel-Zaffran, C.; Maté, M.J.; Vincentelli, R.; Hainaut, M.; Berenger, H.; Pradel, J.; Saurin, A.J.; Ortiz-Lombardía, M.; Graba, Y. A Flexible Extension of the Drosophila Ultrabithorax Homeodomain Defines a Novel Hox/PBC Interaction Mode. *Structure* **2015**, *23*, 270–279. [CrossRef] [PubMed]
19. Livingstone, C.D.; Barton, G.J. Protein sequence alignments: A strategy for the hierarchical analysis of residue conservation. *Comput. Appl. Biosci.* **1993**, *9*, 745–756. [CrossRef] [PubMed]
20. Sievers, F.; Wilm, A.; Dineen, D.; Gibson, T.J.; Karplus, K.; Li, W.; Lopez, R.; McWilliam, H.; Remmert, M.; Söding, J.; *et al.* Fast, scalable generation of high-quality protein multiple sequence alignments using Clustal Omega. *Mol. Syst. Biol.* **2011**, *7*. [CrossRef] [PubMed]
21. Noyes, M.B.; Christensen, R.G.; Wakabayashi, A.; Stormo, G.D.; Brodsky, M.H.; Wolfe, S.A. Analysis of homeodomain specificities allows the family-wide prediction of preferred recognition sites. *Cell* **2008**, *133*, 1277–1289. [CrossRef] [PubMed]
22. Berger, M.F.; Badis, G.; Gehrke, A.R.; Talukder, S.; Philippakis, A.A.; Peña-Castillo, L.; Alleyne, T.M.; Mnaimneh, S.; Botvinnik, O.B.; Chan, E.T.; *et al.* Variation in Homeodomain DNA Binding Revealed by High-Resolution Analysis of Sequence Preferences. *Cell* **2008**, *133*, 1266–1276. [CrossRef] [PubMed]
23. Slattery, M.; Riley, T.; Liu, P.; Abe, N.; Gomez-Alcala, P.; Dror, I.; Zhou, T.; Rohs, R.; Honig, B.; Bussemaker, H.J.; *et al.* Cofactor Binding Evokes Latent Differences in DNA Binding Specificity between Hox Proteins. *Cell* **2011**, *147*, 1270–1282. [CrossRef] [PubMed]
24. Zhao, J.J.; Lazzarini, R.A.; Pick, L. Functional dissection of the mouse Hox-a5 gene. *EMBO J.* **1996**, *15*, 1313–1322. [PubMed]
25. Tan, X.-X.; Bondos, S.; Li, L.; Matthews, K.S. Transcription Activation by Ultrabithorax Ib Protein Requires a Predicted α-Helical Region. *Biochemistry* **2002**, *41*, 2774–2785. [CrossRef] [PubMed]
26. Williams, T.M.; Williams, M.E.; Innis, J.W. Range of HOX/TALE superclass associations and protein domain requirements for HOXA13:MEIS interaction. *Dev. Biol.* **2005**, *277*, 457–471. [CrossRef] [PubMed]
27. Hsiao, H.-C.; Gonzalez, K.L.; Catanese, D.J., Jr.; Jordy, K.E.; Matthews, K.S.; Bondos, S.E. The Intrinsically Disordered Regions of the *Drosophila* melanogaster Hox Protein Ultrabithorax Select Interacting Proteins Based on Partner Topology. *PLoS ONE* **2014**, *9*, e108217. [CrossRef] [PubMed]

28. Regulski, M.; Harding, K.; Kostriken, R.; Karch, F.; Levine, M.; McGinnis, W. Homeo box genes of the Antennapedia and Bithorax Complexes of Drosophila. *Cell* **1985**, *43*, 71–80. [CrossRef]
29. Rezsohazy, R. Non-transcriptional interactions of Hox proteins: inventory, facts, and future directions. *Dev. Dyn.* **2014**, *243*, 117–131. [CrossRef] [PubMed]
30. Weirauch, M.T.; Yang, A.; Albu, M.; Cote, A.G.; Montenegro-Montero, A.; Drewe, P.; Najafabadi, H.S.; Lambert, S.A.; Mann, I.; Cook, K.; *et al.* Determination and Inference of Eukaryotic Transcription Factor Sequence Specificity. *Cell* **2014**, *158*, 1431–1443. [CrossRef] [PubMed]
31. Martin, A.; Serano, J.M.; Jarvis, E.; Bruce, H.S.; Wang, J.; Ray, S.; Barker, C.A.; O'Connell, L.C.; Patel, N.H. CRISPR/Cas9 Mutagenesis Reveals Versatile Roles of Hox Genes in Crustacean Limb Specification and Evolution. *Curr. Biol.* **2016**, *26*, 14–26. [CrossRef] [PubMed]
32. Niu, W.; Lu, Z.J.; Zhong, M.; Sarov, M.; Murray, J.I.; Brdlik, C.M.; Janette, J.; Chen, C.; Alves, P.; Preston, E.; *et al.* Diverse transcription factor binding features revealed by genome-wide ChIP-seq in *C. elegans*. *Genome Res.* **2011**, *21*, 245–254. [CrossRef] [PubMed]
33. Bach, C.; Buhl, S.; Mueller, D.; García-Cuéllar, M.-P.; Maethner, E.; Slany, R.K. Leukemogenic transformation by HOXA cluster genes. *Blood* **2010**, *115*, 2910–2918. [CrossRef] [PubMed]
34. Misra, M.; Sours, E.; Lance-Jones, C. Hox transcription factors influence motoneuron identity through the integrated actions of both homeodomain and non-homeodomain regions. *Dev. Dyn.* **2012**, *241*, 718–731. [CrossRef] [PubMed]
35. Gutierrez, A.; Knoch, L.; Witte, H.; Sommer, R.J. Functional specificity of the nematode Hox gene mab-5. *Development* **2003**, *130*, 983–993. [CrossRef] [PubMed]
36. Hueber, S.D.; Bezdan, D.; Henz, S.R.; Blank, M.; Wu, H.; Lohmann, I. Comparative analysis of Hox downstream genes in Drosophila. *Development* **2007**, *134*, 381–392. [CrossRef] [PubMed]
37. Pearson, J.C.; Lemons, D.; McGinnis, W. Modulating Hox gene functions during animal body patterning. *Nat. Rev. Genet.* **2005**, *6*, 893–904. [CrossRef] [PubMed]
38. Mann, R.S.; Lelli, K.M.; Joshi, R. Hox Specificity: Unique Roles for Cofactors and Collaborators. *Curr. Top. Dev. Biol.* **2009**, *88*, 63–101. [PubMed]
39. Gebelein, B.; Culi, J.; Ryoo, H.D.; Zhang, W.; Mann, R.S. Specificity of Distalless repression and limb primordia development by abdominal Hox proteins. *Dev. Cell* **2002**, *3*, 487–498. [CrossRef]
40. Li-Kroeger, D.; Witt, L.; Grimes, H.; Cook, T.A.; Gebelein, B. Hox and Senseless Antagonism Functions as a Molecular Switch to Regulate EGF Secretion in the *Drosophila* PNS. *Dev. Cell* **2008**, *15*, 298–308. [CrossRef] [PubMed]
41. Crocker, J.; Abe, N.; Rinaldi, L.; McGregor, A.P.; Frankel, N.; Wang, S.; Alsawadi, A.; Valenti, P.; Plaza, S.; Payre, F.; *et al.* Low affinity binding site clusters confer hox specificity and regulatory robustness. *Cell* **2015**, *160*, 191–203. [CrossRef] [PubMed]
42. Hirth, F.; Loop, T.; Egger, B.; Miller, D.F.B.; Kaufman, T.C.; Reichert, H. Functional equivalence of Hox gene products in the specification of the tritocerebrum during embryonic brain development of Drosophila. *Development* **2001**, *128*, 4781–4788. [PubMed]
43. Raines, A.M.; Magella, B.; Adam, M.; Potter, S.S. Key pathways regulated by HoxA9,10,11/HoxD9,10,11 during limb development. *BMC Dev. Biol.* **2015**, *15*, 28. [CrossRef] [PubMed]
44. Shen, W.F.; Montgomery, J.C.; Rozenfeld, S.; Moskow, J.J.; Lawrence, H.J.; Buchberg, A.M.; Largman, C. AbdB-like Hox proteins stabilize DNA binding by the Meis1 homeodomain proteins. *Mol. Cell. Biol.* **1997**, *17*, 6448–6458. [CrossRef] [PubMed]
45. Jolma, A.; Yin, Y.; Nitta, K.R.; Dave, K.; Popov, A.; Taipale, M.; Enge, M.; Kivioja, T.; Morgunova, E.; Taipale, J. DNA-dependent formation of transcription factor pairs alters their binding specificity. *Nature* **2015**, *527*, 384–388. [CrossRef] [PubMed]
46. Gebelein, B.; McKay, D.J.; Mann, R.S. Direct integration of Hox and segmentation gene inputs during *Drosophila* development. *Nature* **2004**, *431*, 653–659. [CrossRef] [PubMed]
47. Li-Kroeger, D.; Cook, T.A.; Gebelein, B. Integration of an abdominal Hox complex with Pax2 yields cell-specific EGF secretion from Drosophila sensory precursor cells. *Development* **2012**, *139*, 1611–1619. [CrossRef] [PubMed]
48. Wu, X.; Ellmann, S.; Rubin, E.; Gil, M.; Jin, K.; Han, L.; Chen, H.; Kwon, E.M.; Guo, J.; Ha, H.C.; *et al.* ADP ribosylation by PARP-1 suppresses HOXB7 transcriptional activity. *PLoS ONE* **2012**, *7*, e40644. [CrossRef] [PubMed]

49. Eklund, E.A.; Jalava, A.; Kakar, R. Tyrosine Phosphorylation of HoxA10 Decreases DNA Binding and Transcriptional Repression during Interferon γ-induced Differentiation of Myeloid Leukemia Cell Lines. *J. Biol. Chem.* **2000**, *275*, 20117–20126. [PubMed]
50. Vijapurkar, U.; Fischbach, N.; Shen, W.; Brandts, C.; Stokoe, D.; Lawrence, H.J.; Largman, C. Protein kinase C-mediated phosphorylation of the leukemia-associated HOXA9 protein impairs its DNA binding ability and induces myeloid differentiation. *Mol. Cell. Biol.* **2004**, *24*, 3827–3837. [CrossRef] [PubMed]
51. Yaron, Y.; McAdara, J.K.; Lynch, M.; Hughes, E.; Gasson, J.C. Identification of novel functional regions important for the activity of HOXB7 in mammalian cells. *J. Immunol.* **2001**, *166*, 5058–5067. [CrossRef] [PubMed]
52. Peifer, M.; Wieschaus, E. Mutations in the *Drosophila* gene extradenticle affect the way specific homeo domain proteins regulate segmental identity. *Genes Dev.* **1990**, *4*, 1209–1223. [CrossRef] [PubMed]
53. Nourse, J.; Mellentin, J.D.; Galili, N.; Wilkinson, J.; Stanbridge, E.; Smith, S.D.; Cleary, M.L. Chromosomal translocation t(1;19) results in synthesis of a homeobox fusion mRNA that codes for a potential chimeric transcription factor. *Cell* **1990**, *60*, 535–545. [CrossRef]
54. Kamps, M.P.; Murre, C.; Sun, X.; Baltimore, D. A new homeobox gene contributes the DNA binding domain of the t(1;19) translocation protein in pre-B all. *Cell* **1990**, *60*, 547–555. [CrossRef]
55. Bürglin, T.R. Analysis of TALE superclass homeobox genes (MEIS, PBC, KNOX, Iroquois, TGIF) reveals a novel domain conserved between plants and animals. *Nucleic Acids Res.* **1997**, *25*, 4173–4180. [CrossRef] [PubMed]
56. Chan, S.-K.; Jaffe, L.; Capovilla, M.; Botas, J.; Mann, R.S. The DNA binding specificity of ultrabithorax is modulated by cooperative interactions with extradenticle, another homeoprotein. *Cell* **1994**, *78*, 603–615. [CrossRef]
57. Chang, C.P.; Brocchieri, L.; Shen, W.F.; Largman, C.; Cleary, M.L. Pbx modulation of Hox homeodomain amino-terminal arms establishes different DNA-binding specificities across the Hox locus. *Mol. Cell. Biol.* **1996**, *16*, 1734–1745. [CrossRef] [PubMed]
58. Pöpperl, H.; Bienz, M.; Studer, M.; Chan, S.K.; Aparicio, S.; Brenner, S.; Mann, R.S.; Krumlauf, R. Segmental expression of Hoxb-1 is controlled by a highly conserved autoregulatory loop dependent upon exd/pbx. *Cell* **1995**, *81*, 1031–1042. [CrossRef]
59. Rauskolb, C.; Peifer, M.; Wieschaus, E. Extradenticle, a regulator of homeotic gene activity, is a homolog of the homeobox-containing human proto-oncogene pbx1. *Cell* **1993**, *74*, 1101–1112. [CrossRef]
60. Rauskolb, C.; Wieschaus, E. Coordinate regulation of downstream genes by extradenticle and the homeotic selector proteins. *EMBO J.* **1994**, *13*, 3561–3569. [PubMed]
61. Sun, B.; Hursh, D.A.; Jackson, D.; Beachy, P.A. Ultrabithorax protein is necessary but not sufficient for full activation of decapentaplegic expression in the visceral mesoderm. *EMBO J.* **1995**, *14*, 520–535. [PubMed]
62. Izpisúa-Belmonte, J.C.; Falkenstein, H.; Dollé, P.; Renucci, A.; Duboule, D. Murine genes related to the Drosophila AbdB homeotic genes are sequentially expressed during development of the posterior part of the body. *EMBO J.* **1991**, *10*, 2279–2289. [PubMed]
63. Van Dijk, M.A.; Murre, C. Extradenticle Raises the DNA binding specificity of homeotic selector gene products. *Cell* **1994**, *78*, 617–624. [CrossRef]
64. Ebner, A.; Cabernard, C.; Affolter, M.; Merabet, S. Recognition of distinct target sites by a unique Labial/Extradenticle/Homothorax complex. *Development* **2005**, *132*, 1591–1600. [CrossRef] [PubMed]
65. Ryoo, H.D.; Marty, T.; Casares, F.; Affolter, M.; Mann, R.S. Regulation of Hox target genes by a DNA bound Homothorax/Hox/Extradenticle complex. *Development* **1999**, *126*, 5137–5148. [PubMed]
66. Ferretti, E.; Cambronero, F.; Tümpel, S.; Longobardi, E.; Wiedemann, L.M.; Blasi, F.; Krumlauf, R. Hoxb1 Enhancer and Control of Rhombomere 4 Expression: Complex Interplay between PREP1-PBX1-HOXB1 Binding Sites. *Mol. Cell. Biol.* **2005**, *25*, 8541–8552. [CrossRef] [PubMed]
67. Ferretti, E.; Marshall, H.; Popperl, H.; Maconochie, M.; Krumlauf, R.; Blasi, F. Segmental expression of Hoxb2 in r4 requires two separate sites that integrate cooperative interactions between Prep1, Pbx and Hox proteins. *Development* **2000**, *127*, 155–166. [PubMed]
68. Tümpel, S.; Cambronero, F.; Ferretti, E.; Blasi, F.; Wiedemann, L.M.; Krumlauf, R. Expression of Hoxa2 in rhombomere 4 is regulated by a conserved cross-regulatory mechanism dependent upon Hoxb1. *Dev. Biol.* **2007**, *302*, 646–660. [CrossRef] [PubMed]

69. Manzanares, M.; Bel-Vialar, S.; Ariza-McNaughton, L.; Ferretti, E.; Marshall, H.; Maconochie, M.M.; Blasi, F.; Krumlauf, R. Independent regulation of initiation and maintenance phases of Hoxa3 expression in the vertebrate hindbrain involve auto- and cross-regulatory mechanisms. *Development* **2001**, *128*, 3595–3607. [PubMed]
70. Uhl, J.D.; Zandvakili, A.; Gebelein, B. A Hox Transcription Factor Collective Binds a Highly Conserved Distal-less *cis*-Regulatory Module to Generate Robust Transcriptional Outcomes. *PLoS Genet* **2016**, *12*, e1005981. [CrossRef] [PubMed]
71. Joshi, R.; Passner, J.M.; Rohs, R.; Jain, R.; Sosinsky, A.; Crickmore, M.A.; Jacob, V.; Aggarwal, A.K.; Honig, B.; Mann, R.S. Functional Specificity of a Hox Protein Mediated by the Recognition of Minor Groove Structure. *Cell* **2007**, *131*, 530–543. [CrossRef] [PubMed]
72. Kim, S.; Broströmer, E.; Xing, D.; Jin, J.; Chong, S.; Ge, H.; Wang, S.; Gu, C.; Yang, L.; Gao, Y.Q.; *et al.* Probing Allostery Through DNA. *Science* **2013**, *339*, 816–819. [CrossRef] [PubMed]
73. Narasimhan, K.; Pillay, S.; Huang, Y.-H.; Jayabal, S.; Udayasuryan, B.; Veerapandian, V.; Kolatkar, P.; Cojocaru, V.; Pervushin, K.; Jauch, R. DNA-mediated cooperativity facilitates the co-selection of cryptic enhancer sequences by SOX2 and PAX6 transcription factors. *Nucleic Acids Res.* **2015**, *43*, 1513–1528. [CrossRef] [PubMed]
74. Chang, C.P.; Shen, W.F.; Rozenfeld, S.; Lawrence, H.J.; Largman, C.; Cleary, M.L. Pbx proteins display hexapeptide-dependent cooperative DNA binding with a subset of Hox proteins. *Genes Dev.* **1995**, *9*, 663–674. [CrossRef] [PubMed]
75. Lu, Q.; Kamps, M.P. Heterodimerization of Hox proteins with Pbx1 and oncoprotein E2a-Pbx1 generates unique DNA-binding specifities at nucleotides predicted to contact the N-terminal arm of the Hox homeodomain–demonstration of Hox-dependent targeting of E2a-Pbx1 *in vivo*. *Oncogene* **1997**, *14*, 75–83. [CrossRef] [PubMed]
76. Abe, N.; Dror, I.; Yang, L.; Slattery, M.; Zhou, T.; Bussemaker, H.J.; Rohs, R.; Mann, R.S. Deconvolving the recognition of DNA shape from sequence. *Cell* **2015**, *161*, 307–318. [CrossRef] [PubMed]
77. Tanay, A. Extensive low-affinity transcriptional interactions in the yeast genome. *Genome Res.* **2006**, *16*, 962–972. [CrossRef] [PubMed]
78. Scardigli, R.; Bäumer, N.; Gruss, P.; Guillemot, F.; Roux, I.L. Direct and concentration-dependent regulation of the proneural gene Neurogenin2 by Pax6. *Development* **2003**, *130*, 3269–3281. [CrossRef] [PubMed]
79. Ramos, A.I.; Barolo, S. Low-affinity transcription factor binding sites shape morphogen responses and enhancer evolution. *Phil. Trans. R. Soc. B* **2013**, *368*, 20130018. [CrossRef] [PubMed]
80. Evans, N.C.; Swanson, C.I.; Barolo, S. Sparkling insights into enhancer structure, function, and evolution. *Curr. Top. Dev. Biol.* **2012**, *98*, 97–120. [PubMed]
81. Crocker, J.; Noon, E.P.-B.; Stern, D.L. The Soft Touch: Low-Affinity Transcription Factor Binding Sites in Development and Evolution. *Curr. Top. Dev. Biol.* **2016**, *17*, 455–469.
82. Chan, S.K.; Pöpperl, H.; Krumlauf, R.; Mann, R.S. An extradenticle-induced conformational change in a HOX protein overcomes an inhibitory function of the conserved hexapeptide motif. *EMBO J.* **1996**, *15*, 2476–2487. [PubMed]
83. Galant, R.; Walsh, C.M.; Carroll, S.B. Hox repression of a target gene: Extradenticle-independent, additive action through multiple monomer binding sites. *Development* **2002**, *129*, 3115–3126. [PubMed]
84. Lelli, K.M.; Noro, B.; Mann, R.S. Variable motif utilization in homeotic selector (Hox)–cofactor complex formation controls specificity. *Proc. Natl. Acad. Sci. USA* **2011**, *108*, 21122–21127. [CrossRef] [PubMed]
85. Merabet, S.; Kambris, Z.; Capovilla, M.; Bérenger, H.; Pradel, J.; Graba, Y. The Hexapeptide and Linker Regions of the AbdA Hox Protein Regulate Its Activating and Repressive Functions. *Dev. Cell* **2003**, *4*, 761–768. [CrossRef]
86. Merabet, S.; Saadaoui, M.; Sambrani, N.; Hudry, B.; Pradel, J.; Affolter, M.; Graba, Y. A unique Extradenticle recruitment mode in the Drosophila Hox protein Ultrabithorax. *Proc. Natl. Acad. Sci. USA* **2007**, *104*, 16946–16951. [CrossRef] [PubMed]
87. Tour, E.; Hittinger, C.T.; McGinnis, W. Evolutionarily conserved domains required for activation and repression functions of the Drosophila Hox protein Ultrabithorax. *Development* **2005**, *132*, 5271–5281. [CrossRef] [PubMed]

88. Hudry, B.; Remacle, S.; Delfini, M.-C.; Rezsohazy, R.; Graba, Y.; Merabet, S. Hox Proteins Display a Common and Ancestral Ability to Diversify Their Interaction Mode with the PBC Class Cofactors. *PLoS Biol.* **2012**, *10*, e1001351. [CrossRef] [PubMed]
89. Jürgens, G.; Wieschaus, E.; Nüsslein-Volhard, C.; Kluding, H. Mutations affecting the pattern of the larval cuticle inDrosophila melanogaster. *Wilhelm Rouxs Arch. Dev. Biol.* **1984**, *193*, 283–295. [CrossRef]
90. Moskow, J.J.; Bullrich, F.; Huebner, K.; Daar, I.O.; Buchberg, A.M. Meis1, a PBX1-related homeobox gene involved in myeloid leukemia in BXH-2 mice. *Mol. Cell. Biol.* **1995**, *15*, 5434–5443. [CrossRef] [PubMed]
91. Rieckhof, G.E.; Casares, F.; Ryoo, H.D.; Abu-Shaar, M.; Mann, R.S. Nuclear Translocation of Extradenticle Requires homothorax, which Encodes an Extradenticle-Related Homeodomain Protein. *Cell* **1997**, *91*, 171–183. [CrossRef]
92. Abu-Shaar, M.; Ryoo, H.D.; Mann, R.S. Control of the nuclear localization of extradenticle by competing nuclear import and export signals. *Genes Dev.* **1999**, *13*, 935–945. [CrossRef] [PubMed]
93. Berthelsen, J.; Kilstrup-Nielsen, C.; Blasi, F.; Mavilio, F.; Zappavigna, V. The subcellular localization of PBX1 and EXD proteins depends on nuclear import and export signals and is modulated by association with PREP1 and HTH. *Genes Dev.* **1999**, *13*, 946–953. [CrossRef] [PubMed]
94. Jaw, T.J.; You, L.R.; Knoepfler, P.S.; Yao, L.C.; Pai, C.Y.; Tang, C.Y.; Chang, L.P.; Berthelsen, J.; Blasi, F.; Kamps, M.P.; *et al.* Direct interaction of two homeoproteins, homothorax and extradenticle, is essential for EXD nuclear localization and function. *Mech. Dev.* **2000**, *91*, 279–291. [CrossRef]
95. Pai, C.Y.; Kuo, T.S.; Jaw, T.J.; Kurant, E.; Chen, C.T.; Bessarab, D.A.; Salzberg, A.; Sun, Y.H. The Homothorax homeoprotein activates the nuclear localization of another homeoprotein, extradenticle, and suppresses eye development in *Drosophila*. *Genes Dev.* **1998**, *12*, 435–446. [CrossRef] [PubMed]
96. Hudry, B.; Thomas-Chollier, M.; Volovik, Y.; Duffraisse, M.; Dard, A.; Frank, D.; Technau, U.; Merabet, S. Molecular insights into the origin of the Hox-TALE patterning system. *eLife* **2014**, *3*, e01939. [CrossRef] [PubMed]
97. Berthelsen, J.; Zappavigna, V.; Ferretti, E.; Mavilio, F.; Blasi, F. The novel homeoprotein Prep1 modulates Pbx–Hox protein cooperativity. *EMBO J.* **1998**, *17*, 1434–1445. [CrossRef] [PubMed]
98. Jacobs, Y.; Schnabel, C.A.; Cleary, M.L. Trimeric Association of Hox and TALE Homeodomain Proteins Mediates Hoxb2 Hindbrain Enhancer Activity. *Mol. Cell. Biol.* **1999**, *19*, 5134–5142. [CrossRef] [PubMed]
99. Shanmugam, K.; Green, N.C.; Rambaldi, I.; Saragovi, H.U.; Featherstone, M.S. PBX and MEIS as Non-DNA-Binding Partners in Trimeric Complexes with HOX Proteins. *Mol. Cell. Biol.* **1999**, *19*, 7577–7588. [CrossRef] [PubMed]
100. Shen, W.F.; Rozenfeld, S.; Kwong, A.; Kömüves, L.G.; Lawrence, H.J.; Largman, C. HOXA9 forms triple complexes with PBX2 and MEIS1 in myeloid cells. *Mol. Cell. Biol.* **1999**, *19*, 3051–3061. [CrossRef] [PubMed]
101. Uhl, J.D.; Cook, T.A.; Gebelein, B. Comparing anterior and posterior Hox complex formation reveals guidelines for predicting *cis*-regulatory elements. *Dev. Biol.* **2010**, *343*, 154–166. [CrossRef] [PubMed]
102. Grice, J.; Noyvert, B.; Doglio, L.; Elgar, G. A Simple Predictive Enhancer Syntax for Hindbrain Patterning Is Conserved in Vertebrate Genomes. *PLoS ONE* **2015**, *10*, e0130413. [CrossRef] [PubMed]
103. Baëza, M.; Viala, S.; Heim, M.; Dard, A.; Hudry, B.; Duffraisse, M.; Rogulja-Ortmann, A.; Brun, C.; Merabet, S. Inhibitory activities of short linear motifs underlie Hox interactome specificity *in vivo*. *eLife* **2015**, *4*, e06034. [CrossRef] [PubMed]
104. Gong, K.-Q.; Yallowitz, A.R.; Sun, H.; Dressler, G.R.; Wellik, D.M. A Hox-Eya-Pax complex regulates early kidney developmental gene expression. *Mol. Cell. Biol.* **2007**, *27*, 7661–7668. [CrossRef] [PubMed]
105. Gavis, E.R.; Hogness, D.S. Phosphorylation, expression and function of the Ultrabithorax protein family in Drosophila melanogaster. *Development* **1991**, *112*, 1077–1093. [PubMed]
106. Hsia, C.C.; Paré, A.C.; Hannon, M.; Ronshaugen, M.; McGinnis, W. Silencing of an abdominal Hox gene during early development is correlated with limb development in a crustacean trunk. *Evol. Dev.* **2010**, *12*, 131–143. [CrossRef] [PubMed]
107. Jaffe, L.; Ryoo, H.D.; Mann, R.S. A role for phosphorylation by casein kinase II in modulating Antennapedia activity in *Drosophila*. *Genes Dev.* **1997**, *11*, 1327–1340. [CrossRef] [PubMed]
108. Taghli-Lamallem, O.; Hsia, C.; Ronshaugen, M.; McGinnis, W. Context-dependent regulation of Hox protein functions by CK2 phosphorylation sites. *Dev. Genes Evol.* **2008**, *218*, 321–332. [CrossRef] [PubMed]

109. Eklund, E.A.; Goldenberg, I.; Lu, Y.; Andrejic, J.; Kakar, R. SHP1 protein-tyrosine phosphatase regulates HoxA10 DNA binding and transcriptional repression activity in undifferentiated myeloid cells. *J. Biol. Chem.* **2002**, *277*, 36878–36888. [CrossRef] [PubMed]
110. Lindsey, S.; Huang, W.; Wang, H.; Horvath, E.; Zhu, C.; Eklund, E.A. Activation of SHP2 protein-tyrosine phosphatase increases HoxA10-induced repression of the genes encoding gp91(PHOX) and p67(PHOX). *J. Biol. Chem.* **2007**, *282*, 2237–2249. [CrossRef] [PubMed]
111. Wang, H.; Lindsey, S.; Konieczna, I.; Bei, L.; Horvath, E.; Huang, W.; Saberwal, G.; Eklund, E.A. Constitutively active SHP2 cooperates with HoxA10 overexpression to induce acute myeloid leukemia. *J. Biol. Chem.* **2009**, *284*, 2549–2567. [CrossRef] [PubMed]
112. Basnet, H.; Su, X.B.; Tan, Y.; Meisenhelder, J.; Merkurjev, D.; Ohgi, K.A.; Hunter, T.; Pillus, L.; Rosenfeld, M.G. Tyrosine phosphorylation of histone H2A by CK2 regulates transcriptional elongation. *Nature* **2014**, *516*, 267–271. [CrossRef] [PubMed]
113. Kraus, W.L.; Hottiger, M.O. PARP-1 and gene regulation: Progress and puzzles. *Mol. Aspects Med.* **2013**, *34*, 1109–1123. [CrossRef] [PubMed]
114. Wang, S.; Jones, K.A. CK2 Controls the Recruitment of Wnt Regulators to Target Genes *in Vivo*. *Curr. Biol.* **2006**, *16*, 2239–2244. [CrossRef] [PubMed]
115. Biggin, M.D. Animal Transcription Networks as Highly Connected, Quantitative Continua. *Dev. Cell* **2011**, *21*, 611–626. [CrossRef] [PubMed]
116. Carroll, S.B.; Grenier, J.K.; Weatherbee, S.D. *From DNA to Diversity: Molecular Genetics and the Evolution of Animal Design*, 2nd ed.; Blackwell Pub: Malden, MA, USA, 2005.
117. Casares, F.; Calleja, M.; Sánchez-Herrero, E. Functional similarity in appendage specification by the Ultrabithorax and abdominal-A Drosophila HOX genes. *EMBO J.* **1996**, *15*, 3934–3942. [PubMed]
118. McKay, D.J.; Lieb, J.D. A Common Set of DNA Regulatory Elements Shapes Drosophila Appendages. *Dev. Cell* **2013**, *27*, 306–318. [CrossRef] [PubMed]
119. Beh, C.Y.; El-Sharnouby, S.; Chatzipli, A.; Russell, S.; Choo, S.W.; White, R. Roles of cofactors and chromatin accessibility in Hox protein target specificity. *Epigenetics Chromatin* **2016**, *9*. [CrossRef] [PubMed]
120. Choe, S.-K.; Ladam, F.; Sagerström, C.G. TALE factors poise promoters for activation by Hox proteins. *Dev. Cell* **2014**, *28*, 203–211. [CrossRef] [PubMed]
121. Choe, S.-K.; Sagerström, C.G. Variable Meis-dependence among paralog group-1 Hox proteins. *Biochem. Biophys. Res. Commun.* **2005**, *331*, 1384–1391. [CrossRef] [PubMed]
122. Shogren-Knaak, M.; Ishii, H.; Sun, J.-M.; Pazin, M.J.; Davie, J.R.; Peterson, C.L. Histone H4-K16 Acetylation Controls Chromatin Structure and Protein Interactions. *Science* **2006**, *311*, 844–847. [CrossRef] [PubMed]
123. Amin, S.; Donaldson, I.J.; Zannino, D.A.; Hensman, J.; Rattray, M.; Losa, M.; Spitz, F.; Ladam, F.; Sagerström, C.; Bobola, N. Hoxa2 Selectively Enhances Meis Binding to Change a Branchial Arch Ground State. *Dev. Cell* **2015**, *32*, 265–277. [CrossRef] [PubMed]
124. Huang, Y.; Sitwala, K.; Bronstein, J.; Sanders, D.; Dandekar, M.; Collins, C.; Robertson, G.; MacDonald, J.; Cezard, T.; Bilenky, M.; *et al.* Identification and characterization of Hoxa9 binding sites in hematopoietic cells. *Blood* **2012**, *119*, 388–398. [CrossRef] [PubMed]
125. Choo, S.W.; White, R.; Russell, S. Genome-Wide Analysis of the Binding of the Hox Protein Ultrabithorax and the Hox Cofactor Homothorax in *Drosophila*. *PLoS ONE* **2011**, *6*, e14778. [CrossRef] [PubMed]
126. Ladam, F.; Sagerström, C.G. Hox regulation of transcription: More complex(es). *Dev. Dyn.* **2014**, *243*, 4–15. [CrossRef] [PubMed]
127. Collins, C.; Wang, J.; Miao, H.; Bronstein, J.; Nawer, H.; Xu, T.; Figueroa, M.; Muntean, A.G.; Hess, J.L. C/EBPα is an essential collaborator in Hoxa9/Meis1-mediated leukemogenesis. *Proc. Natl. Acad. Sci. USA* **2014**, *111*, 9899–9904. [CrossRef] [PubMed]
128. Gordon, J.A.R.; Hassan, M.Q.; Saini, S.; Montecino, M.; van Wijnen, A.J.; Stein, G.S.; Stein, J.L.; Lian, J.B. Pbx1 represses osteoblastogenesis by blocking Hoxa10-mediated recruitment of chromatin remodeling factors. *Mol. Cell. Biol.* **2010**, *30*, 3531–3541. [CrossRef] [PubMed]
129. Kim, H.-H.; Park, S.J.; Han, J.-H.; Pathak, C.; Cheong, H.-K.; Lee, B.-J. Structural insight into the interaction between the Hox and HMGB1 and understanding of the HMGB1-enhancing effect of Hox-DNA binding. *Biochim. Biophys. Acta* **2015**, *1854*, 449–459. [CrossRef] [PubMed]

130. Lu, Y.; Goldenberg, I.; Bei, L.; Andrejic, J.; Eklund, E.A. HoxA10 represses gene transcription in undifferentiated myeloid cells by interaction with histone deacetylase 2. *J. Biol. Chem.* **2003**, *278*, 47792–47802. [CrossRef] [PubMed]
131. Zappavigna, V.; Falciola, L.; Helmer-Citterich, M.; Mavilio, F.; Bianchi, M.E. HMG1 interacts with HOX proteins and enhances their DNA binding and transcriptional activation. *EMBO J.* **1996**, *15*, 4981–4991. [PubMed]
132. Hassan, M.Q.; Tare, R.; Lee, S.H.; Mandeville, M.; Weiner, B.; Montecino, M.; van Wijnen, A.J.; Stein, J.L.; Stein, G.S.; Lian, J.B. HOXA10 controls osteoblastogenesis by directly activating bone regulatory and phenotypic genes. *Mol. Cell. Biol.* **2007**, *27*, 3337–3352. [CrossRef] [PubMed]
133. Gordon, J.A.R.; Hassan, M.Q.; Koss, M.; Montecino, M.; Selleri, L.; van Wijnen, A.J.; Stein, J.L.; Stein, G.S.; Lian, J.B. Epigenetic regulation of early osteogenesis and mineralized tissue formation by a HOXA10-PBX1-associated complex. *Cells Tissues Organs* **2011**, *194*, 146–150. [CrossRef] [PubMed]
134. Andrioli, L.P.; Oberstein, A.L.; Corado, M.S.G.; Yu, D.; Small, S. Groucho-dependent repression by sloppy-paired 1 differentially positions anterior pair-rule stripes in the Drosophila embryo. *Dev. Biol.* **2004**, *276*, 541–551. [CrossRef] [PubMed]
135. Roth, M.; Bonev, B.; Lindsay, J.; Lea, R.; Panagiotaki, N.; Houart, C.; Papalopulu, N. FoxG1 and TLE2 act cooperatively to regulate ventral telencephalon formation. *Development* **2010**, *137*, 1553–1562. [CrossRef] [PubMed]
136. Verginelli, F.; Perin, A.; Dali, R.; Fung, K.H.; Lo, R.; Longatti, P.; Guiot, M.-C.; del Maestro, R.F.; Rossi, S.; di Porzio, U.; *et al.* Transcription factors FOXG1 and Groucho/TLE promote glioblastoma growth. *Nat. Commun.* **2013**, *4*, 2956. [CrossRef] [PubMed]
137. Cohen, S.M.; Brönner, G.; Küttner, F.; Jürgens, G.; Jäckle, H. Distal-less encodes a homoeodomain protein required for limb development in *Drosophila. Nature* **1989**, *338*, 432–434. [CrossRef] [PubMed]
138. Vachon, G.; Cohen, B.; Pfeifle, C.; McGuffin, M.E.; Botas, J.; Cohen, S.M. Homeotic genes of the bithorax complex repress limb development in the abdomen of the Drosophila embryo through the target gene Distal-less. *Cell* **1992**, *71*, 437–450. [CrossRef]
139. Sambrani, N.; Hudry, B.; Maurel-Zaffran, C.; Zouaz, A.; Mishra, R.; Merabet, S.; Graba, Y. Distinct Molecular Strategies for Hox-Mediated Limb Suppression in *Drosophila*: From Cooperativity to Dispensability/Antagonism in TALE Partnership. *PLoS Genet* **2013**, *9*, e1003307. [CrossRef] [PubMed]
140. Lawrence, P.A. The present status of the parasegment. *Development* **1988**, *104*, 61–65.
141. Cadigan, K.M.; Grossniklaus, U.; Gehring, W.J. Localized expression of sloppy paired protein maintains the polarity of Drosophila parasegments. *Genes Dev.* **1994**, *8*, 899–913. [CrossRef] [PubMed]
142. Grossniklaus, U.; Pearson, R.K.; Gehring, W.J. The *Drosophila* sloppy paired locus encodes two proteins involved in segmentation that show homology to mammalian transcription factors. *Genes Dev.* **1992**, *6*, 1030–1051. [CrossRef] [PubMed]
143. Agelopoulos, M.; McKay, D.J.; Mann, R.S. Developmental regulation of chromatin conformation by Hox proteins in *Drosophila. Cell Rep.* **2012**, *1*, 350–359. [CrossRef] [PubMed]
144. Merabet, S.; Hudry, B. Hox transcriptional specificity despite a single class of cofactors: are flexible interaction modes the key? Plasticity in Hox/PBC interaction modes as a common molecular strategy for shaping Hox transcriptional activities. *BioEssays* **2013**, *35*, 88–92. [CrossRef] [PubMed]
145. Grenier, J.K.; Carroll, S.B. Functional evolution of the Ultrabithorax protein. *Proc. Natl. Acad. Sci. USA* **2000**, *97*, 704–709. [CrossRef] [PubMed]
146. Ronshaugen, M.; McGinnis, N.; McGinnis, W. Hox protein mutation and macroevolution of the insect body plan. *Nature* **2002**, *415*, 914–917. [CrossRef] [PubMed]
147. Averof, M.; Akam, M. Hox genes and the diversification of insect and crustacean body plans. *Nature* **1995**, *376*, 420–423. [CrossRef] [PubMed]
148. Averof, M.; Patel, N.H. Crustacean appendage evolution associated with changes in Hox gene expression. *Nature* **1997**, *388*, 682–686. [PubMed]

10

Hox Genes in Cardiovascular Development and Diseases

Marine Roux [1,2,3] **and Stéphane Zaffran** [1,2,*]

1 Aix Marseille Université, GMGF UMRS910, Faculté de Médecine, 27 Bd Jean Moulin, Marseille 13385, France
2 Inserm U910, Faculté de Médecine, 27 Bd Jean Moulin, Marseille 13005, France
3 Laboratory of Genetics and Development, Institut de Recherches Cliniques de Montréal (IRCM), Université de Montréal, Montréal, QC H3C 3T5, Canada; marine.roux@ircm.qc.ca
* Correspondence: stephane.zaffran@univ-amu.fr

Academic Editors: Vincenzo Zappavigna and Simon J. Conway

Abstract: Congenital heart defects (CHD) are the leading cause of death in the first year of life. Over the past 20 years, much effort has been focused on unraveling the genetic bases of CHD. In particular, studies in human genetics coupled with those of model organisms have provided valuable insights into the gene regulatory networks underlying CHD pathogenesis. *Hox* genes encode transcription factors that are required for the patterning of the anterior–posterior axis in the embryo. In this review, we focus on the emerging role of anteriorly expressed *Hox* genes (*Hoxa1*, *Hoxb1*, and *Hoxa3*) in cardiac development, specifically their contribution to patterning of cardiac progenitor cells and formation of the great arteries. Recent evidence regarding the cooperative regulation of heart development by Hox proteins with members of the TALE-class of homeodomain proteins such as Pbx and Meis transcription factors is also discussed. These findings are highly relevant to human pathologies as they pinpoint new genes that increase susceptibility to cardiac anomalies and provide novel mechanistic insights into CHD.

Keywords: Hox; heart development; cardiogenesis; second heart field; pharyngeal aortic arches; transcription factors

1. Introduction

In vertebrates, the roles of the cardiovascular system, consisting of the heart and blood vessels, are diverse and include important functions such as the transport of nutrients and waste, and the distribution of oxygen during both fetal and post-natal life. Cardiovascular development is a complex but ordered process that is spatially and temporally regulated [1]. This process includes dynamic remodeling of the pharyngeal arch arteries (PAAs) and coordinated septation of the outflow tract (OFT). Hence, a slight perturbation in this process may result in a spectrum of cardiovascular defects. In humans, variably abnormal cardiovascular development is reflected in the high incidence of congenital heart diseases (CHD), affecting nearly 1%–2% of all live births [2]. The cause of CHD is often difficult to determine with certainty; nonetheless, studying factors that control cardiovascular development can help us better understand the etiology of these defects. For a detailed summary of many other molecular pathways guiding heart formation, we suggest reviews [3–7].

Embryonic heart development begins at the cardiac crescent stage, around embryonic day (E) 7.5 in the mouse, when differentiation of cardiac precursors to cardiomyocytes is first observed. The linear heart tube that is subsequently formed is a transient structure composed of an inner layer, the endocardium, and covered by a myocardial layer. The heart tube rapidly elongates by progressive addition of cells at the arterial and venous poles during looping morphogenesis. The source of these new cardiomyocytes is located in a region of pharyngeal mesoderm termed the second heart field (SHF) [5–7]. As SHF progenitor cells are added to the forming heart, they are exposed to a number of surrounding signals [5,7]. After rightward looping, the heart is shaped by expansion of the myocardium, which leads to the formation of four cardiac chambers, two atria and two ventricles (Figure 1). The forming heart is connected to the bilateral dorsal aorta by the aortic sac and the PAAs. Septation of the cardiac chambers and the OFT is required for separation of the oxygenated and carbonated blood flows. Simultaneous development of the great arteries that supply the head, neck, and upper limbs is also crucial for the developing embryo. This process involves extensive vascular remodeling of the five pharyngeal arch arteries to form a unique arterial architecture (Figure 1A). Although initially symmetric, the most rostral PAAs in the 1st and 2nd arches mostly regress, whereas the caudal PAAs are remodeled into adult vascular structures. Development of the OFT and PAAs requires a specific subpopulation of neural crest cells (NCCs) originating in the posterior hindbrain. These cardiac NCCs migrate to the pharyngeal arches and heart to regulate the remodeling of the PAAs and the septation of the OFT, respectively [8]. They remain as smooth muscle cells of the proximal aorta, carotid arteries, and jugular veins [9].

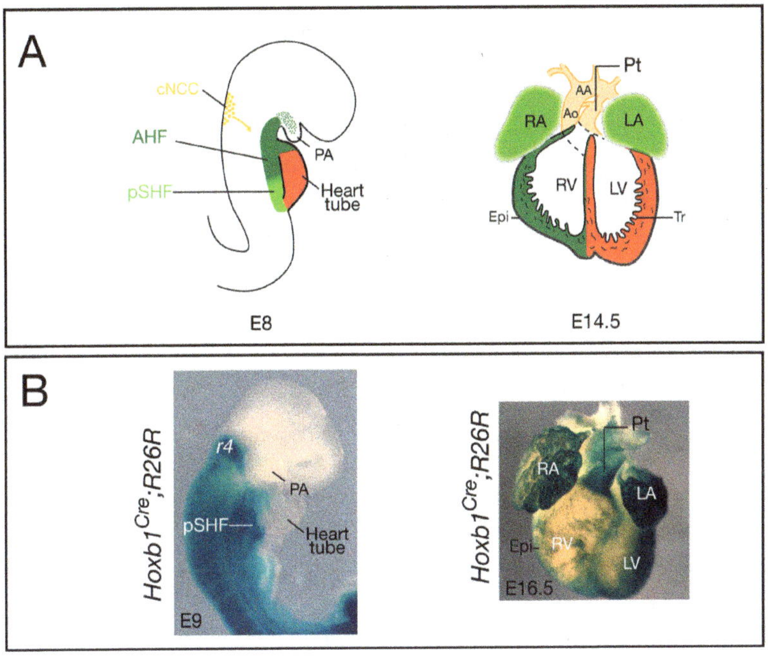

Figure 1. Cardiovascular development and contribution of *Hoxb1*^{Cre}-labeled cells in the mouse. (**A**) The location and contribution of the SHF is shown in green, with the anterior heart field (AHF) subdomain in dark green and the posterior SHF (pSHF) in light green. Cardiac neural crest cells (cNCC) are in orange. At E8, the early cardiac tube forms at the midline and it subsequently undergoes looping; at E14.5, the chambers are separated by the inter-ventricular septum and are connected to the pulmonary trunk (Pt) and aorta (Ao); (**B**) Cre lineage visualized by X-gal staining of *Hoxb1*^{Cre}; *R26R-lacZ* embryo (lateral view), and heart (ventral view), respectively. At E9, X-gal staining shows *Hoxb1*^{cre}+ cells in the pSHF; at E16.5, β-galactosidase activity is detected in both atria and the sub-pulmonary myocardium of *Hoxb1*^{Cre}; *R26R-lacZ* heart. *Hoxb1*^{cre}+ cells also contribute to the epicardium (Epi), which is located at the surface of the heart. AA: Aortic arch; LA: Left atrium; LV: Left ventricle; PA: Pharyngeal arch; RA: Right atrium; RV: Right ventricle; Tr: Trabeculae.

Hox genes encode for highly conserved homeodomain transcription factors. During vertebrate development, Hox proteins establish positional identity along the anterior–posterior (A–P) axis [10]. In mammals, *Hox* genes are organized in analogous clusters on four distinct chromosomes and are expressed in defined and often overlapping domains along the body axis in a manner corresponding to their position along the chromosome. This propriety is known as spatial collinearity [11]. Hox genes also show temporal collinearity, meaning that genes expressed more rostrally are produced before those at the caudal end. Major signaling pathways, such as Fibroblast growth factors (Fgf), retinoic acid (RA), and Wnt, play important roles in the establishment of ordered domains of *Hox* expression. In particular, exogenous RA can induce anterior expansions of *Hox* expression domains. Consistently, several *Hox* genes have well characterized RA-response elements (RAREs) in their enhancers or proximal promoters [12].

Several studies clearly identified that *Hox* function is required during heart development. In the invertebrate *Drosophila melanogaster*, four *Hox* genes are expressed in the organ equivalent of the heart, the dorsal vessel, in essentially non-overlapping domains: *Antennapedia* (*Antp*, from segment T3 to A1), *Ultrabithorax* (*Ubx*, from segment A1 to A4), *abdominal-A* (*abdA*, from segment A4 to A7) and *Abdominal-B* (*AbdB*, the last pair of cardiomyocytes of A7) [13,14]. Furthermore, functional experiments have demonstrated that these genes play critical roles in multiple phases of heart development in *Drosophila*: in the choice of cell lineages within the cardiac mesoderm, in the differentiation of cardiac cells, and in the remodeling of the larval cardiac tube into the adult heart [13,14]. A role for *Hox* gene expression in the vertebrate heart was first suggested by evidence that vitamin A deficiency causes various cardiac anomalies [15]. Vitamin A is the requisite source of RA in the body. Expression of *hoxd-3, hoxa-4,* and *hoxd-4* genes is increased in chick cardiac explants upon RA treatment [16]. In zebrafish, injection of a hyperactive VP16-tagged *hoxb5b* mRNA caused severe cardiac reduction, a highly reminiscent effect of ectopic RA on the formation of atrial and ventricular chambers [17].

In this review we will discuss studies establishing the role of *Hox* genes in several aspects of cardiovascular system development in mammals, from early cardiac specification and heart patterning to the remodeling of the PAAs. Recent studies reported a novel role of the *Hox* genes in the SHF related to regulation by RA signaling. This population of cardiac progenitor cells contributes to the major components of the heart including cardiac OFT, right ventricular, and atrial myocardium. Therefore, altering this early role of the *Hox* genes has later consequences on the onset of malformations of the heart and great arteries such as those described in CHD.

2. *Hox* Genes and Lineage Commitment

During gastrulation, the cardiac progenitors are derivatives of the mesoderm, which emerges from the primitive streak. Lineage tracing experiments in the mouse have shown that the first mesodermal cell lineage to emerge is Flk1+ [18]. At the same time, activation of genes along the Hox clusters occurs sequentially in the primitive streak throughout axis elongation [10]. Several signaling pathways, including BMP, FGF, and Wnt, are required for the specification of cardiogenic mesoderm [19]. A recent study showed that coordination of these pathways induces patterning of the mesoderm and specification by activating Cdx/Hox genes, emphasizing the role of Hox factors during this process [20]. Migration of cells expressing *Mesp1*, the earliest molecular marker for cardiac progenitors, through the primitive streak is triggered by Wnt3a signaling [21]. Interestingly, Lengerke *et al.* [20] demonstrated that addition of Wnt3a to embryonic stem (ES) cells is sufficient to induce expression of posterior *Hox* genes. This was the first study to link Wnt3a to Hox genes. These findings suggest that a Mesp1+ migrating cell that also expressed *Hox* genes may depend on Wnt3a signaling.

How *Hox* gene expression is coordinated during lineage differentiation is still poorly understood. However, long intergenic non-coding RNAs, in addition to Trithorax and Polycomb complexes, have been shown to be important for their regulation [22]. Deng *et al.* [23] have recently reported the role of HoxBlinc in mesoderm specification and *HoxB* gene activation. This study shows that HoxBlinc RNA specifies Flk1+ mesoderm toward cardiac and hematopoietic lineages. It recruits Setd1a/MLL1 complexes and facilitates the organization of a chromatin structure that activates the anterior Hoxb genes to control lineage-specific transcription. This work demonstrated that loss of HoxBlinc RNA leads to a strong decrease in *Hoxb1-b6* gene expression and a downregulation of key transcription factors of cardiac development such as Islet1, Nkx2-5, Mef2c, and Gata4 [23]. Therefore, this study reveals a novel cooperation between lncRNA and *Hox* genes during the specification and differentiation of a specific lineage.

3. *Hox* Genes and Patterning of the Second Heart Field

Elongation of the cardiac tube occurs by addition of SHF progenitor cells at the arterial and venous poles of the forming heart. Recent studies have clarified the origin of these cardiac progenitors and their regulation (see [3,5–7]). SHF cells are characterized by elevated proliferation and delayed differentiation, and by expression of the transcription factors Islet 1 [24], Nkx2-5 [25], and Tbx1 [26]. Since its discovery, there has been accumulating evidence showing that the SHF is pre-patterned [27,28]. Anterior and posterior SHF populations have been identified that contribute to the formation of OFT and right ventricular myocardium at the arterial pole and a large part of the atrial myocardium at the venous pole, respectively [3] (Figure 1A). However, the factors required for the A–P patterning of the SHF were, until recently, unknown. Analysis of the expression pattern of several *Hox* genes in the early embryo demonstrated that in the mesoderm of the SHF, the most anterior expression domains of *Hoxa1*, *Hoxb1*, and *Hoxa3* overlap with the posterior limit of *Islet1* and *Tbx5* [29]. Genetic lineage tracing confirmed that *Hox*-expressing SHF cells contribute to the heart (Figure 1B). Surprisingly, this study revealed that $Hoxb1^{Cre}$-labeled cells contribute to both poles of the heart tube: the inferior wall of the OFT, which gives rise to the sub-pulmonary myocardium at the arterial pole, and the atrio-ventricular canal and atrial myocardium at the venous pole [29]. Similarly, lineage tracing of *Hoxa1-enhIII-Cre-* and $Hoxa3^{Cre}$-labeled cells showed contributions of *Hoxa1*- and *Hoxa3*-expressing cells to both the atria and the distal part of the OFT [29,30]. Consistent with these observations, a recent retrospective clonal analysis confirmed that sub-pulmonary and atrial myocardial cells are clonally related [31].

The pre-patterning of the SHF into domains contributing to both the arterial and venous pole of the heart begs the question of how it is established. Studies of several animal models have shown that discrete levels of RA signaling are required for the A–P patterning of the heart tube (see [15]). In the mouse, loss of function of retinaldehyde dehydrogenase 2 (*Raldh2*), which encodes a critical enzyme for the synthesis of endogenous RA in the early embryo, results in an abnormal heart, with hypoplastic inflow tract region [32]. Further investigations of *Raldh2* mutant mouse embryos have shown that RA signaling is required to define the posterior boundary of the SHF [33,34]. Expression domains of a number of SHF genes, including *Islet1*, *Tbx1*, *Fgf8*, and *Fgf10*, are caudally expanded in $Raldh2^{-/-}$. Additionally, perturbation of RA signaling causes a failure in the deployment of SHF-derived cells, which contributes to the inferior wall of OFT myocardium [29]. In zebrafish, RA signaling has also been shown to regulate the size of the cardiac field through indirect regulation of *hoxb5b* expression in the adjacent forelimb field [35].

Many *Hox* loci contain RA-response elements that control aspects of their expression [36,37]. Consistently, several novel enhancers containing RA-responsive elements have been identified and characterized in the *HoxA* and *HoxB* gene clusters as well as the *Hoxa3* gene [38,39]. When mouse embryos are treated with a teratogenic dose of trans-RA, an anterior shift of the rostral border of *Hoxa1*, *Hoxb1*, and *Hoxa3* expression domains, including the SHF, is observed [29]. Together, these findings suggest that *Hox* gene expression is sensitive to RA dosage and that the influence of RA on heart development may be mediated through its effects on *Hox* genes.

Perturbation of the SHF during elongation of the forming heart tube results in a spectrum of conotruncal congenital defects, ranging from OFT misalignment, including double outlet right ventricle (DORV), overriding aorta, and tetralogy of Fallot (ToF), to ventricular septal defects (VSD) [40]. Several studies showed that human patients carrying a homozygous truncating mutation in *HOXA1* have OFT malformations [41,42]. Interestingly, similar defects have recently been reported for *Hoxa1*-deficient mice (Table 1), demonstrating that *Hoxa1* is required for patterning of the OFT of the heart [43]. More recently, we identified a previously unknown role for the *Hoxb1* gene in the forming heart [44]. Mouse embryos carrying a single *Hoxb1* null allele have VSD and OFT defects (Table 1). These malformations are the consequence of a failure in the deployment of SHF cells during the formation of the OFT [44].

Interestingly, it is the balance between proliferation and differentiation in the SHF that is disrupted in embryos lacking *Hoxb1*. *Hox* genes might also play a role in the differentiation of cardiac progenitor cells. Indeed, reduction of the length of the OFT and increased incidence of heart defects were observed in compound $Hoxa1^{-/-}$;$Hoxb1^{+/-}$ embryos compared to $Hoxa1^{-/-}$ mutants, demonstrating an overlap in function between these two genes during the formation of the OFT [44]. In 2015, Dupays *et al.* [45] reassessed the expression pattern of *Meis1* gene, which encodes a member of the TALE-class factors in the early embryo, and showed that it is strongly expressed in both the anterior region of the SHF and in the distal part of the OFT. Targeted mutations in the *Meis1* locus lead to early lethality and cardiac anomalies including OFT defects [46,47]. Altogether these data suggest that anterior Hox proteins, probably with cofactors, are important for the formation of the OFT through the development of the SHF.

Table 1. Cardiac phenotypes associated with *Hox*, *Pbx*, and *Meis* loss-of-function.

Gene	Mutants	Anomalies	References
Hoxa1	$Hoxa\text{-}1.6^{-/-}$	No cardiac phenotype VSD	[48]
	$Hoxa\text{-}1.6^{-/-}$;$Hoxb1^{GFP/+}$	OFT defects, VSD	[44]
	$Hoxa1^{GFPneo/GFPneo}$	No cardiac phenotype	[49]
	$Hoxa1^{-/-}$	IAA-B, ASC, RAA, VSD, ToF	[43]
Hoxa3	$Hox\text{-}1.5^{-/-}$	No cardiac phenotype	[50]
	$Hoxa3^{-/-}$	Degeneration of the 3rd arch artery Malformation of the carotid artery system	[51,52]
Hoxb1	$Hoxb1^{GFP/GFP}$	OFT defects, VSD	[44]
HoxA/HoxB	$Hoxa^{-/-}$;$Hoxb^{-/-}$	Heart looping defects	[53]
Pbx1	$Pbx1^{-/-}$	Die around E15; PTA and VSD	[47]
	$Pbx1^{+/-}$;$Pbx2^{+/-}$;$Pbx3^{+/-}$	Bicuspid aortic valve (BAV)	[47]
	$Pbx1^{+/-}$;$Pbx2^{-/-}$	Overriding aorta, VSD, BAV, bicuspid pulmonary valve	[47]
	$Pbx1^{+/-}$;$Pbx2^{-/-}$;$Pbx3^{+/-}$	ToF	[47]
Pbx2	$Pbx2^{-/-}$	No cardiac phenotype	[47]
Pbx3	$Pbx3^{-/-}$	No cardiac phenotype	[47]
Pbx4	$pbx4^{b557-/-}$ (zebrafish)	OFT defects	[54]
Meis1	$Meis1^{-/-}$	Overriding aorta, VSD	[47]
	$Meis1^{ECFP/ECFP}$	Die around E14; VSD	[46]
	α-MHC-Cre;$Meis1^{f/f}$	Increased postnatal cardiomyocyte proliferation	[55]
Meis2	*meis2b*-MO (zebrafish)	Heart looping defects	[56]
	$Meis2^{-/-}$	Lethality by E13.5-E15; PTA and VSD	[57]
	$AP2\alpha$-Cre;$Meis2^{f/f}$	DORV, abnormal semilunar valves	[57]

4. Role of *Hox* Genes in Pharyngeal Arch Development and Patterning

During development, PAAs form asymmetrically, following a rostro-caudal and temporal gradient [3]. Remodeling begins around E11.5 when the blood flow at the level of the left 6th arch increases and as the 1st and 2nd arch arteries have already regressed. This results from the rotation of the myocardium and the preferential development of the aorta on the left side to the detriment of the right side [58]. Segments of the dorsal aorta will then regress, allowing individualization of the future common carotid arteries, derived from the 3rd arches. At E14.5, the left 4th arch contributes to the segment of the aortic arch between the carotid and left subclavian arteries, whereas the right one forms a segment connecting the right subclavian arteries to the arterial brachiocephalic trunk. Finally, the right 6th arch is not maintained when the left contributes to the ductus arteriosus that will be shut at birth to allow the establishment of the pulmonary circulation. Anomalies in the patterning or the remodeling of the PAAs, especially of the 4th arch, are critical in the outcome of aortic arch defects such as interruption of the aortic arch (IAA) or subclavian anomalies. NCCs have been shown to be essential for this response.

The NCCs represent a population of multipotent cells with high migratory capacity that will differentiate into a wide range of cell types [59]. A subpopulation of cranial NCCs located in the neural plate between the otic placode and the third pair of somites has been renamed the "cardiac neural crest" for its important contribution to heart development [8,60]. However, a wider subpopulation participates in vascular remodeling between the heart and the face [9]. NCCs migrate through the pharyngeal (also called branchial) arches 3, 4, and 6 and invade the heart by the arterial pole. They contribute to the smooth muscles of the arteries, to the OFT endocardial cushions as well as to the aortico-pulmonary septum separating the aorta and the pulmonary trunk [8,59]. Genetic lineage tracing of NCCs using the $Wnt1^{Cre}$ mouse revealed their contribution to ascending aorta smooth muscle, to the aortic arch, proximal carotid, and coronary arteries [61]. The first cells invading the OFT participate to the formation of the aortico-pulmonary septum and those migrating later contribute to the remodeling of the PAAs, in particular to the smooth muscle wall of the arteries themselves [62].

Much of our early knowledge about the role of cardiac neural crest in heart development comes from chick ablation studies [8]. Mechanical ablation of premigratory NCCs leads to a number of cardiovascular and non-cardiovascular defects. In the neural crest-ablated embryos, the PAAs are abnormally patterned. When cardiac NCCs are ablated, the PAAs form normally but regress or persist in an unpredictable manner [63,64]. These results suggest that NCCs are not required for the initial formation of the endothelial channels that form the PAAs, but they are required for their remodeling and maintenance.

Cardiac NCCs express a specific set of *Hox* genes, depending on their rostro-caudal origin. Expression of the *Hox* genes in the posterior hindbrain undergoes a spatial and temporal collinear regulation by RA. Experiments in the chick embryo showed that *Hoxa1* and *Hoxb1* participate in neural crest specification and epithelial-to-mesenchymal transition by interacting with neural crest-inducing signaling pathways and regulating the expression of key genes involved in these processes, *Snail2* and *Msx1/2* [65]. They also contribute to the establishment and maintaining of *Hoxb1* expression in NCCs derived from rhombomere 4 [66].

The first model showing abnormal great arteries in absence of Hox function was a knockout mouse model for *Hoxa3* with common carotid artery patterning and remodeling defects [50,52]. NCCs invading the third pharyngeal arch normally express *Hoxa3*. Consistently, the use of a transgenic mouse line *Hoxa3-lacZ* revealed reporter expression in the cardiac NCCs [38]. In *Hoxa3* homozygous null mutant mice, the third arch artery degenerates bilaterally at E11.5, resulting in the malformation of the carotid artery system (Table 1) [51].

HOXA1 plays a direct role in human cardiovascular system development. Patients with Bosley–Salih–Alorainy syndrome (BSAS) have homozygous truncating mutations in the *HOXA1* locus and exhibit cardiac and cerebrovascular malformations [42,67]. The cardiovascular malformations included interrupted aortic arch type B (IAA-B), aberrant subclavian artery, VSD, and ToF. Genetic lineage tracing analysis in mice shows that *Hoxa1* is expressed in the NCCs that contribute to OFT cushions [29]. Two studies have also reported heart defects including VSD in *Hoxa1* mutant mice, which is coherent with an early role of *Hoxa1* in the NCCs lineage [43,44]. In 2012, Makki and Cappechi [43] demonstrated that *Hoxa1* is required for NCCs specification by acting upstream of the *Zic1* and *Foxd3* transcription factors, two major players in neural crest development.

5. Hox Gene Specificity: Cofactors in Heart Development

Hox functions are versatile and their activity seems to be context-dependent. Several examples show that the same Hox can act as an activator or a repressor depending on the circumstances [68]. The high degree of conservation of the homeodomain results in poor sequence specificity, which can be improved through binding with different cofactors. An example is provided by the Hox and TALE (three amino acid loop extension) families [69]. Some Hox interact with the Extradenticle/Pbx family of proteins, whereas others interact preferentially with the Homothorax/Meis family members. Hox proteins also present variable sequences outside the homeodomain, allowing them to interact with an important combination of cofactors. The fact that two factors use the same interaction site, such that only one of the factors can be bound at a time, suggests that there may exist several "Hox complexes," which vary in subunit composition [70], and most likely in function. Moreover, a recent study found that Hox proteins operate as tissue-specific factors to modulate the ground state binding of TALE cofactors to instruct anatomic identity [71].

In vertebrates, there are four *Pbx* (pre B cell leukemia homeobox) and three *Meis* (myeloid ecotropic viral integration site 1 homolog) genes. PBX proteins also bind to other factors such as HDACs (Histone deacetylase) and HATs (Histone acetyltransferase), suggesting a role in the regulation of the complex activity. A number of groups have reported on the formation of trimeric complexes involving proteins of the HOX, PBX, and MEIS families [72]. Such interaction is likely to modulate the transcriptional activity of the Hox complex [70]. HOX:PBX binds to HDACs, acting as a repressor of transcription that can be reversed by introducing the MEIS protein [73]. Thus, MEIS proteins compete with HDACs for binding to PBX and modulate accessibility of HDACs. Meis1 is able to interact with the co-activator of transcription CBP, which acts as an acetyltransferase, favoring an open position of the chromatin and therefore transcription. HDACs, which interact with PBX, on the contrary, act as a co-repressor.

Recently, Choe and colleagues [74] examined *hoxb1a* regulation in zebrafish embryos and showed that Pbx/Meis association occurs on its promoter but is not enough for transcriptional activation. Recruitment of hoxb1b releases "poised" RNA-PolII, allowing transcription to start. Thus, the authors suggest that TALE factors access promoters during early embryogenesis to poise them for activation but that Hox proteins are required to trigger efficient transcription.

Recent studies have involved Pbx and their Meis partners as critical regulators of cardiovascular development [54,55,57,69,75,76]. *Pbx1* homozygous mutants show great vessel anomalies as well as OFT septation defects [75]. This was confirmed by another study, which also showed a genetic interaction between *Pbx1, Pbx2,* and *Pbx3,* where combinatorial mutants for these genes had additional cardiac malformations [47]. Nevertheless, *Pbx1* displayed the most severe phenotype, with PTA (persistent truncus arteriosus) (Table 1). These studies showed that Pbx proteins act in NCCs to promote Pax3 expression, which is necessary for OFT development [75]. Recent study in zebrafish demonstrated that *pbx4* is required for establishing a proper OFT, probably through the formation of the SHF [54]. Interestingly, *Meis1* disruption results in cardiac anomalies that resemble *Pbx* mutations [47]. This result led the authors to hypothesize that Pbx1 interacts with Hox and/or Meis1 proteins to control a subset of target genes important for OFT development. Meis1 has also been described as a key factor for proliferation of postnatal cardiomyocytes [55]. It remains to be elucidated what genes are direct targets of Pbx/Meis factors.

As for the other Meis proteins, Machon *et al.* [57] showed that *Meis2* is expressed in the NCCs and that its absence results in abnormal OFT development. Heart defects have also been described in a conditional deletion using a NCC-specific Cre; however, no PTAs were observed, suggesting a role for non-crest Meis2-expressing cells in OFT formation (Table 1). Recently, a number of studies reported *MEIS2* mutations in patients with cardiac defects such as VSD [77,78]. In 2012, Paige and colleagues [56] have studied the temporal chromatin alterations in human ES cells as these cells differentiate into cardiomyocytes, and provided a model to determine key regulator genes of heart development. This work characterized those "key" genes as tightly regulated since the consequence of their inappropriate expression could lead to extremely severe defects caused by cell-fate change. *MEIS2* was among the genes identified. Using the zebrafish model, these authors demonstrated that *MEIS2* is critical for cardiac development. Indeed, embryos that received splice-blocking morpholino oligonucleotide directed against *meis2b* had defective cardiac morphogenesis [56].

Despite these studies, the function of TALE factors in cardiovascular development has not yet been fully addressed.

6. Conclusions

In this review, we have presented and discussed recent data demonstrating the role of Hox factors and TALE-class members in cardiovascular development. The question of functional redundancy among TALE factors has also been described for *Hoxa1* and *Hoxb1*, which synergize during SHF development [44]. Further investigation involving tissue-specific deletions will help to address this question. In addition, it is likely that Pbx/Meis proteins genetically interact with Hox genes during heart development given the similar phenotypes of $Hoxa1^{-/-}$, $Pbx1^{-/-}$, and $Meis1^{-/-}$ mutants. Thus, genetic approaches aimed at deleting one or more alleles of the *Hox*, *Pbx*, and *Meis* genes will provide valuable information regarding the intersection of these genes in OFT development. While recent studies have uncovered the role of Hox and TALE proteins in heart development, little is known about downstream target genes they activate or inhibit. Thus, it will be crucial to identify common target genes, thereby providing insight into potential mechanisms underlying CHDs.

Acknowledgments: We thank Heather Etchevers for critical reading of the manuscript. Grant support was provided by the Association Française contre les Myopathies (NMH-Decrypt Project), the Agence Nationale pour la Recherche (ANR-13-BSV2-0003-01), and the Institut National de la Santé et de la Recherche Médicale to Stéphane Zaffran.

Author Contributions: M.R. and S. Z. wrote the paper. S.Z. made the artwork.

Conflicts of Interest: The authors declare no conflict of interest.

References

1. Poelmann, R.E.; Gittenberger-de Groot, A.C.; Hierck, B.P. The development of the heart and microcirculation: Role of shear stress. *Med. Biol. Eng. Comput.* **2008**, *46*, 479–484. [CrossRef] [PubMed]
2. Hoffman, J.I.; Kaplan, S. The incidence of congenital heart disease. *J. Am. Coll. Cardiol.* **2002**, *39*, 1890–1900. [CrossRef]
3. Vincent, S.D.; Buckingham, M.E. How to make a heart: The origin and regulation of cardiac progenitor cells. *Curr. Top. Dev. Biol.* **2010**, *90*, 1–41. [PubMed]
4. Srivastava, D. Making or breaking the heart: From lineage determination to morphogenesis. *Cell* **2006**, *126*, 1037–1048. [CrossRef] [PubMed]
5. Zaffran, S.; Kelly, R.G. New developments in the second heart field. *Differentiation* **2012**, *84*, 17–24. [CrossRef] [PubMed]
6. Buckingham, M.; Meilhac, S.; Zaffran, S. Building the mammalian heart from two sources of myocardial cells. *Nat. Rev. Genet.* **2005**, *6*, 826–835. [CrossRef] [PubMed]
7. Rochais, F.; Mesbah, K.; Kelly, R.G. Signaling pathways controlling second heart field development. *Circ. Res.* **2009**, *104*, 933–942. [CrossRef] [PubMed]
8. Kirby, M.L.; Gale, T.F.; Stewart, D.E. Neural crest cells contribute to normal aorticopulmonary septation. *Science* **1983**, *220*, 1059–1061. [CrossRef] [PubMed]
9. Etchevers, H.C.; Vincent, C.; le Douarin, N.M.; Couly, G.F. The cephalic neural crest provides pericytes and smooth muscle cells to all blood vessels of the face and forebrain. *Development* **2001**, *128*, 1059–1068. [PubMed]
10. Deschamps, J.; van Nes, J. Developmental regulation of the hox genes during axial morphogenesis in the mouse. *Development* **2005**, *132*, 2931–2942. [CrossRef] [PubMed]
11. Duboule, D.; Dolle, P. The structural and functional organization of the murine hox gene family resembles that of drosophila homeotic genes. *EMBO J.* **1989**, *8*, 1497–1505. [PubMed]
12. Alexander, T.; Nolte, C.; Krumlauf, R. Hox genes and segmentation of the hindbrain and axial skeleton. *Annu. Rev. Cell Dev. Biol.* **2009**, *25*, 431–456. [CrossRef] [PubMed]
13. Lo, P.C.; Frasch, M. Establishing A-P polarity in the embryonic heart tube. A conserved function of Hox genes in *Drosophila* and vertebrates? *Trends Cardiovasc. Med.* **2003**, *13*, 182–187. [CrossRef]
14. Monier, B.; Tevy, M.F.; Perrin, L.; Capovilla, M.; Semeriva, M. Downstream of homeotic genes: In the heart of hox function. *Fly* **2007**, *1*, 59–67. [CrossRef] [PubMed]
15. Zaffran, S.; Niederreither, K. Retinoic acid signaling and heart development. In *The Retinoids*; Dolle, P., Niederreither, K., Eds.; Wiley-Blackwell: Hoboken, NJ, USA, 2015.

16. Searcy, R.D.; Yutzey, K.E. Analysis of Hox gene expression during early avian heart development. *Dev. Dyn* **1998**, *213*, 82–91. [CrossRef]
17. Waxman, J.S.; Yelon, D. Increased Hox activity mimics the teratogenic effects of excess retinoic acid signaling. *Dev. Dyn.* **2009**, *238*, 1207–1213. [CrossRef] [PubMed]
18. Van Vliet, P.; Wu, S.M.; Zaffran, S.; Puceat, M. Early cardiac development: A view from stem cells to embryos. *Cardiovasc. Res.* **2012**, *96*, 352–362. [CrossRef] [PubMed]
19. Zaffran, S.; Frasch, M. Early signals in cardiac development. *Circ. Res.* **2002**, *91*, 457–469. [CrossRef] [PubMed]
20. Lengerke, C.; Schmitt, S.; Bowman, T.V.; Jang, I.H.; Maouche-Chretien, L.; McKinney-Freeman, S.; Davidson, A.J.; Hammerschmidt, M.; Rentzsch, F.; Green, J.B.; *et al.* Bmp and Wnt specify hematopoietic fate by activation of the Cdx-Hox pathway. *Cell Stem Cell* **2008**, *2*, 72–82. [CrossRef] [PubMed]
21. Yue, Q.; Wagstaff, L.; Yang, X.; Weijer, C.; Munsterberg, A. Wnt3a-mediated chemorepulsion controls movement patterns of cardiac progenitors and requires RhoA function. *Development* **2008**, *135*, 1029–1037. [CrossRef] [PubMed]
22. Soshnikova, N. Hox genes regulation in vertebrates. *Dev. Dyn.* **2014**, *243*, 49–58. [CrossRef] [PubMed]
23. Deng, C.; Li, Y.; Zhou, L.; Cho, J.; Patel, B.; Terada, N.; Li, Y.; Bungert, J.; Qiu, Y.; Huang, S. HoxBlinc RNA recruits Set1/MLL complexes to activate Hox gene expression patterns and mesoderm lineage development. *Cell Rep.* **2016**, *14*, 103–114. [CrossRef] [PubMed]
24. Cai, C.L.; Liang, X.; Shi, Y.; Chu, P.H.; Pfaff, S.L.; Chen, J.; Evans, S. Isl1 identifies a cardiac progenitor population that proliferates prior to differentiation and contributes a majority of cells to the heart. *Dev. Cell* **2003**, *5*, 877–889. [CrossRef]
25. Prall, O.W.; Menon, M.K.; Solloway, M.J.; Watanabe, Y.; Zaffran, S.; Bajolle, F.; Biben, C.; McBride, J.J.; Robertson, B.R.; Chaulet, H.; *et al.* An Nkx2–5/Bmp2/Smad1 negative feedback loop controls heart progenitor specification and proliferation. *Cell* **2007**, *128*, 947–959. [CrossRef] [PubMed]
26. Baldini, A. Dissecting contiguous gene defects: Tbx1. *Curr. Opin. Genet. Dev.* **2005**, *15*, 279–284. [CrossRef] [PubMed]
27. Galli, D.; Dominguez, J.N.; Zaffran, S.; Munk, A.; Brown, N.A.; Buckingham, M.E. Atrial myocardium derives from the posterior region of the second heart field, which acquires left-right identity as Pitx2c is expressed. *Development* **2008**, *135*, 1157–1167. [CrossRef] [PubMed]
28. Zaffran, S.; Kelly, R.G.; Meilhac, S.M.; Buckingham, M.E.; Brown, N.A. Right ventricular myocardium derives from the anterior heart field. *Circ. Res.* **2004**, *95*, 261–268. [CrossRef] [PubMed]
29. Bertrand, N.; Roux, M.; Ryckebusch, L.; Niederreither, K.; Dolle, P.; Moon, A.; Capecchi, M.; Zaffran, S. Hox genes define distinct progenitor sub-domains within the second heart field. *Dev. Biol.* **2011**, *353*, 266–274. [CrossRef] [PubMed]
30. Makki, N.; Capecchi, M.R. Hoxa1 lineage tracing indicates a direct role for Hoxa1 in the development of the inner ear, the heart, and the third rhombomere. *Dev. Biol.* **2010**, *341*, 499–509. [CrossRef] [PubMed]
31. Lescroart, F.; Mohun, T.; Meilhac, S.M.; Bennett, M.; Buckingham, M. Lineage tree for the venous pole of the heart: Clonal analysis clarifies controversial genealogy based on genetic tracing. *Circ. Res.* **2012**, *111*, 1313–1322. [CrossRef] [PubMed]
32. Niederreither, K.; Subbarayan, V.; Dolle, P.; Chambon, P. Embryonic retinoic acid synthesis is essential for early mouse post-implantation development. *Nat. Genet.* **1999**, *21*, 444–448. [CrossRef] [PubMed]
33. Ryckebusch, L.; Wang, Z.; Bertrand, N.; Lin, S.C.; Chi, X.; Schwartz, R.; Zaffran, S.; Niederreither, K. Retinoic acid deficiency alters second heart field formation. *Proc. Natl. Acad. Sci. USA* **2008**, *105*, 2913–2918. [CrossRef] [PubMed]
34. Sirbu, I.O.; Zhao, X.; Duester, G. Retinoic acid controls heart anteroposterior patterning by down-regulating Isl1 through the Fgf8 pathway. *Dev. Dyn.* **2008**, *237*, 1627–1635. [CrossRef] [PubMed]
35. Waxman, J.S.; Keegan, B.R.; Roberts, R.W.; Poss, K.D.; Yelon, D. Hoxb5b acts downstream of retinoic acid signaling in the forelimb field to restrict heart field potential in zebrafish. *Dev. Cell* **2008**, *15*, 923–934. [CrossRef] [PubMed]
36. Duester, G. Retinoic acid synthesis and signaling during early organogenesis. *Cell* **2008**, *134*, 921–931. [CrossRef] [PubMed]

37. Niederreither, K.; Dolle, P. Retinoids and heart development. In *Heart Development*; Rosenthal, N., Harvey, R.P., Eds.; Academic Press: San Diego, CA, USA, 2008.
38. Diman, N.Y.; Remacle, S.; Bertrand, N.; Picard, J.J.; Zaffran, S.; Rezsohazy, R. A retinoic acid responsive Hoxa3 transgene expressed in embryonic pharyngeal endoderm, cardiac neural crest and a subdomain of the second heart field. *PLoS One* **2011**, *6*, e27624. [CrossRef] [PubMed]
39. Nolte, C.; Jinks, T.; Wang, X.; Martinez Pastor, M.T.; Krumlauf, R. Shadow enhancers flanking the HoxB cluster direct dynamic hox expression in early heart and endoderm development. *Dev. Biol.* **2013**, *383*, 158–173. [CrossRef] [PubMed]
40. Ward, C.; Stadt, H.; Hutson, M.; Kirby, M.L. Ablation of the secondary heart field leads to tetralogy of fallot and pulmonary atresia. *Dev. Biol.* **2005**, *284*, 72–83. [CrossRef] [PubMed]
41. Holve, S.; Friedman, B.; Hoyme, H.E.; Tarby, T.J.; Johnstone, S.J.; Erickson, R.P.; Clericuzio, C.L.; Cunniff, C. Athabascan brainstem dysgenesis syndrome. *Am. J. Med. Genet. Part A* **2003**, *120A*, 169–173. [CrossRef] [PubMed]
42. Tischfield, M.A.; Bosley, T.M.; Salih, M.A.; Alorainy, I.A.; Sener, E.C.; Nester, M.J.; Oystreck, D.T.; Chan, W.M.; Andrews, C.; Erickson, R.P.; *et al.* Homozygous hoxa1 mutations disrupt human brainstem, inner ear, cardiovascular and cognitive development. *Nat. Genet.* **2005**, *37*, 1035–1037. [CrossRef] [PubMed]
43. Makki, N.; Capecchi, M.R. Cardiovascular defects in a mouse model of hoxa1 syndrome. *Hum. Mol. Genet.* **2012**, *21*, 26–31. [CrossRef] [PubMed]
44. Roux, M.; Laforest, B.; Capecchi, M.; Bertrand, N.; Zaffran, S. Hoxb1 regulates proliferation and differentiation of second heart field progenitors in pharyngeal mesoderm and genetically interacts with Hoxa1 during cardiac outflow tract development. *Dev. Biol.* **2015**, *406*, 247–258. [CrossRef] [PubMed]
45. Dupays, L.; Shang, C.; Wilson, R.; Kotecha, S.; Wood, S.; Towers, N.; Mohun, T. Sequential binding of MEIS1 and NKX2–5 on the Popdc2 gene: A mechanism for spatiotemporal regulation of enhancers during cardiogenesis. *Cell Rep.* **2015**, *13*, 183–195. [CrossRef] [PubMed]
46. Gonzalez-Lazaro, M.; Rosello-Diez, A.; Delgado, I.; Carramolino, L.; Sanguino, M.A.; Giovinazzo, G.; Torres, M. Two new targeted alleles for the comprehensive analysis of Meis1 functions in the mouse. *Genesis* **2014**, *52*, 967–975. [CrossRef] [PubMed]
47. Stankunas, K.; Shang, C.; Twu, K.Y.; Kao, S.C.; Jenkins, N.A.; Copeland, N.G.; Sanyal, M.; Selleri, L.; Cleary, M.L.; Chang, C.P. Pbx/Meis deficiencies demonstrate multigenetic origins of congenital heart disease. *Circ. Res.* **2008**, *103*, 702–709. [CrossRef] [PubMed]
48. Lufkin, T.; Dierich, A.; LeMeur, M.; Mark, M.; Chambon, P. Disruption of the Hox-1.6 homeobox gene results in defects in a region corresponding to its rostral domain of expression. *Cell* **1991**, *66*, 1105–1119. [CrossRef]
49. Godwin, A.R.; Stadler, H.S.; Nakamura, K.; Capecchi, M.R. Detection of targeted GFP-Hox gene fusions during mouse embryogenesis. *Proc. Natl. Acad. Sci. USA* **1998**, *95*, 13042–13047. [CrossRef] [PubMed]
50. Chisaka, O.; Capecchi, M.R. Regionally restricted developmental defects resulting from targeted disruption of the mouse homeobox gene Hox-1.5. *Nature* **1991**, *350*, 473–479. [CrossRef] [PubMed]
51. Chisaka, O.; Kameda, Y. Hoxa3 regulates the proliferation and differentiation of the third pharyngeal arch mesenchyme in mice. *Cell Tissue Res.* **2005**, *320*, 77–89. [CrossRef] [PubMed]
52. Kameda, Y.; Watari-Goshima, N.; Nishimaki, T.; Chisaka, O. Disruption of the hoxa3 homeobox gene results in anomalies of the carotid artery system and the arterial baroreceptors. *Cell Tissue Res.* **2003**, *311*, 343–352. [PubMed]
53. Soshnikova, N.; Dewaele, R.; Janvier, P.; Krumlauf, R.; Duboule, D. Duplications of Hox gene clusters and the emergence of vertebrates. *Dev. Biol.* **2013**, *378*, 194–199. [CrossRef] [PubMed]
54. Kao, R.M.; Rurik, J.G.; Farr, G.H., 3rd; Dong, X.R.; Majesky, M.W.; Maves, L. Pbx4 is required for the temporal onset of zebrafish myocardial differentiation. *J. Dev. Biol.* **2015**, *3*, 93–111. [CrossRef] [PubMed]
55. Mahmoud, A.I.; Kocabas, F.; Muralidhar, S.A.; Kimura, W.; Koura, A.S.; Thet, S.; Porrello, E.R.; Sadek, H.A. Meis1 regulates postnatal cardiomyocyte cell cycle arrest. *Nature* **2013**, *497*, 249–253. [CrossRef] [PubMed]
56. Paige, S.L.; Thomas, S.; Stoick-Cooper, C.L.; Wang, H.; Maves, L.; Sandstrom, R.; Pabon, L.; Reinecke, H.; Pratt, G.; Keller, G.; *et al.* A temporal chromatin signature in human embryonic stem cells identifies regulators of cardiac development. *Cell* **2012**, *151*, 221–232. [CrossRef] [PubMed]

57. Machon, O.; Masek, J.; Machonova, O.; Krauss, S.; Kozmik, Z. Meis2 is essential for cranial and cardiac neural crest development. *BMC Dev. Biol.* **2015**, *15*, 40. [CrossRef] [PubMed]
58. Yashiro, K.; Shiratori, H.; Hamada, H. Haemodynamics determined by a genetic programme govern asymmetric development of the aortic arch. *Nature* **2007**, *450*, 285–288. [CrossRef] [PubMed]
59. Le Lievre, C.S.; le Douarin, N.M. Mesenchymal derivatives of the neural crest: Analysis of chimaeric quail and chick embryos. *J. Embryol Exp. Morphol.* **1975**, *34*, 125–154. [PubMed]
60. Kirby, M.L.; Waldo, K.L. Neural crest and cardiovascular patterning. *Circ. Res.* **1995**, *77*, 211–215. [CrossRef] [PubMed]
61. Jiang, X.; Choudhary, B.; Merki, E.; Chien, K.R.; Maxson, R.E.; Sucov, H.M. Normal fate and altered function of the cardiac neural crest cell lineage in retinoic acid receptor mutant embryos. *Mech. Dev.* **2002**, *117*, 115–122. [CrossRef]
62. Boot, M.J.; Gittenberger-De Groot, A.C.; van Iperen, L.; Hierck, B.P.; Poelmann, R.E. Spatiotemporally separated cardiac neural crest subpopulations that target the outflow tract septum and pharyngeal arch arteries. *Anat. Rec. Part A Discov. Mol. Cell Evol. Biol.* **2003**, *275*, 1009–1018. [CrossRef] [PubMed]
63. Waldo, K.L.; Kumiski, D.; Kirby, M.L. Cardiac neural crest is essential for the persistence rather than the formation of an arch artery. *Dev. Dyn* **1996**, *205*, 281–292. [CrossRef]
64. Hutson, M.R.; Kirby, M.L. Model systems for the study of heart development and disease cardiac neural crest and conotruncal malformations. *Semin. Cell Dev. Biol.* **2007**, *18*, 101–110. [CrossRef] [PubMed]
65. Gouti, M.; Briscoe, J.; Gavalas, A. Anterior Hox genes interact with components of the neural crest specification network to induce neural crest fates. *Stem Cells* **2011**, *29*, 858–870. [CrossRef] [PubMed]
66. Studer, M.; Gavalas, A.; Marshall, H.; Ariza-McNaughton, L.; Rijli, F.M.; Chambon, P.; Krumlauf, R. Genetic interactions between Hoxa1 and Hoxb1 reveal new roles in regulation of early hindbrain patterning. *Development* **1998**, *125*, 1025–1036. [PubMed]
67. Bosley, T.M.; Alorainy, I.A.; Salih, M.A.; Aldhalaan, H.M.; Abu-Amero, K.K.; Oystreck, D.T.; Tischfield, M.A.; Engle, E.C.; Erickson, R.P. The clinical spectrum of homozygous hoxa1 mutations. *Am. J. Med. Genet. Part A* **2008**, *146A*, 1235–1240. [CrossRef] [PubMed]
68. Krasnow, M.A.; Saffman, E.E.; Kornfeld, K.; Hogness, D.S. Transcriptional activation and repression by ultrabithorax proteins in cultured drosophila cells. *Cell* **1989**, *57*, 1031–1043. [CrossRef]
69. Moens, C.B.; Selleri, L. Hox cofactors in vertebrate development. *Dev. Biol.* **2006**, *291*, 193–206. [CrossRef] [PubMed]
70. Ladam, F.; Sagerstrom, C.G. Hox regulation of transcription: More complex(es). *Dev. Dyn.* **2014**, *243*, 4–15. [CrossRef] [PubMed]
71. Amin, S.; Donaldson, I.J.; Zannino, D.A.; Hensman, J.; Rattray, M.; Losa, M.; Spitz, F.; Ladam, F.; Sagerstrom, C.; Bobola, N. Hoxa2 selectively enhances meis binding to change a branchial arch ground state. *Dev. Cell* **2015**, *32*, 265–277. [CrossRef] [PubMed]
72. Ferretti, E.; Marshall, H.; Popperl, H.; Maconochie, M.; Krumlauf, R.; Blasi, F. Segmental expression of Hoxb2 in r4 requires two separate sites that integrate cooperative interactions between Prep1, Pbx and Hox proteins. *Development* **2000**, *127*, 155–166. [PubMed]
73. Choe, S.K.; Lu, P.; Nakamura, M.; Lee, J.; Sagerstrom, C.G. Meis cofactors control HDAC and CBP accessibility at Hox-regulated promoters during zebrafish embryogenesis. *Dev. Cell* **2009**, *17*, 561–567. [CrossRef] [PubMed]
74. Choe, S.K.; Ladam, F.; Sagerstrom, C.G. Tale factors poise promoters for activation by Hox proteins. *Dev. Cell* **2014**, *28*, 203–211. [CrossRef] [PubMed]
75. Chang, C.P.; Stankunas, K.; Shang, C.; Kao, S.C.; Twu, K.Y.; Cleary, M.L. Pbx1 functions in distinct regulatory networks to pattern the great arteries and cardiac outflow tract. *Development* **2008**, *135*, 3577–3586. [CrossRef] [PubMed]
76. Maves, L.; Tyler, A.; Moens, C.B.; Tapscott, S.J. Pbx acts with Hand2 in early myocardial differentiation. *Dev. Biol.* **2009**, *333*, 409–418. [CrossRef] [PubMed]

77. Crowley, M.A.; Conlin, L.K.; Zackai, E.H.; Deardorff, M.A.; Thiel, B.D.; Spinner, N.B. Further evidence for the possible role of MEIS2 in the development of cleft palate and cardiac septum. *Am. J. Med. Genet. Part A* **2010**, *152A*, 1326–1327. [CrossRef] [PubMed]
78. Louw, J.J.; Corveleyn, A.; Jia, Y.; Hens, G.; Gewillig, M.; Devriendt, K. MEIS2 involvement in cardiac development, cleft palate, and intellectual disability. *Am. J. Med. Genet. Part A* **2015**, *167A*, 1142–1146. [CrossRef] [PubMed]

11

HoxA Genes and the Fin-to-Limb Transition in Vertebrates

João Leite-Castro [1], Vanessa Beviano [1,2], Pedro Nuno Rodrigues [1,2,3] and Renata Freitas [1,2,*]

[1] IBMC—Instituto de Biologia Celular e Molecular, Oporto, Portugal; newtonjplc@hotmail.com (J.L.-C.); vanessa.beviano@hotmail.com (V.B.); PRodrigu@ibmc.up.pt (P.N.R.)
[2] I3S—Instituto de Investigação e Inovação em Saúde, Rua Alfredo Allen, 208, 4200-135 Porto, Portugal
[3] ICBAS—Instituto de Ciências Biomédicas Abel Salazar, University of Oporto, Rua de Jorge Viterbo Ferreira n°. 228, 4050-313 Oporto, Portugal
* Correspondence: renata.freitas@ibmc.up.pt

Academic Editors: Vincenzo Zappavigna and Andy Wessels

Abstract: *HoxA* genes encode for important DNA-binding transcription factors that act during limb development, regulating primarily gene expression and, consequently, morphogenesis and skeletal differentiation. Within these genes, *HoxA11* and *HoxA13* were proposed to have played an essential role in the enigmatic evolutionary transition from fish fins to tetrapod limbs. Indeed, comparative gene expression analyses led to the suggestion that changes in their regulation might have been essential for the diversification of vertebrates' appendages. In this review, we highlight three potential modifications in the regulation and function of these genes that may have boosted appendage evolution: (1) the expansion of polyalanine repeats in the HoxA11 and HoxA13 proteins; (2) the origin of +a novel long-non-coding RNA with a possible inhibitory function on *HoxA11*; and (3) the acquisition of *cis*-regulatory elements modulating 5′ *HoxA* transcription. We discuss the relevance of these mechanisms for appendage diversification reviewing the current state of the art and performing additional comparative analyses to characterize, in a phylogenetic framework, *HoxA11* and *HoxA13* expression, alanine composition within the encoded proteins, long-non-coding RNAs and *cis*-regulatory elements.

Keywords: *HoxA* genes; transcriptional regulation; development; evolution; limb; fin

---

1. Introduction

The autopod, the multi-fingered extremity located at the end of a hindlimb or forelimb, is a specific characteristic of tetrapods [1–3]. The fossil record suggests that this anatomical structure evolved from sarcopterygian fins by sequential expansion and elaboration of their endochondral bones and concomitant reduction of the apical dermoskeleton [1–3]. Regarding the number of digits, both paleontological and developmental data suggest an evolutionary sequence that went from an adactyl state detectable up to the divergence of sarcopterygians [4], then sidestepping by digit-like radials in stem sarcopterygians [5], polydactyly in stem tetrapod and culminating in the fixation of pentadactyl [6–8]. From a developmental point of view, understanding the molecular mechanisms that enabled the fin-to-limb transition is a puzzling question that has long held the attention of the scientific community and holds the potential of enlightening molecular biologists on how gene regulatory networks and interactions evolved.

The patterning of the skeletal elements, in both fin and limb, largely relies on the action of signaling centers operating from the onset of their outgrowth: the apical ectodermal ridge (AER) and the zone of polarizing activity (ZPA) [9]. However, there are marked differences in the morphology and development of the most distal part of fins and limbs. In tetrapods, the influence of the AER is maintained throughout limb development, being important for the differentiation of the stylopod, zeugopod and autopod [10–12]. In fish, however, this structure is quickly converted into a finfold (FF) during fin development that then gives rise to the apical dermoskeleton [13]. However, its function seems to be conserved during fish fin development and relates to the coordination of the proximo-distal patterning [14]. Recent studies performed in zebrafish suggest that the timing of the AER-FF transition may have mediated the differences found between fins and limbs and that the acquisition of a mechanism repressing the formation of the FF may have been crucial in the development of tetrapod limbs [15].

The patterning of the endoskeleton structure has been proven to rely on the expression of 5′ *HoxA* and *HoxD* genes in both fins and limbs, as demonstrated by different studies performed on a wide variety of model systems [16–20]. Indeed, several lines of evidence highlight the role of these transcriptional factors as key intervenients in limb development. For example, loss-of-function mutations in genes located at the 5′ end of the mouse HoxA and HoxD clusters lead to the appearance of limbs with atavistic characters [8,18,20]. The impact that these genes have on the patterning of the limb, proven by loss-of-function assays, led Duboule and colleagues to suggest that modifications on Hox regulation may have been a source of morphological variation during the evolution of tetrapod limbs [21]. Moreover, they suggest that increased levels of these transcription factors during fin development may have caused extra-proliferation distally, setting the grounds for the formation of additional endoskeleton elements, which occurred concomitantly with the reduction of the apical dermoskeleton. To make the proof of principle of these ideas, Freitas and colleagues overexpressed the zebrafish counterpart of *HoxD13* (*hoxd13a*), generating a fin phenotype characterized by additional endochondral tissue and reduction or ablation of the apical finfold [22]. Thus, *hoxd13a* overexpression induces a phenotype that mirrors two of the major morphological alterations marked as crucial in limb evolution by the fossil record: endochondral expansion and dermoskeleton retraction [1,22].

Taking into account the results obtained by Zákány *et al.* [8] and Freitas *et al.* [22], it seems that both HoxA and HoxD clusters contributed to the fin-to-limb transition, although at distinct time points and by regulating different cellular events. Zákány and colleagues analyzed the phenotypes of several compound mouse mutants with distal *HoxA* and *HoxD* loss-of-function alleles and proposed that the *HoxA13* gene was the initial/primordial major contributor for the determination of the autopod identity [8]. This was probably followed, in the course of evolution, by the recruitment of distal *HoxD* genes (*HoxD11*, *HoxD12* and *HoxD13*), whose expression is associated with finger elongation and fixation of [8,23]. Further evidences that supports this sequence of events comes from the observation that *HoxA13* is expressed earlier than *HoxD13* in the distal mesenchyme during limb development [23], being probably an earlier specifier of the autopod identity. Moreover, the expression of the *Hox13* paralogs in zebrafish is detected in the most distal mesenchyme of the developing fins, closely mirroring the expression patterns described in tetrapods [17]. This suggests that *HoxA13* was already essential to induce a distal appendage identity even prior to the origin of the autopod in the tetrapod lineage. Taken together, these data suggest that alterations in the expression and function of HoxA13 may have been the primary mechanisms leading to the formation of the autopod in the tetrapod lineage [24,25].

Taking into account the conserved *hoxa13* expression patterns found distally in developing fish fins, why do they not give rise to autopod-like structures? Information gathered from developmental, phylogenetic and gene function studies suggests that, in the course of evolution, two factors had to meet to allow the formation of an autopod domain, those being the alteration of the proximo-distal patterning leading to the formation of novel distal territories, characterized by a specific transcriptome, that co-occurred with, or was followed by, an increase in the levels of *HoxA13* transcripts [8,26].

Regarding the proximo-distal patterning, striking differences have been found in the expression of 5′ *HoxA* genes in fish and tetrapod models that may underlie an increased endoskeletal complexity throughout fin/limb evolution (Figure 1). This process counts with the combined action of three genes encoding master regulators of transcription: *Meis1*, which seems to be essential for stylopod specification; *HoxA11* required for proper zeugopod development; and *HoxA13*, which contributes to an autopod identity [27–31]. Interestingly, during tetrapod limb development, the initially overlapping expression domains of *HoxA11* and *HoxA13* split up to a point in which *HoxA11* is exclusively expressed in the prospective zeugopod, while *HoxA13* is restricted to the future autopod (Figure 1) [16,32,33]. The expression patterns of these genes during fish fin development are, however, quite different. In zebrafish, a full separation of the *hoxa11* and *hoxa13* expression domains has never been reported in the developing fins [17,21,34]. In other fish representing distinct phylogenetic positions, such as chondrichthyans (catshark; [35]) and basal actinopterygians (paddlefish; [26,36]), there is only a transient separation of *hoxa11* and *hoxa13* expression domains during early stages of fin development. These data suggest that the decoupling of *HoxA13/HoxA11* expression domains may have occurred in the common ancestor of gnathostomes and was secondarily lost in a particular lineage of actinopterygians, the teleosts.

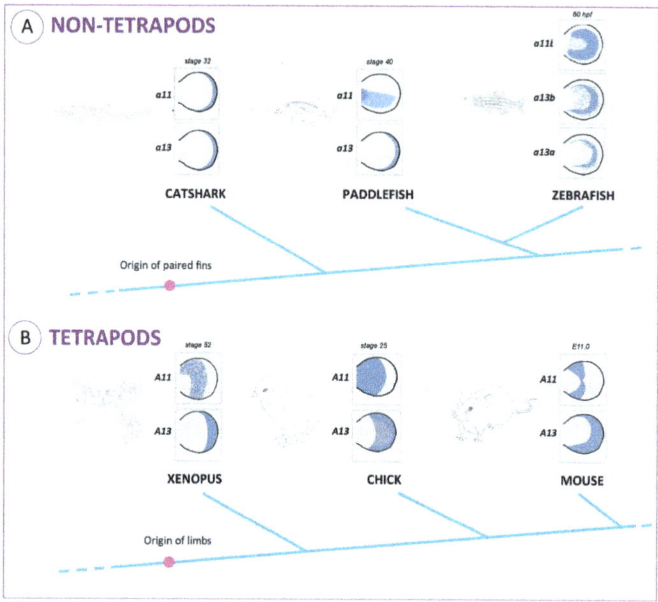

Figure 1. Expression patterns of 5′ *HoxA* genes during the development of paired appendages in vertebrates. The selection of stages took into account data from: *Xenopus* [16]; chick [23]; mouse [31]; catshark [35]; paddlefish [36]; zebrafish [17]. (**A**) Expression domains of *HoxA11* and *HoxA13* in developing fins from fish representing phylogenetic groups that diverged prior to the origin of tetrapods; (**B**) Expression domains of *HoxA11* and *HoxA13* in developing limbs in representatives of tetrapod lineages (in blue).

The origin of a distal domain exclusively expressing *HoxA13* may have influenced the formation of novel endoskeleton structures [26,37,38]. Further support of this hypothesis is provided by regeneration studies in *Xenopus*, where non-separation of *HoxA11* and *HoxA13* expression patterns, making distal blastemal cells *HoxA11* and *HoxA13* double positive, appeared to be associated with the inability to regenerate an autopod [16]. Interestingly, *HoxA11* and *HoxA13* mRNA are co-expressed in the developing limbs of axolotls, but the translated HoxA11 and HoxA13 proteins are distributed in distinct domains consistent with the formation of a zeugopod-autopod border [39]. Questions that remain to be answered are, for example: How was the separation of HoxA11 and HoxA13 domains achieved, and what were the implications for the evolution of the limb morphology? What were the molecular mechanisms associated with this alterations? Or, how did alterations in HoxA11 and HoxA13 protein levels contribute to the autopod origin?

From a gene regulatory perspective, there are several mechanisms that could have contributed to the establishment of a *HoxA13* single positive distal appendage domain that allowed the formation and posterior evolution of the autopod. These may include: (1) alteration in the signaling pathways that act upstream of *HoxA13*, regulating its expression and/or the expression of *HoxA11*; (2) remodeling/alteration or elaboration of the *cis*-regulatory motifs that assume the spatial-temporal control of the expression for these genes; (3) acquisition/loss of elements of the post-transcriptional gene regulatory network; (4) alteration of the protein-protein or protein-DNA interactions through alteration of the coding sequence.

By analyzing the reports published to date, there are three mechanisms that may uphold an evolutionary relevance in the context of the fin-to-limb transition, those being the acquisition/expansion of *HoxA11* and *HoxA13* homopolymeric repeats, acquisition of a new long-non-coding RNA with possible *HoxA11* inhibitory function and the origin of novel *cis*-regulatory regions acting on *HoxA11* and *HoxA13* [40–43]. Here, we will explore the evolutionary/developmental relevance of these three mechanisms, reviewing the current state of the art concerning *HoxA11* and *HoxA13* regulation throughout development in species at different phylogenetic positions and discussing how these developmental regulatory networks may have evolved, being at the core of the morphological diversification found in vertebrate appendages.

2. Polyalanine Repeats and the Fin-to-Limb Transition

Amino acid repeats are common among eukaryote proteins, and their length and frequency cannot be explained solely by random chance, suggesting a biological function and positive selection [44,45]. Interestingly, prokaryote proteins are impoverished in amino acid repeats, when compared to their eukaryotic homologues, indicating that evolution of such amino acid repeats happened after establishment of the initial protein structures and possibly concomitantly with the appearance of novel functionalities in eukaryotes [44]. Indeed, numerous proteins with a variety of different biological functions display amino acid repeats, such as protein kinases, signaling proteins, membrane transporters and transcription factors [44]. Strikingly, the majority of amino acid repeats detected in eukaryotes are present in proteins involved in the formation of large protein-protein or protein-nucleic acid complexes, such as transcriptional factors, possibly influencing their performance [44]. Thus, changes in these particular sequences during evolution may have produced a significant impact, leading to the genesis of novel morphological structures, for example.

Interestingly, alterations in the length of amino acid tandem repeats have long been suggested as a major source of phenotypic variation in evolution [46]. This hypothesis was constructed based on the observation that: (1) amino acid tandem repeats are abundant and frequently conserved in the coding sequence of genes involved in development; (2) repeat length variation, both expansion and contraction, occurs in a locus-specific manner 100,000-times more frequently than point mutations; and (3) transcription, mRNA processing, protein synthesis and morphology are affected by length variation in the amino acid repeats [46]. There are, indeed, some examples that offer support to this hypothesis. For example, within dog breeds, there is a clear association between the contraction of glutamine repeats in the *Alx-4* gene and the development of bilateral polydactyly that characterizes the Great Pyrenees breed [46]. Moreover, a correlation was found between the length of the polyglutamine and polyalanine tracts in the *Runx-2* gene and the facial morphology of different dog breeds [46], showing the potential that amino acid repeat variability may assume in the diversification of phenotypes.

Polyalanine repeats are types of amino acid repeats characterized by tandem repeated alanine residues [44]. These regions can modulate protein function through various mechanisms, such as enabling or increasing a transcriptional factor capacity to act as a repressor, modifying the interactome of a protein or acting as spacers, merely providing a structural component that enhances the performance of its functional domains [40,47–49]. Interestingly, alterations in the alanine tracts of particular Hox proteins have been proven to impact limb development, to a greater or lesser extent [50,51]. For instance, deletion of one of the HoxD13 alanine polymers in mouse caused fusion of the forelimb sesamoid bones [51,52]. Moreover, increased length of HOXD13 polyalanine repeats has been associated with several cases of synpolydactylism [51,53].

HoxA11 and HoxA13 are among the transcriptional factors with crucial function during limb development, where polyalanine repeats have been identified [40,41]. For the human HOXA13, in particular, clear associations were established between the expansion of particular polyalanine tracts (I, III and V) and the development of autopod malformations associated with the human hand-foot-genital syndrome (HFGS) [54,55]. Similar phenotypes were identified in mutant mice with an expanded polyalanine tract [56].

Comparative analyses using HoxA11 amino acid sequences from vertebrates reveal that this protein is composed of three major domains [41]: Domains I and III correspond to the N- and C-terminal regions respectively; Domain II is encoded by exon 1 and corresponds to a variable and overall hydrophilic region (Figure 2A). Characteristic reconstructions of Domains I and III suggest that the rates of coding sequence evolution have not changed significantly in tetrapods (frog and chick) relative to lobe-finned fish (coelacanth). However accelerated rates of coding sequence evolution were observed for the mammalian and newt lineages and shown to be a gene-specific phenomenon [41]. Interestingly, when the hydrophilic region of the HoxA11 proteins is compared, stretches of more than three consecutive alanine residues are identified exclusively in the tetrapod sequences, together with an increase in the length of this region (Figure 2B). Thus, the molecular evolution of this transcription factor might have involved an increase of the hydrophilic region by the addition of polyalanine tracts. An hypothetical scenario is that the alanine repeats present in the common ancestor of tetrapods may have acted as a "seed" for the expansion of polyalanine stretches that then contributed to an increased plasticity of HoxA11, potentiating novel functionalities in this lineage, ultimately contributing to the developmental events involved in the fin-to-limb transition [41].

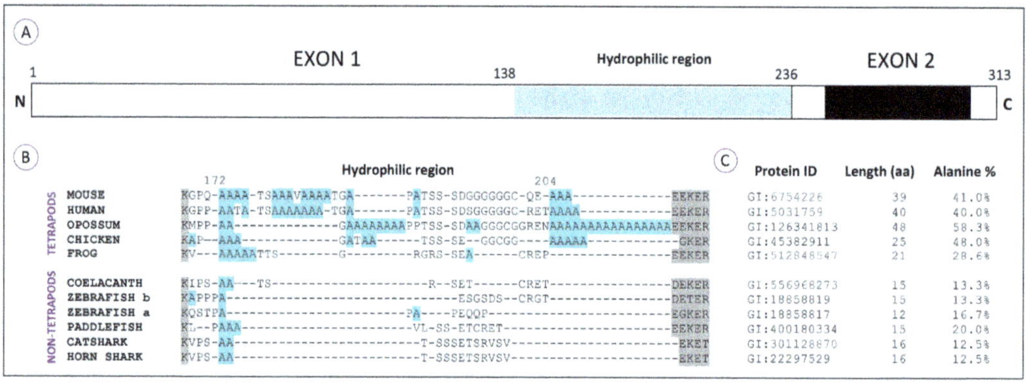

Figure 2. Structure and alanine content of the HoxA11 proteins from vertebrates. (**A**) Schematic representation of the HoxA11 protein, which has 313 aa (amino acid) based on mouse [41,57]. Exons 1 and 2 are represented by white rectangles; the homeobox is in black; and N and C terminal ends are indicated. The variable overall hydrophilic region (98 aa) identified in the N-terminal region is represented by a dotted bar; (**B**) Amino acid sequence alignment of the presumptive hydrophilic region of HoxA11 proteins in vertebrates: mouse (*Mus musculus*), human (*Homo sapiens*), opossum (*Monodelphis domestica*), chicken (*Gallus gallus*), frog (*Xenopus tropicalis*), coelacanth (*Latimeria chalumnae*), zebrafish (*Danio rerio*), paddlefish (*Polyodon spathula*), catshark (*Scyliorhinus canicula*) and horn shark (*Heterodontus francisci*). Alignments were performed with Clustal X [58]. Conserved amino acid residues, flanking the region, are shadowed in grey and alanine residues in blue. Note the presence of polyalanine tracts, with more than four alanine residues clustered, exclusively in tetrapod organisms; (**C**) Identification of the HoxA11 proteins used in the alignment (Protein ID) and the corresponding alanine percentage within the presumptive hydrophilic regions. The sequences used to calculate the alanine percentage correspond to blocks located between extremely conserved residues (in grey). The length of the compared blocks is also indicated here (aa). Note a significant increase of these blocks and a higher percentage of alanine residues in tetrapod organisms, particularly evident in amniotes. Within non-tetrapods, the higher percentage of alanine is found in paddlefish.

Moreover, the overall alanine content within the hydrophilic region is higher in tetrapods than in non-tetrapod organisms. This is especially evident in amniotes, animals that develop a paddle-like hand plate prior to digit differentiation. In fact, the amphibian's hydrophilic region lacks polyalanine tracts that appear to be conserved in other tetrapod lineages, and coincidently, the formation of their autopods is not preceded by the development of a hand plate [59,60]. Thus, the addition of alanine residues in the hydrophilic region may have led to novel functionalities of the HoxA11 protein that may have contributed to the origin of the hand plate. Alternatively, the lower alanine content found in the hydrophilic region of amphibians, when compared to other tetrapod linages, may reflect their own evolutionary history that seems to have led to a highly derived morphology [61].

Interestingly, in non-tetrapod organisms, the higher percentage of alanine residues in the hydrophilic region is found in the paddlefish, which in fact presents a subtle separation of *HoxA11* and *HoxA13* expression domains at early stages of fin development (Figure 1A; [36]). Thus, hypothetically, higher alanine content in the HoxA11 hydrophilic regions may influence the expression pattern of this protein. Evolutionarily, this could have been the process by which *HoxA11* expression patterns evolved, probably contributing to the fin-to-limb transition.

Given that the repressor domains of transcription factors are commonly small, unstructured, hydrophilic regions with high alanine content [62], these observations suggest that expanded alanine stretches might be enhancers of repressor domains that were instrumental in setting tetrapod-specific features, such as the autopods, as proposed by Chiu and colleagues [41]. However, the search of such a repressor function found no evidence in mouse [63]. Nevertheless, there is still grounds for Chiu and colleagues' hypotheses, given that the studies performed in mouse just revealed results for a partial deletion of the alanine residues outside the homeodomain, and there is no indication of their location. Thus, just a complete deletion of the alanine residues within the hydrophilic region would be informative. Moreover, the alanine stretches have other predictable roles, such as increasing the plasticity of HoxA11 function, modifying its interactome and/or acting as spacers, providing a structural component crucial to the performance of its functional domains, as suggested by several authors with respect to the function of polyalanine tracts [40,47–49].

Taking together these analyses, we hypothesize that a sequential increase of the alanine content within the HoxA11 hydrophilic regions may have contributed to: (1) the evolution of *HoxA11* expression patterns during fin development; (2) the establishment of zeugopod/autopod borders; and (3) the formation of the hand plate in the common ancestor of amniotes.

Regarding the HoxA13 proteins, a total of seven polyalanine tracts (I–VII), characterized by a minimum of four clustered alanine residues, has been identified in a variety of vertebrate species at distinct phylogenetic positions (Figure 3A,B; [40,47]). Comparative amino acid analyses reveal that these sequences are characteristic of amniotes, being particularly conserved in mammals (Figure 3B). In fact, Tract III seems to be the only one conserved between mammalian and avian genomes, which suggests that most polyalanine tracts from the HoxA13 proteins arose after the divergence of birds from the lineage that would culminate in the appearance of mammals [40,47]. The incorporation of polyalanine tracts together with flanking segments rich in proline, serine and glycine seems to have increased the N-terminus of mammals approximately 35%, when compared to the fish counterparts [40]. Indeed, while tetrapod HoxA13 proteins have more than 20% alanine content, the zebrafish orthologs have less than 10% (Figure 3C). Within other organisms representing groups that derived prior to the origin of amniotes (frog, coelacanth, zebrafish, paddlefish and sharks), there are alanine residues that align with the polyalanine tracts of amniotes and may represent the primordial elements that gave rise to them. In agreement with this idea, Mortlock and colleagues suggested that the origin of large alanine repeats in mammals might be explained by a dynamic process of recurring replication slippage and point mutation within alanine repeat codons [40].

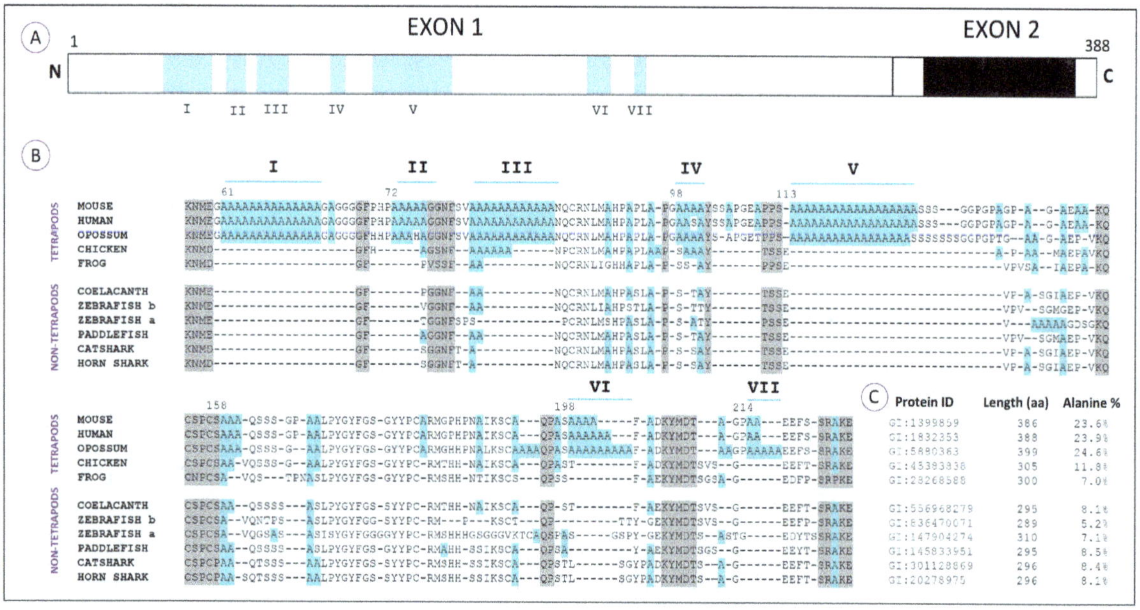

Figure 3. Structure and alanine content of the HoxA13 proteins from vertebrates. (**A**) Schematic representation of the HoxA13 protein, which has 383 aa (amino acid) based on mouse [47,64]. Exons 1 and 2 are represented by white rectangles; the homeobox is in black; and the N and C terminal ends are indicated. Alanine tracts found in the N-terminal region are represented in blue (I–VII); (**B**) Amino acid sequence alignment showing the alanine residues (in blue) detected in the N-terminal region of HoxA13 proteins in vertebrates: mouse (*Mus musculus*), human (*Homo sapiens*), opossum (*Monodelphis domestica*), chicken (*Gallus gallus*), frog (*Xenopus tropicalis*), coelacanth (*Latimeria chalumnae*), zebrafish (*Danio rerio*), paddlefish (*Polyodon spathula*), catshark (*Scyliorhinus canicula*) and horn shark (*Heterodontus francisci*). Note the presence of seven polyalanine tracts (I–VII) exclusively in the mammalian Hoxa13 proteins and conservation of Tract III among amniotes; (**C**) Identification of the HoxA13 proteins used in the alignment (Protein ID) and the corresponding alanine percentage. Note the higher percentage of alanine residues in amniotes, in particular in mammals.

Taking into consideration the potential impact of these polyalanine tracts in the functional plasticity of the proteins, it is possible that their incorporation within the HoxA13 transcription factor may have enabled novel developmental roles contributing to the evolution of the limb structure. However, the absence of such sequences in the HoxA13 of amphibians is intriguing in this scenario. As for HoxA11, we hypothesize that the substantial difference found in the alanine content of the amphibian HoxA13 proteins may explain the lack of a hand plate prior to digit differentiation, which typically characterize the development of amphibian limbs [59]. In order to test this idea, it would be ideal to gain insight into the functional role of Region III, which is conserved in the remaining tetrapod lineages. Several experimental approaches might be relevant to test this hypothesis, including harboring an equivalent region into zebrafish HoxA13, evaluating its capacity to generate a hand plate-like phenotype. Deletions of this particular region in mouse may also be informative to link its functional role to the formation of the hand plate structure.

As for HoxA11, a possible outcome of the modifications in the alanine content of HoxA13 proteins may be the enhancement of their role as repressors. An example of such a type of repressive function is well described for HoxA7, in which N-terminal polyalanine repeats are fundamental to enhance its repressive function [65]. This might have enabled this protein to act as a repressor of the *HoxA11* gene during limb development, leading to the establishment of an ancestral zeugopod/autopod boundary and the development of a hand plate. Moreover, dissociation of *HoxA11* and *HoxA13* expression domains may have potentiated the expansion of the skeletogenic process that culminated in the formation of additional endoskeleton elements, such as digits (Figure 4).

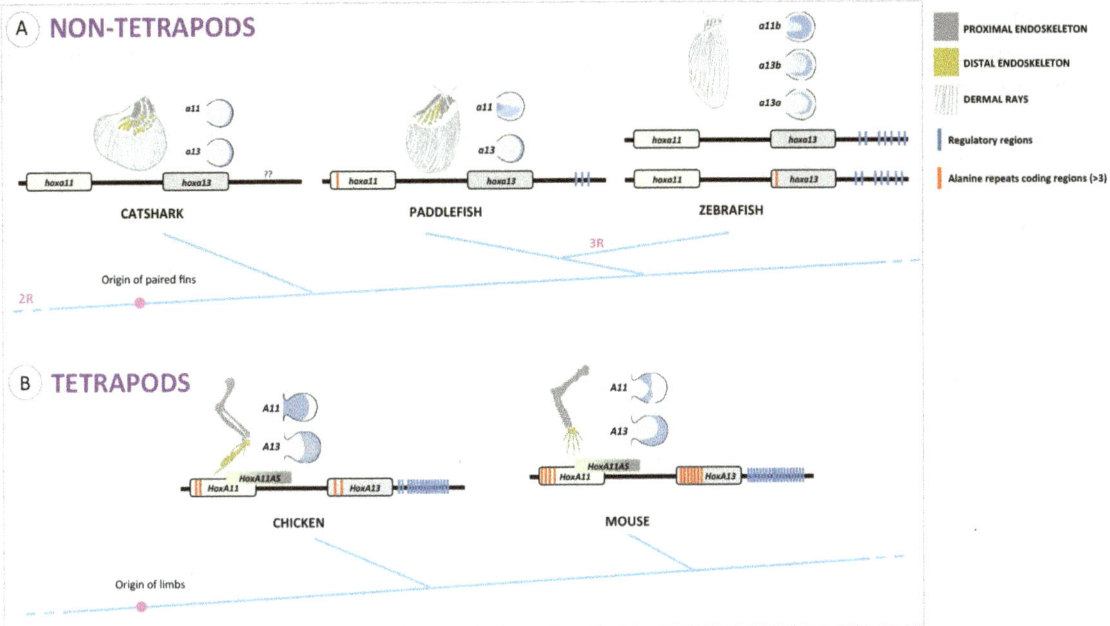

Figure 4. Overview of the morphological, developmental and genomic changes that may have preceded the fin-to-limb transition in vertebrates, comparing non-tetrapod (**A**) and tetrapod (**B**) organisms. Schematic drawing of appendicular skeletons from vertebrates representing distinct phylogenetic positions: basal gnathostomes (catshark); basal actinopterygians (paddlefish); teleosts (zebrafish) and tetrapods (mouse and chicken). 2R and 3R refer to the number of whole-genome duplications that took place prior to the divergence of gnathostomes and teleosts, respectively. Panels adjacent to the appendicular skeletons display *HoxA11* and *HoxA13* expression during appendage development (in blue) based on data from: *Xenopus* [16]; chick [23]; mouse [31]; catshark [35]; paddlefish [36]; zebrafish [17]. The genomic region thought to contain *HoxA11* and *HoxA13* genes in each type of organism is represented under these panels. Note the regions encoding for alanine clusters, with a minimum of three residues (in red) and putative regulatory regions in blue. *HoxA11AS* are the long non-coding *HoxA11* RNAs thought to have an inhibitory effect on *HoxA11* transcription. Question marks in the basal gnathostome condition represent the lack of knowledge of putative regulatory regions operating on 5′ *HoxA* genes. The basal actinopterygian condition was assembled based on data from spotted gar (*Lepisosteus oculatus*). Figure assembly based on [1,40,41,43,57,66,67].

Thus, there is in fact grounds to speculate that the incorporation of polyalanine repeats in HoxA11 and HoxA13 proteins may have been an important step for the evolution of limbs. Although a remark has to be made regarding the alanine content of HoxA11 and HoxA13 proteins between fish, amphibians and birds as the comparative alanine content does not support a clear correlation between higher alanine percentage and the presence of an autopod (Figures 2 and 3). Two hypotheses may explain this observation. The first one is that the relatively similar alanine content of amphibians and birds in comparison with fish, but lower in comparison with mammals, results from the different evolutionary paths that those groups experienced [68]. Indeed, some reptiles and amphibians have a remarkable evolutionary history with reference to the evolution of autopod, with squamates having multiple autopod losses, estimated to have occurred at least 25 times independently [69]. Thus, for amphibians, reptiles and birds, a different molecular signature may be correlated with the evolution of the autopod [68]. The second hypothesis is that the contribution of alanines for autopod evolution is not merely quantitative, but also qualitative. However, functional assays are mandatory to make the proof of principle of these hypotheses, as the importance of alanine repeats for limb evolution remains largely unexplored. Taking advantage of the genome editing capacity of the CRISPR-Cas9 technology [70] and the lack of polyalanine repeats in zebrafish [40,41], it would be possible to introduce polyalanine repeats within the zebrafish HoxA11 or HoxA13 and assess how such sequences impact fin development and/or alter the way these transcription factors interact with their downstream targets. The realization of such experiments holds the potential of exploring how homopolymeric amino acid repeats may have impacted the fin-to-limb transition at quantitative and qualitative levels and, on a broader scale, provides evidence of how such repeat expansions may have contributed to the morphological plasticity in vertebrates.

3. Non-Coding RNAs and the Fin-to-Limb Transition

In the past decade, the scientific community has discovered that RNA's biological function is far more complex than the initially proposed role as simple mediators in the DNA-to-protein information flow [71]. Recently, several classes of non-coding RNAs (ncRNAs) with proven biological function have been discovered, such as: long-non-coding RNAs (lncRNAs), microRNAs (miRNAs), Piwi-interacting RNAs (piRNAs), small interfering RNAs (siRNAs), enhancer RNAs (eRNAs) and promoter-associated RNAs (PARs) [72]. Those ncRNAs have various functions, such as suppression of transposon activity and post-transcriptional gene silencing. They may also act as gene transcriptional activators or repressors and be involved in gene epigenetic control [72–74].

Although exceptions are known, overall, those RNAs seem to have two characteristics in common: a low transcriptional level and a high evolutionary divergence, with their numbers increasing in the genomes along with the species' evolutionary history [72,75]. For example, a recent study demonstrates that the majority of long-non-coding RNAs in each species are lineage specific [75]. This finding, along with the discovery that ncRNAs are important for proper gene regulation during development in some species [74,76], designates the ncRNAs as possible drivers of species morphological innovation [74].

Interestingly, the *HOX* loci produce a considerable repertoire of ncRNAs belonging to two major classes: miRNAs (e.g., *mir-10*, *mir-196*, *mir-99*, *mir-126*) and lncRNAs (e.g., *HOTAIR*, *HOTTIP*, *Mistral*, *HOTATRM1*) [77]. Those ncRNAs have been implicated in the regulation of *HOX* gene expression at the transcriptional, post-transcriptional and translational level [77], and a particularly interesting example of the contribution of such RNA species in the modulation of morphological design comes from mir-196, a well-conserved miRNA in vertebrates [76,78]. McGlinn and colleagues reported that knocking down *mir-196* expression in chicken embryos caused a morphologic alteration of the last cervical vertebrae into a thoracic-like identity [76]. The phenotypic alteration seems to result partially from an expansion of the *Hoxb8* expression domain, caused by the reduction of *mir-196* transcripts [76]. This example shows how relevant miRNAs, and on a broader scale the ncRNAs, are for the regulation of important patterning genes during development [76], with possible evolutionary implications in the establishment of specific morphological traits.

In addition, these ncRNAs may also participate in the epigenetic control of transcription. The lncRNAs HOTAIR was the first one to be discovered to play that role, alerting the importance of RNAs as controllers of the transcriptional activity that may have been highly influential in the determination of the cell fate and, ultimately, in the evolution of animal morphology [79,80]. However, the versatility of functions that these lncRNAs perform during development and disease remains largely unknown. Distinct sources of information (such as conservation, coding potential patterns and anatomical properties) are currently being used to identify the main lncRNA gene families [80]. Regarding the anatomical properties, lncRNAs have been grouped, for example, as: (1) antisense lncRNAs that overlap protein-coding genes; (2) intronic lncRNAs that are encoded within introns of protein-coding genes; (3) divergent lncRNAs encoded in the opposite direction of a protein coding gene, but sharing its promoter; (4) intergenic lncRNAs that are encoded within the intergenic space between protein-coding loci (lincRNAs).

Among the Hox genes found to be involved in limb development, the *HoxA11* locus stands out as capable of generating several antisense lncRNAs (*HoxA11AS*) in a variety of tetrapods [42,57,74,81] that were found to be polyadenylated and alternately processed [42,57]. Moreover, comparison of these antisense *HoxA11* transcripts revealed tissue-specific differences, as well as aspects of RNA processing conserved during evolution, which strongly suggest that the antisense RNAs from the *Hoxa 11* locus are functionally significant [42].

Two of these *HoxA11AS* have overlapping complementary sequences with the first exon of HoxA11 (ENST00000520360.4 and ENST00000522863.1; Figure 4). In contrast, others seem to have sequence complementary with the enhancer/promoter region of *HoxA11* and appear to lack sequence conservation between fish and mammals (e.g., ENST00000522674.1). In order to evaluate the antisense transcript expression of *HoxA11AS*, Hsieh-Li and colleagues performed *in situ* hybridizations using sense and antisense probes designed for the first exon of *HoxA11* [57]. Surprisingly, they found complementary expression patterns of these probes in mouse embryonic limbs. At E9.5, the *HoxA11* gene is transcribed throughout the distal limb bud region and then gets progressively proximally restricted, up to the point where the *HoxA11* gene is only expressed in the prospective zeugopod region [42,57]. This progressive distal repression of the *HoxA11* gene coincides with the appearance of the representatives of *HoxA11AS* transcripts that begin to be expressed in the most distal limb bud region at E10.5. This antisense expression domain then expands proximally, occupying the prospective autopod region by E11.5 [42,57]. The complementary expression patterns found between the sense and antisense probes designed against the *HoxA11* first exon suggest that production of *HoxA11AS* transcripts during limb development may conduct expression inhibition of *HoxA11* [42,57].

Cross species alignments using the "Comparative Genomic" tool from Ensemble (release 80, May 2015, [82]) suggest that the putative *HoxA11AS* encoding regions have been subjected to a different selective pressure during vertebrates' evolution. Indeed, the regions that overlap with the complementary sequence of *HoxA11*, in the first exon, seem to be significantly conserved among tetrapods. However, another *HoxA11AS* encoding region, which seems to be complementary to the *HoxA11* enhancer/promoter region (ENST00000522674.1), is apparently less conserved [74]. Within mammals, this particular *HoxA11* antisense is detectable in eutherians and is expressed during embryonic development, but also in the endometrium during pregnancy [74]. The potential role of this *HoxA11AS* in mammal reproduction may justify its generalized presence within eutherians' genomes. In non-mammal organisms, such as in the frog, the degree of sequence conservation drops to less than 50% [74]. Thus, this amphibian lncRNA, if functional, might be exclusively associated with the regulation of limb development and, as such, under lower selective pressure. Moreover, cross species alignment also revealed an absence of significant similarity between human *HOXA11AS* gene and the syntenic zebrafish, stickleback and coelacanth regions. Moreover, when Bartel and colleagues characterized the lncRNAs during zebrafish development using chromatin marks, poly(A)-site mapping and RNA-Seq data, they did not find any potential homologues of *HOXA11AS*, either by sequence conservation or genomic location [83]. This might be due to the striking lack of conservation of the teleost HoxA intergenic regions, when compared to the ones from other gnathostome lineages [84]. In fact, teleosts have duplicated and more compact HoxA clusters, with shorter intergenic regions [84], which may suggest that they lack most of the genomic region able to transcribe *HoxA11AS*. It remains to be uncovered if fish species that diverge prior to teleost radiation and where the length of the HoxA intergenic regions more closely resemble the ones from the tetrapods have the capacity to produce these lncRNAs (e.g., shark, paddlefish).

Given the possible inhibitory effect of *HoxA11AS* on *HoxA11* and what seems to be a tetrapod-specific characteristic of *HoxA11AS*, we present the hypothesis that the evolutionary origin of *HoxA11AS* was one of the molecular events that might have contributed to the evolution of the fin/limb morphology. According to this hypothesis, the appearance of *HoxA11AS* expression in the distal limb region inhibits *HoxA11* expression in that limb domain, thus uncoupling *HoxA11* and *HoxA13* limb expression domains and enabling the definition of a distal *HoxA13* single positive autopod domain. The apparent lack of *HoxA11AS* homologues in the zebrafish, according to our hypothesis, may help to explain why the expression domains of *HoxA11/HoxA13* do not split during fin development. In contrast, the transient separation of these domains detected during shark and paddlefish fin development [26,35,36] suggests that these organisms may produce *HoxA11AS*. However, these *HoxA11AS* are probably transcribed in the antisense direction within their own *HoxA11* loci; therefore, they may not be conserved among vertebrates. Taken as inspiration the work from Hsieh-Li and colleagues [57], it would be interesting to analyze the expression pattern of sense and antisense *HoxA11* transcripts in these organisms.

An important step to gain insight into our preferential hypothesis, which relates the origin of *HoxA11AS* with the evolution of vertebrate fins, is to functionally characterize these transcripts to better understand their biological relevance to limb development. To achieve this goal, a possible strategy would be to take advantage of the chick model, which seems to have conserved *HoxA11AS*, and induce its overexpression, knockdown and knockout *in ovo*, following the strategy used by McGlinn and colleagues [76]. This would allow testing the impact of *HoxA11AS* in the definition of the *HoxA11* expression domain during the formation of limbs. Another important step to explore our hypothesis is to assess the evolutionary importance of these lncRNAs during vertebrate fin/limb development. To this end, an innovative strategy would be the knock-in of *HoxA11AS* in zebrafish and the characterization of its impact during fin development, assuming that the capacity to metabolize these lncRNAs was assembled prior to the divergence of tetrapods.

4. *Cis*-Regulation of 5′ *HoxA* Genes and the Fin-to-Limb Transition

According to the previously-presented hypothesis, the definition of the autopod domain is associated with the proximal restriction of *HoxA11* expression during the development of appendicular structures in stem groups of tetrapods [26]. However, the autopod morphology, regarding digit size and number, acquired modifications during evolution probably due to an increase in the transcription of 5′ *HoxA* and *HoxD* genes [8,85,86]. Thus, recruitment of novel non-coding regulatory elements, which are able to modulate the transcription of these genes, seems to have been a crucial step for the evolution of the autopod [85,86]. The enhancer elements that govern the expression of 5′ *HoxD* have been characterized during the development of mouse limbs [85,86]. These reports highlighted that a particular combination of regulatory elements is required for the specification of the appropriate time, dosage and pattern of 5′ *HoxD* gene expression.

Regarding the *HoxA11* and *HoxA13* genes, despite their crucial role for the correct development and proximo-distal patterning of the limbs, information on their *cis*-regulatory network is still scarce, although the research teams led by Innis [47], Kmita [66], Gómez-Skarmeta and Shubin [43] have shed some light into the enhancers that drive *HoxA13* expression in the developing limbs and their possible impact in the fin-to-limb transition. From the works published to date on mice, the enhancer network that regulates distal *HoxA13* expression in the developing limbs is comprised of at least nineteen different elements located upstream of, and excluding, the *HoxA* cluster (Figure 4) [66]. Some of those enhancers are present within genes located upstream of the *HoxA* cluster, namely the *Hibadh*, *Tax1bp1* and *Jazf1* genes [66]. A curious observation is that the enhancers driving *HoxA13* distal expression in mouse limb form several sub-megabase topological domains (sub-TADs) that contact each other in a fashion that, in some cases, seems to be independent of the enhancer's transcriptional activity. This suggests that the sub-TADs organization is governed by a developmental structure/function interplay instead of a mere clustering of active enhancers [66].

When the conservation of the mouse *HoxA13* enhancer network was assayed, using sequence homology and ATAC-Seq (assay for transposase-accessible chromatin using sequencing) in zebrafish and gar, only the gar genome showed conservation of three (e10, e13 and e16) of the 19 enhancers that regulate *HoxA13* transcription in mouse limb, and at least the gar e16 region induces, in zebrafish and mouse transgene reporter assays, an expression pattern that is very similar to the one driven by mouse e16 [43]. Taking into account that lack of sequence conservation is not necessarily sufficient to exclude the presence of enhancers in fish, these data nevertheless suggest two possible circumstances. Firstly, the zebrafish *hoxa13a* and *hoxa13b* enhancer network, if present, represents a derived state, as no sequence homology was detected between zebrafish, gar and mouse *HoxA* enhancers [43]. This observation may result from extensive remodeling of *HoxA* regulatory elements caused by the occurrence of a third whole genome duplication in teleosts followed by gene sub-functionalization, thus leading to the inability of the zebrafish *hoxa13* paralogs to respond to a "distal limb program" during development [43]. Secondly, the enhancer network that regulates *HoxA13*'s late phase expression suffered extensive elaboration/remodeling over the course of the vertebrates' evolution, although some of its elements might have already been present in the last common ancestor of bony vertebrates [43]. The extensive elaboration of *HoxA13* enhancer elements throughout vertebrate evolution suggests that the fish *HoxA13* transcriptional level during fin development is too low to elicit an adequate "distal limb identity", thus suggesting that *HoxA13* gene dosage increase was a requirement for the fin-to-limb transition. This hypothesis finds support in the work published, in 1997, by Zákány and colleagues [8], where by analyzing a set of distal *HoxA* and *HoxD* compound mutants, this team found that overall autopod morphology, and in particular digit number and length, greatly varies in response to *Hox* dosage [8] and to a minor extent to the qualitative nature of the expressed HoxA and HoxD proteins (*Hox* code) [85]. Moreover, overexpression of zebrafish *hoxd13a* was shown to cause changes in the morphogenesis of the fins towards a limb-like identity [22]. Testing out the hypothesis that an increase of *HoxA13* dosage was an evolutionary requirement for the fin-to-limb transition could be easily achieved in zebrafish following a strategy previously described by Freitas and colleagues [22]. Performing such a proof-of-principle study not only upholds the potential to disclose how increasing levels of HoxA13 proteins may have contributed to the morphological transition between fish fins and tetrapod limbs, but may also shed light on how regulation and function of *Hox13* paralogs genes may have set the grounds for autopod evolution.

5. Conclusions and Future Directions

In this review, we attempted to survey the evolution of HoxA11 and HoxA13 transcription factors and their regulatory control in vertebrates. Evaluating the current state of the art and performing complementary analyses, using information currently available for distinct vertebrate lineages, we support the hypotheses implicating changes in the alanine content, capacity to generate long-non-coding RNAs and incorporation of a novel *cis*-regulatory region in the developmental mechanisms involved in the fin-to-limb transition in vertebrates (Figure 4).

Conserved polyalanine tracts with a minimum of four residues are common within the HoxA11 and HoxA13 proteins of tetrapods (Figures 2–4). However, these types of sequences are undetectable in the orthologs of fish. In the HoxA11 proteins, these enriched alanine domains co-localize with the hydrophilic region, being higher the percentages of alanines detectable in amniotes. Among the fish analyzed, the hydrophilic region of paddlefish seems to be the one that has a higher content of alanines. These observations point us to two hypotheses: (1) an increased alanine content within HoxA11 proteins may have impacted the formation of specific amniote limb structures during evolution, such as the hand plates that precede digit differentiation; (2) incorporation of extra alanine residues into the hydrophilic regions of HoxA11 protein probably started prior to the tetrapod divergence, in the common ancestor of actinopterygians and sarcopterygians. In the HoxA13 proteins, most polyalanine tracts (I–VII) are conserved in mammals (Figures 3 and 4). However, the polyalanine Tract III seems to be present in the chick HoxA13 proteins, which may suggest a particular role of these tracts in the formation of amniote-specific limb structures, the hand plates.

Taking into consideration the functional potential found for this type of amino acid repeats [40,47,49], their integration into the HoxA11 and Hoxa13 proteins during evolution might have modified the way they interact with their downstream targets, contributing to the phenotypic variation found in the evolutionary transition from fish fins to tetrapod limbs. For example, the polyalanine repeats found in the HoxA11 and HoxA13 proteins might have enhanced their repressor role upon other molecules, helping to establish novel domains of cell identity, leading to the differentiation of the hand plate in amniotes. The repressor potential of HoxA13 in particular might have inhibited the expression of *HoxA11* in the most distal mesenchyme of the appendages, generating a novel field of cell identity, which might have potentiated the formation of additional skeletal elements. However, the proof of principle of these hypotheses is lacking. Here, we suggest, as a future working perspective, the use of CRISPR-Cas9 technology [70] to harbor polyalanine repeats within the zebrafish *HoxA11* or *HoxA13* in order to evaluate the phenotypic consequences of such sequences during appendage development. These experiments would certainly add to our knowledge of how changes in the alanine content of transcription factors impacted the diversification of vertebrate structures.

In addition, we show here that the capacity to generate *HoxA11AS*, which are long-non-coding RNAs, seems to be exclusive to tetrapods (Figure 4) [42,57,74,81]. Moreover the expression of particular *HoxA11* antisense RNAs was found to be complementary to *HoxA11*, suggesting that while *HoxA11* is progressively repressed distally, *HoxA11AS* expression expands proximally. This observation points to a repressor role of *HoxA11AS* on *HoxA11* that might have contributed to split the *HoxA11/HoxA13* expression domains, potentiating the formation of the zeugopod and the autopod. Given that during catshark and paddlefish fin development, the expression domains of *hoxa11* and *hoxa13* also split transiently into two non-overlapping territories, it would be important to search for *HoxA11AS* homologues in these organisms and evaluate the potential repressor role on *HoxA11*. These ideas also lack functional validation, and a first step would be to gain more insight into the expression of HoxA11 antisense in fish representing distinct phylogenetic positions. Moreover, it would be important to better characterize the role of *HoxA11AS* during tetrapod limb development, to test their developmental and evolutionary potential harboring of *HoxA11AS* in zebrafish and to characterize the impact during appendage development.

Finally, we suggest that the transcriptional levels of *HoxA11* and *HoxA13* might have also influenced appendage evolution, as has been proposed for *HoxD13* [22]. Indeed, the autopod morphology in mouse, in particular digit number and length, is influenced by the *Hox* dosage [8], which often depends on the enhancer capacity of *cis*-regulatory regions within their genomic landscape [87]. Marked differences in the putative regulatory regions of the 5′ *HoxA* genes were identified in fish species representing distinct phylogenetic positions, such as the spotted gar and the zebrafish (Figure 4). Thus, *cis*-regulatory changes may have interfered with the differential fin morphology of distinct fish groups (Figure 4). Within tetrapods, subtle changes were also identified in the putative regulatory regions of these genes that may have contributed to the distinct autopod anatomy found in mammals and birds, for example [43]. Additionally, fish *versus* tetrapod comparisons suggest a significant increase of *cis*-regulatory modules in tetrapods [43], a phenomena that might have been crucial to achieve the appropriate levels of Hox expression required for the origin of the autopod. To further explore the consequences of increased levels of 5′ HoxA proteins and the addition of novel regulatory modules for the developmental processes resulting in the evolution of limbs, we emphasize the advantages of the zebrafish model to perform the appropriate proof of principle studies due to its current technical repertoire, allowing a variety of functional assays [22].

Acknowledgments: We are grateful to Mafalda Araújo for aid in phylogenetic analyses. We also acknowledge the financial support of "Fundação para a Ciência e Tecnologia" (FCT–Portugal) through the research project EXPL/BEX-BID/0801/2013.

Author Contributions: J.L.C. conceived of and wrote the manuscript. V.B. conceived of the artwork and collaborated in figure assembly. P.N.R. contributed to the discussion. R.F. participated in the writing, figure assembly and overall supervision of the work. All authors commented on the manuscript at distinct stages.

Conflicts of Interest: The authors declare no conflict of interest.

References

1. Shubin, N.; Tabin, C.; Carroll, S. Deep homology and the origins of evolutionary novelty. *Nature* **2009**, *457*, 818–823. [CrossRef] [PubMed]
2. Coates, M.I.; Jeffery, J.E.; Rut, M. Fins to limbs: What the fossils say. *Evol. Dev.* **2002**, *4*, 390–401. [CrossRef] [PubMed]
3. Shubin, N. The evolution of paired fins and the origin of tetrapod limbs—Phylogenetic and transformational approaches. *Evol. Biol.* **1995**, *28*, 39–86.

4. Janvier, P. *Early Vertebrates*; Oxford University Press: Oxford, UK, 1996.
5. Boisvert, C.A.; Mark-Kurik, E.; Ahlberg, P.E. The pectoral fin of Panderichthys and the origin of digits. *Nature* **2008**, *456*, 636–638. [CrossRef] [PubMed]
6. Coates, M.I.; Clack, J.A. Polydactyly in the earliest known tetrapod limbs. *Nature* **1990**, *347*, 66–68. [CrossRef]
7. Coates, M.I. The origin of vertebrate limbs. *Dev. Suppl.* **1994**, 169–180.
8. Zakany, J.; Fromental-Ramain, C.; Warot, X.; Duboule, D. Regulation of number and size of digits by posterior Hox genes: A dose-dependent mechanism with potential evolutionary implications. *Proc. Natl. Acad. Sci. USA* **1997**, *94*, 13695–13700. [CrossRef] [PubMed]
9. Delgado, I.; Torres, M. Gradients, waves and timers, an overview of limb patterning models. *Semin. Cell Dev. Biol.* **2016**. [CrossRef] [PubMed]
10. Zeller, R.; Lopez-Rios, J.; Zuniga, A. Vertebrate limb bud development: Moving towards integrative analysis of organogenesis. *Nat. Rev. Genet.* **2009**, *10*, 845–858. [CrossRef] [PubMed]
11. Mariani, F.V.; Ahn, C.P.; Martin, G.R. Genetic evidence that FGFs have an instructive role in limb proximal-distal patterning. *Nature* **2008**, *453*, 401–456. [CrossRef] [PubMed]
12. Cooper, K.L.; Hu, J.K.; ten Berge, D.; Fernandez-Teran, M.; Ros, M.A.; Tabin, C.J. Initiation of proximal-distal patterning in the vertebrate limb by signals and growth. *Science* **2011**, *332*, 1083–1086. [CrossRef] [PubMed]
13. Grandel, H.; Schulte-Merker, S. The development of the paired fins in the zebrafish (*Danio rerio*). *Mech. Dev.* **1998**, *79*, 99–120. [CrossRef]
14. Nakamura, T.; Klomp, J.; Pieretti, J.; Schneider, I.; Gehrke, A.R.; Shubin, N.H. Molecular mechanisms underlying the exceptional adaptations of batoid fins. *Proc. Natl. Acad. Sci. USA* **2015**, *112*, 15940–15945. [CrossRef] [PubMed]
15. Yano, T.; Abe, G.; Yokoyama, H.; Kawakami, K.; Tamura, K. Mechanism of pectoral fin outgrowth in zebrafish development. *Development* **2012**, *139*, 2916–2925. [CrossRef] [PubMed]
16. Ohgo, S.; Itoh, A.; Suzuki, M.; Satoh, A.; Yokoyama, H.; Tamura, K. Analysis of hoxa11 and hoxa13 expression during patternless limb regeneration in Xenopus. *Dev. Biol.* **2010**, *338*, 148–157. [CrossRef] [PubMed]
17. Ahn, D.; Ho, R.K. Tri-phasic expression of posterior Hox genes during development of pectoral fins in zebrafish: Implications for the evolution of vertebrate paired appendages. *Dev. Biol.* **2008**, *322*, 220–233. [CrossRef] [PubMed]
18. Sheth, R.; Marcon, L.; Bastida, M.F.; Junco, M.; Quintana, L.; Dahn, R.; Kmita, M.; Sharpe, J.; Ros, M.A. Hox genes regulate digit patterning by controlling the wavelength of a turing-type mechanism. *Science* **2012**, *338*, 1476–1480. [CrossRef]
19. Tarchini, B.; Duboule, D.; Kmita, M. Regulatory constraints in the evolution of the tetrapod limb anterior-posterior polarity. *Nature* **2006**, *443*, 985–988. [CrossRef] [PubMed]
20. Dolle, P.; Dierich, A.; Lemeur, M.; Schimmang, T.; Schuhbaur, B.; Chambon, P.; Duboule, D. Disruption of the Hoxd-13 gene induces localized heterochrony leading to mice with neotenic limbs. *Cell* **1993**, *75*, 431–441. [CrossRef]
21. Sordino, P.; van der Hoeven, F.; Duboule, D. Hox gene expression in teleost fins and the origin of vertebrate digits. *Nature* **1995**, *375*, 678–681. [CrossRef] [PubMed]
22. Freitas, R.; Gomez-Marin, C.; Wilson, J.M.; Casares, F.; Gomez-Skarmeta, J.L. Hoxd13 contribution to the evolution of vertebrate appendages. *Dev. Cell.* **2012**, *23*, 1219–1229. [CrossRef] [PubMed]
23. Nelson, C.E.; Morgan, B.A.; Burke, A.C.; Laufer, E.; DiMambro, E.; Murtaugh, L.C.; Gonzales, E.; Tessarollo, L.; Parada, L.F.; Tabin, C. Analysis of Hox gene expression in the chick limb bud. *Development* **1996**, *122*, 1449–1466. [PubMed]
24. Perez, W.D.; Weller, C.R.; Shou, S.M.; Stadler, H.S. Survival of Hoxa13 homozygous mutants reveals a novel role in digit patterning and appendicular skeletal development. *Dev. Dyn.* **2010**, *239*, 446–457. [CrossRef] [PubMed]
25. Knosp, W.M.; Scott, V.; Bachinger, H.P.; Stadler, H.S. HOXA13 regulates the expression of bone morphogenetic proteins 2 and 7 to control distal limb morphogenesis. *Development* **2004**, *131*, 4581–4592. [CrossRef] [PubMed]
26. Metscher, B.D.; Takahashi, K.; Crow, K.; Amemiya, C.; Nonaka, D.F.; Wagner, G.P. Expression of Hoxa-11 and Hoxa-13 in the pectoral fin of a basal ray-finned fish, Polyodon spathula: Implications for the origin of tetrapod limbs. *Evol. Dev.* **2005**, *7*, 186–195. [CrossRef] [PubMed]

27. Yokouchi, Y.; Nakazato, S.; Yamamoto, M.; Goto, Y.; Kameda, T.; Iba, H.; Kuroiwa, A. Misexpression of Hoxa-13 induces cartilage homeotic transformation and changes cell adhesiveness in chick limb buds. *Genes Dev.* **1995**, *9*, 2509–2522. [CrossRef] [PubMed]
28. Yokouchi, Y.; Ohsugi, K.; Sasaki, H.; Kuroiwa, A. Chicken Homeobox gene Msx-1—Structure, expression in limb buds and effect of retinoic acid. *Development* **1991**, *113*, 431–444. [PubMed]
29. Yokouchi, Y.; Sasaki, H.; Kuroiwa, A. Homeobox gene-expression correlated with the bifurcation process of limb cartilage development. *Nature* **1991**, *353*, 443–445. [CrossRef]
30. Mercader, N.; Leonardo, E.; Azpiazu, N.; Serrano, A.; Morata, G.; Martínez, C.; Torres, M. Conserved regulation of proximodistal limb axis development by Meis1/Hth. *Nature* **1999**, *402*, 425–429. [PubMed]
31. Mercader, N.; Selleri, L.; Criado, L.M.; Pallares, P.; Parras, C.; Cleary, M.L.; Torres, M. Ectopic Meis1 expression in the mouse limb bud alters P-D patterning in a Pbx1-independent manner. *Int. J. Dev. Biol.* **2009**, *53*, 1483–1494. [CrossRef] [PubMed]
32. Sato, K.; Koizumi, Y.; Takahashi, M.; Kuroiwa, A.; Tamura, K. Specification of cell fate along the proximal-distal axis in the developing chick limb bud. *Development* **2007**, *134*, 1397–1406. [CrossRef] [PubMed]
33. Tabin, C.; Wolpert, L. Rethinking the proximodistal axis of the vertebrate limb in the molecular era. *Genes Dev.* **2007**, *21*, 1433–1442. [CrossRef] [PubMed]
34. Neumann, C.J.; Grandel, H.; Gaffield, W.; Schulte-Merker, S.; Nusslein-Volhard, C. Transient establishment of anteroposterior polarity in the zebrafish pectoral fin bud in the absence of sonic hedgehog activity. *Development* **1999**, *126*, 4817–4826. [PubMed]
35. Sakamoto, K.; Onimaru, K.; Munakata, K.; Suda, N.; Tamura, M.; Ochi, H.; Tanaka, M. Heterochronic shift in Hox-mediated activation of sonic hedgehog leads to morphological changes during fin development. *PLoS ONE* **2009**, *4*, e5121. [CrossRef] [PubMed]
36. Davis, M.C.; Dahn, R.D.; Shubin, N.H. An autopodial-like pattern of Hox expression in the fins of a basal actinopterygian fish. *Nature* **2007**, *447*, 473–476. [CrossRef] [PubMed]
37. Yano, T.; Tamura, K. The making of differences between fins and limbs. *J. Anat.* **2013**, *222*, 100–113. [CrossRef]
38. Freitas, R.; Gomez-Skarmeta, J.L.; Rodrigues, P.N. New frontiers in the evolution of fin development. *J. Exp. Zool. B Mol. Dev. Evol.* **2014**, *322*, 540–552. [CrossRef] [PubMed]
39. Roensch, K.; Tazaki, A.; Chara, O.; Tanaka, E.M. Progressive specification rather than intercalation of segments during limb regeneration. *Science* **2013**, *342*, 1375–1379. [CrossRef] [PubMed]
40. Mortlock, D.P.; Sateesh, P.; Innis, J.W. Evolution of N-terminal sequences of the vertebrate HOXA13 protein. *Mamm. Genome* **2000**, *11*, 151–158. [CrossRef]
41. Chiu, C.H.; Nonaka, D.; Xue, L.; Amemiya, C.T.; Wagner, G.P. Evolution of Hoxa-11 in lineages phylogenetically positioned along the fin-limb transition. *Mol. Phylogenet. Evol.* **2000**, *17*, 305–316. [CrossRef] [PubMed]
42. Potter, S.S.; Branford, W.W. Evolutionary conservation and tissue-specific processing of Hoxa 11 antisense transcripts. *Mamm. Genome* **1998**, *9*, 799–806. [CrossRef] [PubMed]
43. Gehrke, A.R.; Schneider, I.; de la Calle-Mustienes, E.; Tena, J.J.; Gomez-Marin, C.; Chandran, M.; Nakamura, T.; Braasch, I.; Postlethwait, J.H.; Gómez-Skarmeta, J.L.; *et al.* Deep conservation of wrist and digit enhancers in fish. *Proc. Natl. Acad. Sci. USA* **2015**, *112*, 803–808. [CrossRef]
44. Rorick, M.M.; Wagner, G.P. The origin of conserved protein domains and amino acid repeats via adaptive competition for control over amino acid residues. *J. Mol. Evol.* **2010**, *70*, 29–43. [CrossRef] [PubMed]
45. Mularoni, L.; Ledda, A.; Toll-Riera, M.; Alba, M.M. Natural selection drives the accumulation of amino acid tandem repeats in human proteins. *Genome Res.* **2010**, *20*, 745–754. [CrossRef] [PubMed]
46. Fondon, J.W.; Garner, H.R. Molecular origins of rapid and continuous morphological evolution. *Proc. Natl. Acad. Sci. USA* **2004**, *101*, 18058–18063. [CrossRef] [PubMed]
47. Lehoczky, J.A.; Innis, J.W. Expanded HOXA13 polyalanine tracts in a monotreme. *Evol. Dev.* **2008**, *10*, 433–438. [CrossRef] [PubMed]
48. HannaRose, W.; Hansen, U. Active repression mechanisms of eukaryotic transcription repressors. *Trends Genet.* **1996**, *12*, 229–234. [CrossRef]
49. Lavoie, H.; Debeane, F.; Trinh, Q.D.; Turcotte, J.F.; Corbeil-Girard, L.P.; Dicaire, M.J.; Saint-Denis, A.; Pagé, M.; Rouleau, G.A.; Brais, B. Polymorphism, shared functions and convergent evolution of genes with sequences coding for polyalanine domains. *Hum. Mol. Genet.* **2003**, *12*, 2967–2979. [CrossRef] [PubMed]

50. Frisen, L.; Lagerstedt, K.; Tapper-Persson, M.; Kockum, I.; Nordenskjold, A. A novel duplication in the HOXA13 gene in a family with atypical hand-foot-genital syndrome. *J. Med. Genet.* **2003**, *40*, e49. [CrossRef]
51. Anan, K.; Yoshida, N.; Kataoka, Y.; Sato, M.; Ichise, H.; Nasu, M.; Ueda, S. Morphological change caused by loss of the taxon-specific polyalanine tract in Hoxd-13. *Mol. Biol. Evol.* **2007**, *24*, 281–287. [CrossRef] [PubMed]
52. Johnson, K.R.; Sweet, H.O.; Donahue, L.R.; Ward-Bailey, P.; Bronson, R.T.; Davisson, M.T. A new spontaneous mouse mutation of Hoxd13 with a polyalanine expansion and phenotype similar to human synpolydactyly. *Hum. Mol. Genet.* **1998**, *7*, 1033–1038. [CrossRef] [PubMed]
53. Bruneau, S.; Johnson, K.R.; Yamamoto, M.; Kuroiwa, A.; Duboule, D. The mouse Hoxd13(spdh) mutation, a polyalanine expansion similar to human type II synpolydactyly (SPD), disrupts the function but not the expression of other Hoxd genes. *Dev. Biol.* **2001**, *237*, 345–353. [CrossRef] [PubMed]
54. Goodman, F.R. Limb malformations and the human HOX genes. *Am. J. Med. Genet.* **2002**, *112*, 256–265. [CrossRef]
55. Mortlock, D.P.; Innis, J.W. Mutation of HOXA13 in hand-foot-genital syndrome. *Nat. Genet.* **1997**, *15*, 179–180. [CrossRef] [PubMed]
56. Innis, J.W.; Mortlock, D.; Chen, Z.; Ludwig, M.; Williams, M.E.; Williams, T.M.; Doyle, C.D.; Shao, Z.; Glynn, M.; Mikulic, D.; *et al.* Polyalanine expansion in HOXA13: Three new affected families and the molecular consequences in a mouse model. *Hum. Mol. Genet.* **2004**, *13*, 2841–2851. [CrossRef] [PubMed]
57. Hsieh-Li, H.M.; Witte, D.P.; Weinstein, M.; Branford, W.; Li, H.; Small, K.; Potter, S.S. Hoxa 11 structure, extensive antisense transcription, and function in male and female fertility. *Development* **1995**, *121*, 1373–1385. [PubMed]
58. Larkin, M.A.; Blackshields, G.; Brown, N.P.; Chenna, R.; McGettigan, P.A.; McWilliam, H.; Valentin, F.; Wallace, I.M.; Wilm, A.; Lopez, R.; *et al.* Clustal W and Clustal X version 2.0. *Bioinformatics* **2007**, *23*, 2947–2948. [CrossRef] [PubMed]
59. Torok, M.A.; Gardiner, D.M.; Shubin, N.H.; Bryant, S.V. Expression of HoxD genes in developing and regenerating axolotl limbs. *Dev. Biol.* **1998**, *200*, 225–233. [CrossRef]
60. Keenan, S.R.; Beck, C.W. Xenopus limb bud morphogenesis. *Dev. Dyn.* **2016**, *245*, 233–243. [CrossRef] [PubMed]
61. Handrigan, G.R.; Wassersug, R.J. The anuran Bauplan: A review of the adaptive, developmental, and genetic underpinnings of frog and tadpole morphology. *Biol. Rev. Camb. Philos. Soc.* **2007**, *82*, 1–25. [CrossRef] [PubMed]
62. Mitchell, P.J.; Tjian, R. Transcriptional regulation in mammalian-cells by sequence-specific dna-binding proteins. *Science* **1989**, *245*, 371–378. [CrossRef]
63. Roth, J.J.; Breitenbach, M.; Wagner, G.P. Repressor domain and nuclear localization signal of the murine Hoxa-11 protein are located in the homeodomain, no evidence for role of poly alanine stretches in transcriptional repression. *J. Exp. Zool. B Mol. Dev. Evol.* **2005**, *304*, 468–475. [CrossRef] [PubMed]
64. Mock, B.A.; D'Hoostelaere, L.A.; Matthai, R.; Huppi, K. A mouse homeo box gene, Hox-1.5, and the morphological locus, Hd, map to within 1 cM on chromosome 6. *Genetics* **1987**, *116*, 607–612. [PubMed]
65. Schnabel, C.A.; AbateShen, C. Repression by HoxA7 is mediated by the homeodomain and the modulatory action of its N-terminal-arm residues. *Mol. Cell. Biol.* **1996**, *16*, 2678–2688. [CrossRef] [PubMed]
66. Berlivet, S.; Paquette, D.; Dumouchel, A.; Langlais, D.; Dostie, J.; Kmita, M. Clustering of tissue-specific sub-TADs accompanies the regulation of HoxA genes in developing limbs. *PLoS Genet.* **2013**, *9*, e1004018. [CrossRef] [PubMed]
67. Schneider, I.; Shubin, N.H. The origin of the tetrapod limb: From expeditions to enhancers. *Trends Genet.* **2013**, *29*, 419–426. [CrossRef] [PubMed]
68. Kohlsdorf, T.; Cummings, M.P.; Lynch, V.J.; Stopper, G.F.; Takahashi, K.; Wagner, G.P. A molecular footprint of limb loss: Sequence variation of the autopodial identity gene Hoxa-13. *J. Mol. Evol.* **2008**, *67*, 581–593. [CrossRef] [PubMed]
69. Wiens, J.J.; Brandley, M.C.; Reeder, T.W. Why does a trait evolve multiple times within a clade? Repeated evolution of snakelike body form in squamate reptiles. *Evolution* **2006**, *60*, 123–141. [CrossRef] [PubMed]
70. Jao, L.E.; Wente, S.R.; Chen, W. Efficient multiplex biallelic zebrafish genome editing using a CRISPR nuclease system. *Proc. Natl. Acad. Sci. USA* **2013**, *110*, 13904–13909. [CrossRef] [PubMed]

71. Ponting, C.P.; Oliver, P.L.; Reik, W. Evolution and functions of long noncoding RNAs. *Cell* **2009**, *136*, 629–641. [CrossRef] [PubMed]
72. Kaikkonen, M.U.; Lam, M.T.Y.; Glass, C.K. Non-coding RNAs as regulators of gene expression and epigenetics. *Cardiovasc. Res.* **2011**, *90*, 430–440. [CrossRef] [PubMed]
73. Mattick, J.S. Non-coding RNAs: The architects of eukaryotic complexity. *EMBO Rep.* **2001**, *2*, 986–991. [CrossRef] [PubMed]
74. Yu, H.S.; Lindsay, J.; Feng, Z.P.; Frankenberg, S.; Hu, Y.; Carone, D.; Shaw, G.; Pask, A.J.; O'Neill, R.; Papenfuss, A.T.; *et al.* Evolution of coding and non-coding genes in HOX clusters of a marsupial. *BMC Genomics* **2012**, *13*. [CrossRef] [PubMed]
75. Hezroni, H.; Koppstein, D.; Schwartz, M.G.; Avrutin, A.; Bartel, D.P.; Ulitsky, I. Principles of long noncoding RNA evolution derived from direct comparison of transcriptomes in 17 species. *Cell Rep.* **2015**, *11*, 1110–1122. [CrossRef]
76. McGlinn, E.; Yekta, S.; Mansfield, J.H.; Soutschek, J.; Bartel, D.P.; Tabin, C.J. *In ovo* application of antagomiRs indicates a role for miR-196 in patterning the chick axial skeleton through Hox gene regulation. *Proc. Natl. Acad. Sci. USA* **2009**, *106*, 18610–18615. [CrossRef] [PubMed]
77. Leite-Castro, J.; Rodrigues, P.N.; Freitas, R. Hox gene regulation in vertebrates. *Trends Dev. Biol.* **2014**, *8*, 77–95.
78. Olena, A.F.; Patton, J.G. Genomic organization of microRNAs. *J. Cell. Physiol.* **2010**, *222*, 540–545. [CrossRef] [PubMed]
79. Rinn, J.L.; Kertesz, M.; Wang, J.K.; Squazzo, S.L.; Xu, X.; Brugmann, S.A.; Goodnough, L.H.; Helms, J.A.; Farnham, P.J.; Segal, E.; *et al.* Functional demarcation of active and silent chromatin domains in human HOX loci by noncoding RNAs. *Cell* **2007**, *129*, 1311–1323. [CrossRef] [PubMed]
80. Rinn, J.L.; Chang, H.Y. Genome regulation by long noncoding RNAs. *Annu. Rev. Biochem.* **2012**, *81*, 145–166. [CrossRef] [PubMed]
81. Bodenmiller, D.M.; Baxter, C.S.; Hansen, D.V.; Potter, S.S. Phylogenetic analysis of Hoxa 11 sequences reveals absence of transposable elements, conservation of transcription factor binding sites, and suggests antisense coding function. *DNA Sequence* **2002**, *13*, 77–83. [CrossRef] [PubMed]
82. Flicek, P.; Amode, M.R.; Barrell, D.; Beal, K.; Billis, K.; Brent, S.; Carvalho-Silva, D.; Clapham, P.; Coates, G.; Fitzgerald, S.; *et al.* Ensembl 2014. *Nucleic Acids Res.* **2014**, *42*, D749–D755.
83. Ulitsky, I.; Shkumatava, A.; Jan, C.H.; Sive, H.; Bartel, D.P. Conserved function of lincRNAs in vertebrate embryonic development despite rapid sequence evolution. *Cell* **2011**, *147*, 1537–1550. [CrossRef] [PubMed]
84. Prohaska, S.J.; Fried, C.; Flamm, C.; Wagner, G.P.; Stadler, P.F. Surveying phylogenetic footprints in large gene clusters: Applications to Hox cluster duplications. *Mol. Phylogenet. Evol.* **2004**, *31*, 581–604. [CrossRef] [PubMed]
85. Montavon, T.; le Garrec, J.F.; Kerszberg, M.; Duboule, D. Modeling Hox gene regulation in digits: Reverse collinearity and the molecular origin of thumbness. *Genes Dev.* **2008**, *22*, 346–359. [CrossRef] [PubMed]
86. Woltering, J.M.; Noordermeer, D.; Leleu, M.; Duboule, D. Conservation and divergence of regulatory strategies at Hox Loci and the origin of tetrapod digits. *PLoS Biol.* **2014**, *12*, e1001773. [CrossRef] [PubMed]
87. Andrey, G.; Duboule, D. SnapShot: Hox gene regulation. *Cell* **2014**, *156*. [CrossRef] [PubMed]

12

The Many Hats of Sonic Hedgehog Signaling in Nervous System Development and Disease

Yesser H. Belgacem *, Andrew M. Hamilton, Sangwoo Shim, Kira A. Spencer and Laura N. Borodinsky *

Department of Physiology & Membrane Biology and Institute for Pediatric Regenerative Medicine, Shriners Hospital for Children Northern California, University of California Davis School of Medicine, Sacramento, CA 95817, USA; andrewmichaelhamilton@gmail.com (A.M.H.); sshim4@gmail.com (S.S.); kaspencer@ucdavis.edu (K.A.S.)

* Correspondence: yesser.belgacem-tellier@inserm.fr (Y.H.B.); lnborodinsky@ucdavis.edu (L.N.B.)

Academic Editors: Henk Roelink and Simon J. Conway

Abstract: Sonic hedgehog (Shh) signaling occurs concurrently with the many processes that constitute nervous system development. Although Shh is mostly known for its proliferative and morphogenic action through its effects on neural stem cells and progenitors, it also contributes to neuronal differentiation, axonal pathfinding and synapse formation and function. To participate in these diverse events, Shh signaling manifests differently depending on the maturational state of the responsive cell, on the other signaling pathways regulating neural cell function and the environmental cues that surround target cells. Shh signaling is particularly dynamic in the nervous system, ranging from canonical transcription-dependent, to non-canonical and localized to axonal growth cones. Here, we review the variety of Shh functions in the developing nervous system and their consequences for neurodevelopmental diseases and neural regeneration, with particular emphasis on the signaling mechanisms underlying Shh action.

Keywords: non-canonical Sonic hedgehog signaling; neuronal differentiation; calcium signaling; synapse formation; neurodevelopmental disorders; medulloblastoma; neural regeneration

---

1. Introduction

The highly complex nature of the central nervous system is not only due to the copious number of cells it is composed of, but also to the number and diversity of connections between these cells. Moreover, neural function relies on dynamic changes in synapse connectivity and strength. This dynamic complexity of the nervous system is present throughout development and into adulthood. Nevertheless, different developmental stages are characterized by sets of specific cellular events, and the temporal and spatial boundaries of the developmental processes that create the nervous system are critically important. These boundaries are established by tight regulation of neural cell proliferation, neural progenitor specification, neuronal differentiation, synapse formation and neural plasticity. Intrinsic and extrinsic factors regulate these events. One potentially simple mechanism of switching from one cellular process to the next would be to switch factors by upregulating the expression of a "new" one while downregulating the expression of the "old" one. However, this paradigm does not always seem to apply; Sonic hedgehog (Shh) is a prominent example of a protein that is expressed in both the incipient nervous system and in the adult brain, but whose role varies widely throughout development. Its function in the nervous system was once thought to be restricted to the promotion of neural stem cell proliferation and neural progenitor specification. More recent studies have challenged this idea by showing that Shh signaling targets many different types of neural cells and by providing evidence of the diversity of processes regulated by Shh. Indeed, Shh appears to serve diverse functions depending on the developmental stage and maturity of the neural cells in question.

Here, we review the many roles of Shh during nervous system development and the signaling mechanisms associated with them. We propose a model of Shh signal switching throughout development to accommodate the wide variety of tasks and adapt to the dynamic changes in neural cell function as the nervous system develops and matures.

2. Neural Cell Proliferation

Nervous system morphogenesis relies on the expansion of stem cells to provide the critical cell mass necessary for neural function. Production of neural cells is achieved by active proliferation of neural stem cells during development, an event promoted by canonical Shh signaling (Figure 1 and Table 1). This pathway is initiated when Hedgehog (Hh) binding to the receptor Patched (Ptch) [1,2] releases inhibition on the co-receptor Smoothened (Smo) [3–5], which leads to upregulation of the activity of the transcriptional activators glioma-associated oncogene (Gli) factors [6,7]. Ectopic expression of *Shh* in the mouse embryonic spinal cord approximately doubles cell proliferation in the dorsal neural tube [8], while an increase in *Gli1* levels in *Nestin*-expressing mouse neural progenitors increases precursor cell numbers in vivo, resulting in a larger brain and broader progenitor domains [9]. Shh/Gli proliferative activity is also apparent in the early postnatal mouse cerebellum, where Shh secreted by Purkinje cells promotes cerebellar granule cell precursor proliferation and prevents their differentiation [10]. In addition to Ptch, Shh proliferative action requires the Hh-binding protein Gas1 (growth arrest specific 1) and the co-receptor Boc (biregional cell-adhesion-molecule-related/downregulated by oncogenes (Cdon)-binding protein) in cerebellar granule neuron progenitors from three-day-old mice [11]. In certain structures of the postnatal and adult brain, neural stem cells continue to produce neurons. Proliferation of adult neural stem cells in mouse hippocampus depends on Shh [12], and Shh also inhibits programmed cell death, contributing to the maintenance of adult neural stem cell niches in the subventricular zone, the dentate gyrus and olfactory bulb [13]. Shh acting through Smo at the primary cilium is also necessary for the recruitment of dentate gyrus postnatal progenitors [14]. However, ablation of primary cilia during embryonic development does not disrupt neurogenesis in the ventricular-subventricular zone of neonatal mice, with the exception of the anterior ventral neural cell population [15], suggesting very specific zones of influence for Shh-regulated neuroproliferation. In the adult brain, neural stem cell proliferation and neurogenesis also appear to depend more on restricting the repressor activity of Gli3R than on the activator character of Gli transcription factors according to double-mutant experiments for *Gli2* and *Gli3* [16].

The mechanisms of Shh/Gli-driven promotion of neural cell proliferation involve the regulation of cell cycle kinetics and cell survival. Shh canonical signaling regulates the length of G1 phase and expression of the anti-apoptotic factor *Bcl2* in the ventral neural tube of chick embryos [17]. Similarly, reciprocal inhibitory interaction between Gli1 and p53 in neural progenitors controls neural stem cell number by enhancing proliferation and restricting stem cell expansion, respectively [9]. Shh promotes neuroepithelial cell proliferation in the embryonic chick and mouse neural tube by controlling progression into the G2 phase through regulating the expression of cyclins E, A and B [18] (Table 1). Additionally, Gli activity is necessary for Wnt-dependent control of G1 phase length in neural progenitors through upregulation of cyclin D1 expression by tcf3/4 [18].

However, regulation of neural cell proliferation by Shh is not always positive. For instance, in the developing *Xenopus* retina, Hh signaling speeds up the cell cycle by reducing the length of G1 and G2 phases. This may contribute to regulating the conversion from slow-cycling stem cells to fast-cycling transient amplifying progenitors [19]. Similarly, in *Drosophila* postembryonic brain neuroblasts, high levels of Hh signaling cause premature neuroblast cell cycle exit and underproliferation, which has consequences on the timing of neurogenesis [20].

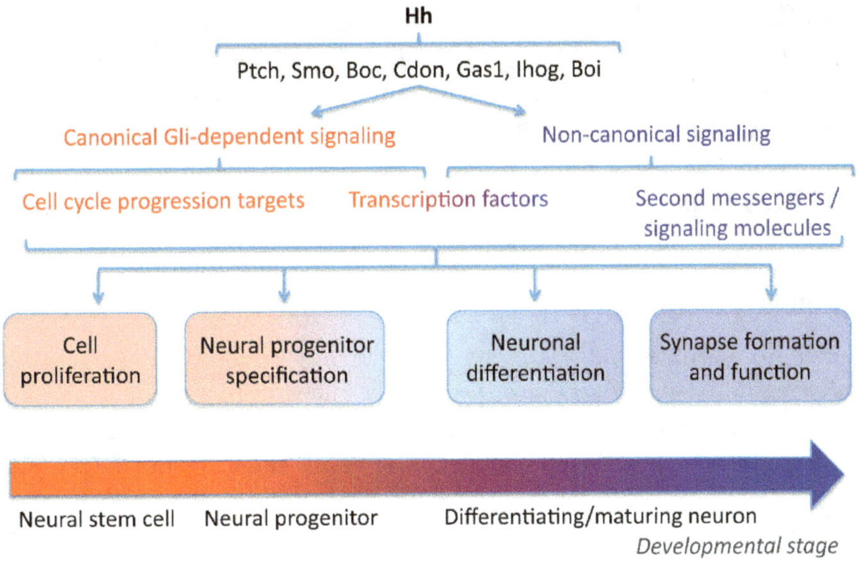

Figure 1. Hedgehog (Hh) signaling spans nervous system development. Hh interacts with different receptors, co-receptors and interacting proteins to recruit canonical, glioma-associated oncogene (Gli)-dependent or non-canonical signaling pathways regulating the cellular events that encompass neural development and function. Boc: brother of cell adhesion molecule-related/down-regulated by oncogenes (Cdon); Boi: brother of interference hedgehog (Ihog); Gas1: growth-arrest specific gene-1; Ptch: Patched; Smo: Smoothened.

Table 1. Mechanisms and functions of canonical and non-canonical Hedgehog (Hh) signaling in neural development.

	Canonical			Non-Canonical		
	Neural Cell Proliferation	**Neural Progenitor Specification**	**Axon Guidance**	**Neuronal Differentiation**	**Axon Guidance**	**Axon Guidance**
Receptors	Ptch, Smo (Gas1, Boc)	Ptch, Smo	Ptch, Smo	Ptch, Smo	Smo, Boc	(1) Smo (mouse), Hhip (chick) (2) Boc (3) Smo, Gas1
Second messengers				IP3, Ca^{2+}		(1) 14-4-3, cAMP (3) Gnaz
Transcription factors	Gli	Gli	Gli	cJun	Transcription independent	Transcription independent
Targets	Bcl2, P53, cyclin A, B, E, D1	Nkx2.2, Nkx6.1, Pax6, Evx1, Phox2A, Gata2, Fox2A, etc.	(1) Slit (2) Stromal cell-derived factor 1	Tlx3	Src	
Subcellular needs for Hh signaling	Nucleus Primary cilium?	Nucleus Primary cilium	Nucleus	Nucleus Primary cilium?	Axonal growth cone	Axonal growth cone
Main roles	Regulation of cell cycle progression and cell survival	Spinal cord and brain patterning during morphogenesis	(1) Midline crossing forebrain commissural axons (2) Retinal ganglion cell axon guidance	Specification of spinal neuron transmitter phenotype	Attractant guidance commissural spinal axons	(1) Repulsive guidance commissural spinal axons (2) Repulsive retinal ganglion cell axon guidance (3) Repulsive guidance enteric axons

?: Not yet determined or neural structure dependent. Bcl2: B-cell lymphoma 2; Boc: brother of cell adhesion molecule-related/down-regulated by oncogenes; cAMP: cyclic adenosine monophosphate; Evx1: even-skipped homeobox 1; Fox2A: forkhead box protein A2; Gas1: growth arrest specific 1; Gnaz: G Protein Subunit α Z; Hhip: hedgehog interacting protein; IP3: inositol 1,4,5-trisphosphate receptor type 3; Pax6: paired box 6; Phox2A: paired like homeobox 2A; Ptch: Patched; Smo: Smoothened; Src: proto-oncogene tyrosine-protein kinase; Tlx3: T-cell leukemia homeobox 3.

The differences in Shh action on neural cell proliferation prompt the notion that the signaling pathways recruited by Shh are diverse and that the cellular outcome varies depending on the target cell and its maturational status, as well as on the molecular partners available at a given space and time. Indeed, Shh mitogenic and patterning functions are uncoupled when proteoglycan binding of Shh is perturbed. While this interaction is necessary for the proliferation of neural stem cells, it is not required for neural tube ventral patterning [21], providing an example of the versatility of Shh action depending on interactions with additional signaling molecules.

3. Neural Progenitor Specification

The transition from proliferation to differentiation in the developing nervous system occurs in a spatiotemporally-regulated manner (Figure 1). Whether Shh signaling participates in this transition remains unclear; however, profuse evidence has demonstrated that Shh function spans beyond its mitogenic role. The competence of the cell and its maturational status restrict the proliferative activity of Shh signaling, as suggested by the aforementioned study in which ectopic Shh expression in early fetal stages elicited enhanced neural cell proliferation, but not when Shh overexpression was implemented in the late stages of embryonic mouse nervous system formation [8]. Nevertheless, overexpression of Shh in late fetal stages, although not mitogenic, still halts the neural precursors in an undifferentiated state [8], indicating that Shh signaling may regulate the transitions between different neural cell maturational states.

One of the first functions attributed to the Shh pathway was its patterning role during tissue morphogenesis. In particular, Hh canonical signaling is responsible for specifying neural progenitors that in turn will give rise to different neuronal phenotypes in the developing nervous system. The patterning of ventral identities in the embryonic spinal cord has become a paradigm of the precise regulation exerted by the Shh canonical pathway [22–27]. Early studies demonstrated that blocking Shh signaling impairs the generation of motor neurons through its action during two critical periods; one in which neural plate cells are specified into ventralized progenitors and a later period during which Shh drives the differentiation of ventralized progenitors into motor neurons or interneurons [25]. The decision to adopt a certain neuronal identity is driven by a spatiotemporal code of transcription factor expression triggered by Shh levels and sensed by neural progenitors depending on their position relative to the ventrally-secreted morphogen. For instance, expression of *Nkx2.2* in ventral neural progenitors is necessary for the neurogenesis of V3 ventral interneurons in the embryonic mouse spinal cord, while its expression limits the ventral boundaries for the generation of the motor neuron domain [24], which in turn depends on the ventral neural progenitor expression of Pax6 [26] (Table 1). However, neural progenitors are not static, as they migrate along the main axes of the neural tube, suggesting that they may sense different levels of Shh throughout development. In zebrafish embryos, this results in heterogeneous populations of specified progenitors that are afterwards sorted into sharply-bordered domains, including that for motor neuron progenitors [28]. Expression of specific cell adhesion molecules facilitates this cell sorting independently of canonical Shh signaling [28]. These findings agree with the discovery that translation of the Shh gradient into patterns of transcription factor expression for specifying neural progenitors occurs through the integration of Shh levels and signaling duration [29]. The interpretation of Shh gradient dynamics relies on a transcriptional network that can store the commands coming from integrated Shh signaling to define the specialization of ventral spinal cord progenitors [30]. Expression of many of these transcription factors is regulated by Gli activity directly [31–35], accounting for the central role of the canonical Shh pathway in the specification of neural progenitors. An interesting aspect of the activity of these transcription factors is that they are transcriptional repressors and as such are believed to repress the expression of alternative phenotypes, thus favoring the acquisition of a specific identity [31].

Glial progenitor cell specification is also dependent on Shh. The most prominent example is the oligodendrocyte lineage, for which the embryonic specification of their precursors depends on Shh derived from the floor plate and the notochord [36,37]. The basic helix-loop-helix transcription factors that are targeted in Shh-driven specification of oligodendrocyte progenitors are Olig1 and Olig2 in the mouse embryonic spinal cord [38]. Intriguingly, Olig2 is required for both motor neuron and oligodendrocyte progenitor specification, the former being specified earlier in the embryonic spinal cord than the latter [39]. A phosphorylation in Olig2 presumably mediated by cyclic adenosine monophosphate (cAMP)-dependent protein kinase A (PKA) participates in the switch from motor neuron to oligodendrocyte precursor [40].

In addition to the spinal cord, the developing brain is also patterned by Shh signaling. Shh is necessary and sufficient to define the size and shape of the midbrain arcs characterized by the expression of different combinations of the transcription factors Phox2A, Gata2, FoxA2 and the stripes expressing Pax6 or Evx1 in chick embryos [41] (Table 1). High rostroventral and low caudodorsal Shh specifies thalamic progenitor cells [42–45]. Interestingly, in the medial ganglionic eminence of mouse embryos, Shh signaling participates in a feed-forward mechanism triggered by the transcription factors Lhx6 and Lhx8, upregulating the expression of Shh, which in turn drives progenitors towards the phenotype of pallial interneurons and components of the septum and bed nucleus stria terminalis [46]. During corticogenesis, Shh canonical signaling regulates the specification of different subtypes of neural progenitors that will eventually provide the differentiated neurons to populate the cortical layers [47]. However, the level of Shh signaling is tightly regulated, and the predominant Gli form is the repressor Gli3R, the primary Gli antagonist of canonical Shh signaling [48–51]. This clamping of Shh canonical signaling at low levels is crucial and appears to be dependent on the negative regulator Suppressor of Fused (SuFu), since neural progenitor cells null for SuFu fail to maintain neural progenitor identity, and this conditional SuFu knockout mouse exhibits impaired corticogenesis [52].

4. Neuronal and Glial Differentiation

Pioneering studies on the invertebrate model organism *Drosophila* took the lead at demonstrating the role of Hh signaling in neuronal differentiation beyond its action in cell proliferation and progenitor specialization (Figure 1). Hh secreted by differentiated neurons triggers the differentiation of the cells next in line, propagating a wave of neuronal differentiation in the developing retina [53–55]. This mechanism is conserved in vertebrates like zebrafish, in which Shh derived from the first retinal neurons elicits a wave of neuronal differentiation extending across the vertebrate retina [56].

In the developing spinal cord, Shh-driven expression of the homeobox gene MNR2 is sufficient to initiate the motor neuron differentiation program that leads to the cholinergic phenotype and characteristic axonal routing [57]. Although the specification of the motor neuron phenotype starts in the motor neuron progenitor, some aspects of motor neuron differentiation continue in the post-mitotic neuron, such as acquiring motor neuron-specific axonal trajectory [58,59], which is activity dependent [59]. Ventral neurons in the forebrain are also responsive to Shh signaling through the expression of the transcription factor Islet-1 [60].

More recent studies have demonstrated roles for Shh in the aspects of neuronal identity that develop further down the path of progenitor specification, in the postmitotic neuron. Altering Shh signaling modifies the neural activity present in developing spinal neurons of *Xenopus laevis* embryos [61] (Figure 2a and Table 1). This activity manifests as Ca^{2+} spikes that last for 10 h during development concurrent with neuronal specialization [61–64]. Intriguingly, the frequency of Ca^{2+} spike activity is higher in ventral spinal neurons compared to their dorsal counterparts [62], resembling the ventrodorsal gradient for Shh in the embryonic spinal cord and opposing the dorsoventral gradient of bone morphogenetic protein (BMP), another morphogen acting inversely to Shh. Our studies reveal that Shh and BMPs regulate Ca^{2+} spike activity in an opposing manner; Shh increases Ca^{2+} spike frequency in spinal neurons, while BMPs decrease it [61,64]. In turn, electrical activity levels regulate the expression of neurotransmitter phenotype [62,65–68]. By recruiting Ca^{2+} spike activity as part of non-canonical signaling, Shh modifies neurotransmitter specification in developing spinal neurons [61]. Enhancing Shh signaling results in greater numbers of gamma-aminobutyric acid-ergic (GABAergic) spinal neurons, while blocking Shh signaling leads to fewer GABAergic neurons. These effects are prevented by suppressing Shh signaling-induced changes in Ca^{2+} spike activity, suggesting that Ca^{2+} signaling acts downstream of Shh action on neurotransmitter specification in developing neurons [61]. These findings suggest that the interaction between Shh and Ca^{2+}-dependent electrical activity adds plasticity to the process of spinal neuron differentiation.

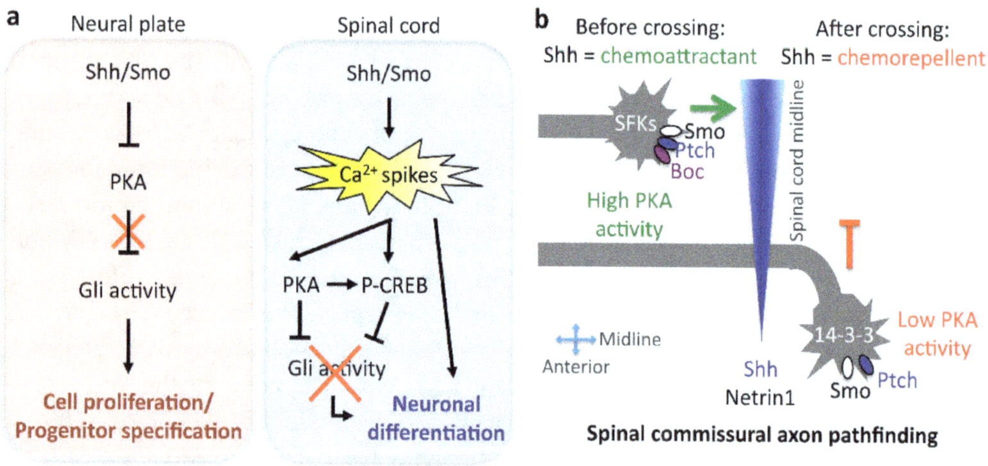

Figure 2. Mechanisms of Sonic hedgehog (Shh) signaling conversion during neural development. (**a**) Canonical Gli signaling is recruited in the developing neural plate, leading to neural stem cell proliferation and neural progenitor specification. The transition from neural plate to spinal cord in *Xenopus laevis* embryos is accompanied by a switch in Shh signaling from Gli to Ca^{2+} spike activity dependent, which results in the recruitment of protein kinase A (PKA) and the shutting off of Gli transcriptional activity in the differentiating neuron. (**b**) Shh action changes when commissural spinal axons cross ventrally the midline from chemoattractant to chemorepellent due to differential, growth cone-localized signaling in mouse embryos. The mediolateral and posteroanterior Shh gradient is represented in blue. P-CREB: phosphorylated cyclic adenosin monophosphate response element-binding protein.

Shh action is also relevant for glial differentiation as shown in a recent study in the developing and adult cerebellar cortex of the mouse. There, Shh produced by the neuronal populations of Purkinje and cerebellar granule cells determines the acquisition of Bergmann glial cell or velate astrocyte identities, through the regulation of the expression of specific glutamate transporters and receptors and the potassium channel Kir4.1 [69]. This study further challenges the concept that both neurons and glial cell differentiation are fully determined by a hardwired developmental program. Instead, this demonstrates that neurons, through Shh signaling, regulate their own cellular environment by dynamically changing the physiological properties of glial cells [69].

5. Axon Guidance

Another important functional feature of differentiating neurons is their axonal trajectory, which enables the establishment of appropriate connections with target cells. In zebrafish, Shh signaling regulates the expression of Netrin1, a known axon guidance molecule expressed by the floor plate and anterior ventral neural tube [70]. Shh effect on axon guidance is not limited to transcriptional regulation of expression of guidance cues; rather, Shh itself functions as a guidance cue by acting as a chemoattractant for commissural spinal axons in their trajectory towards the floor plate before midline crossing [71]. Attraction of commissural axons by Shh is transcription independent and Smo and Boc dependent, and leads to the activation of Src family kinases through a non-canonical signaling pathway [72,73]. Interestingly, this signaling is recruited at the growth cone of the commissural axon in a synergistic manner when both guidance cues, Netrin1 and Shh, are present, serving as a signaling mechanism for the integration of multiple cues [74]. Spinal commissural axons need to travel rostrally after crossing the midline, and this trajectory is also regulated by Shh, although instead of acting as a chemoattractant, here it acts as a chemorepellent for post-commissural spinal axons [75,76]. In chick embryos, the repulsive postcommissural axon guidance is driven through a Ptch- and Smo-independent mechanism mediated by hedgehog interacting protein (Hhip) [75]. In contrast, in the mouse embryonic spinal cord, genetic experiments showed that Hhip is dispensable for Shh-mediated repulsion, while Smo is required for Shh chemorepulsive action [76]. Moreover, the switch in commissural axon responsiveness to Shh is dependent on the developmentally-regulated expression of 14-3-3 proteins that regulate the activity of binding partners, such as PKA, resulting in decreased PKA activity in post-crossing axonal growth cones [76] (Figure 2b).

The Shh effects on axon guidance are also apparent in retinal ganglion cells. First, it was shown that axonal growth was negatively affected by Shh when growing towards the diencephalic ventral midline in the chick embryo [77]. Further studies showed that in vitro exposure to low and high levels of Shh attracts or repels retinal ganglion cell axons from chick embryos, and in vivo overexpression or inhibition of Shh both result in loss of centrally-directed retinal ganglion cell axon projections towards the optic disc [78]. In mice, Shh repels retinal ganglion cell axons, an action mediated by Boc and necessary for the normal segregation of these projections at the optic chiasm [79]. Shh-driven axon guidance can also be indirect, through regulating the expression of factors directly involved in axon guidance; these factors include Slit guidance molecules, which regulate the zebrafish forebrain midline crossing of postoptic commissural axons [80], and the chemokine pathway protein stromal cell-derived factor 1, which is expressed in the optic stalk in a Shh-dependent manner and is necessary for regulating the guidance of retinal ganglion cell axons [81].

In addition to central nervous system axons, Shh regulates the axonal trajectory of enteric neurons that innervate the smooth muscle in the gastrointestinal system. Here, Shh action consists of repelling enteric axons from projecting towards the central villi, allowing for tubular innervation of the gut wall in mouse embryos [82]. In this case, the signaling downstream of Shh includes the binding partner Gas1 and Smo expressed in enteric neurons, which in turn recruit the G protein α Z [82].

All of these studies demonstrate that Shh signaling through noncanonical mechanisms targets postmitotic differentiating neurons to regulate a number of traits, including neurotransmitter phenotype specialization, axonal morphology and axon guidance.

6. Synapse Formation and Plasticity

Once axons reach their targets, they establish synapses, which constitute the circuitry that enables neural network function (Figure 1). During *Drosophila* visual system development, Hh is delivered from photoreceptor axons to the lamina ganglion layer, inducing differentiation of their synaptic partners, the lamina neurons. Hedgehog is localized to axons, growth cones and synapses by a conserved motif in the hedgehog C-terminus [83]. The establishment of the retinotopic map through the formation of specific synapses relies on Hh-induced expression of the transcription factor Single-minded in lamina neurons, which enables association between the pre- and post-synapse by as yet unclear mechanisms [84]. Another subset of synapses in which partners are matched through Hh signaling intervention are the connections between olfactory receptor neurons and their glomerular targets in the *Drosophila* brain. First, Hh leads to differential Ptch expression in peripheral sensory organ precursors, which leads to olfactory neurons with different responsiveness to hedgehog derived from the brain [85]. Hence, in a second step of Hh intervention, only olfactory neurons expressing low levels of Ptch will depend on Hh derived from brain neurons for their axonal targeting through recruitment of Smo and the coreceptor Interference hedgehog (Ihog), a mechanism not available in high Ptch-expressing olfactory neurons due to strong inhibition of Smo [85].

Hh signaling components present in axonal terminals and postsynaptic specializations seem to be an aspect conserved across species. In addition to *Drosophila*, Smo is found presynaptically in close proximity to synaptic vesicles in the hippocampal mossy fibers of adult mice [86], and in the rat hippocampus, Ptch and Smo localize to processes and growth cones in developing neurons and in dendrites, dendritic spines and postsynaptic sites of the mature hippocampus [87]. Shh is present in inhibitory and excitatory synapses in the region I and III of hippocampus proper (CA1 and CA3) and dentate gyrus of the adult rat hippocampus, in pre- and post-synaptic neurons in the vicinity of the active zone, but in distinctly extrasynaptic sites [88]. Shh is also expressed in layer V corticofugal projection neurons, and its coreceptor Boc is expressed in local and callosal projection neurons of layer II/III, whose axons synapse onto subcortical projection neurons [89]. Moreover, the number of synapses formed onto layer V neurons and the strength of these synapses depend on expression of Shh and Boc, respectively [89], indicating the relevance of Shh signaling for synapse formation and function in vertebrates. In agreement with these findings, presynaptic terminals of cultured hippocampal neurons increase in size upon Shh exposure, resulting in increased frequency of miniature postsynaptic currents [90].

7. Interactions of Shh with Other Signaling Pathways in the Nervous System

Shh as a developmental factor for the nervous system is not spatiotemporally isolated, but rather acts concurrently with numerous cues and signaling pathways that dynamically change as the neural tissue develops and matures. Hence, Shh signaling is subjected to modulation by other developmentally-regulated pathways, which may even change Shh signaling's own action in the developing nervous system. Indeed, for Shh to participate seamlessly in the diverse, and sometimes contrasting, cellular events referred to in the above sections as nervous system development progresses, Shh activity needs to change with the maturational status of the neural cell. Here, we review some of the most prominent interactions of Shh with other signaling pathways and the modifications that result in the mechanisms of action for Shh signaling itself during nervous system development (Figure 3).

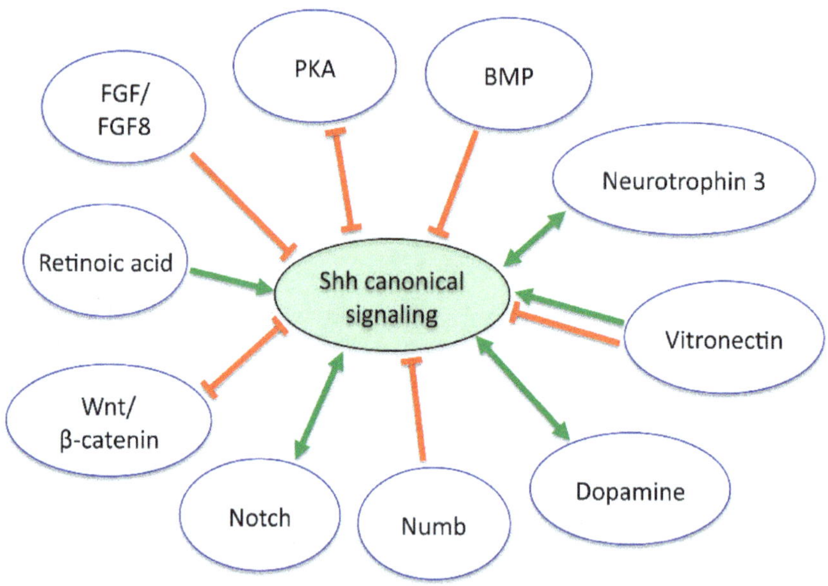

Figure 3. Multiple interactions between Shh and concurrent developmental cues modulate the Shh canonical signaling. Red and green arrows represent antagonistic and synergistic interactions, respectively. Double arrows represent feedback loops. See the text for details.

Rostrally to vertebrate mid-diencephalon, BMP7 and Shh are expressed from the same prechordal mesoderm and act on neural cells coordinately to induce the differentiation of rostral diencephalic ventral midline cells. BMP7 modifies Shh action by shifting the specification towards rostral phenotypes that otherwise, in the absence of BMP7, differentiate into Shh-producing floor plate cells [91]. In chick neural plate explants, BMP signaling modifies the responses of neural cells to Shh by inducing a ventral-to-dorsal switch, antagonistic activity that is modulated in vivo by the expression of BMP inhibitors along with Shh in structures ventral to the neural tube, like the notochord [92]. Another example of signaling interaction demonstrated through in vitro and ex vivo studies from mouse embryonic spinal cord is the synergy between Shh and neurotrophin 3 on motor neuron differentiation, through the regulation of Islet1 expression [93]. Accordingly, neurotrophin 3 knockout mice lack the subclass of fusimotor neurons, which make up 30%–40% of total spinal motor neurons [94]. These findings suggest that the differentiation of some subclasses of motor neurons requires interaction between Shh and neurotrophin 3 [93].

Since most of the transcription factors regulated by Shh are repressors of gene transcription, it is essential to consider that additional signaling pathways must be needed for the induction of a specific phenotype. Indeed, retinoic acid signaling is required for Shh-dependent motor neuron differentiation, while FGF8 downregulates the expression of Shh-induced repressors [95]. Another synergistic partner of Shh is the extracellular matrix protein Vitronectin, which is coexpressed with Shh in the chick notochord, floor plate and ventral neural tube and enhances the differentiation into motor neurons of cultured neuroepithelial cells only when coincubated with Shh, presumably by optimizing Shh presentation to target cells [96]. In contrast, in cerebellar granule cells grown in vitro, Vitronectin overrides the Shh proliferative signal to induce neuronal differentiation by recruiting the transcription factor cyclic-AMP responsive element-binding protein (CREB) [97]. Similarly, FGF promotes differentiation of cerebellar granule neurons in vivo and in vitro by recruiting ERK and JNK kinases and preventing Gli activity, thus abolishing Shh-induced proliferation. Moreover, FGF stops proliferation of *patched* mutant mice-derived tumor cells [98]. In some cases, the interaction between Shh and other signaling factors follows parallel pathways until they intersect at the level of transcriptional regulation, as illustrated by the establishment of thalamic neuron identity defined by a balance of transcription factor expression; thalamic Wnt/β-catenin signaling results in the expression of Neurogenin1/2, while its repressor Nkx2.2 is driven by Shh–Gli signaling, defining the domains of thalamic neuron progenitors through mutual regulation of the transcriptional network [99].

Canonical Gli-dependent Shh signaling is also negatively regulated by PKA, and in the mouse neural tube, this regulation is implemented at the basal body of the primary cilium [100]; thus, pathways that inhibit PKA activity may enhance Gli activity. One interesting example of this interaction is the dopaminergic innervation from the brainstem that promotes motor neuron specification at the expense of V2 interneurons in the embryonic zebrafish spinal cord. This interaction between dopaminergic and Shh canonical signaling happens through activation of D4a dopamine receptors in pre-motor neuron progenitors that inhibit adenylate cyclase, lowering cAMP levels and inhibiting PKA. This results in Gli2 upregulation and increases the number of spinal motor neurons in the developing zebrafish [101]. A similar mechanism is activated upon injury when upregulated dopamine signaling promotes motor neuron regeneration by enhancing Shh signaling [101].

Another interesting type of interaction of signaling pathways with Shh emerges from regulating the localization of Shh signaling components (Table 1). Notch signaling inhibits Ptch localization to the primary cilium, allowing for Smo translocation to this subcellular compartment for full activation of Gli transcription factors, synergizing with Shh activity in regulating neural tube cell specification [102]. In contrast, Numb exerts an antagonistic role on Gli transcriptional activity by binding and targeting Gli to the proteasome-dependent degradation pathway through E3-ubiquitin ligase Itch during cerebellar granule cell progenitor differentiation [103]. Additionally, Notch is known to maintain neural stem cell character and prevent neuronal differentiation and is also inhibited by Numb through ubiquitin-dependent processing of Notch1 [104]. Thus, the obligatory role of Numb in neurogenesis seems to rely on inhibiting Shh proliferative action and Notch signaling [102,105]. Multiple functions of Shh in nervous system proliferation, differentiation and patterning may be facilitated by coordinated regulation and crosstalk between other signaling pathways that include Notch, Numb, BMP and Wnt (Figure 3).

The examples of interacting signaling pathways reviewed thus far result in promoting or preventing the canonical Gli-driven Shh pathway. However, considering that Shh expression persists throughout development and adulthood exerting diverse functions, there is ample support for the model of Shh signaling switching from one pathway that promotes proliferation to others that enable differentiation, maturation and survival of neural cells. In addition to dynamic changes in canonical Shh signaling, Shh becomes engaged in other non-canonical signaling cascades as development progresses, a concept that we recently demonstrated in the embryonic *Xenopus* spinal cord. Gli expression and activity decrease abruptly during spinal cord development when transitioning from the neural plate to spinal cord in *Xenopus* embryos [106,107] and in mice, approximately 30 h post-headfold [30] in nine-day-old embryos [106]. This downregulation of Gli activity is concurrent with the appearance of Ca^{2+} spikes in embryonic *Xenopus* spinal neurons, the frequency of which is enhanced by Shh in a transcription-independent, phospholipase C- and transient receptor potential channel-dependent manner [61]. In turn, Ca^{2+} spikes enhance PKA activity, thus switching Shh action from inhibiting PKA activity in the neural plate to enhancing it in differentiating spinal neurons [107]. This results in a paradoxical turning off of Shh's own canonical Gli-driven signaling during spinal cord development and a switch in Shh signaling to a non-canonical electrical activity and PKA-mediated pathway. The mechanisms of this switch operate on several levels, through inhibiting Gli2 nuclear translocation, favoring Gli2 and Gli3 processing towards repressor forms, and by Shh-dependent activation of CREB, which in turn represses Gli1 expression [107] (Figure 2a).

8. Neural Disease and Regeneration

Given the many important roles that Shh plays during nervous system development, it is not surprising to find that altering Shh signaling contributes to serious neurodevelopmental diseases. Indeed, reduced canonical Shh signaling leads to holoprosencephaly [108], likely due to aberrant patterning and deficient neural cell proliferation. On the other hand, constitutive activation of this pathway by deficient Ptch function [109] underlies the occurrence of several cancers, including a subgroup of the pediatric brain tumor medulloblastoma [108,110]. An increase in Gli1 levels in neural progenitors enables the expansion of neural precursors and neural stem cells and represses expression of p53, an inhibitor of Gli1 activity [9]. The cerebellar granule neuron precursors seem to be responsible for originating the Shh-dependent medulloblastoma [111,112]. In addition to N-myc, a common proto-oncogene [113], another transcription factor that participates in the canonical Shh signaling-driven promotion of neuronal precursor expansion is Atoh1 [114]; Shh prevents Atoh1 degradation by the E3 ubiquitin ligase Huwe1, which results in medulloblastoma when knocked out in mice [115]. Opposing the Shh signaling-driven medulloblastoma cell proliferation and progression is the G protein Gαs, which localizes to the primary cilium in granule neuron precursors and downregulates canonical Shh signaling through a cAMP-driven pathway [116]. One of the hallmarks and critical events in the progression of medulloblastoma is the loss of Ptch1 expression, which deregulates Shh signaling [117–119]. The Shh binding protein Boc is upregulated in medulloblastomas and contributes to the cerebellar granule cell precursor expansion by facilitating aberrantly high Shh signaling, which leads to increased CyclinD1 activity and Ptch1 loss of heterozygosity [120].

In stark contrast to the Shh-dependent overproliferation leading to cerebellar tumors lies the phenotype afflicting Down syndrome patients and the segmentally trisomic postnatal mouse model for Down syndrome (Ts65Dn), which exhibit a smaller cerebellum due to a reduction in the number of granule neurons [121–123]. In Ts65Dn mice, the reduction in granule neurons results from deficient Shh-dependent mitogenic signaling in cerebellar granule neuron precursors during early postnatal development [124]. In an in vitro model, neural precursor cells from the subventricular zone and hippocampus of Ts65Dn mice show reduced canonical Shh signaling, which appears to be rooted in overexpressed Ptch1 due to transcriptional activation of Ptch1 expression by the amyloid precursor protein intracellular domain (AICD), a transcription-promoting fragment of amyloid precursor protein [125]. Moreover, treating newborn trisomic mice with the Smo agonist SAG restores cerebellar granule cell proliferation in one-week-old pups [124] and normal cerebellar morphology in adults [126]. Remarkably, SAG treatment also rescues behavioral deficits dependent on hippocampal function in the Down syndrome mouse model and partially restores theta burst stimulation-induced long-term potentiation in hippocampal slices from these animals [126]. It follows that Down syndrome, as well as other neurodevelopmental disorders, may be responsive to therapeutic intervention targeted to Shh signaling.

Regeneration of damaged neural tissue relies on many cellular events characteristic of neural development. Thus, it is not unexpected to find that Shh appears to be a vital regulator of post-injury neural regeneration. Indeed, Shh and Smo expression are upregulated in facial motor neurons after facial nerve axotomy in adult rats, and cyclopamine, a Smo antagonist, leads to a decrease in the number of surviving motor neurons after injury [127]. Similarly, after zebrafish spinal cord transection, there is an increase in Shh expression in the floor plate and upregulation of Ptch1 and Smo in ependymoradial glial cells, as well as the number of cells expressing the pre-motor neuron markers Olig2, Nkx6.1 and Pax6, which results in motor neuron regeneration, an effect that is prevented by Smo inhibitor cyclopamine [128]. In mice, invasive injury to the somatosensory cortex results in a regenerative response from astrocytes as a consequence of an inflammation-induced reactive gliosis that upregulates Shh signaling, which, in turn, leads to the generation of neural stem cells in the site of injury [129,130]. Moreover, enhancing Shh signaling by treating post-ischemic stroke mice with the Smo agonist purmorphamine is neuroprotective and improves recovery [131]. Intriguingly, recruitment of adult neural stem cells from the subventricular zone to demyelinating regions in the mouse forebrain is enhanced when the canonical Shh effector Gli1 is inhibited, while eliminating Smo-mediated signaling is detrimental for oligodendrocyte differentiation and remyelination [132]. These findings suggest that similarly to development, appropriate regeneration may require a switch from canonical to non-canonical Shh signaling.

9. Conclusions

Shh is a fascinating signaling molecule that is paramount in regulating nervous system development and function. Its relevance casts it as a target of interest for therapeutics of several neurodevelopmental diseases. However, every aspect of its signaling has some intricacy that demands close investigation, from its delivery and receptor molecules, to the subcellular localization of its signaling components, including the participation of the primary cilium in orchestrating Shh signaling in vertebrates, and the downstream pathways. Future challenges include understanding how Shh signaling transitions between functions during nervous system development and identifying the mechanisms that are targeted in disease and the regeneration of neural tissue. Taking into account that there is far more to Shh signaling than its mitogenic action and neural progenitor specification will allow for further discoveries of novel mechanisms mediated by Shh during neural development, disease and regeneration.

Acknowledgments: The work in the Borodinsky lab has been supported by the Basil O'Connor Starter Scholar Research Award Grant 5-FY09-131 from the March of Dimes Foundation, Klingenstein Foundation Award in Neuroscience, NSF 1120796, NIH-NINDS R01NS073055, and Shriners Hospital for Children 86500-NCA, 85220-NCA, 85300-NCA grants to Laura N. Borodinsky and by Shriners Hospital for Children postdoctoral fellowships to Yesser H. Belgacem and Andrew M. Hamilton.

Conflicts of Interest: The authors declare no conflict of interest.

References

1. Goodrich, L.V.; Milenkovic, L.; Higgins, K.M.; Scott, M.P. Altered neural cell fates and medulloblastoma in mouse *patched* mutants. *Science* **1997**, *277*, 1109–1113. [CrossRef] [PubMed]
2. Ingham, P.W.; Taylor, A.M.; Nakano, Y. Role of the *Drosophila patched* gene in positional signalling. *Nature* **1991**, *353*, 184–187. [CrossRef] [PubMed]
3. Alcedo, J.; Ayzenzon, M.; von Ohlen, T.; Noll, M.; Hooper, J.E. The *Drosophila smoothened* gene encodes a seven-pass membrane protein, a putative receptor for the Hedgehog signal. *Cell* **1996**, *86*, 221–232. [CrossRef]

4. Van den Heuvel, M.; Ingham, P.W. *Smoothened* encodes a receptor-like serpentine protein required for Hedgehog signalling. *Nature* **1996**, *382*, 547–551. [CrossRef] [PubMed]
5. Chen, Y.; Struhl, G. Dual roles for Patched in sequestering and transducing Hedgehog. *Cell* **1996**, *87*, 553–563. [CrossRef]
6. Hynes, M.; Stone, D.M.; Dowd, M.; Pitts-Meek, S.; Goddard, A.; Gurney, A.; Rosenthal, A. Control of cell pattern in the neural tube by the zinc finger transcription factor and oncogene *Gli-1*. *Neuron* **1997**, *19*, 15–26. [CrossRef]
7. Dominguez, M.; Brunner, M.; Hafen, E.; Basler, K. Sending and receiving the Hedgehog signal: Control by the *Drosophila* Gli protein Cubitus interruptus. *Science* **1996**, *272*, 1621–1625. [CrossRef] [PubMed]
8. Rowitch, D.H.; St.-Jaques, B.; Lee, S.M.; Flax, J.D.; Snyder, E.Y.; McMahon, A.P. *Sonic hedgehog* regulates proliferation and inhibits differentiation of CNS precursor cells. *J. Neurosci.* **1999**, *19*, 8954–8965.
9. Stecca, B.; Ruiz i Altaba, A. A Gli1–p53 inhibitory loop controls neural stem cell and tumour cell numbers. *EMBO J.* **2009**, *28*, 663–676.
10. Wechsler-Reya, R.J.; Scott, M.P. Control of neuronal precursor proliferation in the cerebellum by Sonic hedgehog. *Neuron* **1999**, *22*, 103–114. [CrossRef]
11. Izzi, L.; Levesque, M.; Morin, S.; Laniel, D.; Wilkes, B.C.; Mille, F.; Krauss, R.S.; McMahon, A.P.; Allen, B.L.; Charron, F. Boc and Gas1 each form distinct Shh receptor complexes with Ptch1 and are required for Shh-mediated cell proliferation. *Dev. Cell* **2011**, *20*, 788–801. [CrossRef] [PubMed]
12. Lai, K.; Kaspar, B.K.; Gage, F.H.; Schaffer, D.V. Sonic hedgehog regulates adult neural progenitor proliferation in vitro and in vivo. *Nat. Neurosci.* **2003**, *6*, 21–27. [CrossRef] [PubMed]
13. Machold, R.; Hayashi, S.; Rutlin, M.; Muzumdar, M.D.; Nery, S.; Corbin, J.G.; Gritli-Linde, A.; Dellovade, T.; Porter, J.A.; Rubin, L.L.; et al. Sonic hedgehog is required for progenitor cell maintenance in telencephalic stem cell niches. *Neuron* **2003**, *39*, 937–950. [CrossRef]
14. Han, Y.G.; Spassky, N.; Romaguera-Ros, M.; Garcia-Verdugo, J.M.; Aguilar, A.; Schneider-Maunoury, S.; Alvarez-Buylla, A. Hedgehog signaling and primary cilia are required for the formation of adult neural stem cells. *Nat. Neurosci.* **2008**, *11*, 277–284. [CrossRef] [PubMed]
15. Tong, C.K.; Han, Y.G.; Shah, J.K.; Obernier, K.; Guinto, C.D.; Alvarez-Buylla, A. Primary cilia are required in a unique subpopulation of neural progenitors. *Proc. Natl. Acad. Sci. USA* **2014**, *111*, 12438–12443. [CrossRef] [PubMed]
16. Petrova, R.; Garcia, A.D.; Joyner, A.L. Titration of GLI3 repressor activity by sonic hedgehog signaling is critical for maintaining multiple adult neural stem cell and astrocyte functions. *J. Neurosci.* **2013**, *33*, 17490–17505. [CrossRef] [PubMed]
17. Cayuso, J.; Ulloa, F.; Cox, B.; Briscoe, J.; Marti, E. The Sonic hedgehog pathway independently controls the patterning, proliferation and survival of neuroepithelial cells by regulating Gli activity. *Development* **2006**, *133*, 517–528. [CrossRef] [PubMed]
18. Alvarez-Medina, R.; Le Dreau, G.; Ros, M.; Marti, E. Hedgehog activation is required upstream of Wnt signalling to control neural progenitor proliferation. *Development* **2009**, *136*, 3301–3309. [CrossRef] [PubMed]
19. Locker, M.; Agathocleous, M.; Amato, M.A.; Parain, K.; Harris, W.A.; Perron, M. Hedgehog signaling and the retina: Insights into the mechanisms controlling the proliferative properties of neural precursors. *Genes Dev.* **2006**, *20*, 3036–3048. [CrossRef] [PubMed]
20. Chai, P.C.; Liu, Z.; Chia, W.; Cai, Y. Hedgehog signaling acts with the temporal cascade to promote neuroblast cell cycle exit. *PLoS Biol.* **2013**, *11*, e1001494. [CrossRef] [PubMed]
21. Chan, J.A.; Balasubramanian, S.; Witt, R.M.; Nazemi, K.J.; Choi, Y.; Pazyra-Murphy, M.F.; Walsh, C.O.; Thompson, M.; Segal, R.A. Proteoglycan interactions with Sonic Hedgehog specify mitogenic responses. *Nat. Neurosci.* **2009**, *12*, 409–417. [CrossRef] [PubMed]
22. Wijgerde, M.; McMahon, J.A.; Rule, M.; McMahon, A.P. A direct requirement for Hedgehog signaling for normal specification of all ventral progenitor domains in the presumptive mammalian spinal cord. *Genes Dev.* **2002**, *16*, 2849–2864. [CrossRef] [PubMed]
23. Briscoe, J.; Pierani, A.; Jessell, T.M.; Ericson, J. A homeodomain protein code specifies progenitor cell identity and neuronal fate in the ventral neural tube. *Cell* **2000**, *101*, 435–445. [CrossRef]
24. Briscoe, J.; Sussel, L.; Serup, P.; Hartigan-O'Connor, D.; Jessell, T.M.; Rubenstein, J.L.; Ericson, J. Homeobox gene Nkx2.2 and specification of neuronal identity by graded Sonic hedgehog signalling. *Nature* **1999**, *398*, 622–627. [PubMed]

25. Ericson, J.; Morton, S.; Kawakami, A.; Roelink, H.; Jessell, T.M. Two critical periods of Sonic hedgehog signaling required for the specification of motor neuron identity. *Cell* **1996**, *87*, 661–673. [CrossRef]
26. Ericson, J.; Rashbass, P.; Schedl, A.; Brenner-Morton, S.; Kawakami, A.; van Heyningen, V.; Jessell, T.M.; Briscoe, J. Pax6 controls progenitor cell identity and neuronal fate in response to graded Shh signaling. *Cell* **1997**, *90*, 169–180. [CrossRef]
27. Ericson, J.; Thor, S.; Edlund, T.; Jessell, T.M.; Yamada, T. Early stages of motor neuron differentiation revealed by expression of homeobox gene Islet-1. *Science* **1992**, *256*, 1555–1560. [CrossRef] [PubMed]
28. Xiong, F.; Tentner, A.R.; Huang, P.; Gelas, A.; Mosaliganti, K.R.; Souhait, L.; Rannou, N.; Swinburne, I.A.; Obholzer, N.D.; Cowgill, P.D.; et al. Specified neural progenitors sort to form sharp domains after noisy Shh signaling. *Cell* **2013**, *153*, 550–561. [CrossRef] [PubMed]
29. Dessaud, E.; Ribes, V.; Balaskas, N.; Yang, L.L.; Pierani, A.; Kicheva, A.; Novitch, B.G.; Briscoe, J.; Sasai, N. Dynamic assignment and maintenance of positional identity in the ventral neural tube by the morphogen Sonic hedgehog. *PLoS Biol.* **2010**, *8*, e1000382. [CrossRef] [PubMed]
30. Balaskas, N.; Ribeiro, A.; Panovska, J.; Dessaud, E.; Sasai, N.; Page, K.M.; Briscoe, J.; Ribes, V. Gene regulatory logic for reading the Sonic Hedgehog signaling gradient in the vertebrate neural tube. *Cell* **2012**, *148*, 273–284. [CrossRef] [PubMed]
31. Nishi, Y.; Zhang, X.; Jeong, J.; Peterson, K.A.; Vedenko, A.; Bulyk, M.L.; Hide, W.A.; McMahon, A.P. A direct fate exclusion mechanism by Sonic hedgehog-regulated transcriptional repressors. *Development* **2015**, *142*, 3286–3293. [CrossRef] [PubMed]
32. Peterson, K.A.; Nishi, Y.; Ma, W.; Vedenko, A.; Shokri, L.; Zhang, X.; McFarlane, M.; Baizabal, J.M.; Junker, J.P.; van Oudenaarden, A.; et al. Neural-specific Sox2 input and differential Gli-binding affinity provide context and positional information in Shh-directed neural patterning. *Genes Dev.* **2012**, *26*, 2802–2816. [CrossRef] [PubMed]
33. Lei, Q.; Jeong, Y.; Misra, K.; Li, S.; Zelman, A.K.; Epstein, D.J.; Matise, M.P. Wnt signaling inhibitors regulate the transcriptional response to morphogenetic Shh-Gli signaling in the neural tube. *Dev. Cell* **2006**, *11*, 325–337. [CrossRef] [PubMed]
34. Oosterveen, T.; Kurdija, S.; Alekseenko, Z.; Uhde, C.W.; Bergsland, M.; Sandberg, M.; Andersson, E.; Dias, J.M.; Muhr, J.; Ericson, J. Mechanistic differences in the transcriptional interpretation of local and long-range Shh morphogen signaling. *Dev. Cell* **2012**, *23*, 1006–1019. [CrossRef] [PubMed]
35. Wang, H.; Lei, Q.; Oosterveen, T.; Ericson, J.; Matise, M.P. Tcf/Lef repressors differentially regulate Shh-Gli target gene activation thresholds to generate progenitor patterning in the developing CNS. *Development* **2011**, *138*, 3711–3721. [CrossRef] [PubMed]
36. Pringle, N.P.; Yu, W.P.; Guthrie, S.; Roelink, H.; Lumsden, A.; Peterson, A.C.; Richardson, W.D. Determination of neuroepithelial cell fate: Induction of the oligodendrocyte lineage by ventral midline cells and Sonic hedgehog. *Dev. Biol.* **1996**, *177*, 30–42. [CrossRef] [PubMed]
37. Orentas, D.M.; Hayes, J.E.; Dyer, K.L.; Miller, R.H. Sonic hedgehog signaling is required during the appearance of spinal cord oligodendrocyte precursors. *Development* **1999**, *126*, 2419–2429. [PubMed]
38. Lu, Q.R.; Yuk, D.; Alberta, J.A.; Zhu, Z.; Pawlitzky, I.; Chan, J.; McMahon, A.P.; Stiles, C.D.; Rowitch, D.H. Sonic hedgehog–regulated oligodendrocyte lineage genes encoding bHLH proteins in the mammalian central nervous system. *Neuron* **2000**, *25*, 317–329. [CrossRef]
39. Takebayashi, H.; Nabeshima, Y.; Yoshida, S.; Chisaka, O.; Ikenaka, K.; Nabeshima, Y. The basic helix-loop-helix factor olig2 is essential for the development of motoneuron and oligodendrocyte lineages. *Curr. Biol.* **2002**, *12*, 1157–1163. [CrossRef]
40. Li, H.; de Faria, J.P.; Andrew, P.; Nitarska, J.; Richardson, W.D. Phosphorylation regulates OLIG2 cofactor choice and the motor neuron-oligodendrocyte fate switch. *Neuron* **2011**, *69*, 918–929. [CrossRef] [PubMed]
41. Agarwala, S.; Sanders, T.A.; Ragsdale, C.W. Sonic hedgehog control of size and shape in midbrain pattern formation. *Science* **2001**, *291*, 2147–2150. [CrossRef] [PubMed]
42. Kiecker, C.; Lumsden, A. Hedgehog signaling from the ZLI regulates diencephalic regional identity. *Nat. Neurosci.* **2004**, *7*, 1242–1249. [CrossRef] [PubMed]
43. Vieira, C.; Garda, A.L.; Shimamura, K.; Martinez, S. Thalamic development induced by Shh in the chick embryo. *Dev. Biol.* **2005**, *284*, 351–363. [CrossRef] [PubMed]

44. Vue, T.Y.; Bluske, K.; Alishahi, A.; Yang, L.L.; Koyano-Nakagawa, N.; Novitch, B.; Nakagawa, Y. Sonic hedgehog signaling controls thalamic progenitor identity and nuclei specification in mice. *J. Neurosci.* **2009**, *29*, 4484–4497. [CrossRef] [PubMed]
45. Jeong, Y.; Dolson, D.K.; Waclaw, R.R.; Matise, M.P.; Sussel, L.; Campbell, K.; Kaestner, K.H.; Epstein, D.J. Spatial and temporal requirements for sonic hedgehog in the regulation of thalamic interneuron identity. *Development* **2011**, *138*, 531–541. [CrossRef] [PubMed]
46. Flandin, P.; Zhao, Y.; Vogt, D.; Jeong, J.; Long, J.; Potter, G.; Westphal, H.; Rubenstein, J.L. *Lhx6* and *Lhx8* coordinately induce neuronal expression of *Shh* that controls the generation of interneuron progenitors. *Neuron* **2011**, *70*, 939–950. [CrossRef] [PubMed]
47. Sousa, V.H.; Fishell, G. Sonic hedgehog functions through dynamic changes in temporal competence in the developing forebrain. *Curr. Opin. Genet. Dev.* **2010**, *20*, 391–399. [CrossRef] [PubMed]
48. Fotaki, V.; Yu, T.; Zaki, P.A.; Mason, J.O.; Price, D.J. Abnormal positioning of diencephalic cell types in neocortical tissue in the dorsal telencephalon of mice lacking functional Gli3. *J. Neurosci.* **2006**, *26*, 9282–9292. [CrossRef] [PubMed]
49. Palma, V.; Ruiz i Altaba, A. Hedgehog-GLI signaling regulates the behavior of cells with stem cell properties in the developing neocortex. *Development* **2004**, *131*, 337–345. [CrossRef] [PubMed]
50. Wang, H.; Ge, G.; Uchida, Y.; Luu, B.; Ahn, S. *Gli3* is required for maintenance and fate specification of cortical progenitors. *J. Neurosci.* **2011**, *31*, 6440–6448. [CrossRef] [PubMed]
51. Wilson, S.L.; Wilson, J.P.; Wang, C.; Wang, B.; McConnell, S.K. Primary cilia and Gli3 activity regulate cerebral cortical size. *Dev. Neurobiol.* **2012**, *72*, 1196–1212. [CrossRef] [PubMed]
52. Yabut, O.R.; Fernandez, G.; Huynh, T.; Yoon, K.; Pleasure, S.J. Suppressor of fused is critical for maintenance of neuronal progenitor identity during corticogenesis. *Cell Rep.* **2015**, *12*, 2021–2034. [CrossRef] [PubMed]
53. Heberlein, U.; Singh, C.M.; Luk, A.Y.; Donohoe, T.J. Growth and differentiation in the *Drosophila* eye coordinated by *hedgehog*. *Nature* **1995**, *373*, 709–711. [CrossRef] [PubMed]
54. Heberlein, U.; Wolff, T.; Rubin, G.M. The TGF beta homolog *dpp* and the segment polarity gene *hedgehog* are required for propagation of a morphogenetic wave in the *Drosophila* retina. *Cell* **1993**, *75*, 913–926. [CrossRef]
55. Ma, C.; Zhou, Y.; Beachy, P.A.; Moses, K. The segment polarity gene *hedgehog* is required for progression of the morphogenetic furrow in the developing *Drosophila* eye. *Cell* **1993**, *75*, 927–938. [CrossRef]
56. Neumann, C.J.; Nuesslein-Volhard, C. Patterning of the zebrafish retina by a wave of Sonic hedgehog activity. *Science* **2000**, *289*, 2137–2139. [CrossRef] [PubMed]
57. Tanabe, Y.; William, C.; Jessell, T.M. Specification of motor neuron identity by the MNR2 homeodomain protein. *Cell* **1998**, *95*, 67–80. [CrossRef]
58. Eisen, J.S. Determination of primary motoneuron identity in developing zebrafish embryos. *Science* **1991**, *252*, 569–572. [CrossRef] [PubMed]
59. Hanson, M.G.; Landmesser, L.T. Normal patterns of spontaneous activity are required for correct motor axon guidance and the expression of specific guidance molecules. *Neuron* **2004**, *43*, 687–701. [CrossRef] [PubMed]
60. Ericson, J.; Muhr, J.; Placzek, M.; Lints, T.; Jessell, T.M.; Edlund, T. Sonic hedgehog induces the differentiation of ventral forebrain neurons: A common signal for ventral patterning within the neural tube. *Cell* **1995**, *81*, 747–756. [CrossRef]
61. Belgacem, Y.H.; Borodinsky, L.N. Sonic hedgehog signaling is decoded by calcium spike activity in the developing spinal cord. *Proc. Natl. Acad. Sci. USA* **2011**, *108*, 4482–4487. [CrossRef] [PubMed]
62. Borodinsky, L.N.; Root, C.M.; Cronin, J.A.; Sann, S.B.; Gu, X.; Spitzer, N.C. Activity-dependent homeostatic specification of transmitter expression in embryonic neurons. *Nature* **2004**, *429*, 523–530. [CrossRef] [PubMed]
63. Root, C.M.; Velazquez-Ulloa, N.A.; Monsalve, G.C.; Minakova, E.; Spitzer, N.C. Embryonically expressed GABA and glutamate drive electrical activity regulating neurotransmitter specification. *J. Neurosci.* **2008**, *28*, 4777–4784. [CrossRef] [PubMed]
64. Swapna, I.; Borodinsky, L.N. Interplay between electrical activity and bone morphogenetic protein signaling regulates spinal neuron differentiation. *Proc. Natl. Acad. Sci. USA* **2012**, *109*, 16336–16341. [CrossRef] [PubMed]
65. Marek, K.W.; Kurtz, L.M.; Spitzer, N.C. cJun integrates calcium activity and *tlx3* expression to regulate neurotransmitter specification. *Nat. Neurosci.* **2010**, *13*, 944–950. [CrossRef] [PubMed]
66. Demarque, M.; Spitzer, N.C. Activity-dependent expression of Lmx1b regulates specification of serotonergic neurons modulating swimming behavior. *Neuron* **2010**, *67*, 321–334. [CrossRef] [PubMed]

67. Dulcis, D.; Spitzer, N.C. Illumination controls differentiation of dopamine neurons regulating behaviour. *Nature* **2008**, *456*, 195–201. [CrossRef] [PubMed]
68. Dulcis, D.; Jamshidi, P.; Leutgeb, S.; Spitzer, N.C. Neurotransmitter switching in the adult brain regulates behavior. *Science* **2013**, *340*, 449–453. [CrossRef] [PubMed]
69. Farmer, W.T.; Abrahamsson, T.; Chierzi, S.; Lui, C.; Zaelzer, C.; Jones, E.V.; Bally, B.P.; Chen, G.G.; Theroux, J.F.; Peng, J.; et al. Neurons diversify astrocytes in the adult brain through Sonic hedgehog signaling. *Science* **2016**, *351*, 849–854. [CrossRef] [PubMed]
70. Strahle, U.; Fischer, N.; Blader, P. Expression and regulation of a *netrin* homologue in the zebrafish embryo. *Mech. Dev.* **1997**, *62*, 147–160. [CrossRef]
71. Charron, F.; Stein, E.; Jeong, J.; McMahon, A.P.; Tessier-Lavigne, M. The morphogen Sonic hedgehog is an axonal chemoattractant that collaborates with Netrin-1 in midline axon guidance. *Cell* **2003**, *113*, 11–23. [CrossRef]
72. Yam, P.T.; Langlois, S.D.; Morin, S.; Charron, F. Sonic hedgehog guides axons through a noncanonical, Src-family-kinase-dependent signaling pathway. *Neuron* **2009**, *62*, 349–362. [CrossRef] [PubMed]
73. Okada, A.; Charron, F.; Morin, S.; Shin, D.S.; Wong, K.; Fabre, P.J.; Tessier-Lavigne, M.; McConnell, S.K. Boc is a receptor for Sonic hedgehog in the guidance of commissural axons. *Nature* **2006**, *444*, 369–373. [CrossRef] [PubMed]
74. Sloan, T.F.; Qasaimeh, M.A.; Juncker, D.; Yam, P.T.; Charron, F. Integration of shallow gradients of Shh and Netrin-1 guides commissural axons. *PLoS Biol.* **2015**, *13*, e1002119. [CrossRef] [PubMed]
75. Bourikas, D.; Pekarik, V.; Baeriswyl, T.; Grunditz, A.; Sadhu, R.; Nardo, M.; Stoeckli, E.T. Sonic hedgehog guides commissural axons along the longitudinal axis of the spinal cord. *Nat. Neurosci.* **2005**, *8*, 297–304. [CrossRef] [PubMed]
76. Yam, P.T.; Kent, C.B.; Morin, S.; Farmer, W.T.; Alchini, R.; Lepelletier, L.; Colman, D.R.; Tessier-Lavigne, M.; Fournier, A.E.; Charron, F. 14-3-3 proteins regulate a cell-intrinsic switch from Sonic hedgehog-mediated commissural axon attraction to repulsion after midline crossing. *Neuron* **2012**, *76*, 735–749. [CrossRef] [PubMed]
77. Trousse, F.; Marti, E.; Gruss, P.; Torres, M.; Bovolenta, P. Control of retinal ganglion cell axon growth: A new role for Sonic hedgehog. *Development* **2001**, *128*, 3927–3936. [PubMed]
78. Kolpak, A.; Zhang, J.; Bao, Z.Z. Sonic hedgehog has a dual effect on the growth of retinal ganglion axons depending on its concentration. *J. Neurosci.* **2005**, *25*, 3432–3441. [CrossRef] [PubMed]
79. Fabre, P.J.; Shimogori, T.; Charron, F. Segregation of ipsilateral retinal ganglion cell axons at the optic chiasm requires the Shh receptor Boc. *J. Neurosci.* **2010**, *30*, 266–275. [CrossRef] [PubMed]
80. Barresi, M.J.; Hutson, L.D.; Chien, C.B.; Karlstrom, R.O. Hedgehog regulated Slit expression determines commissure and glial cell position in the zebrafish forebrain. *Development* **2005**, *132*, 3643–3656. [CrossRef] [PubMed]
81. Stacher Horndli, C.; Chien, C.B. Sonic hedgehog is indirectly required for intraretinal axon pathfinding by regulating chemokine expression in the optic stalk. *Development* **2012**, *139*, 2604–2613. [CrossRef] [PubMed]
82. Jin, S.; Martinelli, D.C.; Zheng, X.; Tessier-Lavigne, M.; Fan, C.M. Gas1 is a receptor for Sonic hedgehog to repel enteric axons. *Proc. Natl. Acad. Sci. USA* **2015**, *112*, E73–E80. [CrossRef] [PubMed]
83. Chu, T.; Chiu, M.; Zhang, E.; Kunes, S. A C-terminal motif targets Hedgehog to axons, coordinating assembly of the *Drosophila* eye and brain. *Dev. Cell* **2006**, *10*, 635–646. [CrossRef] [PubMed]
84. Umetsu, D.; Murakami, S.; Sato, M.; Tabata, T. The highly ordered assembly of retinal axons and their synaptic partners is regulated by Hedgehog/Single-minded in the *Drosophila* visual system. *Development* **2006**, *133*, 791–800. [CrossRef] [PubMed]
85. Chou, Y.H.; Zheng, X.; Beachy, P.A.; Luo, L. Patterning axon targeting of olfactory receptor neurons by coupled hedgehog signaling at two distinct steps. *Cell* **2010**, *142*, 954–966. [CrossRef] [PubMed]
86. Masdeu, C.; Bernard, V.; Faure, H.; Traiffort, E.; Ruat, M. Distribution of Smoothened at hippocampal mossy fiber synapses. *Neuroreport* **2007**, *18*, 395–399. [CrossRef] [PubMed]
87. Petralia, R.S.; Schwartz, C.M.; Wang, Y.X.; Mattson, M.P.; Yao, P.J. Subcellular localization of Patched and Smoothened, the receptors for Sonic hedgehog signaling, in the hippocampal neuron. *J. Comp. Neurol.* **2011**, *519*, 3684–3699. [CrossRef] [PubMed]
88. Petralia, R.S.; Wang, Y.X.; Mattson, M.P.; Yao, P.J. Sonic hedgehog distribution within mature hippocampal neurons. *Commun. Integr. Biol.* **2011**, *4*, 775–777. [CrossRef] [PubMed]

89. Harwell, C.C.; Parker, P.R.; Gee, S.M.; Okada, A.; McConnell, S.K.; Kreitzer, A.C.; Kriegstein, A.R. Sonic hedgehog expression in corticofugal projection neurons directs cortical microcircuit formation. *Neuron* **2012**, *73*, 1116–1126. [CrossRef] [PubMed]
90. Mitchell, N.; Petralia, R.S.; Currier, D.G.; Wang, Y.X.; Kim, A.; Mattson, M.P.; Yao, P.J. Sonic hedgehog regulates presynaptic terminal size, ultrastructure and function in hippocampal neurons. *J. Cell Sci.* **2012**, *125*, 4207–4213. [CrossRef] [PubMed]
91. Dale, J.K.; Vesque, C.; Lints, T.J.; Sampath, T.K.; Furley, A.; Dodd, J.; Placzek, M. Cooperation of BMP7 and SHH in the induction of forebrain ventral midline cells by prechordal mesoderm. *Cell* **1997**, *90*, 257–269. [CrossRef]
92. Liem, K.F., Jr.; Jessell, T.M.; Briscoe, J. Regulation of the neural patterning activity of Sonic hedgehog by secreted BMP inhibitors expressed by notochord and somites. *Development* **2000**, *127*, 4855–4866. [PubMed]
93. Dutton, R.; Yamada, T.; Turnley, A.; Bartlett, P.F.; Murphy, M. Sonic hedgehog promotes neuronal differentiation of murine spinal cord precursors and collaborates with neurotrophin 3 to induce Islet-1. *J. Neurosci.* **1999**, *19*, 2601–2608. [PubMed]
94. Kucera, J.; Fan, G.; Jaenisch, R.; Linnarsson, S.; Ernfors, P. Dependence of developing group Ia afferents on neurotrophin-3. *J. Comp. Neurol.* **1995**, *363*, 307–320. [CrossRef] [PubMed]
95. Novitch, B.G.; Wichterle, H.; Jessell, T.M.; Sockanathan, S. A requirement for retinoic acid-mediated transcriptional activation in ventral neural patterning and motor neuron specification. *Neuron* **2003**, *40*, 81–95. [CrossRef] [PubMed]
96. Pons, S.; Marti, E. Sonic hedgehog synergizes with the extracellular matrix protein vitronectin to induce spinal motor neuron differentiation. *Development* **2000**, *127*, 333–342. [PubMed]
97. Pons, S.; Trejo, J.L.; Martinez-Morales, J.R.; Marti, E. Vitronectin regulates Sonic hedgehog activity during cerebellum development through CREB phosphorylation. *Development* **2001**, *128*, 1481–1492. [PubMed]
98. Fogarty, M.P.; Emmenegger, B.A.; Grasfeder, L.L.; Oliver, T.G.; Wechsler-Reya, R.J. Fibroblast growth factor blocks Sonic hedgehog signaling in neuronal precursors and tumor cells. *Proc. Natl. Acad. Sci. USA* **2007**, *104*, 2973–2978. [CrossRef] [PubMed]
99. Bluske, K.K.; Vue, T.Y.; Kawakami, Y.; Taketo, M.M.; Yoshikawa, K.; Johnson, J.E.; Nakagawa, Y. β-Catenin signaling specifies progenitor cell identity in parallel with Shh signaling in the developing mammalian thalamus. *Development* **2012**, *139*, 2692–2702. [CrossRef] [PubMed]
100. Tuson, M.; He, M.; Anderson, K.V. Protein kinase A acts at the basal body of the primary cilium to prevent Gli2 activation and ventralization of the mouse neural tube. *Development* **2011**, *138*, 4921–4930. [CrossRef] [PubMed]
101. Reimer, M.M.; Norris, A.; Ohnmacht, J.; Patani, R.; Zhong, Z.; Dias, T.B.; Kuscha, V.; Scott, A.L.; Chen, Y.C.; Rozov, S.; et al. Dopamine from the brain promotes spinal motor neuron generation during development and adult regeneration. *Dev. Cell* **2013**, *25*, 478–491. [CrossRef] [PubMed]
102. Kong, J.H.; Yang, L.; Dessaud, E.; Chuang, K.; Moore, D.M.; Rohatgi, R.; Briscoe, J.; Novitch, B.G. Notch activity modulates the responsiveness of neural progenitors to Sonic hedgehog signaling. *Dev. Cell* **2015**, *33*, 373–387. [CrossRef] [PubMed]
103. Di Marcotullio, L.; Ferretti, E.; Greco, A.; de Smaele, E.; Po, A.; Sico, M.A.; Alimandi, M.; Giannini, G.; Maroder, M.; Screpanti, I.; et al. Numb is a suppressor of Hedgehog signalling and targets Gli1 for Itch- dependent ubiquitination. *Nat. Cell Biol.* **2006**, *8*, 1415–1423. [CrossRef] [PubMed]
104. Imayoshi, I.; Sakamoto, M.; Yamaguchi, M.; Mori, K.; Kageyama, R. Essential roles of Notch signaling in maintenance of neural stem cells in developing and adult brains. *J. Neurosci.* **2010**, *30*, 3489–3498. [CrossRef] [PubMed]
105. Stasiulewicz, M.; Gray, S.D.; Mastromina, I.; Silva, J.C.; Bjorklund, M.; Seymour, P.A.; Booth, D.; Thompson, C.; Green, R.J.; Hall, E.A.; et al. A conserved role for Notch signaling in priming the cellular response to Shh through ciliary localisation of the key Shh transducer Smo. *Development* **2015**, *142*, 2291–2303. [CrossRef] [PubMed]
106. Lee, J.; Platt, K.A.; Censullo, P.; Ruiz i Altaba, A. Gli1 is a target of Sonic hedgehog that induces ventral neural tube development. *Development* **1997**, *124*, 2537–2552. [PubMed]
107. Belgacem, Y.H.; Borodinsky, L.N. Inversion of Sonic hedgehog action on its canonical pathway by electrical activity. *Proc. Natl. Acad. Sci. USA* **2015**, *112*, 4140–4145. [CrossRef] [PubMed]

108. Villavicencio, E.H.; Walterhouse, D.O.; Iannaccone, P.M. The Sonic hedgehog–Patched–Gli pathway in human development and disease. *Am. J. Hum. Genet.* **2000**, *67*, 1047–1054. [CrossRef]
109. Pietsch, T.; Waha, A.; Koch, A.; Kraus, J.; Albrecht, S.; Tonn, J.; Sorensen, N.; Berthold, F.; Henk, B.; Schmandt, N.; et al. Medulloblastomas of the desmoplastic variant carry mutations of the human homologue of *Drosophila patched*. *Cancer Res.* **1997**, *57*, 2085–2088. [PubMed]
110. Stecca, B.; Ruiz i Altaba, A. Brain as a paradigm of organ growth: Hedgehog–Gli signaling in neural stem cells and brain tumors. *J. Neurobiol.* **2005**, *64*, 476–490. [CrossRef] [PubMed]
111. Schuller, U.; Heine, V.M.; Mao, J.; Kho, A.T.; Dillon, A.K.; Han, Y.G.; Huillard, E.; Sun, T.; Ligon, A.H.; Qian, Y.; et al. Acquisition of granule neuron precursor identity is a critical determinant of progenitor cell competence to form Shh-induced medulloblastoma. *Cancer Cell* **2008**, *14*, 123–134. [CrossRef] [PubMed]
112. Yang, Z.J.; Ellis, T.; Markant, S.L.; Read, T.A.; Kessler, J.D.; Bourboulas, M.; Schuller, U.; Machold, R.; Fishell, G.; Rowitch, D.H.; et al. Medulloblastoma can be initiated by deletion of Patched in lineage-restricted progenitors or stem cells. *Cancer Cell* **2008**, *14*, 135–145. [CrossRef] [PubMed]
113. Swartling, F.J.; Savov, V.; Persson, A.I.; Chen, J.; Hackett, C.S.; Northcott, P.A.; Grimmer, M.R.; Lau, J.; Chesler, L.; Perry, A.; et al. Distinct neural stem cell populations give rise to disparate brain tumors in response to N-MYC. *Cancer Cell* **2012**, *21*, 601–613. [CrossRef] [PubMed]
114. Flora, A.; Klisch, T.J.; Schuster, G.; Zoghbi, H.Y. Deletion of Atoh1 disrupts Sonic hedgehog signaling in the developing cerebellum and prevents medulloblastoma. *Science* **2009**, *326*, 1424–1427. [CrossRef] [PubMed]
115. Forget, A.; Bihannic, L.; Cigna, S.M.; Lefevre, C.; Remke, M.; Barnat, M.; Dodier, S.; Shirvani, H.; Mercier, A.; Mensah, A.; et al. Shh signaling protects Atoh1 from degradation mediated by the E3 ubiquitin ligase Huwe1 in neural precursors. *Dev. Cell* **2014**, *29*, 649–661. [CrossRef] [PubMed]
116. He, X.; Zhang, L.; Chen, Y.; Remke, M.; Shih, D.; Lu, F.; Wang, H.; Deng, Y.; Yu, Y.; Xia, Y.; et al. The G protein αsubunit Gαs is a tumor suppressor in Sonic hedgehog-driven medulloblastoma. *Nat. Med.* **2014**, *20*, 1035–1042. [CrossRef] [PubMed]
117. Berman, D.M.; Karhadkar, S.S.; Hallahan, A.R.; Pritchard, J.I.; Eberhart, C.G.; Watkins, D.N.; Chen, J.K.; Cooper, M.K.; Taipale, J.; Olson, J.M.; et al. Medulloblastoma growth inhibition by hedgehog pathway blockade. *Science* **2002**, *297*, 1559–1561. [CrossRef] [PubMed]
118. Oliver, T.G.; Read, T.A.; Kessler, J.D.; Mehmeti, A.; Wells, J.F.; Huynh, T.T.; Lin, S.M.; Wechsler-Reya, R.J. Loss of *patched* and disruption of granule cell development in a pre-neoplastic stage of medulloblastoma. *Development* **2005**, *132*, 2425–2439. [CrossRef] [PubMed]
119. Pazzaglia, S.; Tanori, M.; Mancuso, M.; Gessi, M.; Pasquali, E.; Leonardi, S.; Oliva, M.A.; Rebessi, S.; di Majo, V.; Covelli, V.; et al. Two-hit model for progression of medulloblastoma preneoplasia in Patched heterozygous mice. *Oncogene* **2006**, *25*, 5575–5580. [CrossRef] [PubMed]
120. Mille, F.; Tamayo-Orrego, L.; Levesque, M.; Remke, M.; Korshunov, A.; Cardin, J.; Bouchard, N.; Izzi, L.; Kool, M.; Northcott, P.A.; et al. The Shh receptor Boc promotes progression of early medulloblastoma to advanced tumors. *Dev. Cell* **2014**, *31*, 34–47. [CrossRef] [PubMed]
121. Crome, L.; Cowie, V.; Slater, E. Statistical Note on Cerebellar and Brain-Stem Weight in Mongolism. *J. Ment. Defic. Res.* **1966**, *10*, 69–72. [CrossRef]
122. Baxter, L.L.; Moran, T.H.; Richtsmeier, J.T.; Troncoso, J.; Reeves, R.H. Discovery and genetic localization of Down syndrome cerebellar phenotypes using the Ts65Dn mouse. *Hum. Mol. Genet.* **2000**, *9*, 195–202. [CrossRef] [PubMed]
123. Olson, L.E.; Roper, R.J.; Baxter, L.L.; Carlson, E.J.; Epstein, C.J.; Reeves, R.H. Down syndrome mouse models Ts65Dn, Ts1Cje, and Ms1Cje/Ts65Dn exhibit variable severity of cerebellar phenotypes. *Dev. Dyn.* **2004**, *230*, 581–589. [CrossRef] [PubMed]
124. Roper, R.J.; Baxter, L.L.; Saran, N.G.; Klinedinst, D.K.; Beachy, P.A.; Reeves, R.H. Defective cerebellar response to mitogenic Hedgehog signaling in Down's syndrome mice. *Proc. Natl. Acad. Sci. USA* **2006**, *103*, 1452–1456. [CrossRef] [PubMed]
125. Trazzi, S.; Mitrugno, V.M.; Valli, E.; Fuchs, C.; Rizzi, S.; Guidi, S.; Perini, G.; Bartesaghi, R.; Ciani, E. APP-dependent up-regulation of Ptch1 underlies proliferation impairment of neural precursors in Down syndrome. *Hum. Mol. Genet.* **2011**, *20*, 1560–1573. [CrossRef] [PubMed]
126. Das, I.; Park, J.M.; Shin, J.H.; Jeon, S.K.; Lorenzi, H.; Linden, D.J.; Worley, P.F.; Reeves, R.H. Hedgehog agonist therapy corrects structural and cognitive deficits in a Down syndrome mouse model. *Sci. Transl. Med.* **2013**, *5*, 201ra120. [CrossRef] [PubMed]

127. Akazawa, C.; Tsuzuki, H.; Nakamura, Y.; Sasaki, Y.; Ohsaki, K.; Nakamura, S.; Arakawa, Y.; Kohsaka, S. The upregulated expression of Sonic hedgehog in motor neurons after rat facial nerve axotomy. *J. Neurosci.* **2004**, *24*, 7923–7930. [CrossRef] [PubMed]
128. Reimer, M.M.; Kuscha, V.; Wyatt, C.; Sorensen, I.; Frank, R.E.; Knuwer, M.; Becker, T.; Becker, C.G. Sonic hedgehog is a polarized signal for motor neuron regeneration in adult zebrafish. *J. Neurosci.* **2009**, *29*, 15073–15082. [CrossRef] [PubMed]
129. Sirko, S.; Behrendt, G.; Johansson, P.A.; Tripathi, P.; Costa, M.; Bek, S.; Heinrich, C.; Tiedt, S.; Colak, D.; Dichgans, M.; et al. Reactive glia in the injured brain acquire stem cell properties in response to Sonic hedgehog. *Cell Stem Cell* **2013**, *12*, 426–439. [CrossRef] [PubMed]
130. Amankulor, N.M.; Hambardzumyan, D.; Pyonteck, S.M.; Becher, O.J.; Joyce, J.A.; Holland, E.C. Sonic hedgehog pathway activation is induced by acute brain injury and regulated by injury-related inflammation. *J. Neurosci.* **2009**, *29*, 10299–10308. [CrossRef] [PubMed]
131. Chechneva, O.V.; Mayrhofer, F.; Daugherty, D.J.; Krishnamurty, R.G.; Bannerman, P.; Pleasure, D.E.; Deng, W. A Smoothened receptor agonist is neuroprotective and promotes regeneration after ischemic brain injury. *Cell Death Dis.* **2014**, *5*, e1481. [CrossRef] [PubMed]
132. Samanta, J.; Grund, E.M.; Silva, H.M.; Lafaille, J.J.; Fishell, G.; Salzer, J.L. Inhibition of Gli1 mobilizes endogenous neural stem cells for remyelination. *Nature* **2015**, *526*, 448–452. [CrossRef] [PubMed]

13

Histologic Assessment of Drug-Eluting Grafts Related to Implantation Site

Jean-Christophe Tille [1,*], Sarra de Valence [2], Delia Mandracchia [3], Benjamin Nottelet [2], Francesco Innocente [4], Robert Gurny [2], Michael Möller [2] and Beat H. Walpoth [4]

1 Division of Clinical Pathology, University Hospital of Geneva, Geneva 1211, Switzerland
2 School of Pharmaceutical Sciences, University of Geneva, Geneva 1211, Switzerland; sarra.devalence@gmail.com (S.d.V.); benjamin.nottelet@univ-montp1.fr (B.N.); Robert.Gurny@unige.ch (R.G.); Michael.Moeller@unige.ch (M.M.)
3 Department of Pharmacy-Drug Sciences, University of Bari "Aldo Moro", Bari 70125, Italy; delia.mandracchia@uniba.it
4 Department of Cardiovascular Surgery, Faculty of Medicine, University Hospital, Geneva 1211, Switzerland; ufficiostampa@hpg23.it (F.I.); Beat.Walpoth@hcuge.ch (B.H.W.)
* Correspondence: jean-christophe.tille@hcuge.ch

Academic Editors: Robin Muise-Helmericks and Andy Wessels

Abstract: Drug-eluting vascular prostheses represent a new direction in vascular surgery to reduce early thrombosis and late intimal hyperplasia for small calibre grafts. Subcutaneous implantation in rats is a rapid and cost-effective screening model to assess the drug-elution effect and could, to some extent, be useful to forecast results for vascular prostheses. We compared biological and histological responses to scaffolds in different implantation sites. Polycaprolactone (PCL), paclitaxel-loaded PCL (PCL-PTX) and dexamethasone-loaded PCL (PCL-DXM) electrospun scaffolds were implanted subcutaneously and in an infrarenal abdominal aortic model in rats for up to 12 weeks. At the conclusion of the study, a histological analysis was performed. Cellular graft invasion revealed differences in the progression of cellular infiltration between PCL-PTX and PCL/PCL-DXM groups in both models. Cell infiltration increased over time in the aortic model compared to the subcutaneous model for all groups. Cell counting revealed major differences in fibroblast, macrophage and giant cell graft colonisation in all groups and models over time. Macrophages and giant cells increased in the PCL aortic model; whereas in the subcutaneous model these cell types increased only after three weeks or even decreased in the drug-eluting PCL groups. Other major findings were observed only in the aortic replacement such as extracellular matrix deposition and neo-angiogenesis. The subcutaneous implant model can be used for screening, especially when drug-eluting effects are studied. However, major histological differences were observed in cell type reaction and depth of cell penetration compared to the aortic model. Our results demonstrate that the implantation site is a critical determinant of the biological response.

Keywords: vascular prosthesis; tissue engineering; extracellular matrix; drug release; biodegradable polymers; foreign body reaction; pathology

---

1. Introduction

Biomaterial implants are used in a variety of anatomic locations. It has been recognized that maintaining a controlled biological reaction in the peri-implant region is essential. Although final device designs are mostly used in their target anatomical location, the biocompatibility of polymer components is often evaluated subcutaneously or intramuscularly.

The rat subcutaneous tissue implant site has proven to be a high-throughput, relatively low-cost screening technique for testing tissue responses to new materials [1,2]. The healing characteristics around polymer implants in subcutaneous tissues are often extrapolated to other tissues with the expectation that the observed healing will be similar in other locations.

This model could also allow an initial evaluation of the effects of a drug-loaded graft on tissue responses in a more relevant environment than the *in vitro* assay.

Biostable synthetic grafts such as expanded polytetrafluorethylene (ePTFE) have different healing and cellular infiltration characteristics in different implantation sites, e.g., subcutaneous *versus* adipose tissue in the rat [3,4]. Similarly, ePTFE grafts implanted in the abdominal wall showed different cell colonization in the prosthesis compared to vascular implantation [5].

Drug incorporation in fibers using electrospinning of drug-loaded solutions has been described for various compounds and is a promising approach for controlled drug delivery applications related to tissue engineering [6,7].

The aim of this study was to prepare drug-eluting biodegradable grafts with a defined and comparable nanostructure, and perform an *in vivo* assessment in two different anatomical locations, one with a pulsatile flow and the other without, in order to compare the drug-effect and cellular response.

2. Results

2.1. *In Vitro Studies*

2.1.1. Fibre Morphology and Mechanical Properties

A mechanical and morphological evaluation of the PCL-DXM compared to PCL-PTX and non-loaded PCL grafts was carried out. For the same PCL concentration during electrospinning, DXM-loaded grafts have higher maximal stress and strain values than non-loaded grafts (data not shown), as previously reported for PTX-loaded grafts [8]. Fiber diameter (Figure 1A,C) and tensile strength (Figure 1B)were comparable for PCL and PCL-PTX scaffolds. In terms of morphology and as illustrated in Figure 1A, PCL-DMX loaded fibers of the surface present an average smaller diameter but still are in the same range as the PCL and PCL-PTX ones (*i.e.*, around 1 μm, Figure 1C). This difference had little impact on tensile stress as it was of *ca.* 5 MPa for PCL-DXM against *ca.* 4 MPa for PCL-PTX and 5 MPa for PCL. All these values are above the value of 1.4 MPa found for natural vessels.

2.1.2. Evaluation of DXM-Loading and Incorporation Efficiency

Studies were carried out to quantify DXM loading into the graft. PCL-DXM grafts loaded with 1% DXM (*wt/wt*) yielded a high incorporation ranging from 64% to 98%, representing 130–206 mg DXM for a 2.5 cm long graft. The drug incorporation data for the PCL-PTX graft have been reported previously [8].

2.1.3. *In Vitro* DXM Release Profile

The results of DXM release from the electrospun PCL graft *in vitro* show an initial time lag on the first day, probably related to the high interaction of the hydrophobic drug with the polymer matrix. At day 6, approximately 30% of the drug had been released followed by a slower release phase over the following 16 days of 60% of the initially incorporated drug (Figure 2). The release profiles from the PCL-PTX graft have been previously reported [8].

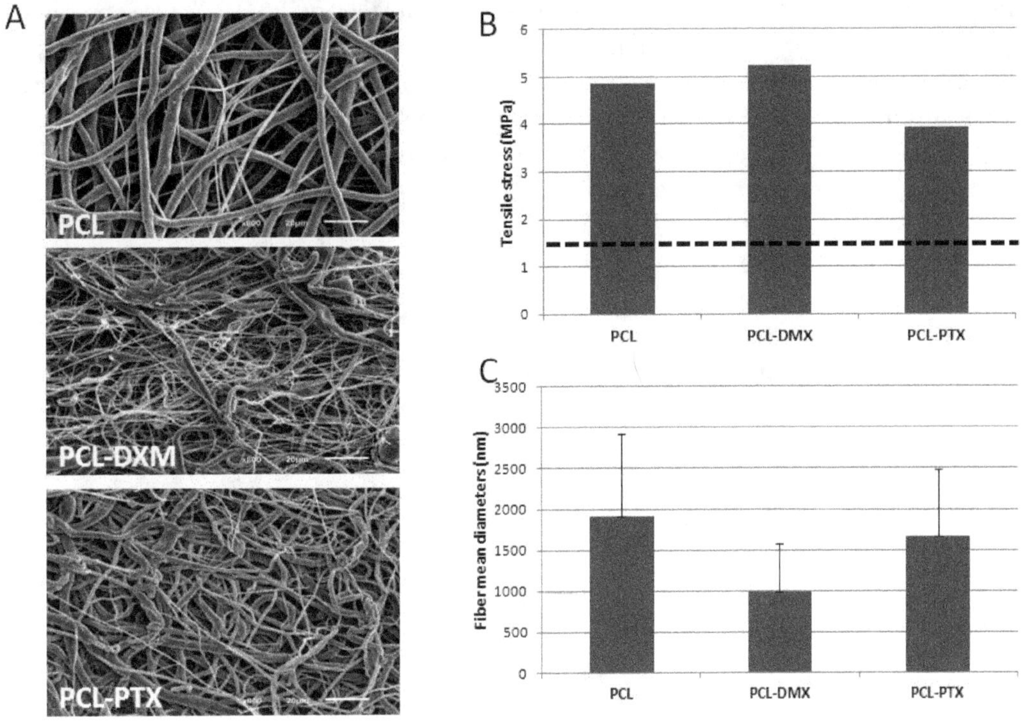

Figure 1. Morphological and mechanical properties of biodegradable polycaprolactone (PCL) electrospun grafts. (**A**) SEM pictures of graft inner surfaces at 700× magnification for PCL (12% *wt/vol*), PCL (9% *wt/vol*)-DXM (1%) and PCL(12% *wt/vol*)-PTX (0.75%); (**B**) Tensile stress values of PCL (12% *wt/vol*), PCL (9% *wt/vol*)-DXM (1%) and PCL (12% *wt/vol*)-PTX (0.75%) gamma sterilized graft (n = 1 for each); dott line correspond to "native" vessels tensile stress of 1.4 MPa; (**C**) Mean fiber diameters of PCL (12% *wt/vol*), PCL (9% *wt/vol*)-DXM 1% and PCL (12% *wt/vol*)-PTX (0.75%) gamma sterilized graft. (Data are expressed as mean ± SD and correspond to fiber diameter mean values calculated from 100 measurements).

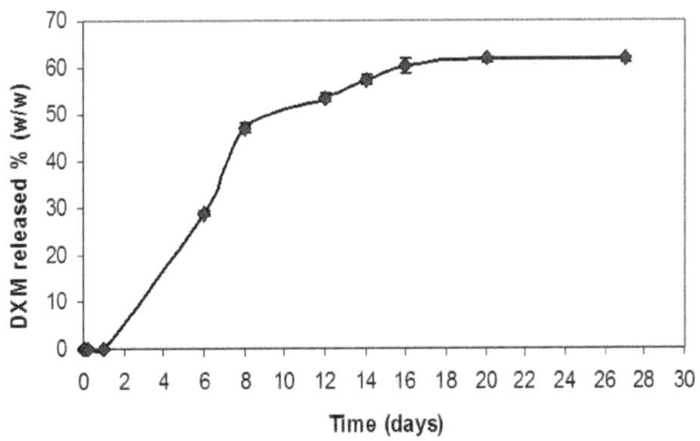

Figure 2. *In vitro* drug release profiles of a PCL graft containing 1% (*wt/wt*) DXM over 16 days. Data are expressed as an average of results obtained from 3 grafts at each time point (mean ± SD).

2.2. *In Vivo Studies*

2.2.1. *In Vivo* Rat Subcutaneous Model

Unloaded-PCL, PCL-PTX (0.5% and 0.75%) or PCL-DXM (1%) grafts were implanted subcutaneously in rats for 1, 3 and 12 weeks to determine the *in vivo* anti-proliferative effect for PTX and anti-inflammatory effect for DXM.

After 1 week of implantation, a major difference between PCL and PCL-PTX was observed. A full histological description of PCL grafts has been previously reported in the aorta [9]. Compared to PCL, PCL-PTX grafts had neither a peri-graft cellular reaction, a macrophage or giant cell reaction, nor cellular infiltration within the grafts (Figure 3a,b). The effect was similar for PCL-PTX-loaded grafts at 0.5% and 0.75% and both groups were combined for further evaluation. On the other hand, PCL-DXM grafts had stronger giant cell and macrophagic reactions compared to PCL (Figure 3c).

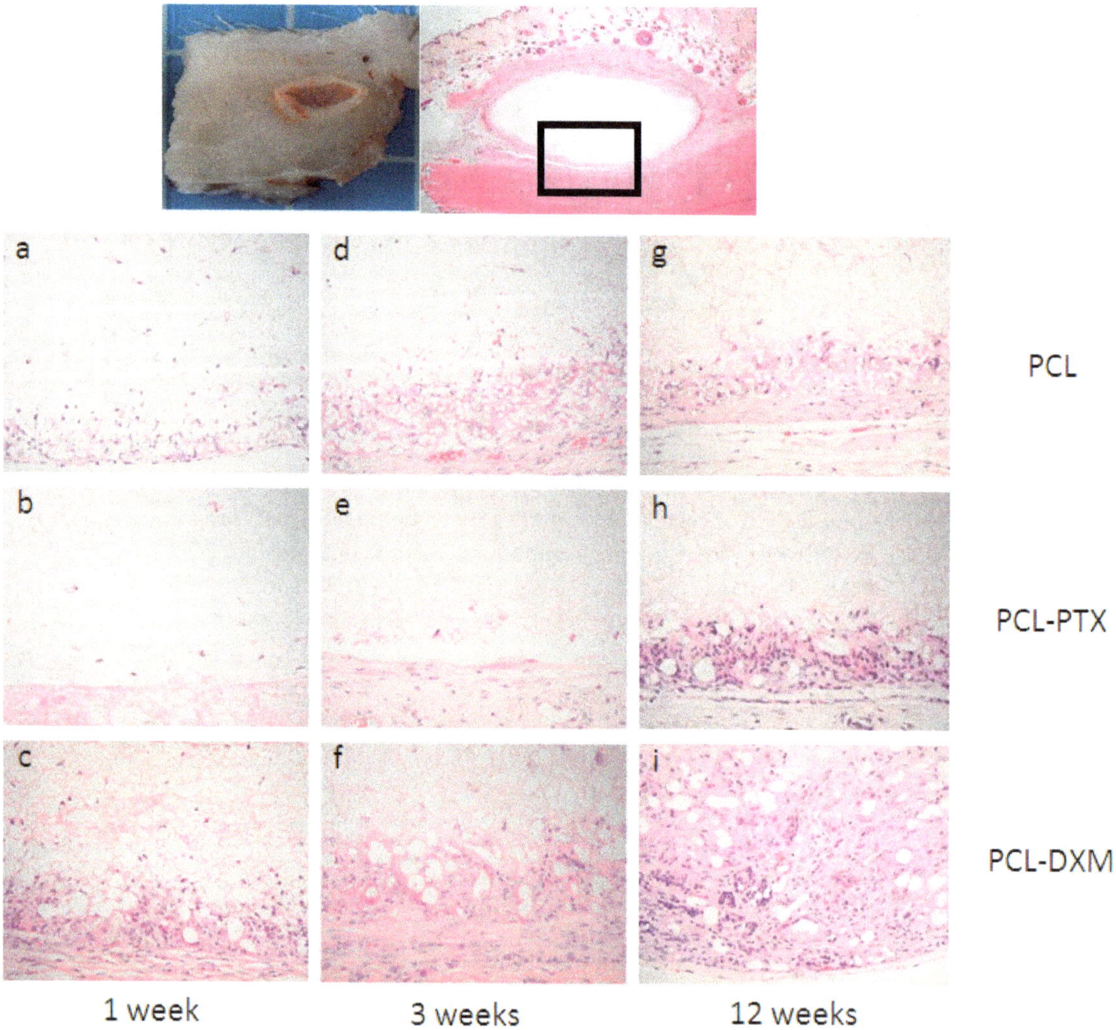

Figure 3. Histological analysis of the subcutaneous implant model. Subcutaneous implants and HE staining. The squares in the first row indicate the localisation of the magnifications below. Implant at 1 week (**a**–**c**), 3 weeks (**d**–**e**) and 12 weeks (**g**–**i**) of PCL grafts (**a**,**d**,**e**), PCL-PTX 0.75% (**b**,**e**,**h**) and PCL-DXM (**c**,**f**,**i**). HE staining, magnification 200×. The luminal side is on the top of the pictures.

Between 1 and 3 weeks after implantation, the cellular infiltration of PCL and PCL-DXM grafts increased with higher numbers of giant cells, macrophages and some fibroblasts (Figure 3d,f). No cellular reaction was observed in the PCL-PTX group (Figure 3e).

After 12 weeks of implantation, all grafts were colonized by host cells from the external part with a giant cell reaction at the periphery of the graft to the internal part with macrophages and some fibroblasts inside the graft wall. At the 12 week time point, no difference could be observed between PCL and PCL-PTX grafts (Figure 3g,h). In PCL-DXM grafts, the giant cells and macrophagic reaction was stronger compared to PCL and PCL-PTX grafts (Figure 3i).

No calcification was observed within the grafts in all groups and at all time-points. A few capillaries (neo-angiogenesis), were found focallyin one 12-week PLC-DXM graft at 12 weeks, but this was not seen in the PCL or PCL-PTX grafts at any of the time points (Figure 4).

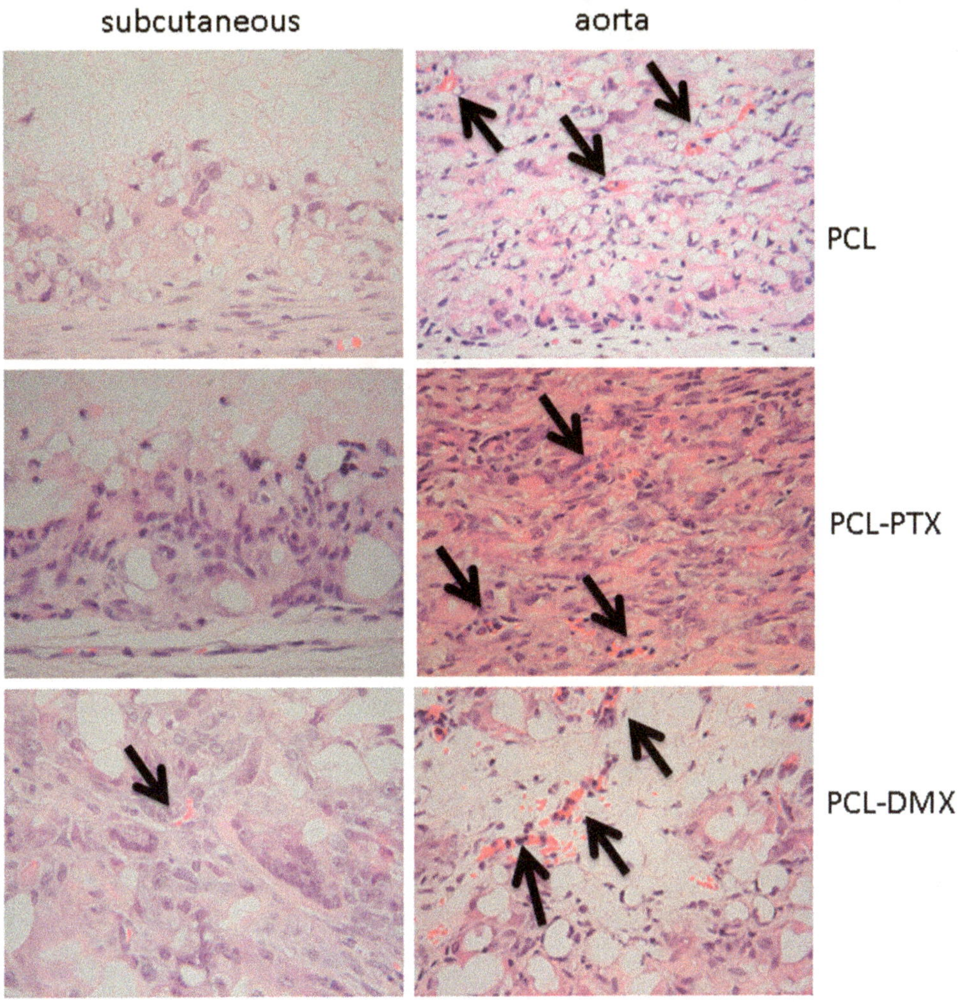

Figure 4. Histological analysis of neoangiogenesis in both models. Implants at 12 weeks in the subcutaneous and aortic models. Arrows indicate capillaries. HE staining, magnification 400×.

2.2.2. *In Vivo* Rat Aorta Interposition Model

We compared *in vivo* aorta implantation studies [8,9] to subcutaneous implantations to first compare the effect of the drugs and, second, the cellular reaction in and around the graft in different *in vivo* implantation sites.

One week after implantation, PCL-PTX explanted aortic grafts had no cellular reactions or infiltration as observed in the subcutaneous model (Figure 5b). The PCL-DXM group demonstrated a stronger cellular reaction with fibroblast infiltration and macrophages compared to the PCL group (Figure 5a,c). Compared to the subcutaneous implant model, the cellular reaction present in the PCL-DXM vascular group appeared to be more pronounced.

After 3 weeks of implantation, some fibroblasts infiltrated the PCL-PTX grafts with no giant cells or macrophages (Figure 5e). PCL-DXM and PCL groups showed the same aspect with giant cells formation, macrophages and some fibroblasts in the graft (Figure 5d,f).

After 12 weeks of implantation, PCL-PTX, PCL-DXM and PCL grafts had identical cellular reactions, which were composed of giant cells, macrophages and fibroblasts. The macrophagic reaction was lighter in PCL-DXM grafts compared to others grafts (Figure 5g–i).

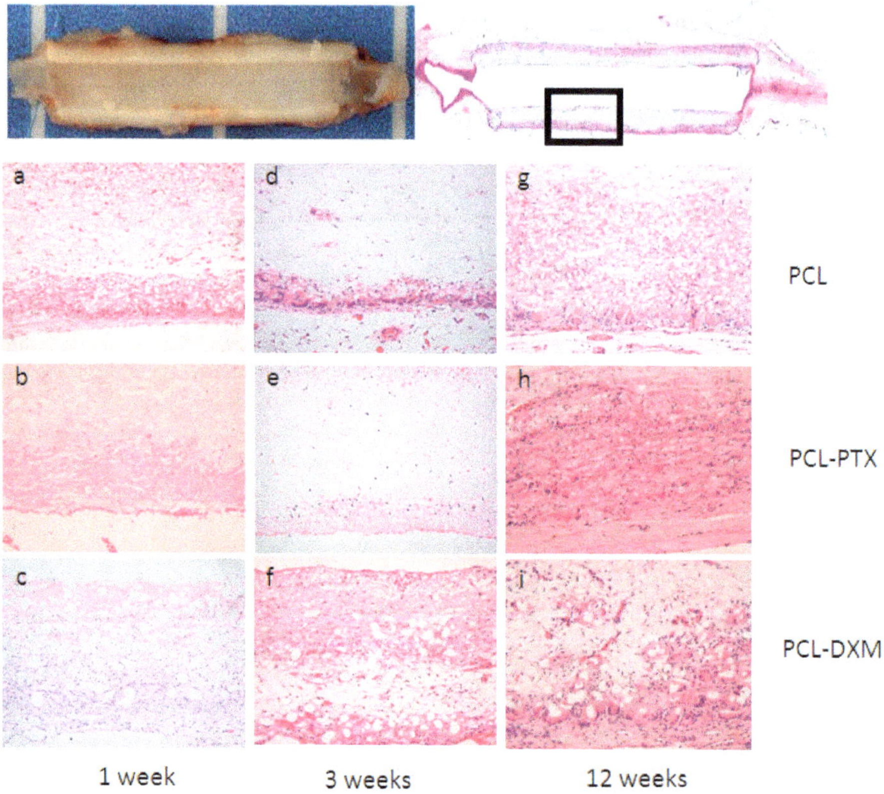

Figure 5. Histological analysis of the aorta graft model. Upper part: representative images of an aortic graft implant, the square indicates the location of the magnifications below. Vascular implant at 1 week (**a**–**c**), 3 weeks (**d**–**e**) and 12 weeks (**g**–**i**) of PCL graft (**a**,**d**,**e**), PCL-PTX 0.75% (**b**,**e**,**h**) and PCL-DXM (**c**,**f**,**i**). HE staining, magnification 200×. The luminal side is on the top.

Neoangiogenesis was observed at 12 weeks in the grafts for all groups (Figure 4). Calcifications in the intima hyperplasia were present in all groups at 12 weeks near the anastomosis or in the middle of the graft and developed through chrondroid metaplasia (data not shown), as previously reported [9].

2.2.3. Morphological Cellular Invasion of the Grafts

Cell type infiltrations, giant cells, macrophages and fibroblasts within the grafts were counted in the subcutaneous and vascular models.

The giant cell reaction was stable in PCL-DXM grafts in the subcutaneous model and had a transient increase from 1 to 3 weeks in the PCL graft. PCL-PTX grafts had an increase in giant cells at 12 weeks (Figure 6, top left). In the aortic location, the number of giant cells increased continuously from 1 to 12 weeks for PCL and PCL-DXM and appeared at 12 weeks for PCL-PTX, similarly to the subcutaneous model (Figure 6, top right). Giant cells numbers at 1 week were higher in the subcutaneous model for PCL-DXM compared to the aortic model and lower in the PLC graft in the subcutaneous model at 1 week compared to the aortic model.

The cellular density of macrophages remained stable in the PCL-DXM graft in the subcutaneous model and increased over time in PCL and PCL-PTX graft (Figure 6, middle left). In the aortic model, the number of macrophages increased over time in all three groups (Figure 6, middle right). Interestingly, the macrophage numbers in PCL and PCL-DXM grafts were significantly higher in the subcutaneous model at 1 and 3 weeks compared to the aortic model ($p < 0.05$).

Fibroblast numbers in the subcutaneous model had the same kinetic as the giant cells for PCL, PCL-DXM and PCL-PTX grafts (Figure 6, lower left). In the aortic model, fibroblast numbers slightly decreased in PCL grafts and remain stable in PCL-DXM grafts from 3 to 12 weeks. In PCL-PTX grafts some fibroblasts infiltrated the graft at 3 weeks and increased up to 12 weeks (Figure 6, lower right). Compared to the subcutaneous model, the fibroblast numbers were higher in the aortic model for the three grafts.

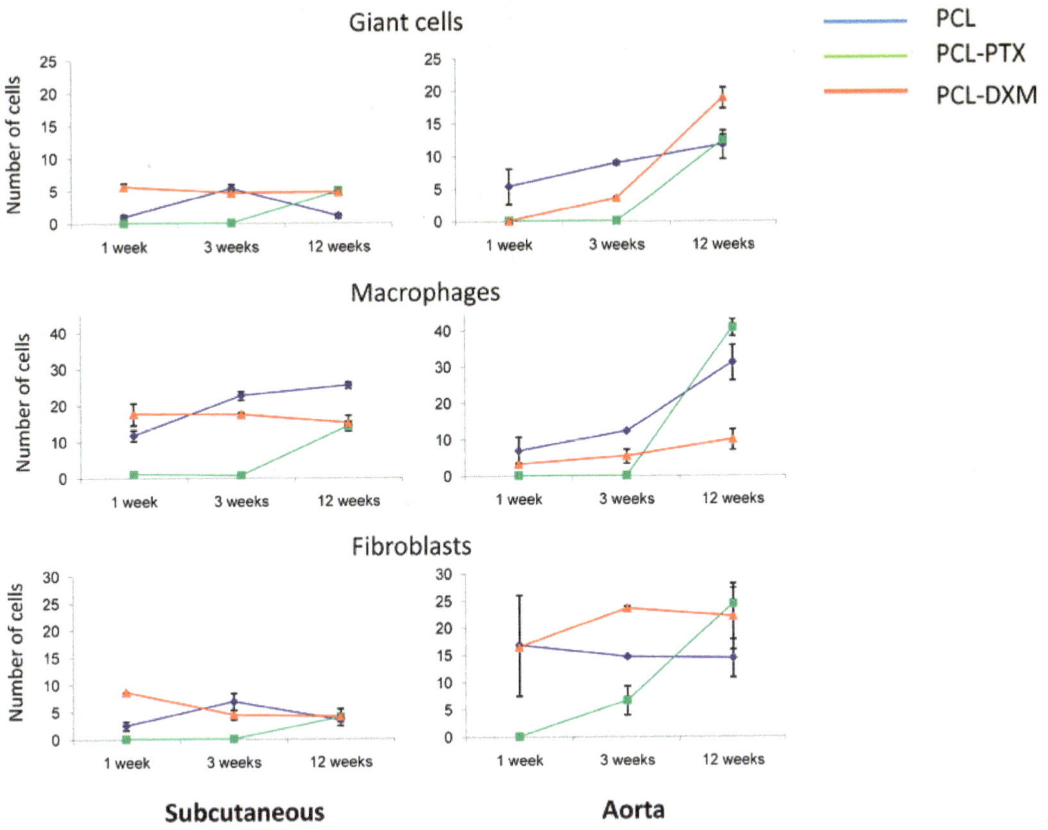

Figure 6. Type of cellular infiltration. The subcutaneous model (left column) and the aortic model (right column). Data are expressed as an average count of various cell types from all grafts at all time points with SEM.

2.2.4. Morphometric Analysis

Quantitative morphometric cellular graft infiltration between PCL, PCL-PTX and PCL-DXM grafts in the subcutaneous and aorta interposition models was compared over time.

In the subcutaneous model, the cellular invasion in PCL grafts increased up to 3 weeks (5.5% at 1 week and 18.8% at 3 weeks) and remained stable up to 12 weeks (16.5%) (Figure 7A). In the PCL-DXM grafts cellular invasion continuously increased from 1 to 12 weeks after implantation (20.9% at 1 week, 33.8% at 3 weeks and 51.1% at 12 weeks) (Figure 7A). The anti-proliferative effect of PTX on cellular infiltration was active at 1 and 3 weeks, then vanished at 12 weeks, with a cellular infiltration similar to PCL grafts (18.2% at 12 weeks) (Figure 7A).

In the aortic model, the cellular invasion of the graft increased continuously over time from 1 to 12 weeks in all grafts (Figure 7B). The delay of the cellular graft infiltration observed for the subcutaneous PCL-PTX graft was also present in the aortic model. An interesting point was that the percentage of cellular infiltration was higher in the aortic model for PCL and PCL-PTX graft and lower in the aortic model for PCL-DXM compared to the subcutaneous model.

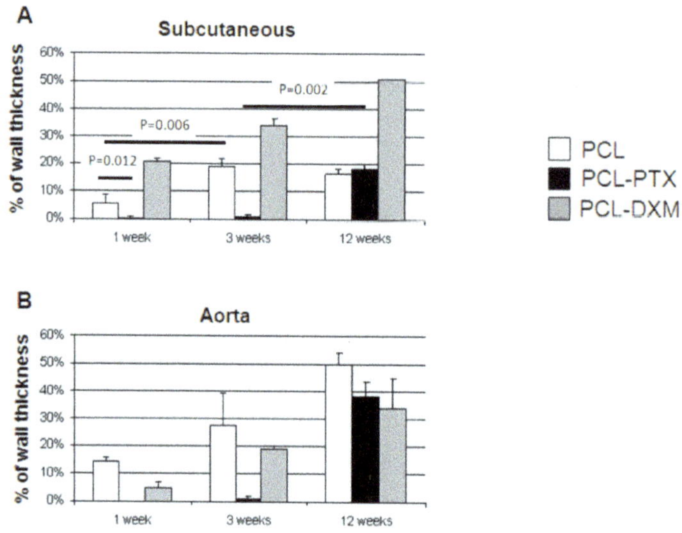

Figure 7. Morphometric cellular infiltration of the grafts. Percentage of cellular ingrowth into the grafts at 1, 3 and 12 weeks in the subcutaneous model (**A**) and the aortic model (**B**) (mean values ± SEM).

2.2.5. Extracellular Matrix Formation

In order to compare the collagen synthesis and deposition in the graft, Miller-Masson staining was performed at all time points in the three different grafts. In the subcutaneous model, there was no collagen deposition from 3 to 12 weeks (Figure 8). On the other hand, in the aortic model, collagen deposition in the graft was present from 6 weeks and increased at 12 weeks in all grafts (Figure 8).

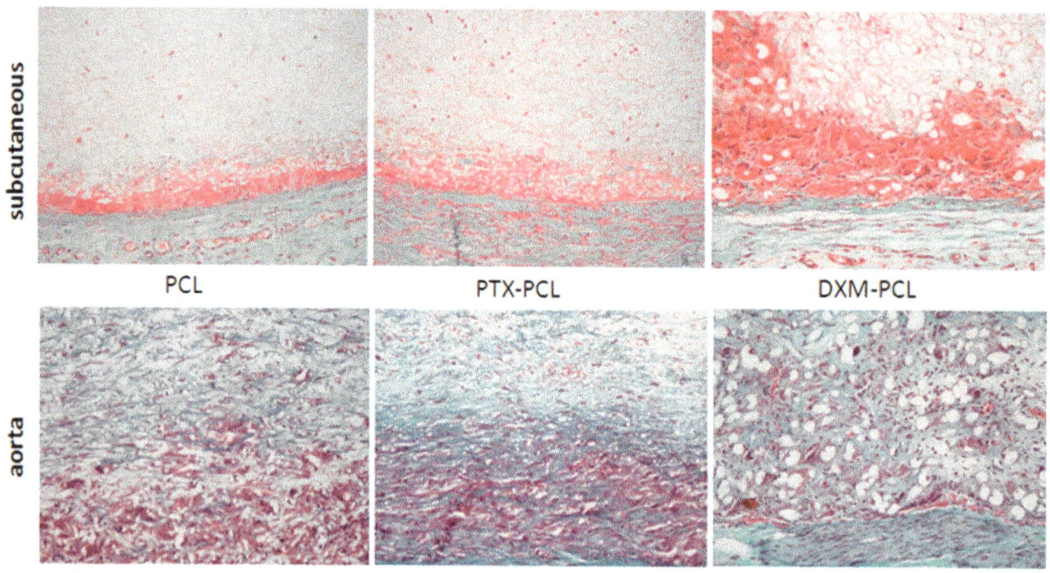

Figure 8. Collagen deposition in the graft. Miller-Masson combined staining of the subcutaneous model (upper panel) and the aortic model (lower panel) at 12 weeks. No collagen deposition (green) was found in the subcutaneous model. Magnification 200×.

3. Discussion

This study served as a comparative evaluation for the drug-eluting biodegradable graft effect and host cellular reaction *in vivo* in a classical subcutaneous model compared to an aortic replacement model used for small vascular grafts. Historically, several extrapolations from subcutaneous implant models had been made when testing new materials for clinical applications, but direct comparisons between anatomical implantation sites have rarely been published.

A detailed assessment of the prepared and investigated PCL, PCL-PTC and PCL-DXM, respectively, proved their comparability in their mechanical and morphological properties, as well as their drug release characteristics.

We observed a marked biological effect of PTX in both subcutaneous and aortic models with a delayed cellular reaction and graft invasion. DXM does not seem to affect the host reaction in a time dependent manner compared to unloaded-PCL, but a stronger giant cell reaction was observed in the subcutaneous model compared to the aortic model. This could be due to the slight differences in fibre surface nanostructures and alignments. The macrophagic reaction was not lower in PCL-DXM grafts, possibly due to the low DXM dosage, below 0.5 mg/kg. Another explanation is that DXM, a well-known immune modulator, had no effect on bone marrow-derived macrophages' differentiation *in vitro* and also decreased macrophages' adhesiveness [10,11]. This mechanism could also be taken into account for the increased macrophage infiltration.

Cellular infiltration of the grafts over time was the main difference between the two *in vivo* models. In the subcutaneous model, the cellular infiltration increased for PCL-PTX and PCL-DXM up to 12 weeks and was stable for three weeks for PCL. Interpretation at the one week time point is limited since only one PCL-PTX graft was retrieved at one week. In contrast, in the aortic model, there was a continuously increased cellular infiltration in all three biodegradable grafts. We did not observe differences in the diameter or orientation of the fibres, indicating that these were not modified during the processing. We know that the scaffold nanostructure architecture is an important point with regard to the healing process and host response *in vivo* [12]. This difference could possibly be related to the vascular environment with a blood flow induced mechanical strain on the grafts promoting scaffold structure modification and cell mobilization within the graft, as shown for ePTFE implants [5].

The magnitude of the macrophage and fibroblast reaction to the graft was also different for the subcutaneous and the aortic models. The numbers of fibroblasts in the grafts were higher in the subcutaneous model and the opposite was seen for macrophages, which infiltrated in lower numbers at early time points in the subcutaneous model. This difference in cell type reaction and infiltration could be explained by the differences between the models: a static subcutaneous model *vs.* a pulsatile aortic model. Extracellular matrix formation in the graft, such as collagen—a key component for mechanical stability—was only present in the aortic model and not observed in the subcutaneous model. This difference in extracellular accumulation was also observed by others in bio-reactor matured models of tissue-engineered vessels. Grafts in a pulsatile *in vitro* flow system contained more matrix deposition compared to static conditions, indicating an effect of pulse on fibroblasts [13,14].

Calcification located only in the intima hyperplasia was only present in the aortic model at 12 weeks in all grafts and never in the subcutaneous model. This may also be linked to the pulsatile conditions.

Neoangiogenesis, accompanying cellular colonization of the graft, is an important factor for tissue remodelling. Neoangiogenesis was observed only focally in one subcutaneously implanted PCL-DXM graft, but was present in all grafts implanted in the aortic model after 12 weeks.

4. Materials and Methods

4.1. Preparation of Unloaded, PTX-Loaded and DXM-Loaded PCL Grafts by Electrospinning

Two mm inner diameter unloaded PCL and paclitaxel-loaded PCL (PCL-PTX, with two different PTX concentration: 0.5% and 0.75%) grafts were prepared according to a process optimized by a factorial design approach that has been reported in detail elsewhere [8,9,15].

In a similar factorial design approach, various optimized dexamethasone-loaded grafts (PCL-DXM) were prepared, and the effects of different DXM concentrations on fibre morphology, graft tensile stress and strain behaviours were investigated. Briefly, the DXM-loaded grafts were prepared in the same manner as the unloaded controls by electrospinning of PCL solutions (7.5%, 9% and 12% [*wt/vol*]) containing different amounts of DXM (1% and 10% [*wt/wt*]). The generated fibres were collected on a rotating (4500 rpm) and translating (200 rpm, 4-cm amplitude) mandrel to form a nonwoven tubular PCL graft. The grafts were dried at 37 °C under vacuum overnight. Residual solvents were analysed with a Headspace Sampler (HP7694, Hewlett-Packard, Palo Alto, CA, USA) coupled to a gas-chromatography system (GC System, 6850 Series, Agilent Technologies, Santa Clara, CA, USA) equipped with an HP-1 capillary column (Agilent Technologies). All grafts were γ-sterilized at 25 kGy before characterization and *in vivo* use. The PCL-DXM graft chosen for *in vivo* studies was made from a PCL 9% [*wt/vol*] PCL solution with DXM 1% because it showed similar fibre morphology and mechanical properties to plain PCL grafts used previously [8,9,14].

4.2. Assessment of DXM-Loading and Distribution in the Graft

The extent of drug loading and the incorporation efficiency were assessed by high performance liquid chromatography (HPLC, Acquity UPLC system, Waters Corporation, Milford, CT, USA) analysis in 2.5 cm long DXM-loaded grafts with a theoretical loading of 1% and 10% [*wt/wt*]. Additionally, the homogeneity of DXM distribution along the graft ($n = 3$) was verified by cutting four 1 cm long pieces from a 4 cm DXM-loaded graft.

4.3. Assessment of DXM Release

Grafts theoretically loaded with 1% of DXM ($n = 3$) were placed in vials with an exact measured amount of release medium of phosphate buffered saline (pH = 7.4) containing 0.05% (*vol/vol*) Tween 20. The vials were kept at 37 °C with rotational mixing at 100 rpm for 1, 5, and 16 h and 1, 6, 8, 12, 14 and 16 days. At each time point, 1 mL aliquots of the release medium were taken for DXM quantification. The remaining release medium was aspirated and fresh release medium was added at each time point to maintain sink conditions.

4.4. Mechanical and Morphological Properties Evaluation

Fibre diameter and morphology of the grafts were investigated by scanning electronic microscopy (SEM, JEOL JSM-6510LV, Tokyo, Japan) as previously reported [8,9,15]. Briefly, tensile stress was measured on complete vascular grafts using a mechanical testing bench from Schenck AG (Nänikon, Dübendorf, Switzerland) using a cross-head speed of 10 mm/min at room temperature. The vascular grafts were fixed by clamps using 5 mm of each end of the graft, so that either a 1 or 3 cm length was used for the measurements corresponding for the 2 and 4 mm, respectively, I.D. grafts.

Statistical Analyses

A statistical analysis was performed only for the cellular invasion in the subcutaneous model with a Mann Whitney U test two-tailed (http://www.socscistatistics.com). Due to a small number of samples, it was not possible to perform a statistical analysis for the PCL-DXM sample in this model and also in the aorta model for all samples. A $p < 0.05$ was considered as significant.

4.5. In Vivo Studies

The experimental protocol was approved by the Animal Experiments Ethical Committee of the University of Geneva (Protocol No: 06/52) and the Veterinary Office of State of Geneva, Switzerland (No: 1081/3232/II), and carried out in conformity with the Guide for Care and Use of the Laboratory Animals (National Research Council, Washington, DC: National Academy Press; 1996).

4.5.1. Rat Subcutaneous Model

In vivo studies were carried out on 60 male Sprague–Dawley rats (375 g) according to the ethical principles of laboratory animal care and Swiss regulations. The animals were kept in individual cages and received food and water *ad libitum*. Anaesthesia was induced with isoflurane 4% and the animals were maintained under deep anesthesia with isoflurane 2%, 40% oxygen through facial masks.

After shaving the abdominal region, we implanted PCL, PCL-PTX, PCL-DXM grafts (10 mm length and 2 mm internal diameter, wall thickness 600 μm) subcutaneously with a maximum of 6 grafts per rat. The animals were followed for 1 week (PCL n = 8, PCL-PTX n = 10 (0.5% = 2, 0.75% = 8), PCL-DXM n = 3), 3 weeks (PCL n = 9, PCL-PTX n = 12 (0.5% = 3, 0.75% = 9), PCL-DXM n = 3) and 12 weeks (PCL n = 6, PCL-PTX n=6 (0.5% = 2, 0.75% = 4), PCL-DXM n = 3). Euthanasia was then performed under deep isoflurane anesthesia by intracardiac injection of KCl. At sacrifice only 2 PCL-DXM and 1 PCL-DXM grafts were found remaining at the implantation site at 1 and 12 weeks, respectively. All other grafts were retrieved.

4.5.2. Rat Aorta Interposition Model

Our experimental rat model, including the anaesthesia and operative techniques as well as the follow-up protocol, was described in detail previously [9]. Briefly, 28 male Sprague-Dawley rats (375 g) were anaesthetized and operated on according to this protocol. All grafts (10 mm length and 2 mm internal diameter, wall thickness 600 μm) were implanted randomly into the rat abdominal aorta. All rats were followed for 1 week (PCL n = 3, PCL-PTX n = 3, PCL-DXM n = 3), 3 weeks (PCL n = 3, PCL-PTX n = 3, PCL-DXM n = 3) and 12 weeks (PCL n = 3, PCL-PTX n = 3, PCL-DXM n = 3). No anticoagulation or anti-platelet drugs were given. At conclusion of the study period, digital subtraction angiography was performed to assess patency and rats were sacrificed after explantation of the vascular graft with the adjacent abdominal aorta segments. All grafts were patents at all time points.

4.5.3. Histological Assessment and Quantitative Morphometric Analysis

For subcutaneous implants, the area of implantation was dissected and fixed in 4% formaldehyde for 24 h. A slice of the subcutaneously implanted graft with adjacent tissue was embedded in paraffin wax. For the aorta, explanted grafts with both anastomoses were fixed in 4% formaldehyde for 24 h, cut into two longitudinal halves, and then embedded into paraffin. Histological sections of 4 μm were stained with Hematoxylin-Eosin (HE) and Miller-Masson for elastin and collagen deposition.

4.5.4. Morphological Cell Quantification Analysis

In the subcutaneous and vascular explants, based on classical morphology, giant cells around the graft, macrophages and fibroblasts within the graft were counted under a light microscope at 400× magnifications over 5 representative fields on HE stains, and the final cell number was per 1 high power field (400× magnification). Morphological identification of the different cells types was done with the following criteria: macrophages were identified by an abundant cytoplasm with an eccentric reniform nucleus; giant cells are large cell with numerous nuclei; fibroblasts have a spindle-shape with an elongated nucleus and little cytoplasm, as illustrated below (Figure 9).

Neoangiogenesis was analyzed on HE stains and capillaries were defined as a small vessel circled by endothelial cells and containing erythrocytes in the lumen. Slides were digitalized with the Mirax scan (Carl Zeiss MicroImaging GmbH, Göttingen, Germany) and images analyzed with ImageJ software (version 1.3, National Institute of Health, Bethesda, MD, USA). Cellular invasion of the graft was defined as the percentage of graft area which was penetrated densely by host cells from the external side towards the inner side.

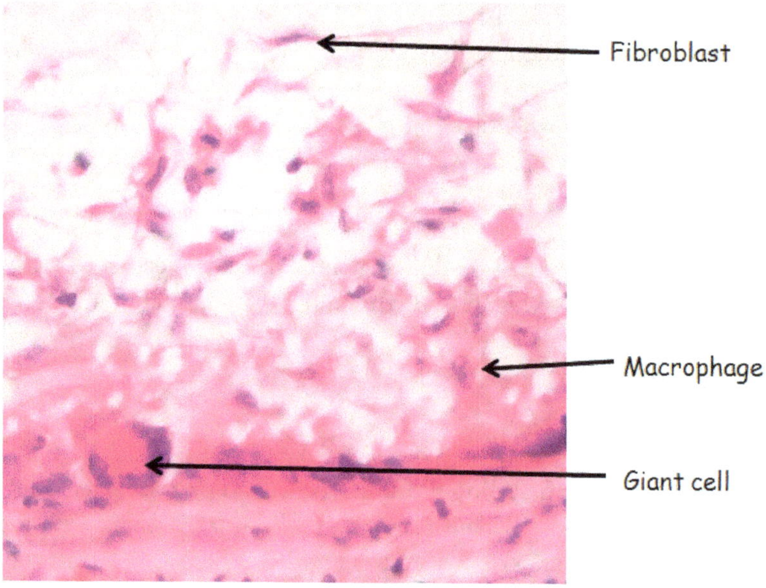

Figure 9. Morphology of the different analysezed cells type. HE stain (magnification 400×).

5. Conclusions

The subcutaneous implant model for testing biodegradable polymers can be used for screening, especially when drug-eluting effects are studied. However, major histological differences in cell reaction, such as the depth of cell colonization compared to the aortic replacement model, indicate some limitations and demonstrate that the results from one model cannot be generalized for later application at a different implantation site. It is therefore highly recommended to perform the implantation at the site of interest in order to confirm initial results obtained in the subcutaneous implant model, especially for drug releasing, biodegradable implants.

Acknowledgments: We would like to thank Unn Lutzen and Patrizia Gindre for their precious support for the stains and immunohistochemistry. This study was supported by a grant from thr Swiss National Science Foundation (320000-119822-1).

Author Contributions:
J-CT: responsible histopathological analysis, integration of the results, writing of the manuscript
SdV: responsible for *in vitro* and morphometric studies
DM: responsible for *in vitro* studies
BN: responsible for *in vitro* studies
FI: responsible for *in vivo* experiments
RG: responsible for the experimental approach design, control and integration of the results and final version writing of the manuscript
MM: responsible for the experimental approach design, control and integration of the results and final version writing of the manuscript
BHW: responsible for the experimental approach design, control and integration of the results and final version writing of the manuscript

Conflicts of Interest: The authors declare no conflict of interest.

References

1. Mako, W.J.; Shah, A.; Vesely, I. Mineralization of glutaraldehyde-fixed porcine aortic valve cusps in the subcutaneous rat model: Analysis of variations in implant site and cuspal quadrants. *J. Biomed. Mater. Res.* **1999**, *45*, 209–213. [CrossRef]
2. Sharkawy, A.A.; Klitzman, B.; Truskey, G.A.; Reichert, W.M. Engineering the tissue which encapsulates subcutaneous implants. I. Diffusion properties. *J. Biomed. Mater. Res.* **1997**, *37*, 401–412. [CrossRef]
3. Salzmann, D.L.; Kleinert, L.B.; Berman, S.S.; Williams, S.K. The effects of porosity on endothelialization of eptfe implanted in subcutaneous and adipose tissue. *J. Biomed. Mater. Res.* **1997**, *34*, 463–476. [CrossRef]

4. Williams, S.K.; Berman, S.S.; Kleinert, L.B. Differential healing and neovascularization of eptfe implants in subcutaneous *versus* adipose tissue. *J. Biomed. Mater. Res.* **1997**, *35*, 473–481. [CrossRef]
5. Bellon, J.M.; Bujan, J.; Contreras, L.A.; Hernando, A.; Jurado, F. Similarity in behavior of polytetrafluoroethylene (ePTFE) prostheses implanted into different interfaces. *J. Biomed. Mater. Res.* **1996**, *31*, 1–9. [CrossRef]
6. Zeng, J.; Xu, X.; Chen, X.; Liang, Q.; Bian, X.; Yang, L.; Jing, X. Biodegradable electrospun fibers for drug delivery. *J. Control. Release* **2003**, *92*, 227–231. [CrossRef]
7. Zeng, J.; Yang, L.; Liang, Q.; Zhang, X.; Guan, H.; Xu, X.; Chen, X.; Jing, X. Influence of the drug compatibility with polymer solution on the release kinetics of electrospun fiber formulation. *J. Control. Release* **2005**, *105*, 43–51. [CrossRef] [PubMed]
8. Innocente, F.; Mandracchia, D.; Pektok, E.; Nottelet, B.; Tille, J.C.; de Valence, S.; Faggian, G.; Mazzucco, A.; Kalangos, A.; Gurny, R.; *et al.* Paclitaxel-eluting biodegradable synthetic vascular prostheses: A step towards reduction of neointima formation? *Circulation* **2009**, *120*, S37–S45. [CrossRef] [PubMed]
9. Pektok, E.; Nottelet, B.; Tille, J.C.; Gurny, R.; Kalangos, A.; Moeller, M.; Walpoth, B.H. Degradation and healing characteristics of small-diameter poly(epsilon-caprolactone) vascular grafts in the rat systemic arterial circulation. *Circulation* **2008**, *118*, 2563–2570. [CrossRef] [PubMed]
10. Lloberas, J.; Soler, C.; Celada, A. Dexamethasone enhances macrophage colony stimulating factor- and granulocyte macrophage colony stimulating factor-stimulated proliferation of bone marrow-derived macrophages. *Int. Immunol.* **1998**, *10*, 593–599. [CrossRef] [PubMed]
11. Roberts, C.P.; Murphy, A.A.; Santanam, N.; Parthasarathy, S. Regulation of monocyte to macrophage differentiation by antiglucocorticoids and antioxidants. *Am. J. Obstet. Gynecol.* **1998**, *179*, 354–362. [CrossRef]
12. Cao, H.; McHugh, K.; Chew, S.Y.; Anderson, J.M. The topographical effect of electrospun nanofibrous scaffolds on the in vivo and in vitro foreign body reaction. *J. Biomed. Mater. Res. A* **2009**, *93*, 1151–1159.
13. Hoerstrup, S.P.; Zund, G.; Sodian, R.; Schnell, A.M.; Grunenfelder, J.; Turina, M.I. Tissue engineering of small caliber vascular grafts. *Eur. J. Cardiothorac. Surg.* **2001**, *20*, 164–169. [CrossRef]
14. Syedain, Z.H.; Meier, L.A.; Bjork, J.W.; Lee, A.; Tranquillo, R.T. Implantable arterial grafts from human fibroblasts and fibrin using a multi-graft pulsed flow-stretch bioreactor with noninvasive strength monitoring. *Biomaterials* **2011**, *32*, 714–722. [CrossRef] [PubMed]
15. Nottelet, B.; Pektok, E.; Mandracchia, D.; Tille, J.C.; Walpoth, B.; Gurny, R.; Moller, M. Factorial design optimization and *in vivo* feasibility of poly(epsilon-caprolactone)-micro- and nanofiber-based small diameter vascular grafts. *J. Biomed. Mater. Res. A* **2009**, *89*, 865–875. [CrossRef] [PubMed]

PERMISSIONS

All chapters in this book were first published in JDB, by MDPI; hereby published with permission under the Creative Commons Attribution License or equivalent. Every chapter published in this book has been scrutinized by our experts. Their significance has been extensively debated. The topics covered herein carry significant findings which will fuel the growth of the discipline. They may even be implemented as practical applications or may be referred to as a beginning point for another development.

The contributors of this book come from diverse backgrounds, making this book a truly international effort. This book will bring forth new frontiers with its revolutionizing research information and detailed analysis of the nascent developments around the world.

We would like to thank all the contributing authors for lending their expertise to make the book truly unique.

They have played a crucial role in the development of this book. Without their invaluable contributions this book wouldn't have been possible. They have made vital efforts to compile up to date information on the varied aspects of this subject to make this book a valuable addition to the collection of many professionals and students.

This book was conceptualized with the vision of imparting up-to-date information and advanced data in this field. To ensure the same, a matchless editorial board was set up. Every individual on the board went through rigorous rounds of assessment to prove their worth. After which they invested a large part of their time researching and compiling the most relevant data for our readers.

The editorial board has been involved in producing this book since its inception. They have spent rigorous hours researching and exploring the diverse topics which have resulted in the successful publishing of this book. They have passed on their knowledge of decades through this book. To expedite this challenging task, the publisher supported the team at every step. A small team of assistant editors was also appointed to further simplify the editing procedure and attain best results for the readers.

Apart from the editorial board, the designing team has also invested a significant amount of their time in understanding the subject and creating the most relevant covers. They scrutinized every image to scout for the most suitable representation of the subject and create an appropriate cover for the book.

The publishing team has been an ardent support to the editorial, designing and production team. Their endless efforts to recruit the best for this project, has resulted in the accomplishment of this book. They are a veteran in the field of academics and their pool of knowledge is as vast as their experience in printing. Their expertise and guidance has proved useful at every step. Their uncompromising quality standards have made this book an exceptional effort. Their encouragement from time to time has been an inspiration for everyone.

The publisher and the editorial board hope that this book will prove to be a valuable piece of knowledge for researchers, students, practitioners and scholars across the globe.

LIST OF CONTRIBUTORS

Matthew R. McCann and Cheryle A. Séguin
Department of Physiology and Pharmacology, Schulich School of Medicine & Dentistry, Western University, London, ON N6A 5C1, Canada
Bone and Joint Institute, Western University, London, ON N6A 5C1, Canada

Jayne Aiken and Jeffrey K. Moore
Department of Cell and Developmental Biology, University of Colorado School of Medicine, MS8108, 12801 E 17th Ave, Aurora, CO 80045, USA

Georgia Buscaglia and Emily A. Bates
Department of Pediatrics, University of Colorado Anschutz Medical Campus, Aurora, CO 80045, USA

Ashley M. Driver and Christopher Shumrick
Division of Human Genetics, Department of Pediatrics, Cincinnati Children's Hospital Medical Center, Cincinnati, OH 45229, USA

Rolf W. Stottmann
Division of Human Genetics, Department of Pediatrics, Cincinnati Children's Hospital Medical Center, Cincinnati, OH 45229, USA
Division of Developmental Biology, Department of Pediatrics, Cincinnati Children's Hospital Medical Center, Cincinnati, OH 45229, USA

Fabio Alexis Lefebvre
Institut de Recherches Cliniques de Montréal (IRCM), Montréal, QC H2W 1R7, Canada
Département de Biochimie, Université de Montréal, Montréal, QC H3T 1J4, Canada

Éric Lécuyer
Institut de Recherches Cliniques de Montréal (IRCM), Montréal, QC H2W 1R7, Canada
Département de Biochimie, Université de Montréal, Montréal, QC H3T 1J4, Canada
Division of Experimental Medicine, McGill University, Montréal, QC H3A 0G4, Canada
IRCM, RNA Biology Laboratory, 110 Avenue des Pins, Ouest, Montréal, QC H2W 1R7, Canada

Elizabeth C. Coffey, Maggie E. Pasquarella and Michelle F. Goody
School of Biology and Ecology, University of Maine, Orono, ME 04469, USA

Clarissa A. Henry
School of Biology and Ecology, University of Maine, Orono, ME 04469, USA
Graduate School of Biomedical Sciences and Engineering, University of Maine, Orono, ME 04469, USA

Giuliana D. Clemente, Matthew R. Hannaford and Jens Januschke
Division of Cell and Developmental Biology, School of Life Sciences, University of Dundee, Dundee DD1 5EH, UK

Hamze Beati and Hans-Arno J. Müller
Division of Cell and Developmental Biology, School of Life Sciences, University of Dundee, Dundee DD1 5EH, UK
Institut für Biologie, Fachgebiet Entwicklungsgenetik, Universität Kassel, Heinrich-Plett Str. 40, 34321 Kassel, Germany

Katja Kapp
Institut für Biologie, Fachgebiet Entwicklungsgenetik, Universität Kassel, Heinrich-Plett Str. 40, 34321 Kassel, Germany

Eric R. Griffis
Centre of Gene Regulation and Expression, School of Life Sciences, University of Dundee, Dundee DD1 5EH, UK

Bin Liu, Hongmei Xiao and Chunjie Zhao
Key Laboratory of Developmental Genes and Human Diseases, MOE, School of Medicine, Southeast University, Nanjing 210009, China

Odessa R. Yabut, Hui Xuan Ng, Gloria Fernandez and Jeremy Kuhn
Department of Neurology, University of California San Francisco, San Francisco, CA 94158, USA

Keejung Yoon
Department of Neurology, University of California San Francisco, San Francisco, CA 94158, USA
College of Biotechnology and Bioengineering, Sungkyunkwan University, Suwon, Gyeonggi-do 16419, Korea

Samuel J. Pleasure
Department of Neurology, University of California San Francisco, San Francisco, CA 94158, USA
Programs in Neuroscience and Developmental Biology, Eli and Edythe Broad Center of Regeneration Medicine and Stem Cell Research, University of California San Francisco, San Francisco, CA 94143, USA

Arya Zandvakili
Molecular and Developmental Biology Graduate Program, Cincinnati Children's Hospital Medical Center, Cincinnati, OH 45229, USA
Medical-Scientist Training Program, University of Cincinnati College of Medicine, Cincinnati, OH 45267, USA

Brian Gebelein
Division of Developmental Biology, Cincinnati Children's Hospital Medical Center, Cincinnati, OH 45229, USA

Stéphane Zaffran
Aix Marseille Université, GMGF UMRS910, Faculté de Médecine, 27 Bd Jean Moulin, Marseille 13385, France
Inserm U910, Faculté de Médecine, 27 Bd Jean Moulin, Marseille 13005, France

Marine Roux
Aix Marseille Université, GMGF UMRS910, Faculté de Médecine, 27 Bd Jean Moulin, Marseille 13385, France
Inserm U910, Faculté de Médecine, 27 Bd Jean Moulin, Marseille 13005, France
Laboratory of Genetics and Development, Institut de Recherches Cliniques de Montréal (IRCM), Université de Montréal, Montréal, QC H3C 3T5, Canada

João Leite-Castro
IBMC—Instituto de Biologia Celular e Molecular, Oporto, Portugal

Vanessa Beviano and Renata Freitas
IBMC—Instituto de Biologia Celular e Molecular, Oporto, Portugal
I3S—Instituto de Investigação e Inovação em Saúde, Rua Alfredo Allen, 208, 4200-135 Porto, Portugal

Pedro Nuno Rodrigues
IBMC—Instituto de Biologia Celular e Molecular, Oporto, Portugal
I3S—Instituto de Investigação e Inovação em Saúde, Rua Alfredo Allen, 208, 4200-135 Porto, Portugal
ICBAS—Instituto de Ciências Biomédicas Abel Salazar, University of Oporto, Rua de Jorge Viterbo Ferreira n°. 228, 4050-313 Oporto, Portugal

Yesser H. Belgacem, Andrew M. Hamilton, Sangwoo Shim, Kira A. Spencer and Laura N. Borodinsky
Department of Physiology & Membrane Biology and Institute for Pediatric Regenerative Medicine, Shriners Hospital for Children Northern California, University of California Davis School of Medicine, Sacramento, CA 95817, USA

Jean-Christophe Tille
Division of Clinical Pathology, University Hospital of Geneva, Geneva 1211, Switzerland

Sarra de Valence, Nottelet, Robert Gurny and Michael Möller
School of Pharmaceutical Sciences, University of Geneva, Geneva 1211, Switzerland

Marco Barucca, Adriana Canapa and Maria Assunta Biscotti
Dipartimento di Scienze della Vita e dell'Ambiente, Università Politecnica delle Marche, via Brecce Bianche, 60131 Ancona, Italy

Index

www.ingramcontent.com/pod-product-compliance
Lightning Source LLC
LaVergne TN
LVHW080957210926
848485LV00004B/841